New Materials for
a Circular Economy

Edited by

Alberto García-Peñas

Gaurav Sharma

Published by **Materials Research Forum LLC**
Millersville, PA 17551, USA

Published as part of the book series
Materials Research Foundations
Volume 149 (2023)
ISSN 2471-8890 (Print)
ISSN 2471-8904 (Online)

Print ISBN 978-1-64490-262-2
eBook ISBN 978-1-64490-263-9

Distributed worldwide by

Materials Research Forum LLC
105 Springdale Lane
Millersville, PA 17551
USA
https://www.mrforum.com

Manufactured in the United States of America
10 9 8 7 6 5 4 3 2 1

We wish to dedicate this book
"To our parents & Family's and specially in remembrance of Pari "

Table of Contents

Keywords
About the editors

Preface

The continues consumption of natural resources for covering the necessities of human society in terms of applications, energy and transport is increasing as world population rises. However, our resources are limited, and effective applications and efficient use of energy will be required to preserve the environment and life. In this sense, the strategies of some countries are focused on a new model based on a circular economy, which could displace the lineal economy used around the world.

This strategy involves all the segments of society, whose main goals are a revolution in terms of cycles of life for products, increase the efficiency of electric and electronic devices and the eco-design. For that purpose, the generation of new materials will be essential, and also its recycling or conversion after use.

New regulation from many countries is affecting the industry which needs to look for alternatives as a response to the rules in terms of sustainability and recycling, but keeping the spectra of properties and rentability of their products. Thus, industries and also governments are investing in research to address the following questions: Can I find an alternative material for my product responding to sustainable development and keeping the rentability? Could I incorporate other features keeping or increasing the costs? Can the product be recycled after use for manufacturing of new devices? Could I improve the efficiency of my battery? Can I develop a competitive car using renewable energy?

This book will gather all the new advances in terms of materials which could contribute to a circular economy, including the new ways of production, management, recycling and conversion of new and regular materials.

The importance of microplastics is addressed in **Chapter 1** where the most common degradation processes of plastics and the techniques used for their quantification and separation processes are exposed. Specifically, the food plastic packaging is growing dramatically, contributing to the increase of microplastics. For that reason, new alternatives like lignocellulose-based materials obtained from biorefinery for food packaging could reduce the problem (**Chapter 2**). On the other hand, the use of solar energy for recycling plastics could contribute to the sustainable development in terms of reduction of the accumulation in land fields and energy efficiency for a sustainable development (**Chapter 3**). The use of biopolymers and derivatives can be a successful way to replace a huge part of the plastics known as commodities (**Chapter 4**). The reduction of the greenhouse gases emission, CO_2 specifically, is another important field where numerous research groups are working on developing polymeric systems for its capture (**Chapter 5**).

The circular economy tries to enlarge the lifetime of the products, devices and structures. One of the most important problems of the metallic structures is the corrosion, and the development of smart anticorrosive polymeric coatings could mitigate this issue (**Chapter 6**). On the other hand, the use of residues as raw materials could be an effective way to promote sustainability such as the use of scrap for new steel (**Chapter 7**).

A generation of new specific nanomaterials can provide multiple benefits to many applications such as energy storage, catalysis, and water remediation (**Chapter 8**). On the other hand, **Chapter 9** involves the advances in the treatment of waste derived from electronic components of cars, which could open a new perspective in the automobile industry.

The importance of a new generation of hydrogels for 3D printing and their possibilities will be treated in **Chapter 10**, whilst the use of lignocellulosic residues as a feedstock for the development of biorefinery systems will be addressed in **Chapter 11**. Finally, the use of polymeric electrolytes in fuel cells will be discussed in **Chapter 12** showing the recent advances in terms of energy storage.

The book will be interesting for the industry and researchers, as they will find the main research lines associated with the circular economy. In addition, this book could be useful for subjects related to new bachelor and master degrees based on sustainability.

Alberto García-Peñas – Gaurav Sharma

New Materials for a Circular Economy   Materials Research Forum LLC
Materials Research Foundations 149 (2023) 1-23   https://doi.org/10.21741/9781644902639-1

Chapter 1

From Plastics to Microplastics: Quantification, Degradation and Mitigation

P. Herrasti[1*], N. Menendez[1], J. Sanchez-Marcos[1], F.J. Recio[1] and E. Mazario[1]

[1]Universidad Autónoma de Madrid, Facultad de Ciencias, Departamento de Química Física Aplicada, 28049 Madrid. España

pilar.herrasti@uam.es

Abstract

This chapter focuses on the transformation of plastics into smaller units such as microplastics. The most common degradation processes and the techniques used for their quantification and separation will be discussed. We will also try to see the most common and harmful processes for the environment that microplastics can produce, especially focusing on the processes of adsorption of pollutants that will be transported together with microplastics as a vector of pollution. Finally, we will indicate the methodologies used for the reuse of plastics, with the interest of some newer processes. We will end with some issues that should and must be addressed in the near future to prevent further contamination of the environment by these products.

Keywords

Plastic, Microplastic, Separation, Degradation, Adsorption

Contents

1. Introduction

Today, people are aware of the excessive consumption of raw materials and energy, this will have significant repercussions for our future. For this reason, there is a need to increase the useful life of what we produce. The emerging circular economy can be defined as "a model of production and consumption that involves sharing, renting, reusing, repairing, refurbishing and recycling existing materials and products as often as possible to create added value". Extending the useful life of consumables leads to a significant reduction in the levels of waste deposited in the environment. This type of economy is born in contrast to the current production system that occupies practically all industrial sectors and is characterized by consuming raw materials, processing them, and converting them into products that reach the market and that, in general, only have a single cycle of use. It is a thoughtful approach that considers the chemistry of the materials, the design of the products, the processes required to manufacture them, and, finally, the systems in the production chain, all in the context of not harming people and the environment. When it comes to plastics, these materials are used globally and are substitutes for other materials such as glass, cardboard, etc. Still, plastic waste is long-lasting and accumulates over time. Impacts on wildlife are widespread, leading to the collective conclusion that plastic in the environment is harmful and causes damage to the environment. The work of scientists is fundamental to understanding the impact of plastics and how to solve the problems that arise from their use, accumulation, and degradation. We can mention three possible solutions to reduce the effects of plastics: (1) identify and quantify the sources of land-based plastics, (2) expand zero waste strategies, and (3) develop new business solutions, both for new developments and for reuse. The bridge between the linear and circular economy consists of material circularity, which requires economic investment and corporate responsibility toward these ends.

The present society cannot be imagined without the presence of plastics; the growth in the production of plastics has been substantial in recent decades, currently amounting to more than 320 million tons per year [1]. Most plastics become solid waste in the trash, and only 6-26% is recycled. Plastics are produced by the polymerization of monomers and additives, and therefore their composition and characteristics differ. While most commonly used plastics are not biodegradable, their environmental half-life varies depending on the type of polymer and environmental conditions but ranges from days to centuries. The most common plastics detected in the environment are polyethylene (PE), polyvinyl chloride (PVC), polypropylene (PP), polyethylene terephthalate (PET), and polystyrene (PS)[2][3]. When these plastics are of macroscopic size can be collected, and their accumulation provides visual damage rather than contaminants. The problem is amplified when these plastics are reduced in size, making them considerably more difficult to recover. So-called microplastics (MP) or nanoplastics (NP) can derive from two sources. The first is from intentionally manufactured, primary MP, for example, microbeads in personal care products or industrial abrasives for delicate surfaces. Microbeads are also used in cleaning products, coatings, paints, drilling fluids in the oil and gas industry, and precursor resins and pellets to manufacture finished plastic products. Secondary MPs are formed from the fragmentation of larger plastics during use (e.g., tire wear particles) in the collection and compaction process or by environmental degradation of macroplastics. Degradation can occur through multiple methods, including hydrolysis, photodegradation by ultraviolet light, mechanical abrasion, and biodegradation[4],[5],[6]. Degradation mechanisms are not uniform for all plastics. For example, PS and PE are more sensitive to UV light than other plastics. MPs are of particular interest because of their characteristics related to their small size: 1) they can be transferred rapidly and over long distances in the environment; 2) they have a large surface area for rapid sorption of contaminants and release of sorbates and constituent chemicals; 3) they can easily enter the food web and move up the trophic levels to reach humans, and 4) they could migrate through animal tissues[7]. Figure 1 shows a summary of where MPs come from, what kind of materials can be adsorbed on them, and their possible degradation or aging and their migration towards adsorption by plants, the last step be ingestion by humans.

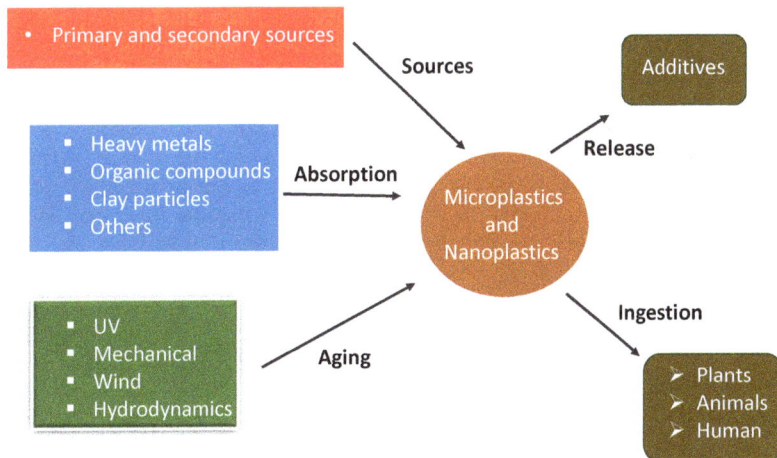

Figure 1. Different processes during the life of microplastics in the environment [8].

There are many difficulties in sampling and analyzing MP contamination, especially in complex ecosystems. The first step in evaluating the amount of microplastics in a given area is to collect a representative sample to know the amount and size distribution of this sample. Countless published studies have focused on sampling specific areas where there are large amounts of plastics, such as the sea surface, coasts, or beaches. Different studies conclude that very different values in the size distribution of the MPs were found [9],[10]. On the other hand, there is difficult to determine these in areas such as soils, where the complexity of these matrices is enormously high. In soils, physical properties such as organic matter content, particle distribution, pH, and bulk density play a critical role in the efficiency of extraction methods [11]. Again, in the literature, very controversial data are found where some studies report recovery efficiencies of 85-100% [12] while others are much lower at 5-75% [13]. From all of the above, we can say that sampling and collection of Mps prior to their possible elimination are fundamental. Therefore, it seems evident that efficient processes are necessary to separate these polymeric particles from both industrial and municipal wastewater flows to prevent their arrival in the natural environment where they accumulate. However, it seems that most research is focused on the characterization and quantification of microplastics and the study of their potential hazards. Although this research is of great importance, the search for feasible separation techniques for large water streams seems to be subordinate to the research on microplastics.

Effective separation is particularly difficult due to two aspects: (1) large volumes of water with low solids concentrations and (2) small particle sizes, in the lower to submicrometer range. Figure 2 [14] shows different centrifugation and sedimentation techniques that can be employed for the separation of different particle sizes. This figure only represents an overview and does not characterize the exact separation limits of each of the processes. It is clear from the figure that a general trend is that with decreasing particle size, the scale of the process and the volume of treated water also decrease. In general, sedimentation devices tend to have problems when the density difference between the solid and liquid phase is not significant, which limits the separation in case of low-density MPs. Of all the techniques in Figure 2, ultrafiltration is an exception, as it is capable of working with large volumes of water, small particle sizes and is effective in dilute suspensions, but although it has many advantages it also has some disadvantages, the most important being the size of the equipment and the additional flow resistance resulting from the pressure drop. There are other methodologies or techniques that are also applied to the separation of microplastics in aqueous media such as chemical coagulation [15], flotation [16] and magnetic seed filtration [17,18].

As can be deduced, the traditional or most widely used methods for the separation of MPs have a number of drawbacks that require the development of new technologies where the aforementioned inherent problems are diminished in order to achieve an efficient separation from the medium in which they are found.

As already mentioned, when the separation of MPs from the medium is not effective, they can cause the adsorption of pollutants and metals that are carried away and cause contamination not only by the accumulation of the MPs but also by the compounds adsorbed on them.

In addition to the recovery of MP, another point to be addressed is the search for new applications for the reuse of the recovered MP and plastics. In industry, these are reused by manufacturing new plastics or materials in which they are incorporated. Giving a new life to these materials is a very important research goal in the coming years.

This chapter will try to give some ideas on the latest advances in the separation, characterization, use and reuse of MP, providing new perspectives and developments in this field. It will also show that a circular economy is of vital importance to avoid pollution and the excessive and unnecessary use of raw materials and energy for the manufacture of plastics.

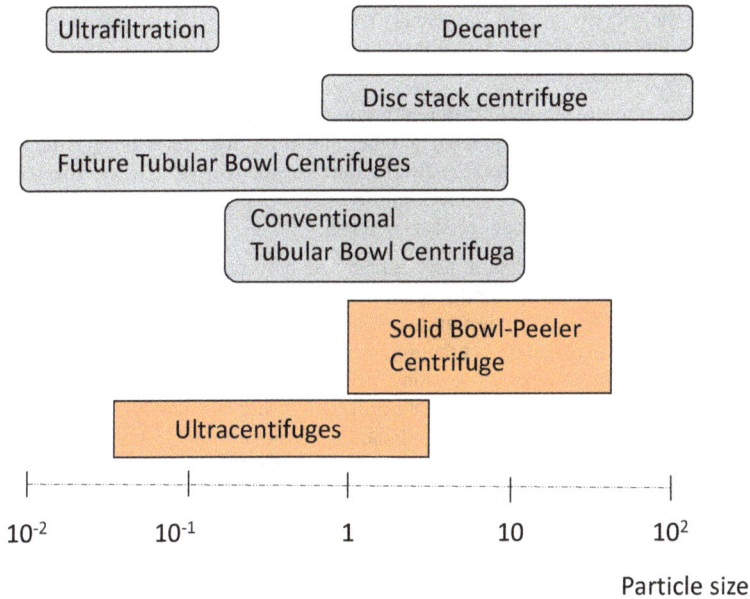

Figure 2. Different types of centrifuges used for the separation of particle size.

2. Microplastic separation

In a first approach, a visual separation of non-plastic materials such as, gravel and organic matter should be performed. In general, this process does not involve any modification of the MPs since no chemicals are added. Once the MPs have been separated, they can be separated by their density. To do this, in general, an aqueous suspension is prepared to which a salt is added, usually NaCl, which increases the medium density and allows the lighter plastics to be separated by flotation [19]. The density of plastics is between 0.8 and 1.6 g cm^{-3}, and this density is greatly influenced by the concentration of additives in the matrix, the type of polymer and even by adsorbed substances and organisms. Most studies use NaCl solutions whose density is 1.2 g cm^{-3}, this density is not too high, so polymers with high density such as polyvinyl chloride and polyethylene terephthalate cannot be easily separated with NaCl [20]. In this case other salts can be used, such as NaI, which has a density of 1.8 g cm^{-3}. One of the problems when using this salt is that it has a high price, so a first separation can be made with NaCl separating a high percentage of low density MPs and then a separation with NaI, with which the amount of NaI needed is much lower [21].

When MPs are found in biological tissues, it is common to perform an acid or basic digestion for their recovery. HNO$_3$ is usually used as acid and KOH or NaOH as base. The problem with this methodology is that many of the acids and bases used can degrade the MPs, resulting in inconsistent recoveries.

Another option is to use sieves of different types and mesh sizes. The materials remaining on the surface of the sieve are collected and the smaller particles pass through the sieve. In general, several sieves are used to help differentiate between different sizes. This technique allows the separation of plastics from 5 mm to 0.2 µm, although filters for microplastics separation with small mesh sizes are also used [6].

When particles are smaller than 1 µm, active and passive chromatographic separation techniques are often used. Among the active separations is field flow fractionation (FFF), in which external fields are applied in microfluidic environments. The technique consists of injecting a fluid suspension by pumping it into a long, narrow channel under a force field perpendicular to the flow direction. This field causes the separation of particles present in the fluid, depending on their mobility under the force field [22]. In contrast, a passive separation occurs with hydrodynamic chromatography (HDC) which uses hydrodynamics and surface forces to separate particles from the fluid. In this methodology the advantages of fluid dynamics and exclusion chromatography are combined and particles can be separated based on size differences [23].

New recovery methodologies have emerged in recent years, mostly related to the use of magnetic materials that can be used to attract microplastics. The use of magnetic materials to recover small particles of plastics relies on the hydrophobicity of most MPs in order to magnetize them. J. Grbic et al. [24] synthesized hydrophobic Fe nanoparticles to be bonded to the MPs, thus allowing their separation from the medium by means of a magnet. By means of this technique they were able to recover percentages between 78 and 93 % of MPs from polyethylene, polyethylene terephthalate, polystyrene, polyurethane, polyvinyl chloride and polypropylene, in seawater. The recovery percentage is highly dependent on the size of the microplastics and decreases as the size of the microplastics increases. This methodology could be incorporated as one of the steps in the extraction of MPs, for example after their separation by density, and could be a methodology to be applied for the purification of drinking water.

Along the same lines as the aforementioned work, the possibility of using magnetized carbon nanotubes (M-CNT) for the recovery of MPs has been studied. Using this methodology, PE, PET and PA (polyamide) have been efficiently recovered by adsorption of the magnetic carbon nanotubes. The adsorption capacities vary from 1600, 1400 and 1000 mg M-CNT/g for PE, PET and PA, respectively. The process was performed with

MPs from a kitchen waste treatment plant. The M-CNTs can be recycled presenting magnetic properties comparable to the original ones. Although it is a new technique and its possibilities have not been tested at the industrial level, it may provide a new insight into the separation of microplastics [25].

A very interesting work was carried out by Wang J. et al. [26] in which different microplastics were coated with magnetite nanoparticles (Fe_3O_4). To perform this synthesis, a coprecipitation method was used with Fe^{2+} and Fe^{3+} salts in a basic medium, which provides a thin layer of magnetic nanoparticles on them. With this, the surface of the plastics is hydrophilized and the flotation separation of PVC in a mixture of MPs was evaluated. Due to the weak interaction of the Fe ions with the PVC surface, the coating on this plastic is much lower, which produces a reverse flotation behavior between the PVC and the rest of the plastics, facilitating in this case their separation. After systematic optimization of the process, flotation separation of PVC was achieved at 100% recovery values.

The preparation of ferrofluids for the recovery of MPs has also been studied recently. These ferrofluids were synthesized without the addition of stabilizing agents or surfactants by using oils and iron oxide powder. A volume ratio of oil: magnetic powder of 1:2.5 was achieved by removing 99% of PET in simulated water, with the percent removal of MPs in real water from laundry water being 64%. Although the methodology seems practical for the recovery of MPs, it must be taken into consideration that the use of high oil content in relation to the plastic content affects the yield and therefore the percentage of MP removal. Another fact demonstrated is that the dosage of magnetic oxides must be high to increase the magnetic field strength and thereby increase the overall extraction yield [27].

Another of the methodologies that have been tested, for the time being at laboratory scale, is the use of micromotors. There are not many studies in this direction, but they can serve as a proof of concept for their subsequent application on a larger scale. Micromotors are compounds within the so-called smart materials [28] and although there are many types and mechanisms by which they can move, we will focus on two examples from the literature, MnO_2 and TiO_2 microparticles, the former are driven mechanically and the latter by light. MnO_2 micromotors were developed by Wang et al.[29] and since this first work an infinite number of MnO_2-based micromotors have been developed, their main applications being biomedical and environmental [30]. In relation to environmental applications is the work developed by Heng Ye et al. [31] in which they synthesized self-propelled and magnetically steerable iron oxide-MnO_2 micromotors. These micromotors were capable of both removing organic dyes and separating MPs from contaminated water.

With respect to TiO_2 micromotors, photocatalytic micromotors of the type (Au@mag@TiO_2, mag = Ni, Fe) have been synthesized for the removal of MPs [32]. It has been demonstrated that this micromotor can move in both peroxide medium and water under UV illumination. In the work two types of micromotors were synthesized, single and assembled chains. With these systems they tried to remove MPs extracted from face cleansing creams, also using a very low concentration of hydrogen peroxide 0.2%. They obtained a removal efficiency of 71%, when no hydrogen peroxide was used the efficiency was reduced to only 67%. This small difference concludes that these micromotors can be used in both environments.

3.   Detection and quantification of microplastics

Due to the difficulty of monitoring MPs in the environment, even decades later, there is still not enough data to obtain a complete picture of the amount and also the available techniques for the detection and quantification of MPs still limited [33]. The first step, which is crucial, is to recognize and detect the type of MPs in the study matrices, and for this it is necessary to collect a representative sample that fits the research objectives. Different reviews have been published on the methods of detection and quantification of MPs [34,35,36]. The detection and subsequent quantification of MPs is challenging in many respects, with analytical determination being one of the main problems. There are different units to quantify their abundance. For water and sediment/sand, the units that have been used in the literature include a) pieces of MPs per unit area (p/m^2), per unit volume (p/m^3) or per unit mass (p/kg) or per sample site (p/beach) and b) mass of MPs per unit area (g/m^2), per unit volume (g/m^3) or per unit mass (g/kg) of sample.

In first approximation, MPs are usually identified visually with a microscope; in general, it is useful when these MPs are between 1 and 5 mm in size. This visualization can be very important to evaluate MPs of different color, shape and light transmission and to some extent to distinguish plastics from non-plastics. However, the method has low reliability, especially in the case of small, transparent and fiber-like particles [37]. In addition, another problem with the use of microscopes alone is that they cannot provide the chemical composition and therefore cannot differentiate between mixtures of synthetic polymers. In order to detect smaller particles, scanning electron microscopy (SEM) is necessary and if coupled with energy dispersive spectroscopy (EDS) can simultaneously determine the morphology of very small materials and also determine their composition. Although it is a powerful technique, it has drawbacks such as the fact that the sample must be conductive, MPs are not conductive and therefore in order to be observed they must be coated with a layer of carbon or gold. This increases the cost of the process, as well as introducing artifacts and interfering with the morphological analysis of the MPs. Another disadvantage to be taken into account is that in order to visualize plastic materials by SEM, a high-energy electron beam is used, which can lead to melting of certain plastics during the measurement process.

To advance in detection and also in quantification, the most widely used techniques today are spectroscopic. One of the most widely used for detection is Fourier transform infrared spectroscopy (FTIR). This technique provides chemical information by detecting the vibrational modes of the molecules in the sample. Each of the absorbance peaks corresponds to specific chemical bonds or molecular vibrations. Many authors have studied MPs using this technique, for example several types of MPs in the water of the Ross Sea have been analyzed [38] and also has been reported the identification of MPs in costal environment [39] and in the artic sea [40]. A very important advance in this field is the development of FTIR micro-spectroscopy (μ-FTIR), which allows a chemical mapping that has produced an important breakthrough in the detection of microplastics. The technique consists of combining IR spectroscopy with an IR microscope. This technique allows the detection of particles below 10 μm. It has been used in different environments such as water [41], sediments [42], atmosphere [43], organisms [44] etc. Another of the spectroscopic techniques also widely used for the detection of MPs is Raman spectroscopy, the technique is based on the inelastic scattering of light that occurs when matter is irradiated by a monochromatic light source. When this light interacts with the sample a part of it changes wavelength, which is called Raman Effect. Raman spectra give a representation of the intensity of the scattered light versus wavenumber. As in the case of IR spectroscopy, many articles have used this technique to identify MPs [45,46]. In order to detect particles down to 1 μm it is necessary to use Raman micro-spectroscopy. S. Karbalaei et al. [47] identified MPs in three Malaysian commercial brands of fishmeal using this technique. A study similar to the one developed by other authors with the μ-FTIR technique but using the μ-Raman technique has been developed by Kazour M. et al. who studied MP contamination in three different matrices (water, sediments and biota) in the Mediterranean [48] . One of the difficulties of these techniques is that the spectra are compared with library spectra to determine the type of MPs, but in many cases these plastics are modified over time and their spectra deviate from those that would correspond to the fresh materials. Currently, some libraries of different MPs that have been deteriorated are being elaborated in order to compare them with those obtained experimentally [49].

Another spectroscopic technique that has been used for the identification of MPs and recently commercialized is the photothermal infrared spectroscopy, which consists of a visible light probe to measure the photothermal response of the target particles after IR irradiation [50].

Other detection methodologies require the destruction of the MPs, for example, by vaporizing them, their components can be analyzed by chromatographic separation and subsequent identification by mass spectroscopy [51], or by thermogravimetry coupled to differential scanning calorimetry [52]. These techniques can undoubtedly detect the presence of MPs, but the destruction of the MPs implies that it is not possible to evaluate their size and shape. This problem can to some extent be solved if prior to pyrolysis a size separation of the microplastics is performed [53].

Spectroscopic techniques have advanced in the quantification and differentiation of MPs. But as in determination, in the case of quantification many assays are required to obtain reliable spectra especially of small and in many cases aged particles. It is therefore desirable to develop alternative methods that facilitate the identification and quantification of MPs. In this regard, Won Joon Shim et al. [54]. developed a method for the identification and quantification of MPs using a Nile red staining method. The method consists of using this fluorescent dye to color the plastics so that they can be determined by green fluorescence. It was used for the determination of MPs such as polyethylene, polypropylene, polystyrene, polycarbonate, polyurethane and polyethylene-vinyl acetate although some MPs such as polyvinyl chloride, polyamide and polyester could not be detected. Although the method is simple and fast and quantifies a large number of polymers, however with an added problem is that the MPs must be separated from organic residues, as co-staining of this organic matter occurs. It is therefore very important to remove this organic matter before staining. This poses a major problem in the use of this methodology as there are no absolutely reliable methods to remove all organic materials other than plastics in the samples.

The quantification of MPs remains one of the many unresolved issues. The biggest problem is the lack of standardization as such quantification is carried out in many ways leading to very different values of MP content.

As a summary of the most common locations where MPs are found, their separation and subsequent characterization and quantification can be seen in Figure 3.

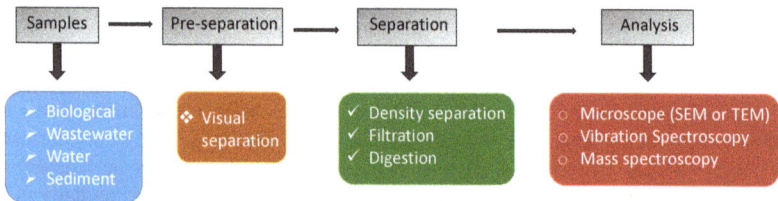

Figure 3. Methods of separation and analysis of microplastics for different samples [55].

4.   Adsorption of contaminants by microplastics

MPs are materials that interact with a large number of chemical substances in the environment, including the adsorption of heavy metals and organic pollutants. Once these toxic substances are adsorbed on MPs they can be transported together with them, which is why MPs are considered to be vectors of hazardous pollutants. This adsorption phenomenon is produced by different types of interactions between the MP and the toxic substance. According to Torres F.G. et al. [56] there are six mechanisms involved in adsorption; these can be seen in Figure 4.

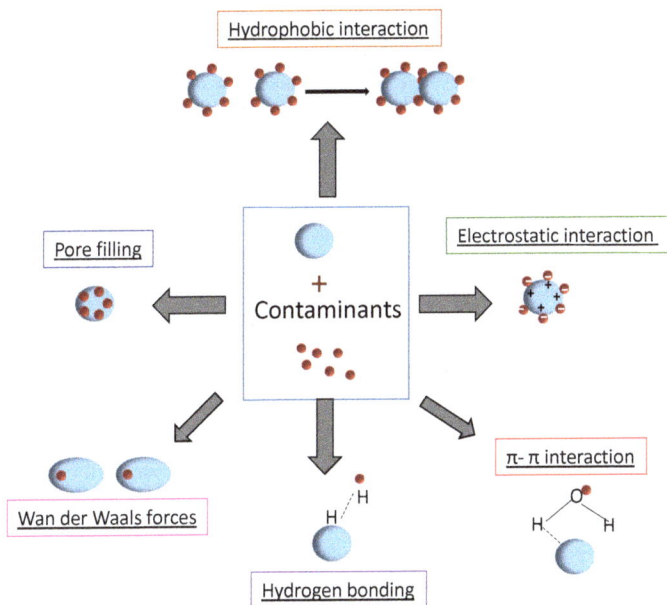

Figure 4. The type(s) of mechanism(s) by which adsorption occurs will depend on the characteristics of the MP and of the chemical contaminant (adapted of the reference [56])

Hydrophobic interactions are due to the attraction between two non-polar substances, which results in the formation of aggregates or clusters, in which case only hydrophobic contaminants will be adsorbed on the surface of the MPs. Electrostatic interactions occur because the MPs have opposite charge to the contaminant. To know the surface charge of the MPs, the measurement of the Z potential, whose value changes with the pH of the medium, is used. There is a pH for which the Z potential is zero, called the zero-charge potential. For pH below this value the surface becomes positively charged and will attract negatively charged chemical compounds, and vice versa if the pH is higher than the zero-charge potential. This type of adsorption by electrostatic interactions also occurs with metal ions such as copper, lead, cadmium, etc., being in this case the most important interaction mechanism.

The pore-filling mechanism consists in the adsorption of the contaminant that is entirely trapped in the pores of the MP [57]. Some examples of this type of mechanism have been observed in the adsorption of oxytetracycline (OTC), a broad-spectrum antibiotic, on virgin and damaged polystyrene microplastics [58], or in the adsorption of pesticides such as DDT and phenentrene on PVC and PE [59]. In general, this type of adsorption occurs on MPs with a mesoporous network with large pore sizes.

Adsorption by hydrogen bonding can occur when both proton donor and acceptor groups are present. This type of mechanism is not very common and does not occur in a wide range of polymers but this type of adsorption has been detected in some cases, as for example in the study by Liu et al who investigated the adsorption of 17 β-Estradiol on different polymers, and concluded that for example it presents a high adsorption on PA due to the presence of hydrogen donors that can form hydrogen bonds with the amide groups of the plastic. This fact was not observed for other MPs such as polycarbonate or polymethylmethacrylate [60] .

The last two types of adsorption that can occur are due to Van der Waals forces or π-π interactions between the MP and the contaminant compound. Although these types of adsorption can occur, they occur to a much lesser extent than the interactions cited above. Velzeboer et al. [61] indicated that the high aromaticity of PS MPs significantly increases the adsorption of PCBs due to π-π and hydrophobic interaction mechanisms, which occur simultaneously

It is interesting to mention a study carried out by Ateia M. et al. [62] where the differences that exist in the literature on the study of adsorption of contaminants in regular plastics in size and purity, compared to the plastics used in a real life natural environment whose shapes are irregular and have a large number of compounds in their composition. The most important conclusion in this work is that in laboratory studies the MPs that are analyzed are few and the adsorption experiments are performed under different experimental conditions, so the data found are very scattered and the classification of the adsorption capacity of MPs is not very realistic and can be misleading. Therefore, to know and to be able to interpret correctly the adsorption of pollutants on plastics it is convenient to use real plastics and to work in the same experimental conditions for the adsorption of pollutants.

There are countless studies on the adsorption behavior of MPs in different types of scenarios. Most studies have focused on laboratory experiments where the MPs used are virgin. However, secondary MPs derived from the degradation of the former are the main source of MPs in the real environment. The processes that produce this degradation are many, such as ultraviolet light (UV), temperature, mechanical friction, etc. The properties of degraded MPs differ considerably from non-degraded ones with respect to their roughness [63], the functional groups they contain [64] and the specific surface area [65]. Other processes that cause the aging of MPs are abiotic and biological processes, although it is a process to be taken into account, there are few studies in this direction. For example, Johansena et al. [66]. found that the formation of a biofilm on microplastics accumulated more adenonucleotides than the same virgin MPs.

Although all these processes occur in nature, the problem is that their effect is low over short periods of time which greatly restricts the understanding of the long-term weathering behavior of MPs. As an alternative to natural processes, accelerated aging processes are carried out in the laboratory using Fenton or Photo-Fenton processes that can give us information about the long-term aging of MPs. This aging is performed by artificially controlling the intensity of UV rays and the concentration of reactive oxygen species (ROS) produced in the aforementioned processes. It is evident that the adsorption of pollutants will be very different in virgin and degraded MPs.

The use of antibiotics has increased enormously in recent decades, so that they can reach the sea by different routes. This has aroused the interest of scientists, since antibiotics induce the expression of resistance genes in microorganisms that affect plants and animals and can reach the food chain. Xiaowe W. et al. [67] studied the adsorption of triclosan (TCS) on different aged polypropylene MPs concluding that the aged MPs showed higher TCS adsorption capacity, especially in high salinity waters. They also showed that adsorption is controlled by the oxygen/carbon ratio and the ionic strength of the water.

Adsorption studies of oxytetracycline (OTC), conducted by Zhang et al. [58] showed that OTC adsorbs on PS foams and this adsorption is much more intense when the PS is aged. The adsorption process changes with the conditions of pH, ionic strength and humic acid concentration. This fact demonstrates that adsorption processes can change drastically with environmental conditions.

In the case of tetracycline (TC), adsorption studies have been carried out on different types of MPs (PE, PS and PVC). The amount of TC adsorbed was shown to depend on the polarity of the type of MPs and that the interparticle and extra particle diffusion processes are major drivers in the adsorption process. The surface behavior of plastics was studied, and substantial changes of carbon-carbon double bonds were found to disappear after TC adsorption. This fact may be due to substitution or to the formation of a free hydrogen bond [68].

Other authors have studied the degree of accelerated PS aging by UV light irradiation in different conditions (air, pure water and seawater). The degree of aging was studied by 2D-FTIR-COS and it was concluded that UV irradiation produces a higher deterioration of the MPs in air, followed by seawater and finally pure water. They also showed that for different heavy metals such as Pb^{2+}, Cu^{2+}, Cd^{2+}, Ni^{2+} and Zn^{2+} adsorption increases with increasing degree of aging [69].

The adsorption of two different pharmaceutical compounds, atorvastatin and amlodipine, on aged PS was also studied using a photo-fenton process [70].These compounds have been frequently detected in natural and contaminated waters at concentrations of ng/L or μg/L. Similarly, the authors demonstrated that the behavior was very different between virgin and aged MPs. In this work, it was observed that by the aging process, intermediate products are generated in high concentration that significantly affect the adsorption of both pharmaceutical compounds. This does not happen when the concentration of the intermediates formed is low. Therefore, it can be deduced that the intermediates during the degradation process play a critical role in the adsorption of the studied compounds. The effect of aging by a Fenton process and adsorption of Cd^{2+} ions on PS MPs has been studied by M. Lang et al. Their results have shown that aging not only provides an increase in the surface area of the MPs, but also the oxidation of functional groups on the PS surface, which produces an increase in the interaction of several orders of magnitude with respect to virgin PS [71].

Turner et al. also studied the adsorption of metals on virgin and aged PE. They again demonstrated an increase in adsorption capacity when these MPs were degraded. The amount of adsorbate varies per gram of MPs and can increase for example from 55-69 in virgin MPs to 160-175 mg/g in the case of Cu^{2+} adsorption [72].

In addition to significantly increasing the adsorption capacity of contaminants, in the course of the plastic degradation process in the environment, many of the additives they contain (plasticizers, flame retardants, antioxidants, lubricants, pigments, etc.) may be released, increasing the toxicity of the system in which they are found [73]. There are few studies on the content and release of metals into the environment by MPs, although there is evidence of the presence of Zn and Cd. There are also studies where it has been suggested that a large part of the trace metals associated with MPs derive from an inherent load, not from subsequent adsorption [74]. For example, Turner et al. [75] have shown in a field study that Pb additives in the MPs themselves, especially PVC, have a greater environmental impact than the Pb adsorbed by them.

5.   Mitigation

Before we talk about MPs mitigation, we need to talk about how to start at a higher order, i.e. how to mitigate the macroplastics problem. A reduction of these in the medium will undoubtedly result in a lower amount of microplastics.

One way to reduce these is to recycle macroplastics, when we talk about recycling, in most cases it is understood that the plastic is mechanically shredded for later reuse, and also this type of recycling can lead to the production of large quantities of MPs, which is undesirable. It must be taken into account that this type of recycling causes a degradation in the properties of plastics after multiple uses. Nevertheless, this kind of recycling has demonstrated a successful reduction of plastic waste creating other products. For example, the production of clothes, shoes, sun-glasses etc. from derelict fishing gear. Other approaches to plastic waste conversion focus on the production of synthesis gas or fuels and naphtha [76].

When mechanical recycling is not possible, as in the case of sheet, contaminated and mixed plastic waste streams, as well as multilayer packaging products, alternative recycling techniques must be employed. All the processes involved are aimed at achieving a circular economy by ensuring that the monomers and oligomers of which plastics are composed can be re-polymerized after purification and used in the same or similar applications as the virgin polymer equivalent, which is produced from fossil fuels. All the techniques in use or emerging are crucial to move towards a circular economy of plastics, producing an increase over the current percentage by weight of recycled plastic globally, while providing solutions.

Some of the most commonly used methodologies for plastics processing are: mechanical processes, pyrolysis, dissolution, solvolysis, incineration and thermal treatment. Figure 5 shows these processes [76].

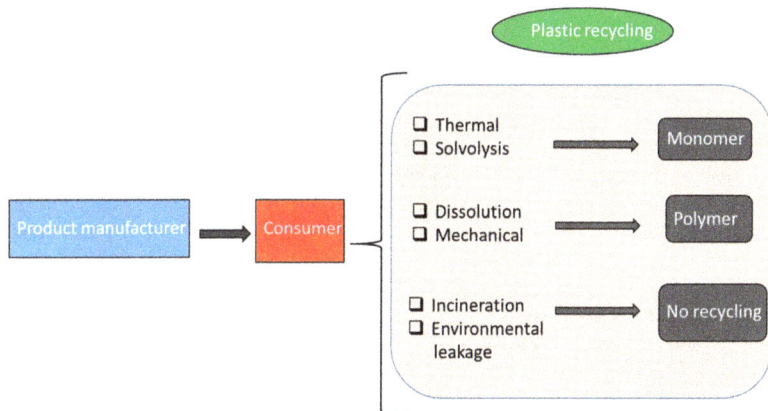

Figure 5. Different recycling methods to obtain monomers, polymers or complete destruction of non-reusable plastic (adapted of reference [76]).

Some of these methodologies cause the total disposal of plastics into the environment either by incineration or direct discharge. Others, such as thermal treatment or solvolysis, produce the decomposition of the plastics into their respective components and finally others cause the reduction in size for their subsequent recyclability in the same or similar products. The use of one or other methodologies depends on the type of plastic to be treated.

There are new approaches, not as widely used but that can solve some of the problems that occur in the above-mentioned methodologies, such as the low thermal conductivity of the plastic and the general problems of heat distribution in industrial reactors. Some approaches use alternative forms of energy supply for depolymerization, such as mechanochemical approaches, photo-reforming, microwave and plasma reactors, which have the potential for more efficient heating. Alternative solvents are also being pursued, such as supercritical fluids, ionic liquids (ILs) and deep eutectic solvents, which yield higher conversion and greater selectivity.

Some recent works try to use plastics in conjunction with other materials such as magnetic nanoparticles to develop multifunctional products [77].

If we focus on MPs, they cannot be reused in the same way as macroplastics because their quantity in certain areas is very low and they cannot be reused or reused again. Therefore, the most important problem is to try to minimize their use in products and to reduce and recycle the plastics we use. It is clear that environmental concentrations may change in the long term, and then a better understanding of the amount of the MPs in different geographical regions will identify areas where the mitigation actions will be most effectives.

Conclusions

Although there has been great progress in the knowledge, quantification and reuse of MPs, there are still many points to be understood and elucidated that should help us in this field to achieve the circular economy mentioned at the beginning of this chapter. In order to understand the problem, we are facing, it is important to understand the complexity of the composition of plastic products. Their number and diversity is so wide that we are faced with materials that may contain an infinite number of components that in the short or long term may cause damage to the environment. The study by means of advanced analytical methods will make it possible to detect not only MPs but also NPs that, due to their surface/volume ratio, can interact more easily with toxic components in the environment and facilitate their transport from one place to another. In the same scenario is the attempt to analyze and understand the effect of particle size on their biodegradability. The complexity of the interactions of MPs with the environment needs to be studied in greater depth, and to this end the toxicity of the great variety of plastics that exist in our environment needs to be analyzed in the laboratory but also in the environment.

In order to solve or at least minimize the future problems that MPs, and plastics in general, chemists and chemical engineers must find new and improved processes for recycling a wide variety of commonly used plastic materials, and much progress can be expected in the coming years. On the other hand, it must be recognized by governments that the accumulation of plastics is a global problem that needs to be tackled as their harm knows no boundaries and these scientific and technological advances will need to be accompanied by improved policy frameworks and platforms that connect all important stakeholders.

Acknowledgements

This research was funded by Ministerio de Ciencia e Innovación, project number PID2021-123431OB-I00

References

[1] S. L. Wright and F. J. Kelly, Plastic and Human Health: A Micro Issue?, Environ. Sci. Technol. 51 (2017) 6634-6647. https://doi.org/10.1021/acs.est.7b00423

[2] C. P. Ward, C. J. Armstrong, A. N. Walsh, J. H. Jackson, and C. M. Reddy, Sunlight Converts Polystyrene to Carbon Dioxide and Dissolved Organic Carbon, Environ. Sci. Technol. Lett. 6 (2019) 669-674. https://doi.org/10.1021/acs.estlett.9b00532

[3] C. M. Rochman, E. Hoh, B. T. Hentschel, and S. Kaye, Long-Term Field Measurement of Sorption of Organic Contaminants to Five Types of Plastic Pellets: Implications for Plastic Marine Debris, Environ. Sci. Technol. 47 (2013) 1646-1654. https://doi.org/10.1021/es303700s

[4] H. K. Imhof, J. Schmid, R. Niessner, N. P. Ivleva, and C. Laforsch, A novel, highly efficient method for the separation and quantification of plastic particles in sediments

of aquatic environments, Limnol. Oceanogr. Methods, 10 (2012) 524-537. https://doi.org/10.4319/lom.2012.10.524

[5] O. S. Ogunola, O. A. Onada, and A. E. Falaye, Mitigation measures to avert the impacts of plastics and microplastics in the marine environment, Environ. Sci. Pollut. Res. 25 (2018) 9293-9310. https://doi.org/10.1007/s11356-018-1499-z

[6] J. Li, H. Liu, and J. Paul Chen, Microplastics in freshwater systems: A review on occurrence, environmental effects, and methods for microplastics detection, Water Res. 137 (2018) 362-374. https://doi.org/10.1016/j.watres.2017.12.056

[7] A. L. Andrady, Microplastics in the marine environment, Mar. Pollut. Bull. 62 (2011) 1596-1605. https://doi.org/10.1016/j.marpolbul.2011.05.030

[8] L. Peng, D. Fu, H. Qi, C. Q. Lan, H. Yu, and C. Ge, Micro- and nano-plastics in marine environment: Source, distribution and threats - A review, Sci. Total Environ. 698 (2020) 134254. https://doi.org/10.1016/j.scitotenv.2019.134254

[9] K. Enders, R. Lenz, C. A. Stedmon, and T. G. Nielsen, Abundance, size and polymer composition of marine microplastics ≥10µm in the Atlantic Ocean and their modelled vertical distribution, Mar. Pollut. Bull. 100 (2015) 70-81. https://doi.org/10.1016/j.marpolbul.2015.09.027

[10] C. Andrés, F. Echevarría, J.I. González-Gordillo and C. M. Duarte, Plastic debris in the open ocean, Proc. Natl. Acad. Sci.vol. 111 (2014) 10239-10244. https://doi.org/10.1073/pnas.1314705111

[11] N. P. Ivleva, A. C. Wiesheu, and R. Niessner, Microplastic in Aquatic Ecosystems, Angew. Chemie Int. Ed. 56 (2017) 1720-1739. https://doi.org/10.1002/anie.201606957

[12] M. Liu, Y. Song, S. Lu, R. Qiu, J. Hu, X. Li, M. Bigalke, H. Shi and D. He, A method for extracting soil microplastics through circulation of sodium bromide solutions, Sci. Total Environ. 691 (2019) 341-347. https://doi.org/10.1016/j.scitotenv.2019.07.144

[13] Z. Wang, S. E. Taylor, P. Sharma, and M. Flury, Poor extraction efficiencies of polystyrene nano- and microplastics from biosolids and soil, PLoS One. 13 (2018) e0208009. https://doi.org/10.1371/journal.pone.0208009

[14] L. E. Spelter, A. Steiwand, and H. Nirschl, Processing of dispersions containing fine particles or biological products in tubular bowl centrifuges, Chem. Eng. Sci. 65 (2010) 4173-4181. https://doi.org/10.1016/j.ces.2010.04.028

[15] M. Lapointe, J. M. Farner, L. M. Hernandez, and N. Tufenkji, Understanding and Improving Microplastic Removal during Water Treatment: Impact of Coagulation and Flocculation, Environ. Sci. Technol. 54 (2020) 8719-8727. https://doi.org/10.1021/acs.est.0c00712

[16] C. Wang, H. Wang, J. Fu, and Y. Liu, Flotation separation of waste plastics for recycling-A review, Waste Manag. 41 (2015) 28-38. https://doi.org/10.1016/j.wasman.2015.03.027

[17] Y. S. Shaikh and P. Kampeis, Development of a novel disposable filter bag for separation of biomolecules with functionalized magnetic particles, Eng. Life Sci. 17 (2017) 817-828. https://doi.org/10.1002/elsc.201600190

[18] F. Rhein, F. Scholl, and H. Nirschl, Magnetic seeded filtration for the separation of fine polymer particles from dilute suspensions: Microplastics, Chem. Eng. Sci. 207 (2019) 1278-1287. https://doi.org/10.1016/j.ces.2019.07.052

[19] F. Radford, L. M. Zapata-Restrepo, A. A. Horton, M. D. Hudson, P. J. Shaw, and I. D. Williams, Developing a systematic method for extraction of microplastics in soils, Anal. Methods. 13 (2021) 1695-1705. https://doi.org/10.1039/D0AY02086A

[20] R. C. Thompson, Y. Olsen, R. P. Mitchell, A. Davis, and et al, Lost at Sea: Where Is All the Plastic?, Science. 304 (2004) 838. https://doi.org/10.1126/science.1094559

[21] B. Quinn, F. Murphy, and C. Ewins, Validation of density separation for the rapid recovery of microplastics from sediment, Anal. Methods. 9 (2017) 1491-1498. https://doi.org/10.1039/C6AY02542K

[22] K.-G. Wahlund, Flow field-flow fractionation: Critical overview, J. Chromatogr. A. 1287 (2013) 97-112. https://doi.org/10.1016/j.chroma.2013.02.028

[23] B. W. J. Pirok, N. Abdulhussain, T. Aalbers, B. Wouters, R. A. H. Peters, and P. J. Schoenmakers, Nanoparticle Analysis by Online Comprehensive Two-Dimensional Liquid Chromatography combining Hydrodynamic Chromatography and Size-Exclusion Chromatography with Intermediate Sample Transformation, Anal. Chem. 89 (2017) 9167-9174. https://doi.org/10.1021/acs.analchem.7b01906

[24] J. Grbic, B. Nguyen, E. Guo, J. B. You, D. Sinton, and C. M. Rochman, Magnetic Extraction of Microplastics from Environmental Samples, Environ. Sci. Technol. Lett. 6 (2019) 68-72. https://doi.org/10.1021/acs.estlett.8b00671

[25] Y. Tang, S. Zhang, Y. Su, D. Wu, Y. Zhao, and B. Xie, Removal of microplastics from aqueous solutions by magnetic carbon nanotubes, Chem. Eng. J. 406 (2021) 126804. https://doi.org/10.1016/j.cej.2020.126804

[26] J. Wang, D. Yue, and H. Wang, In situ Fe3O4 nanoparticles coating of polymers for separating hazardous PVC from microplastic mixtures, Chem. Eng. J. 407 (2021) 127170. https://doi.org/10.1016/j.cej.2020.127170

[27] S. Hamzah, L.Y. Ying, A. Azzura abd, R. Azmi, N.A. Razali, N. Hanis Hayati Hairom, N. A. Mohamad and M. H. Che Harun, Synthesis, characterisation and evaluation on the performance of ferrofluid for microplastic removal from synthetic and actual wastewater, J. Environ. Chem. Eng. 9 (2021) 105894. https://doi.org/10.1016/j.jece.2021.105894

[28] F. Soto, E. Karshalev, F. Zhang, B. Esteban Fernandez de Avila, A. Nourhani, and J. Wang, Smart Materials for Microrobots, Chem. Rev. 122 (2022) 5365-5403. https://doi.org/10.1021/acs.chemrev.0c00999

[29] H. Wang, G. Zhao, and M. Pumera, Beyond Platinum: Bubble-Propelled Micromotors Based on Ag and MnO2 Catalysts, J. Am. Chem. Soc. 136, (2014) 2719-2722. https://doi.org/10.1021/ja411705d

[30] K. Villa, J. Parmar, D. Vilela, and S. Sánchez, Metal-Oxide-Based Microjets for the Simultaneous Removal of Organic Pollutants and Heavy Metals, ACS Appl. Mater. Interfaces. 10 (2018) 20478-20486. https://doi.org/10.1021/acsami.8b04353

[31] H. Ye, Y. Wang, X. Liu, D. Xu, H. Yuan, H. Sun, S. Wang and X. Ma, Magnetically steerable iron oxides-manganese dioxide core-shell micromotors for organic and microplastic removals, J. Colloid Interface Sci. 588 (2021) 510-521. https://doi.org/10.1016/j.jcis.2020.12.097

[32] L. Wang, A. Kaeppler, D. Fischer, and J. Simmchen, Photocatalytic TiO2 Micromotors for Removal of Microplastics and Suspended Matter, ACS Appl. Mater. Interfaces. 36 (2019) 32937-32944. https://doi.org/10.1021/acsami.9b06128

[33] V. Hidalgo-Ruz, L. Gutow, R. C. Thompson, and M. Thiel, Microplastics in the Marine Environment: A Review of the Methods Used for Identification and Quantification, Environ. Sci. Technol. 46 (2012) 3060-3075. https://doi.org/10.1021/es2031505

[34] L. Fok, T. W. L. Lam, H.-X. Li, and X.-R. Xu, A meta-analysis of methodologies adopted by microplastic studies in China, Sci. Total Environ. 718 (2020) 135371. https://doi.org/10.1016/j.scitotenv.2019.135371

[35] A. B. Silva, A. S. Bastos, C. I. L. Justino, J. P. da Costa, A. C. Duarte, and T. A. P. Rocha-Santos, Microplastics in the environment: Challenges in analytical chemistry - A review, Anal. Chim. Acta, 1017 (2018) 1-19. https://doi.org/10.1016/j.aca.2018.02.043

[36] F. Stock, C. Kochleus, B. Bänsch-Baltruschat, N. Brennholt, and G. Reifferscheid, Sampling techniques and preparation methods for microplastic analyses in the aquatic environment - A review, TrAC Trends Anal. Chem. 113 (2019) 84-92. https://doi.org/10.1016/j.trac.2019.01.014

[37] Y. K. Song, S.H. Hong, M. Jang, Gi. M. Han, M. Rani, J. Lee and W. J. Shim, A comparison of microscopic and spectroscopic identification methods for analysis of microplastics in environmental samples, Mar. Pollut. Bull. 93 (2015) 202-209. https://doi.org/10.1016/j.marpolbul.2015.01.015

[38] A. Cincinelli, C. Scopetani, D. chelazzi, E. Lombardini, T. Martellini, A. Katsoyiannis, M.C. Fossi and S. Corsolini, Microplastic in the surface waters of the Ross Sea (Antarctica): Occurrence, distribution and characterization by FTIR,

Chemosphere, 175 (2017) 391-400.
https://doi.org/10.1016/j.chemosphere.2017.02.024

[39] J. P. G. L. Frias, P. Sobral, and A. M. Ferreira, Organic pollutants in microplastics from two beaches of the Portuguese coast, Mar. Pollut. Bull. 60 (2010) 1988-1992. https://doi.org/10.1016/j.marpolbul.2010.07.030

[40] M. A. Browne, P. Crump, S.J. Niven, E. Teuten, A. Tonkin, T. Galloway and R. Thompson, Accumulation of Microplastic on Shorelines Worldwide: Sources and Sinks, Environ. Sci. Technol. 45 (2011) 9175-9179. https://doi.org/10.1021/es201811s

[41] P. Wu, Y. Tang, M. Dang, S. Wang, H. Jin, Y. Liu, H. Jing, C. Zheng, S. Yi and Z. Cai, Spatial-temporal distribution of microplastics in surface water and sediments of Maozhou River within Guangdong-Hong Kong-Macao Greater Bay Area, Sci. Total Environ. 717 (2020) 135187. https://doi.org/10.1016/j.scitotenv.2019.135187

[42] G. Peng, R. Bellerby, F. Zhang, X. Sun, and D. Li, The ocean's ultimate trashcan: Hadal trenches as major depositories for plastic pollution, Water Res. 168 (2020) 115121. https://doi.org/10.1016/j.watres.2019.115121

[43] K. Liu, X. Wang, T. Fang, P. Xu, L. Zhu, and D. Li, Source and potential risk assessment of suspended atmospheric microplastics in Shanghai, Sci. Total Environ. 675 (2019) 462-471. https://doi.org/10.1016/j.scitotenv.2019.04.110

[44] B. Nan, L. Su, C. Kellar, N. J. Craig, M. J. Keough, and V. Pettigrove, Identification of microplastics in surface water and Australian freshwater shrimp Paratya australiensis in Victoria, Australia, Environ. Pollut. 259 (2020) 113865. https://doi.org/10.1016/j.envpol.2019.113865

[45] M. Löder and G. Gerdts, Methodology Used for the Detection and Identification of Microplastics-A Critical Appraisal. Springer, 2015. https://doi.org/10.1007/978-3-319-16510-3_8

[46] S. Zhao, M. Danley, J. E. Ward, D. Li, and T. J. Mincer, An approach for extraction, characterization and quantitation of microplastic in natural marine snow using Raman microscopy, Anal. Methods. 9 (2017) 1470-147. https://doi.org/10.1039/C6AY02302A

[47] S. Karbalaei et al., Analysis and inorganic composition of microplastics in commercial Malaysian fish meals, Mar. Pollut. Bull. 150 (2020) 110687. https://doi.org/10.1016/j.marpolbul.2019.110687

[48] M. Kazour, S. Jemaa, C. Issa, G. Khalaf, and R. Amara, Microplastics pollution along the Lebanese coast (Eastern Mediterranean Basin): Occurrence in surface water, sediments and biota samples, Sci. Total Environ. 696 (2019) 133933. https://doi.org/10.1016/j.scitotenv.2019.133933

[49] A. Käppler, D. Fischer, S. Oberbeckmann, G. Schernewski, M. Labrenz, K.J. Eichhorn and B. Voit, Analysis of environmental microplastics by vibrational

microspectroscopy: FTIR, Raman or both?, Anal. Bioanal. Chem. 408 (2016) 8377-8391. https://doi.org/10.1007/s00216-016-9956-3

[50] Y. Su, X. Hu, H. Tang, K. Lu, H. Li, S. Liu, B. Xing and R. Ji, Steam disinfection releases micro(nano)plastics from silicone-rubber baby teats as examined by optical photothermal infrared microspectroscopy, Nat. Nanotechnol. 17 (2022) 76-85. https://doi.org/10.1038/s41565-021-00998-x

[51] E. Dümichen, P. Eisentraut, C. G. Bannick, A.K. Barthel, R. Senz, and U. Braun, Fast identification of microplastics in complex environmental samples by a thermal degradation method, Chemosphere. 174 (2017) 572-584. https://doi.org/10.1016/j.chemosphere.2017.02.010

[52] P. M. Peacock and C. N. McEwen, Mass Spectrometry of Synthetic Polymers, Anal. Chem. 78 (2006) 3957-3964. https://doi.org/10.1021/ac0606249

[53] H. Luo, Y. Zhao, Y. Li, Y. Xiang, D. He, and X. Pan, Aging of microplastics affects their surface properties, thermal decomposition, additives leaching and interactions in simulated fluids, Sci. Total Environ. 714 (2020) 136862. https://doi.org/10.1016/j.scitotenv.2020.136862

[54] W. J. Shim, Y. K. Song, S. H. Hong, and M. Jang, Identification and quantification of microplastics using Nile Red staining, Mar. Pollut. Bull. 113 (2016) 469-476. https://doi.org/10.1016/j.marpolbul.2016.10.049

[55] B. Nguyen, D. Claveau-Mallet, L. M. Hernandez, E. G. Xu, J. M. Farner, and N. Tufenkji, Separation and Analysis of Microplastics and Nanoplastics in Complex Environmental Samples, Acc. Chem. Res. 52 (2019) 858-866. https://doi.org/10.1021/acs.accounts.8b00602

[56] F. G. Torres, D. C. Dioses-Salinas, C. I. Pizarro-Ortega, and G. E. De-la-Torre, Sorption of chemical contaminants on degradable and non-degradable microplastics: Recent progress and research trends, Sci. Total Environ. 757 (2021) 143875. https://doi.org/10.1016/j.scitotenv.2020.143875

[57] M. Filella, Questions of size and numbers in environmental research on microplastics: methodological and conceptual aspects, Environ. Chem. 12 (2015) 527-538. https://doi.org/10.1071/EN15012

[58] H. Zhang, J. Wang, B. Zhou, Y. Zhou, Z. Dai, Q. Zhou, P. Chriestie and Y. Luo, Enhanced adsorption of oxytetracycline to weathered microplastic polystyrene: Kinetics, isotherms and influencing factors, Environ. Pollut. 243 (2018) 1550-1557. https://doi.org/10.1016/j.envpol.2018.09.122

[59] A. Bakir, S. J. Rowland, and R. C. Thompson, Transport of persistent organic pollutants by microplastics in estuarine conditions, Estuar. Coast. Shelf Sci. 140 (2014) 14-21. https://doi.org/10.1016/j.ecss.2014.01.004

[60] X. Liu, J. Xu, Y. Zhao, H. Shi, and C.-H. Huang, Hydrophobic sorption behaviors of 17β-Estradiol on environmental microplastics, Chemosphere. 226 (2019) 726-735. https://doi.org/10.1016/j.chemosphere.2019.03.162

[61] I. Velzeboer, C. J. A. F. Kwadijk, and A. A. Koelmans, Strong Sorption of PCBs to Nanoplastics, Microplastics, Carbon Nanotubes, and Fullerenes, Environ. Sci. Technol. 48 (2014) 4869-4876. https://doi.org/10.1021/es405721v

[62] M. Ateia, T. Zheng, S. Calace, N. Tharayil, S. Pilla, and T. Karanfil, Sorption behavior of real microplastics (MPs): Insights for organic micropollutants adsorption on a large set of well-characterized MPs, Sci. Total Environ. 720 (2020) 137634. https://doi.org/10.1016/j.scitotenv.2020.137634

[63] A. ter Halle, L. Ladirat, X. Gendre, D. Goudouneche, C. Pusineri, C. Routaboul, C. Tenailleau, B. Duployer and E. Perez, Understanding the Fragmentation Pattern of Marine Plastic Debris, Environ. Sci. Technol. 50 (2016) 5668-5675. https://doi.org/10.1021/acs.est.6b00594

[64] E. Hernandez, B. Nowack, and D. M. Mitrano, Polyester Textiles as a Source of Microplastics from Households: A Mechanistic Study to Understand Microfiber Release During Washing, Environ. Sci. Technol. 51 (2017) 7036-7046. https://doi.org/10.1021/acs.est.7b01750

[65] J. Wang, X. Liu, G. Liu and Z. Zhang, Size effect of polystyrene microplastics on sorption of phenanthrene and nitrobenzene, Ecotoxicol. Environ. Saf. 173 (2019) 331-338,. https://doi.org/10.1016/j.ecoenv.2019.02.037

[66] M. P. Johansen, T. Cresswell, J. Davis, D. L. Howard, N. R. Howell, and E. Prentice, Biofilm-enhanced adsorption of strong and weak cations onto different microplastic sample types: Use of spectroscopy, microscopy and radiotracer methods, Water Res. 158 (2019) 392-400. https://doi.org/10.1016/j.watres.2019.04.029

[67] X. Wu, P. Liu, H. Huang, and S. Gao, Adsorption of triclosan onto different aged polypropylene microplastics: Critical effect of cations, Sci. Total Environ. 717 (2020) 137033. https://doi.org/10.1016/j.scitotenv.2020.137033

[68] F. Yu, C. Yang, G. Huang, T. Zhou, Y. Zhao, and J. Ma, Interfacial interaction between diverse microplastics and tetracycline by adsorption in an aqueous solution, Sci. Total Environ. 721 (2020) 137729. https://doi.org/10.1016/j.scitotenv.2020.137729

[69] R. Mao, M. Lang, X. Yu, R. Wu, X. Yang, and X. Guo, Aging mechanism of microplastics with UV irradiation and its effects on the adsorption of heavy metals, J. Hazard. Mater. 393 (2020) 122515. https://doi.org/10.1016/j.jhazmat.2020.122515

[70] P. Liu, K. Lu, J. Li, X. Wu, L. Qian, M. Wang and S. Gao, Effect of aging on adsorption behavior of polystyrene microplastics for pharmaceuticals: Adsorption mechanism and role of aging intermediates, J. Hazard. Mater. 384 (2019) 121193. https://doi.org/10.1016/j.jhazmat.2019.121193

[71] M. Lang, X. Yu, J. Liu, T. Xia, T. Wang, H. Jia and X. Guo, Fenton aging significantly affects the heavy metal adsorption capacity of polystyrene microplastics, Sci. Total Environ. 722, (2020) 137762. https://doi.org/10.1016/j.scitotenv.2020.137762

[72] Q. Wang, Y. Zhang, X. Wangjin, Y. Wang, G. Meng, and Y. Chen, The adsorption behavior of metals in aqueous solution by microplastics effected by UV radiation, J. Environ. Sci. 87 (2020) 272-280. https://doi.org/10.1016/j.jes.2019.07.006

[73] J. N. Hahladakis, C. A. Velis, R. Weber, E. Iacovidou, and P. Purnell, An overview of chemical additives present in plastics: Migration, release, fate and environmental impact during their use, disposal and recycling, J. Hazard. Mater. 344 (2018) 179-199. https://doi.org/10.1016/j.jhazmat.2017.10.014

[74] J. Wang, J. Peng, Z. Tan, Y. Gao, Z. Zhan, Q. Chen and L. Cai, Microplastics in the surface sediments from the Beijiang River littoral zone: Composition, abundance, surface textures and interaction with heavy metals, Chemosphere. 171 (2017) 248-258. https://doi.org/10.1016/j.chemosphere.2016.12.074

[75] A. Turner, L. Holmes, R. C. Thompson, and A. S. Fisher, Metals and marine microplastics: Adsorption from the environment versus addition during manufacture, exemplified with lead, Water Res. 173 (2020) 115577. https://doi.org/10.1016/j.watres.2020.115577

[76] I. Vollmer, M. J.F. Jenks, M. C.P. Roelands, R.J. White, T. van Harmelen, P. de Wild, G.P. van der Laan, F. Meirer, J.T.F. Keurentjes, and B.M. Weckhuysen, Beyond Mechanical Recycling: Giving New Life to Plastic Waste, Angew. Chem. Int. Ed. Engl. 59 (2020) 15402-15423. https://doi.org/10.1002/anie.201915651

[77] S. Fernández-Velayos, J. Sánchez-Marcos, A. Munoz-Bonilla, P. Herrasti, N. Menéndez, and E. Mazarío, Direct 3D printing of zero valent iron@polylactic acid catalyst for tetracycline degradation with magnetically inducing active persulfate, Sci. Total Environ. 806 (2022) 150917. https://doi.org/10.1016/j.scitotenv.2021.150917

New Materials for a Circular Economy   Materials Research Forum LLC
Materials Research Foundations 149 (2023) 24-69   https://doi.org/10.21741/9781644902639-2

Chapter 2

Lignocellulose-Based Materials for Food Packaging: A Biorefinery Perspective

Miguel Ladero[1]*, Juan M. Bolivar[1], Victoria E. Santos[1], Nuria Gomez[2], Juan C. Villar[2],
Priscilla Vergara[2], Pedro Yustos[1], Maria Isabel Guijarro[1], Jose M. Carbajo[2], Ursulla Fillat[2],
Itziar A. Escanciano[1], Victor Martin Dominguez[1], Jorge Garcia Montalvo[1],
Celia Alvarez Gonzalez[1]

[1]FQPIMA group, Chemical Engineering and Materials Department. Faculty of Chemistry.
Universidad Complutense. 28040 Madrid, Spain

[2]Cellulose and Paper Group, Forest Product Department, Forest Research Centre (INIA, CSIC),
Ctra. de A Coruña km 7.5, 28040. Madrid, Spain

mladerog@ucm.es ; juanmbol@ucm.es; vesantos@ucm.es; nuria@inia.csic.es;
villar@inia.csic.es; vergara.priscilla@inia.csic.es; pyustosc@ucm.es; migg@ucm.es;
chema@inia.csic.es; fillat.ursula@inia.csic.es; itziaria@ucm.es; vmdominguez@ucm.es;
jorgar10@ucm.es; cealva03@ucm.es)

Abstract

To date, food plastic packaging has become widespread because it dramatically increases the shelf life of foods. However, social awareness of the negative effects of such a practice (emission of greenhouse gases, ubiquitous presence of nano- and microplastics, effects on the food chain, etc.) is growing. At the same time, cellulose-based materials like paper and cardboard have the notable advantage of their recyclability and the nature of their source. Other chemical compounds contained in plants and other living beings have great potential as components of cellulose-based packaging, making it possible to improve the mechanical, thermal and barrier properties of this key material in second-generation biorefineries. As the integrated biorefinery concept is a holistic view of the total conversion of biomass into energy, chemicals, materials, food and feed. Successful bio-based food packaging is able to replace the best plastic packaging and can be envisaged as a key factor for the expansion of the bioeconomy and the circular economy beyond the food sector, further integrating human activities in the geochemical cycles and, thus, boosting their sustainability.

Keywords

Packaging, Food, Biorefinery, Renewable, Sustainable, Compostable, Cellulose-Based, Barrier, Functional Material, Active Ingredients

Contents

1. Introduction

A century ago, the fossil fuel-based industry opened the door to modernity, enabling an economic, cultural and scientific growth unprecedented in the history of mankind. This development has permitted the consolidation of the chemical industry, playing a fundamental role in the production of pharmaceuticals, foodstuffs, polymers, fertilizers and fuels, among many others. However, this hyper-growth, due to the massive exploitation of the earth's resources, has led to the misconception a few decades ago that fossil fuels were unlimited and cheap. But nothing could be further from the truth. The chemical industry faces the problem of supplying and guaranteeing the same intensity (quantity and quality) of products to meet the increase in the world's population and the desire for a higher standard of living in many regions of the world. At the same time, global needs should be met while preserving global resources and the environment by finding new forms and sources of energy and material goods without compromising the environment and, thus, harming the per capita growth of mankind.

This situation has been the source of several concepts. Sustainability is envisaged as a highway to meet mankind needs for goods and societal development in a framework of respect towards our planet (the so-called three pillars model, with diverse indicators to indicate progress in economy, society and environment field). Circular economy and bioeconomy emerge from this concept, focusing on holistic management of energy and material resources. This management involves the use of renewable, biobased resources able to capture the sun energy in a most efficient manner through the Calvin cycle (photosynthesis), while new, more efficient materials are sought for batteries and similar devices to store energy. In these days, fossil resources are increasing their market prices due to geopolitical and, in the end, depletion reasons, boosting the search for novel energy chemical vectors (hydrogen, biofuels) and sources (convective movements in air and water –winds, tides, waves-) that reduces the high energy dependence in first-world countries. This policy is also a consequence of the quest for a non-carbon economy in the European Union and other world regions, a perspective that is considered in the annual BP (British Petroleum) outlook report in three different scenarios [1]. The political framework has been created by UN in 2015 through their Sustainable Development Goals (SDGs) in an attempt to focus world attention on the need to end poverty and ensure acceptable standards of living for all using sustainable development as a strategy to solve our needs while not compromising the solutions need by future generations [2].

Biomass is the matter contained or produced by living beings, matter that, in the end, is created through the Calvin cycle and connected biochemical routes in photosynthetic organisms, creating organic matter out of inorganic one (essentially CO_2, but not only) using the energy of the sun. From each 1000 kJ of sun energy reaching these living beings, 46 to 60 kJ are fixed in organic matter [3]. Thus, lignocellulosic biomass (LCB) is a plentiful and renewable resource to be transformed into chemicals, biofuels, food and feed ingredients and materials through biorefinery processes (thermal, physical, chemical and biochemical) [4]. If only agricultural and forest wastes are considered, the annual production exceeds 140 gigatons (Gt). Indeed, this type of waste supposes between 70 and 95% of the total lignocellulosic biomass (LCB) [5], formed by agroforestry residues from a wide variety of sources, such as wood, branches, bark, pruning, cereal straw, etc., which are produced in abundance, continuously, in many cases concentrated in one place and at a time of year, which presents obvious advantages for processing.

Fig. 1 tries to capture the multiple feedstock biomass resources and transformation by all type of processes into diverse materials, ingredients and monomers. Although biorefinery processes can target most mankind needs, from energy to food, the transformation of biomass into materials has been developed from the origins of mankind: wood, cellulose, paper, board, rosin, wool, etc. will continued to be a solution for future needs, though now we are seeking for the replacement of those materials created from fossil fuels since the first years of the last century [6].

Lignocellulosic and algal biomass is plentiful and a potential source of several materials and material ingredients. Food wastes with a high lignocellulose content are created every year (up to 1,300 Mt), while LCB related to food is unavoidable (for example, 1.3-1.4 kg of straw are created per kg of wheat grain and 1.0 kg corn stover comes in hand with 1.0 kg corn grain, figures that are to be considered in the light of world crop production. World corn production 2021/2022 will be 1,210.45 Mt, according to the United States Department of Agriculture (USDA), while wheat production will reach a peak this year at 778.6 Mt) [7]. As for biomass related to plant food wastes (responsible for the 6% of greenhouse gases, in the last term), pomaces and peels are highly produced. For example, apple crop in 2018 amounted to 86 Mt, apple further processing and consumption created about 15 Mt pomace [8]. With a world production of 21.6 Mt of olives in 2018, almost 3 Mt olive pomace and pâté cake were created, together with almost 25 Mt pruning, all agro-residues notably rich in polyphenols, fiber, and lignocellulosic matter [9]. Total citrus production in the world amounts to 120 Mt each year, being a 20% of it processed to juice, which creates about 12 Mt pomace used for feeding, mostly, but abundant in pectin and other polysaccharides [10]. With more than 9.7 million hectares, citrus fruit production is extensive in warm climate regions, and its pruning leads to more than 30 Mt lignocellulosic biomass. As a final example regarding agrowaste related LCB, wine production, responsible for the 0.3% of annual greenhouse gases (GHG) emission, is turning to be ecological, organic and environmentally-friendly, making use of most grape pomace for bioethanol production. In Spain, wine-related residues amounted to 2.7 Mt in 2019, with a production of grape pomace or marc above 470.000 tons and the parallel generation of more than 700.000 tons vine pruning, a waste that nowadays is used for low-value applications, such as the generation of thermal energy by combustion. At world level, 60 Mt grapes are produced yearly, generating 8.49 Mt grape pomace or marc and more than 12 Mt pruning [11]. As for global production of timber (hardwood and softwood), it is rising steadily from 2008 at a constant compound annual growth rate CAGR higher than 20%, being the production in 2018 more than 2.2 billion cubic meters (about 1000 Mt). Perspectives for 2050 indicate an impressive production of 2900 Mt timber [12].

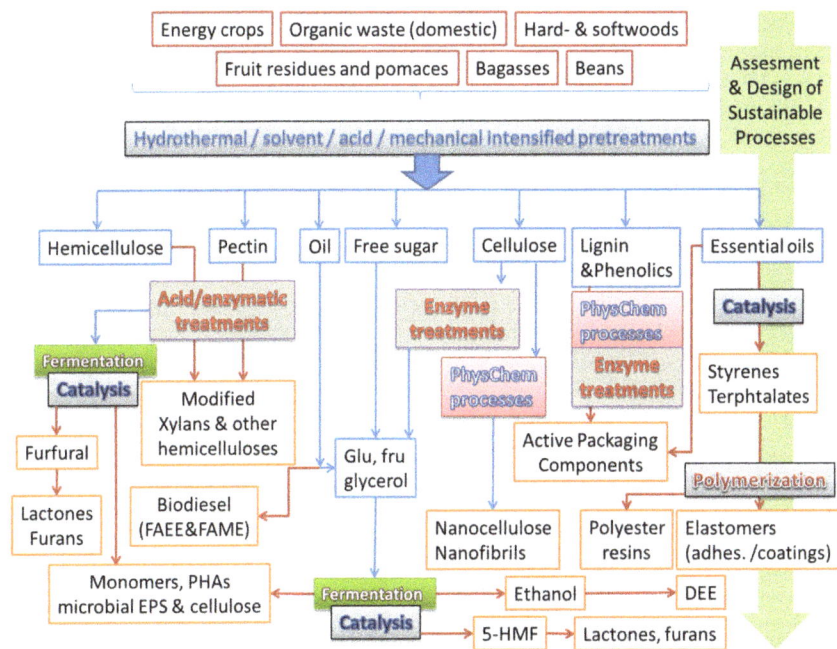

Figure 1. An overall perspective of biorefinery processes towards energy, chemicals and materials.

New Materials for a Circular Economy   Materials Research Forum LLC
Materials Research Foundations 149 (2023) 24-69   https://doi.org/10.21741/9781644902639-2

Recently, boosted by their interest in the bioenergy, food and feed fields, the production of macro- and microalgae is experiencing an upsurge. This biomass, relatively different from LCB, has an enormous potential also as provider for materials (e.g. alginate, agarose) and material monomers and ingredients [13]. Still, as recently as 2015 seaweed and algae production scarcely was 30 Mt, far away from the world production of terrestrial plant biomass (LCB) [14].

2. Packaging: a global need

Packaging is a necessity for the transport and conservation of any type of merchandise. It transports, protects and preserves commercial products and can also be the medium for advertising and for collecting information about their content. In the case of food packaging, some current trends advocate their total or partial elimination to avoid accumulating waste. However, its effect, due to the preservation of fresh foods, is just the contrary: it reduces food waste enormously, a major environmental and social problem of our days [15].

Plastics, after paper and cardboard, are the most used materials in packaging. Its success is due to some of its characteristics such as cost, lightness, versatility, flexibility, transparency, heat sealing and good mechanical and barrier properties [15]. In developed societies, the widespread use of packaging is associated with a high amount of waste. In EU27, during 2017 they accounted for 77.5 Mt, of which 7.5 Mt correspond to Spain [16]. In the case of those corresponding to plastic packaging, this represented 14.5 Mt and 1.6 Mt, for EU27 and Spain, respectively, figures that are clearly lower than the waste of paper and cardboard packaging that represents 31.4 Mt in EU27 and 3.7 Mt nationwide. However, and contrary to what these figures might suggest, the greatest impact is caused by plastic packaging, of which only 11.2 Mt are recovered in EU27 (29.2 Mt of paper packaging), while only 6.1 Mt are recycled (26.9 Mt paper). This is the case of commonly used plastics such as polyethylene or polyethylene terephthalate, which are neither compostable nor biodegradable, while another category of polymers, with wide application in food packaging, are biodegradable such as starch, polyhydroxy-alkanoates, cellulose, chitosan or polylactic acid. The situation in Spain is, broadly speaking, not very different and highlights the difficulty of giving a new use to plastic packaging.

There is a great variety of materials and containers used for packaging. The most widely used worldwide are those made from lignocellulosic material such as paper, cardboard and corrugated cardboard; flexible plastics (mainly low-density polyethylene), rigid plastic (polyethylene-based), high-density polyethylene and polyethylene terephthalate, aluminium and glass packaging [17, 18]. The value of the global packaging market has increased from 2011 till 2020 with a CAGR of almost 4.0%, according to diverse Smithers Packaging reports (Fig. 2). The prospective growth till 2024 indicates a slightly lower CAGR: 2.3%, with a prospective market value higher than 1000 million $ [18].

The disposal of used packaging is having a huge impact on the environment, which is why reducing the carbon footprint of this packaging has been identified as one of the main environmental challenges of modern society. In the specific case of the packaging of perishable products, some current trends advocate their total or partial elimination to avoid the accumulation of waste. However, its success is controversial, because it greatly reduces food waste, due to its conservation [17].

Plastics are the most widely used materials in packaging intended to come into contact with food. Its success is due to some of its characteristics such as low cost, lightness, versatility, flexibility, transparency, resistance and excellent barrier properties, essential for the preservation of this type of food. In addition, plastic has the advantage of being heat-sealable to achieve vacuum packaging and protective atmospheres [15]. In developed societies, the widespread use of packaging is associated with a large amount of waste. Most of these are incinerated or end up in landfills, as most petroleum-based plastics are not biodegradable and have little recyclability. Unfortunately, however, a large proportion of this plastic waste has accumulated in the oceans [19, 20], becoming a threat to human health through the food chain due to the presence of microplastics [21].

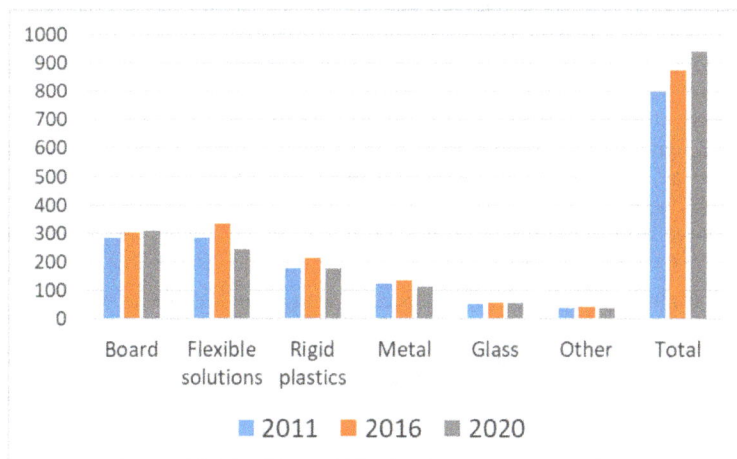

Figure 2. Global market evolution of packaging goods 2011-2022 in billion $ [17, 18].

In this way, it has been estimated that during 2019 the EU27 generated an estimated volume of waste of 79.6 Mt, 2.8% higher than in 2018, mainly due to the increase in glass and plastic packaging [22], of which 8.0 Mt correspond to Spain. In the case of those corresponding to plastic packaging, this represented 15.4 Mt and 1.7 Mt, for EU27 and Spain, respectively. The glass packaging represented 15.2 and 1.5 Mt for EU27 and Spain, respectively. However, the waste of paper and cardboard packaging represents 32.3 Mt in EU27 and 3.9 million ton nationwide.

New Materials for a Circular Economy   Materials Research Forum LLC
Materials Research Foundations 149 (2023) 24-69   https://doi.org/10.21741/9781644902639-2

However, and contrary to what these figures might suggest, the greatest impact is caused by plastic packaging, of which only 11.9 Mt are recovered in EU27 (29.5 Mt of paper packaging), while only 6.3 Mt are recycled (26.5 Mt paper). The situation in Spain is not, broadly speaking, very different from that of the EU27; with a recovery of 1.1 Mt (66.8%) for plastic packaging and 0.9 Mt (51.5%) is recycled.

To achieve sustainable and environmentally friendly food packaging, it is mandatory to implement several joint strategies, some related to the structure and the materials of packaging and others related to the extension of the useful life of the packaged product and, considering the reducing the possibility of food becoming waste [15, 23]. Among the first ones, it is worth highlighting the use lighter and thinner packages, retaining or improving performance using reinforcement bio-materials [24], the optimization of design and the replacement of packaging materials with ones made from natural resources and recycled materials, in addition these materials should be recyclable, biodegradable and, if possible, compostable or, at least, oxo- or photodegradable [25].

Polyethylene or polyethylene terephthalate, materials that are neither compostable nor biodegradable, are ubiquitous in the packaging of perishable foods such as meat, fish, dairy and fatty foods. This is due to its high barrier properties against oxygen, carbon dioxide, moisture, aromas, water, microorganisms and fats. On the other hand, anther biopolymers such as polylactide (PLA), poly(butylene succinate) (PBS), poly(hydroalkanoates) (PHA), starch, and lignocellulose-based biopolymers (cellulose, lignin, hemicellulose, cellulose derivatives) are produced from renewable sources and are biodegradable, also gaining prominence due to their high potential and availability; even some of them are already available on an industrial scale and at competitive prices. Fig. 3 shows the different families of plastics and elastomers classified by their origin and biodegradability [26-29].

Regarding the manufacture of functional packaging to extent shelf life of package and food waste reduction, the so-called active packaging is now being developed, which incorporate different active components that work as absorbers, emitters or releasers of compounds that are able to play a pivotal role in food preservation. Thus, their main target is to prevent microbial and chemical contamination, as well as to maintain visual and organoleptic properties of food [30].

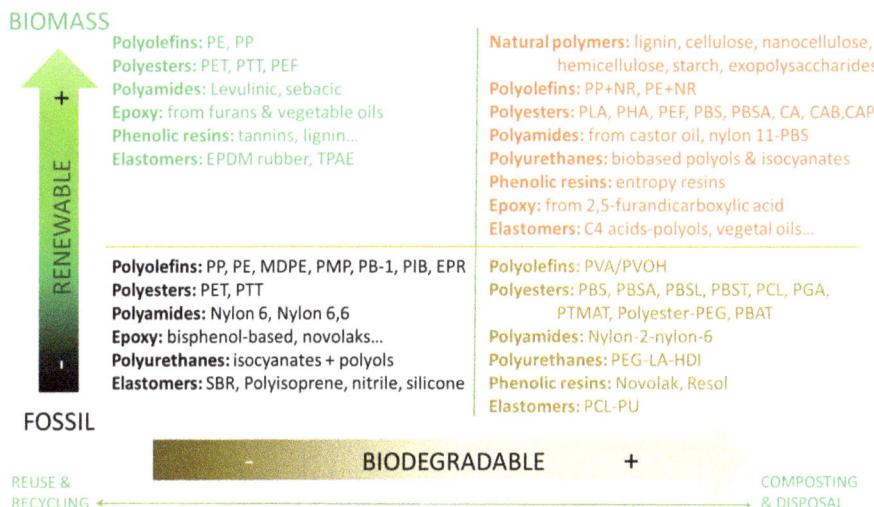

BIOMASS

RENEWABLE

+

Polyolefins: PE, PP
Polyesters: PET, PTT, PEF
Polyamides: Levulinic, sebacic
Epoxy: from furans & vegetable oils
Phenolic resins: tannins, lignin...
Elastomers: EPDM rubber, TPAE

Natural polymers: lignin, cellulose, nanocellulose,
   hemicellulose, starch, exopolysaccharides
Polyolefins: PP+NR, PE+NR
Polyesters: PLA, PHA, PEF, PBS, PBSA, CA, CAB,CAP
Polyamides: from castor oil, nylon 11-PBS
Polyurethanes: biobased polyols & isocyanates
Phenolic resins: entropy resins
Epoxy: from 2,5-furandicarboxylic acid
Elastomers: C4 acids-polyols, vegetal oils...

Polyolefins: PP, PE, MDPE, PMP, PB-1, PIB, EPR
Polyesters: PET, PTT
Polyamides: Nylon 6, Nylon 6,6
Epoxy: bisphenol-based, novolaks...
Polyurethanes: isocyanates + polyols
Elastomers: SBR, Polyisoprene, nitrile, silicone

Polyolefins: PVA/PVOH
Polyesters: PBS, PBSA, PBSL, PBST, PCL, PGA,
   PTMAT, Polyester-PEG, PBAT
Polyamides: Nylon-2-nylon-6
Polyurethanes: PEG-LA-HDI
Phenolic resins: Novolak, Resol
Elastomers: PCL-PU

–

FOSSIL

BIODEGRADABLE   **+**

REUSE &
RECYCLING

COMPOSTING
& DISPOSAL

Figure 3. Polymers (plastics & elastomers) classification regarding origin and biodegradability [18, 26-29].

3. Biomass as a source of food packaging components

Biorefineries are being developed to create diverse products out of biomass and, in particular, lignocellulosic biomass, as it represents the majority of bioresources created by photosynthesis [3]. A key aspect in the use of biomass are the pretreatment, which are needed to make it more prone to physical, chemical and biological modification into products of interest, the efficient use of water and energy are critical for a sustainable implementation [4, 5]. Therefore, the development of low energy input pretreatments for several lignocellulosic residues related to agro and food industry activities (corn stover, wheat straw, cardoon stems, citrus and apple peel wastes, etc.) is of strategic relevance to render a biorefinery feasible from the economical perspective.

A wide techno-scientific literature is evidencing that ethanol-acid water extraction at moderate pressure and temperature conditions (thus, low severity conditions) is enough to obtain a high yield in glucose, lignin and hemicellulosic monosaccharides, outperforming more classical sulfuric acid pretreatments for most agro- and food wastes treated [5]. This can be extended to diverse solvent-based pretreatments, all emerging from the organosolv concept. Moreover, it was also probed that recycling of treatment liquors does not have any deleterious effect on pretreatment efficiency but, on the contrary, reused ethanol-acid water liquors were more efficient in LCB pretreatment, reducing the need for enzymes in the subsequent saccharification stages. However, it is a limited research focused on batch

New Materials for a Circular Economy   Materials Research Forum LLC
Materials Research Foundations 149 (2023) 24-69   https://doi.org/10.21741/9781644902639-2

processing with several classic and advanced technologies [5]. These studies are increasingly important and search for the intensification of polyphenols extraction by enhancing L-S mass transfer, moving to continuous processing and scaling-up extraction processes to render an adequate solid for further fractionation with adequate ethanol-acid water mixtures focused on lignin extraction, hemicellulose hydrolysis and cellulose purification [7].

Acid-alkali pretreatments and the combination of these with different intensified pretreatments (microwaves, ultrasound…) have been also widely studied, proving to be effective technologies for biomass degradation, achieving high glucose yields [5, 7, 31]. However, the use of corrosive chemicals, the high cost of downstream with the processes for sludge neutralization, the high operating conditions, make it necessary to search for new pretreatments and solvents that results in sustainable processes. Moreover, these pretreatments are mainly focused on lignin removal regardless of damage to its structure, minimizing the alteration of the cellulose fraction for subsequent saccharification. In this regard, selective process and greener solvents, as supercritical fluids, ionic liquids, and deep eutectic solvents, represent a promise with high potential for the pretreatment of lignocellulosic biomass allowing the recovery of the lignin for its valorization [32, 33] and the non-generation of bioproducts [34].

The use of supercritical CO_2 improved mass transfer by facilitating penetration into the solid structure due to properties such as high diffusivity and low viscosity. No significant changes in the structure were observed when pre-treated with supercritical CO_2 alone, however, the presence of cosolvents increased the solvation abilities and polarity [34, 35], showing changes in the biomass structure. The use of supercritical CO_2 with a water-ethanol solvent showed higher digestibility, reaching 90% lignin removal using rice husk [36]. The combination of supercritical CO_2 with hot water and hot water-ethanol also showed better digestibility [34]. Milder oxidation process has also been studied, improving the enzymatic digestibility of rice straw by 88% when Fenton's reagent was used and a 100% increase in cellulose content when was combined with NaOH in sugarcane bagasse pretreatment. It was seen that pretreatment by Fenton's reagent allowed biomass fractionation but was incapable by itself of performing selective oxidation, being necessary to combine it with other types of oxidation [37].The ionic liquids (ILs), formed by a large organic cation and a small anion, are very selective, allowing under mild conditions, to extract a large amount of cellulose for subsequent hydrolysis [34,37] and being able to separate the different fractions of hemicellulose and lignin to obtain value-added products if it solvent is designed properly. One of the disadvantages is their high cost and difficult reuse [38]. For this reason, deep eutectic solvents (DESs) are gaining relevance, especially choline chloride, being solvents that have similar physical-chemical properties to ILS (high polarity, low melting points...), but are more economical, non-toxic, biodegradable and easy to prepare [38]. These solvents are prepared by mixing a hydrogen bond donor (HBD) and a hydrogen bond acceptor (HBA) in appropriate ratios at moderate temperatures [34,37]. Lignin oligomers with high purity (96%) were obtained using ChCl-oxalic acid in the pretreatment of wood lignocellulose [39]. DES (deep eutectic solvents) allow the fractionation of biomass, allowing the extraction of high purity lignin to produce a wide range of value-added products, such as phenolic compounds [32], since lignin contains a wide range of phenolic compounds in low concentration [37].

Extraction of polyphenols has experienced notable research upraise in the last years for several reasons: the need to valorize the plentiful biomass residues or reducing the costs of waste management. Valorization implies various step such as material pretreatment, which normally implies a drying and a grinding process, followed by the extraction procedure. Finally, a downstream step to separate the remaining solids and a solvents removal to recover them.

The most applied extraction process of polyphenols is the classical one based on maceration, reflux extraction using a Soxhlet extractor, liquid percolation through the material, among other contact techniques. Those techniques require lower investment as it is a simple method, however has some drawbacks like a high solvent needs, high energy input, lower efficiency and high extraction time [40]. To solve those problems and increase the profitability of the process intensification processes are under development. Those processes should require lower energy, time and solvent inputs and a smaller number of operation units as a combination of chemical and physical operations happens and higher mass and heat transfer rate can be achieved [41]. Most employed intensification methods include microwave (MAE) and ultrasound (UAE) assisted extraction, among others [42]. Being MAE and UAE, the most employed, as its use is easier and more scalable. Microwave assisted extraction implies the application of microwave energy that facilities the diffusion of the solvents and the solubilization of extractants bonded to the matrix. In the case of ultrasound extraction employs waves too, but in this case, waves generate physical disruptions of the matrix facilitating the diffusion of the polyphenols to the solvents with a lower breakdown of extractants [40].

Moreover, apart from extraction method, ideal extraction solvents should be investigated too. The development of advanced green solvents is boosted by UN 2030 Agenda for sustainable development to implement green chemistry processes [43]. Those solvents highlight as non-toxic, non-volatile, recyclable, biodegradable and do not require high energy inputs for their synthesis. There are two main groups of green solvents like bio-based solvents and most promising and selective neoteric solvents as Deep Eutectic Solvents (DES) and Ionic Liquids (ILs). Bio-Based solvents comprise solvents produced by biorefineries highlighting ethanol and glycerol [44]. The main advantage of those solvents is their renewable source, but they share the same disadvantages of classical extraction described before. ILs are a type of salts composed by cations and anions with a lower melting point (<100°C) with exclusive physicochemical properties like low vapor pressure, high stability and a high solubilization capacity, to name a few. So, they are good candidates for extraction of bioactive compounds, gut its use can be limited due to their toxicity [45]. To overcome the toxicity problems of several ILs and also due to economic reasons, DES has been developed. They are eutectic mixtures of acids and bases linked through hydrogen bonds with a reduced melting point. These solvents are easy to produce, and have a higher stability and lower toxicity than ILs. In any case, the most important advantage of neoteric solvents is their tunability, a compositional feature that permits a selective extraction of polyphenols out of very complex mixtures [46].

In addition, cyclodextrins and other organic vehicles allows for the application of supramolecular chemistry to this field and, again, permits a selective extraction of certain compounds out of complex mixtures [47]. Finally, enzymatic-assisted extraction (EAE) can be applied to improve extraction in mild conditions. Useful enzymes include cellulases, pectinases, hemicellulases, lignases and tannases, which breaks the bonds between matrix components and phenolics, so the extraction yield can be improved. EAE presents some advantages over classical extraction. For example, enzymes require milder, fewer extraction steps as liquid-solid extraction is combined with an enzymatic hydrolysis in the same process and the substrate specificity is higher, extracting a larger number of phenolics as some bounded phenolics to the cellular matrix are released. Besides, another advantage is the possible combination with an adequate process intensification technology (MAE or UAE) and advanced solvent. The problems with EAE are that enzymes can achieve an excessive degradation of the matrix, the possible inhibition of enzymes due to the extracted phenolics and a higher cost of the process because of enzyme cost [48].

In any case, an adequate selection and optimization of mass-transfer operation units should be carried out to maximize the extraction yield. The most important variables to study are mechanical treatment and temperature conditions. Mechanical treatment involves the particle size and agitation speed, key parameters that will determine the surface in contact with the solvent, but viscosity problems and filtration troubles can happen if particle size is small. Temperature conditions can improve the mass transfer, but can produce some extractant degradation, so an optimization is required [49]. To achieve this optimization a kinetic model should be developed to understand the dynamic phenomena underlying the process. In almost all cases extraction process comprises two stages, the first one implies extraction of superficial extractants (washing stage), typically a very fast process, and the second one that requires the diffusion of extractant to the medium being a slower step. The most important parameters in empirical models are the concentration or activity of compounds, the extraction order, and a kinetic rate constant. Once the model is fitted an optimal control of the process could be applied [50]. Finally, after optimization of extraction process and kinetic analysis, a downstream process is required to concentrate and purify polyphenols obtained from agro- and food wastes, usually by adsorption and membrane operations. Adsorption process permits the separation of the different analytes in function of them partition coefficient employing different resins (hydrophobic, hydrophylic or ion-exchange). Membrane separation is based on molecular volume differences between desired phenolics and the rest of compounds in the medium [51].

4. Lignin-first biorefineries: the phenolic components of packaging

Bioeconomy needs of the creation of sustainable and integrated processes, based on the use of biomass as raw material [52, 53]. These processes, which are carried out in the second-generation or lignocellulosic biorefineries, are essential to boost the emerging bioeconomy and promote energy and material resources sustainability. The recent consensus is that a sustainable biorefinery depends largely on the deconstruction and recovery of the three main components of lignocellulosic biomass: cellulose, hemicellulose and lignin [54].

The complex structure of lignocellulosic biomass is one of the main obstacles for its conversion into chemical products. Therefore, a pretreatment stage will be required in order to separate and maximize the recovery of each of the main components [5]. Lignin is the second most abundant polymer on Earth after cellulose, representing 15-40% of lignocellulosic biomass and 40% of its energy content, so it has significant potential as a sustainable source for fuel, chemicals and materials production [55, 56]. However, the limitation of the current biorefinery lies in the focus on exploiting carbohydrates for biofuels, leaving lignin as an underutilized waste [54]. Fig. 4 compiles diverse strategies to obtain material ingredients (monomers, crosslinkers, additives) for food packaging, together with the main features desired in these materials from perspectives of functionality and sustainability (mentioning some of the main technical, economic and environmental tools to be applied, as technoeconomic analysis and life cycle assessment).

Lignin from pulping represents 80% of world production (approximately 70 million metric tons per year) and is mainly used as fuel for the production of heat and energy in the combustion and chemical recovery stages of the pulp, or is disposed of in landfills. Only slightly other industrial applications have been developed (2% of the production of the paper industry) such as: obtaining additives for construction, biosurfactants, stabilizing agents, among others [56, 57]. Therefore, its valorization, beyond the calorific value, could substantially improve the economics and environmental performance of the paper industries and future lignocellulosic biorefineries.

Currently, the market for lignin and its derivatives is growing at a rate of 2.28% per year, from $599 million in 2014 to $704 million in 2022 [56]. As a first step for the valorization of lignin, it must be isolated from the lignocellulosic material through different thermal, chemical or biochemical processes. However, the structural complexity due to the diversity and the irregular polymeric structure of the aromatic compounds that form it, are the main barrier for its valorization and application [57].

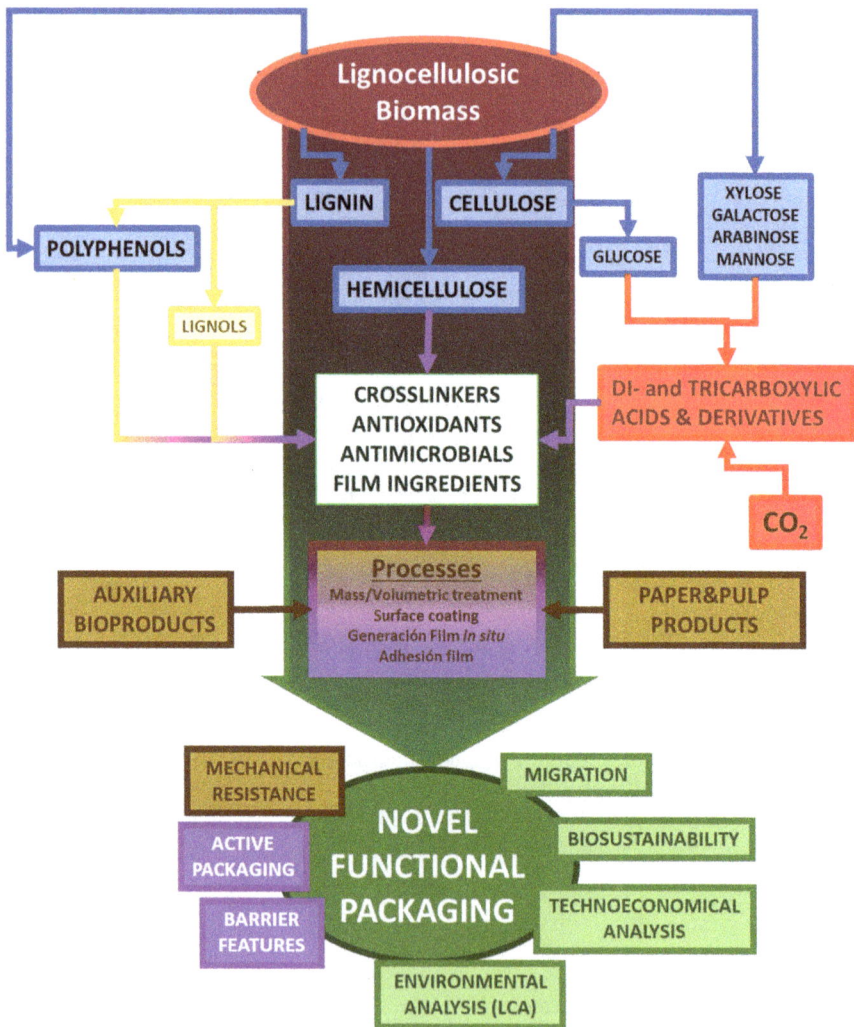

Figure 4- Strategies leading from renewables (plant biomass, carbon dioxide) and inorganic mineral raw materials to cellulose- and lignocellulose-based food packaging.

New Materials for a Circular Economy   Materials Research Forum LLC
Materials Research Foundations 149 (2023) 24-69   https://doi.org/10.21741/9781644902639-2

The processes developed for this purpose can be divided into two groups, according to the solubilized lignocellulosic fraction: (i) processes in which lignin is solubilized from lignocellulosic biomass and removed by separating the solid residue from the spent liquor (Kraft, Organosolv, soda and sulphite pulping), and (ii) processes in which polysaccharides are selectively hydrolyzed, obtaining a solid residue rich in lignin and with some carbohydrate degradation products (enzymatic and Klason methods) [58]. The fractionated lignin obtained will depend largely on the biomass raw material and the extraction method used, so its selection will depend on the desired final application [56, 50].

On the other hand, lignin extraction methods irreversibly affect the conformation of the structure compared to the native form [55]. Even, under severe conditions, it undergoes significant degradation and irreversible condensation is obtained. Cleavage of the aryl or alkyl ether bonds (e.g. β-O-4 and α-O-4), ester bonds and C-C bonds, are the key in lignin depolymerization as well as demethoxylation and alkylation [59]. Therefore, developing simpler methodologies that avoid the condensation and/or repolymerization of lignin, and that are also respectful of the environment in order to obtain a lignin with a less altered native structure, is a great challenge to consider.

"Lignin-first approach"

In recent years, a new strategy has been developed called "lignin-first" which aims to isolate lignin before carbohydrate conversion and obtain a high yield of lignin with minimal condensation through either catalysis or protection-group chemistry [59]. The goal is to improve the conversion rate of lignin into chemicals and increase the promise of the alternative source (aromatics, phenolics, and biogasoline). However, significantly, it is not the same concept as lignin valorization, but rather a method that derives value from lignin, achieving the goal of using lignocellulose biomass sustainably [60].

Within this approach, we find the *Reductive catalytic fractionation* (RCF), a complex process based on heterogeneous catalysis and involves three main steps: solvolysis, depolymerization and reductive stabilization. The development of the method shows that the availability of the main bond between the lignin monomers, B-O-4' (45-84%), is the key. Lignin obtained by this process have preserved or stabilized structures, and the potential to overcome challenges in the direct application of lignin to materials [55].

Therefore, effective valorization of lignin is a pivotal strategy for integrated biorefineries and an opportunity for obtaining components for packaging. On one hand, lignin processing is a critical step in the valorization of lignocellulose biomass. On the other, the suitable lignin depolymerization and transformation is a fantastic opportunity for the production of phenolic-derived chemical in the context of circular bioeconomy and sustainable chemistry [61-63]. As commented, the three fundamental aspects to succeed is this goal are the fractionation and biomass pretreatment, depolymerization process chemistry and the upgrading strategies [61-63]. The chemical upgrading of lignin implies two levels of application, one macromolecular related to the route towards to lignin-derived materials, another related to bio-based phenolic chemicals. In any case, the chemical processing of lignin compiles a reaction network of depolymerization, repolymerization/condensation, and functionalization that can be performed with different chemical and catalytic transformations, e.g. acid-based, reductive, oxidative…Among this palette of chemical routes, lignin oxidative lignin polymerization is an interesting key approach to produce platform chemicals including phenolic compounds, dicarboxylic acids, and quinones in high selectivity and yield [61–63].

Due to biological nature of lignin, its microbial or enzymatic transformations are an opportunity for the lignin-first biorefineries. Diverse oxidative enzymes as peroxidases and laccases from fungal sources naturally act on lignin to convert it into useful feedstock for the host organism. In particular polyphenol oxidases (laccases) display a major role in the lignin degradation and they have been studied with regard to the application for lignin transformation from enzyme -assisted biomass degradation to lignin oxidation, lignin-modification and recently to fine surface modification of lignocellulosic materials [64, 65]. In this last sense grafting of functional molecules to obtain modified materials is really promising. These aspects will be briefly commented as follows.

First technological application of laccases has been related to employment in the modification of lignocellulosic material [63, 66]. Use of laccases has covered oxidative depolymerization of lignin, but also phenol removal and more recently phenol functionalization and polymerization [62, 64, 65, 67]. The basic of the lignin oxidative depolymerization by laccases is well-understood [62, 64, 65, 67]. Laccases cleave mainly phenolic lignin units, being the redox potential the key parameter to evaluate the substrate scope, in presence of artificial or natural mediators (such as such as vanillin and p-coumarate) the oxidation can be target to the side chain of phenolic monomers. However, the exploitation of the catalytic potential for the upgrading of lignin and the design of controlled catalysts and reactions is still under development [67–69].

The potential of laccase-mediated functionalization has been started to be harness as technological tool in synthesis and material engineering. Laccase has the enormous potential to be implemented as the multitasking biocatalyst in whole process of paper making. The enzyme can be utilized effectively for pulping, delignification of pulps either alone or in the combination with other bleaching enzymes [62, 70]. More recently, laccase-mediated grafting on lignocelluloses has gained considerable attention as an environmentally benign method to covalently modify biomaterials [67–69]. Laccase has been evaluated for the biografting of pulp fibers, decolorize and stabilize the effluent of the paper mills, biotransformation of high molecular weight (HMW) lignin to lower molecular weight (LMW) aromatic compounds. The scope of this approach has been further widened by applying mediators or high redox potential laccases and those that modify synthetic polymers and proteins [66, 68,71,72] The grafting of bio-based molecules to bio-based materials for fine tuning of properties using oxidative chemistry of laccases offers two basic possibilities. First is the application of laccase in constructing lignin-based materials via grafting of other molecules to lignin [71]. The other is the grafting of lignin-derived phenolics to carbohydrate-based materials.

Grafting of functional molecules to obtain uniform properties are among the most investigated strategies to modify technical lignin. Laccases and peroxidases have been used to mediate the grafting of water-soluble molecules with carboxylic, sulfonic acid groups, and hydrophilic N–OH onto different types of Kraft lignin [67, 73-75]. Recently the enzyme assisted polymerization of lignosulfonate has been studied for the formulation of adhesives [76, 77]. The application of laccases in the field of biopolymers extends also to the grafting of different compounds to polymers or the functionalization of polymers and material coating. The enzymatic functionalization of lignocellulosic fibers using oxidoreductases was successfully achieved by targeting lignin moieties as grafting sites on the surface [73]. Laccase-mediated grafting on lignocelluloses has gained considerable attention as an environmentally benign method to covalently modify wood, paper and cork. In recent decades this technique has also been employed to modify fibers with a polysaccharide backbone, such as cellulose or chitosan, to tune surface properties as color or functional properties as antimicrobial activity or antioxidant activity to the material [66, 70, 78]. Recently, a coating liquid suitable for coating food packing paper was patented and contains laccase-grafted modified chitosan and an antibacterial palm wax emulsion. The laccase-mediated grafting very often involves oligomerization of phenols. Mechanical strength, hydrophobicity, antimicrobial activity or antioxidant activity are the properties that improve chemical or mechanical stability of lignocelluloses by laccase-assisted treatments. In recent years, efforts had been made to endow the collagen-based materials with some functional activities by the addition of phenolic compounds [79–81].

These emerging applications of laccases represent an important repertoire of opportunities for both laccases-mediated synthesis and laccase-mediated material functionalization. Understanding of the substrate scope, reaction engineering, and reaction kinetics is fundamental for the successful application of laccases. However, there is a lack of fundamental understanding of how to control and direct these oxidation reactions towards selectively generating valuable end tubed materials from lignin and lignin-derived phenols. Integration of enzyme engineering, catalysis, material sciences and process engineering will be a crucial strategy for the implementation of novel processes and the creation of neoteric lignin-based materials.

5.   Microbial and catalytic routes to bio-based products for material synthesis

Several catalytic and microbial processes to tri- and dicarboxylic acids are being intensely researched with the aim of obtaining bio-based monomers and crosslinkers, in addition to their applications as food and feed ingredients [82, 83]. These polyacids are considered sustainable biomonomers for condensation polymerization to polyesters and polyamides [82]. Moreover, the metabolic routes led to C4 diacids include a CO_2 molecule per acid molecule, contributing to CO_2 capture and fixation into chemicals and, in this case, polymeric materials, contributing to the reduction of GHG emissions [82, 83]. When using microbial routes, it can be necessary to redirect the carbon flow to the target molecules to optimize the bioprocesses. This fact involves non-trivial assumptions, as the cells must be able to transform a given biomass into a specific compound, achieving a maximum concentration, yield and productivity, while taking into account that the selling price should be around 2 euros per kilo maximum [82, 83].

Within the top 12 chemical building blocks designated by the US Department of Energy [84], two prominent diacids are produced by fungal fermentations: Itaconic and fumaric acids. Both of them have promising applications on polymer industry [85], having a large potential for the production of unsaturated polyesters, coating, painting, and resin industries. Itaconic acid is a good candidate to substitute acrylic acid on chemical industry. Moreover, these two acids have several applications on pharma and medical industries; for example, fumaric acid esters can be used as a treatment for diseases such as psoriasis or multiple sclerosis; while poly(propylene fumarate) main applications lay in orthopedics and tissue engineering [86]. Also, several derivates from both acids have promising applications as antimicrobial, anti-inflammatory, or antitumoral agents [87, 88].

The bio-based routes for production involves fungal fermentation. Fungal fermentations are slow bioprocesses that proceed through several stages: sporulation, inoculum, and production. The first stages are focused on fungal reproduction and initial growth, to develop proper metabolic states and an adequate, usually critical, fungal morphology. Indeed, this morphology must be properly maintained along the production stage, being cell immobilization helpful, as it allows to conserve the adequate physiology for production [89, 90].

Also, oxygen transfer and consumption are essential: a defect in oxygen hinders fungal growth; while an excess boosts the total oxidation of substrates to CO_2 and H_2O [86]. These facts, together with oxidative and hydrodynamic stresses, indicate that agitation must be well balanced to reach appropriate oxygen levels in the liquid, being necessary a constant and not interrupted input in many cases [91].

Nowadays, fumaric acid has a global production of 2.55 kton/year (2021), productivity that is expected to increase with a CAGR of 3.88% in the period 2022-2027 [92]. This acid is produced through petrochemical isomerization of maleic anhydride, obtained from benzene. Before the advent of petrochemical industry, fumaric acid was produced in the USA by fermentative processes with the fungi *Rhizopus arrhizus* or *oryzae*. Reductive Tricarboxylic acids pathway is their main source of fumaric acid, involving CO_2 fixation to convert pyruvate (C3) to oxalacetate (C4) [86, 93].

The stable fumaric acid production at high productivity and/or low resource cost is still a challenge, as diverse factors affect the morphology of *Rhyzopus* spp (developed on first stages). Controlling pH with $CaCO_3$ is a usual strategy that boosts productivity; however, it increases energy input and difficulties the acid purification. Exists a patent of DuPont with the highest titter and productivity reported till now: 130 $g \cdot L^{-1}$ in 6 days.

In addition, certain limiting conditions promotes fumarase expression, which increases the final fumaric acid production. Most usual limiting conditions are related with nitrogen source [94, 95] and oxygen transfer rate, which indicates that a balance on process conditions must be found, looking for optimum values for maximum productivities. These fungi can grow and produce on several alternative media: xylose-containing hydrolysates, glycerol, food wastes, apple pomace hydrolysates or brewery wastewater [86, 88].

Concerning itaconic acid, it had a world production of about 407 kt/year in 2020, meaning a market of US$ 500 million [96]. Reasoned by the good efficiency of this compound, it is estimated that the annual production capacity will increase by 5.5% between 2016 and 2023 [90]. Its price is about 2 US$/kg [14]; however, has been demonstrate that using renewable raw materials its price could be reduced on a 50% [90].

This dicarboxylic acid has been produced by a fermentative process for many years, carried out by *Aspergillus terreus*. Being a consolidated production process, the influencing parameters like media composition, pH, or morphology influences are well known. However, the productivity is very low, having very limited benefits. Despite of this, biotechnological production is clearly more efficient and profitable than chemical production process: citric acid pyrolysis. For all these reasons, the process is in constant improvement, one of the proposed modifications is using alternative microorganisms as *Ustilago maydis* [97].

The metabolic pathway for itaconic acid production on producer microorganisms is linked to TCA cycle or tricarboxylic acid cycle. Cis-aconitate is produced inside the mitochondria and excreted to cytosol, for being converted to itaconic acid and transported outside the cell [90]; this metabolic pathway can be initiated by different sugars such as glucose, xylose, sucrose, arabinose [90], or mannose [98]. Actually, is being under study the use of different kinds of biomass, for converting this process on a biorefinery: sugarcane molasses, starch hydrolysates, or steam exploded corn stover are proved to provide acceptable performances, with potential for implementation [99]. However, these raw materials alter metabolism, driven to by-products formation [90].

Finally, exist studies about coproduction of both acids, carried out by *Aspergillus oryzae* from lignocellulosic materials, employing different process regimes and providing promising results [89, 100, 101].

A most relevant C4 diacid is the succinic acid, which is a specialty chemical traditionally produced by catalytic hydrogenation of maleic acid or anhydride in refineries [102]. However, due to its high production costs and serious problems of environmental contamination, in recent years microbial synthesis processes have been developed from residues rich in sugar that allow production costs to be reduced [102]. This acid has many applications in the food and pharmaceutical industry and has a high potential in chemical industry processes for detergents, surfactants, chelators, solvents and biodegradable polymers [104]. According to US Department of Energy, is a key building block and the most important C4 in the structure of biorefineries [105].

In 2017, the bio-derived succinic acid market was around US$ 175.7 million, as a 20% CAGR was predicted, it is projected to reach a value of US$ 900 million by 2026 [106]. In fact, succinic acid generated by biological processes was reported to be the fastest-growing bio-based market in 2015. In 2013, bio-based succinic acid had a market price of US$ 2.86

per kg, while if it was obtained from fossil resources the price was reduced to US$ 2.00 per kg. So, it is essential to reduce production costs and for this it is necessary to solve the problem of availability of raw material in the long term, increase productivity in fermentation and reduce costs associated with purification processes [107].

Among the succinic acid-producing microorganisms, bacteria isolated from the rumen of ruminae such as *Actinobacillus. succinogenes*, *Anaerobiospirillum succiniciproducens* or *Mannheimia succiniciproducens* are the most promising candidates due to its ability to naturally use CO_2 to produce high amounts of this acid [108].

With the aim of solving the problem of the availability of substrate biomass, in recent years researchers have tried to find methods to use different wastes as a carbon source as Ferone *et al.* [109] who produced 24.1 $g·L^{-1}$ of succinic acid from high sugar content beverages by *Actinobacillus succinogenes* operating in batch. To overcome the challenge of low productivity in the processes, different types and operating conditions have been studied. Flippi *et al.* [110] managed to obtain 40.2 $g L^{-1}$ of succinic with a high productivity of 0.79 $g L^{-1} h^{-1}$ from grace pomace and stalks thanks to a fed-batch type operation. Given that the cost of purifying the succinic acid obtained in a bioprocess is 60% of the total process costs, Escanciano *et al.* [111] studied the process of producing succinic acid with *Actinobacillus succinogenes* cells in a resting state, managing to minimize the generation of by-products by 27.5% compared to the same operation carried out with cells in a growth state.

Citric acid, on its side, is one of the most in-demand polycarboxylic acids. In 2020, the production of this compound in China, the world's largest producer, was around 2 Mt. It is estimated that by 2025, the world market for this acid could grow by more than 5.24%, reaching US$ 3.6 billion [112]. In recent years, a multitude of investigations have been carried out on the production of citric acid through fermentation processes, taking advantage of residues as carbon sources thanks to the biocatalyst action of microorganisms such as *Aspergillus niger* and *Yarrowia lypolytica* [106, 113, 114].

Citric acid is widely used in in the food, pharmaceutical, and personal care industries. However, it is worth noting its use as a crosslinker, application that is gaining a notable attention in recent years due to the fervent need to avoid the use of non-biodegradable based-petroleum materials [114, 115]. Wen et al. [115] used citric acid as multifunctional cross-linkers as well as an effective reinforcers to improve mechanical and antibacterial properties of a biodegradable film for food packaging with anti-fog properties. Hassan *et al.* [116] produced biodegradable foams composed of biodegradable starch and cellulose crosslinked with 5 % (w/w) citric acid, observing its crosslinking action notably improved the thermal stability of the composite foams. Recently, Ponnusamy *et al.* [117] improved the water resistance and water vapor permeability properties of nanofibrils chitosan-cellulose composite films thanks to their preparation with in situ crosslinking of citric acid.

Polycarboxylic acid and their derivatives as crosslinkers: some examples

Fumaric acid can be used as a crosslinking agent in natural polymeric structures to form hydrogels. Natural polymer-based hydrogels are 3D hydrophilic network structures made up of two or more natural polymers, crosslinker(s) and solvent(s) [118]. Generally natural polymers are cellulose, chitin, and their derivatives carboxymethyl cellulose and chitosan respectively, which have been widely studied due to their non-toxicity, biocompatibility, extracellular matrix structure, and abundance [119, 120]. The hydrogels formed from natural polymers can be used for theoretically inexhaustible applications in renewable energy (as energy carriers in textile, thermoelectric material or agents in anti-microbial energy devices/textiles). This is due to their abundance and sustainability compared with depleting petroleum-based synthetic counterparts [118, 121-123].

Recent research has studied the role of fumaric acid as a crosslinker of these structures: for example, the used of fumaric acid in carboxymethyl cellulose-based hydrogels produces structures that respond to a stimulus-response system [124]. Also, carboxymethylcellulose-polyvinyl alcohol hydrogels were synthesized by crosslinking in presence of fumaric acid [125]. The crosslinking with fumaric acid provides the optimal structure, suitable chemical properties and improved the thermal stability of the hydrogels, making it as a promising green and sustainable hydrogel matrix [126]. The effect of crosslinking with fumaric acid on an individualized natural polymer has also been studied. A novel biodegradable carboxymethylcellulose and fumaric acid-based hydrogel was synthesized and characterized [124]. This study probed that crosslinking carboxymethyl cellulose with fumaric acid induces a more ordered structure in the hydrogel. In addition, the material was more stable to temperature and biodegradation at high fumaric acid contents.

A wide range of succinic acid ester products can be used in the chemical (dimethyl, diethyl, di-isobutyl, and dioctyl succinates as green solvents and mono or dibutyl, didecyl, diamyl and diisoamyl succinates in plastic and fuel additives) [127]. Among these materials, the most interesting and developed are the aliphatic poly(esters) obtained from the condensation polymerization of dicarboxylic acids (succinic acid) which can react with several diols (1,2-ethanediol or 1,3-propanediol or 1,4-butanediol). The properties of these copolyesters will depend on their structure due to the combination used between these acids and diols. Aliphatic polyesters are the most likely to biodegrade under certain conditions. An important bioaliphatic polyester is polybutylene succinate (PBS) with properties comparable to polypropylene (PP) or polyethylene terephthalate (PET) [128]. On the other hand, the introduction of aromatic units in the synthesis of polyesters leads to an improvement of the mechanical properties, together with a modular biodegradability [127]. The presence of the aromatic component in the polyester chain is important to obtain polymers with adequate crystallization rates and high melting temperature [129].

The synthesis of poly(esters) based on succinic acid has been carried out by different means, enzymatic polymerizations [130, 131] (*Candida antartica* lipase B or CALB), supported zirconia catalysts [132], fermentations, using different bacterial strains [128], using a heterogeneous catalyst composed of catalyst composed of silico-tungstic acid (SiW12) and MCM-22 (SiW12-MCM-22) [133], microwave assisted esterification of succinic acid using esterification in the absence of solvents [134].

New Materials for a Circular Economy   Materials Research Forum LLC
Materials Research Foundations 149 (2023) 24-69   https://doi.org/10.21741/9781644902639-2

In recent years, improving the mechanical properties of aliphatic polyesters has become a challenge for both researchers and industry. On the one hand, aliphatic monomers such as 1,4-butanediol, ethanediol, succinic acid and adipic acid, which are necessary to synthesize aliphatic polyesters, have already been produced from bio-based feedstocks [135]. Different strategies have been studied to improve the properties of some polyesters: copolymerization modifies the physicochemical, mechanical and gas barrier properties. The addition of plasticizers can also help providing an increase in ductility and a reduction in viscosity, improving the processability of the material. In this case, the use of crosslinking agents that improve the miscibility between the different components.

Plasticizers facilitate the processability and increase the flexibility and toughness of industrial plastics. Organic phthalate esters are the most widely used plasticizers in the PVC (polyvinylchloride) industry. However, significant concerns over the use of phthalates have arisen in recent years owing to environmental and health issues [128, 131, 135].

The synergistic effect of BCNW reinforcement and diacids crosslinking towards PVA properties has been studied. The composite films were prepared by simple mechanical dispersion of reinforcement material, bacterial cellulose nanowhiskers (BCNW) into PVA solution followed by crosslinking with diacids, succinic acid (SA), and adipic acid (AdA), subsequently characterized [136].

The BCNW reinforcement mainly increased the strength of the PVA whereas crosslinking improved swelling, thermal stability and strength of the PVA. The BCNW–SA-crosslinked PVA films showed the highest tensile strength and least water uptake in comparison to their parent cross-linked and noncrosslinked PVA samples. Diacids crosslinked samples have marginally less water uptake in comparison to PVA. SA, crosslinker with short length made PVA more rigid and crystalline in comparison to long crosslinker AdA. SA and AdA crosslinked samples are stable at a higher temperature. Fig. 5 depicts the step-by-step LCB fractionation and transformation towards some of the main bio-based acid monomers and their final application as material ingredients.

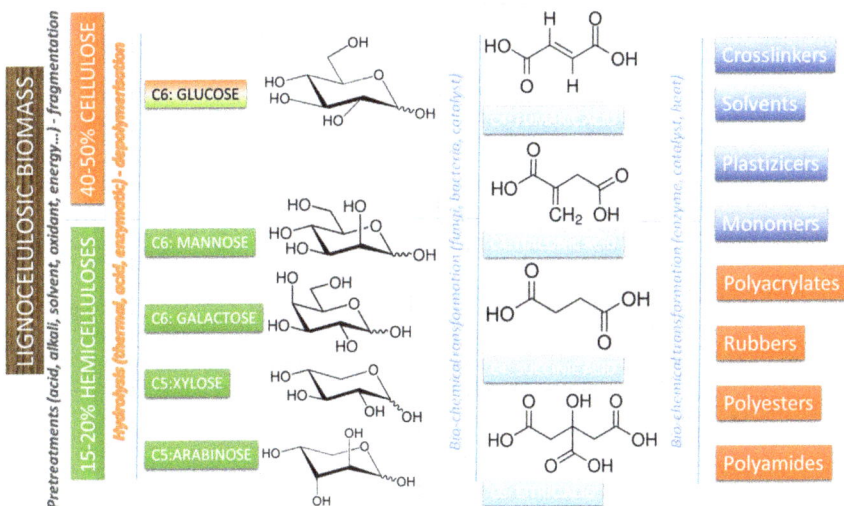

Figure 5. LCB strategies towards polyacids and their uses as material ingredients.

6. Novel packaging alternatives to plastics

Food packaging materials usually consist of a coating or multi-layer of different types of polymers that contains additives for a better performance (fillers, plasticizers, colorants…). This complexity makes more difficult and expensive packaging recycling and landfilling is often the preferred alternative. This supposes a high environmental impact and the solutions proposes different approaches to reduce the packaging weight or to substitute its materials by other easier to recycle or, at least, compostable and/or biodegradable.

Reducing the material thickness (weight) could be compensated with the application of a barrier to give strength to the material or to create an obstacle to substances transference or migration (very relevant in packaging in contact with foodstuff). Packaging could also contribute with other functionalities additional to its main purpose. In these cases, it is possible to distinguish among intelligent or smart packaging, active packaging or functional packaging. The first type includes possibilities as monitoring the conditions of food and include elements as labels or tags. Active packaging acts on the packaging environment, for instance by removing oxidants or contaminants. Functional packaging is designed to impart new functionalities not associated to the packaging materials (plastics or paper) as hydrophobicity or microbial resistance. These solutions contribute to enlarge the self-life of food and obviously reduce the impact of food losses, but also introduce complexity in the packaging design and could be an obstacle to packaging recyclability.

New Materials for a Circular Economy   Materials Research Forum LLC
Materials Research Foundations 149 (2023) 24-69   https://doi.org/10.21741/9781644902639-2

One-third of all food produced each year is squandered or spoiled before it can be consumed. In the developed countries, this waste happens during the food distribution chain and in the households when it spoils in our fridges and cabinets. It is believed that a major factor in preventing food loss lies in the use of quality packaging that prevents moisture and oxygen transmission between products and their environment [23, 135]. For this reason, the packaging materials intended to meet perishable foodstuffs such as meat, fish, dairy and fatty foods, usually consist of a flexible plastic film (sac) or rigid plastic thermoformed containers that contains additives to extending the shelf life of food [20, 30].

Various studies have assessed the life cycle environmental impact of different food products, part of which have highlighted the contribution of packaging to overall environmental load of commercial product [15, 138]. In fact, as previously discussed, it is unquestionable that plastic packages are responsible for high environmental impact associated with their life cycles, especially due to the difficult and expensive packaging recycling and landfilling is often the preferred alternative [19, 20]. The solutions proposes different approaches to reduce the packaging weight or to substitute its materials by other easier to recycle or, at least, compostable and/or biodegradable.

In a first approach, glass and tinplate could be a suitable alternative to plastic materials in packaging, as they have excellent barrier properties and their reuse and recycling systems are well established in modern society. However, its packaging relative environmental impact overall environmental load of food are higher than 50 % evaluated as percentage of global warming potential irrespective of the food category. While, carton-based containers allow a dramatic reduction of the package impact, which amounts to just 9.7-12.1% of global warming potential [15].

Therefore, paper and board packaging could be the alternative with the lowest carbon footprint, because they are manufactured from natural sources and are recyclable and biodegradable. However, paper and cardboard show a clear disadvantage from the other packaging materials that is the lack of barrier properties. To solve this problem, the paper and cardboard have been laminated with flexible aluminum and synthetic polymers foils and consequently the problems of their recyclability and biodegradability remain [139, 140].

In addition, the cellulose fiber-based food packaging has the disadvantage of the remarkable strength loss suffered due to high relative humidity generated by refrigerated storage and transport of perishable food products [24]. On the other hand, is of concern the uncertainty of safety for food contact, particularly for papers manufactured from recycled fiber, which reduces the guarantees to ensure the absence of migration of substances into food.

Biopolymers can be used to replace non-biodegradable plastics, they can be obtained from agro-forest wastes thereby reducing environmental impact and allowing the packaging to be biodegradable or compostable. Examples of bio-materials are bioplastics such as polyethylene and polyethyleneterephtalate, polylactic acid or chitosan which has been suggested for packaging solutions [139].

New Materials for a Circular Economy   Materials Research Forum LLC
Materials Research Foundations 149 (2023) 24-69   https://doi.org/10.21741/9781644902639-2

Biopolymers can be classified into three categories depending on the origin and synthesis processes (Fig. 6): (i) from biomass and natural sources, polysaccharides (cellulose, starch, and chitosan), proteins (whey, collagen) and lipids (bees and plant waxes), (ii) synthetic biodegradable copolymers from biomass as polylactic acid and (iii) those obtained by microorganism as hydroxyalkanoates or bacterial cellulose.

Biopolymers can be used as edible packaging that can be consumed along with the food itself [141], as films [142] and as surface treatments in coatings [143]. The use of biopolymers in paper products can improve their recyclability and allow for more ecological and plastic-free packaging [144].

Starch is a polysaccharide composed mainly by linear amylose (poly-α-1, 4-d-glucopyranoside), and branched amylopectin (poly-α-1, 4-d-glucopyranoside and α-1, 6-d-glucopyranoside). It can be obtained from corn, potato, waxy maize, wheat and tapioca [145]. Native starch is rarely used in the paper industry due to its insolubility, high viscosity and tendency to retrogradation [146]. Nevertheless, modified starches as oxidized starch, starch ester, cationic starch, enzyme degraded starch, carboxymethyl starch and thermoplastic starch have been widely used as binders, surface sizing and coatings in paper manufacturing [146].

In nature, cellulose derived from biomass shows a hierarchical structure that encompass different levels from a macro to a nano-scale [147]. Cellulose microfibrils are produced by mechanical disintegration [148]. Some mechanical, enzymatic and chemical pre-treatments (as TEMPO mediated oxidation and carboxymethylation) have been applied to reduce the high amount of energy consumed [149]. These pretreatments also allow to obtain cellulose nanofibrils (CNF) with a diameter of 2 to 50 nm and lengths up to several micrometers [150]. CNF have been studied to form films [151] and in papermaking to improve mechanical strength [152] and gas barrier properties [153-155]. Besides that, barrier properties of cellulose derivatives as cellulose esters, carboxymethyl cellulose, cellulose acetate, ethyl cellulose films and coatings have been assessed for packaging applications [17].

Chitosan is a linear polymer consisting of β-(1–4)-D-glucosamine and N-acetyl-D-glucosamine produced by de-acetylation of chitin, which is a polysaccharide obtained from crustacean shells. Chitosan films exhibit good mechanical properties and has been coated in paper increasing its barrier properties [143].

Other polysaccharides can be produced by microorganism as pullulan, gellan gum, xanthan gum, bacterial cellulose, or bacterial alginates and can be applied in food packaging [156]. Proteins as gelatin and wheat gluten have been used as edible films in food industry and nowadays several studies are performed in order to enhance its mechanical and barrier properties [142].

In general, films and coatings of polysaccharides display good gas barrier properties [143]. However, due to its hydrophilicity poor moisture barrier properties are achieved [151]. To

New Materials for a Circular Economy   Materials Research Forum LLC
Materials Research Foundations 149 (2023) 24-69   https://doi.org/10.21741/9781644902639-2

overcome this limitation composite films or coatings can be used when adding lipids as waxes or fatty acids in the hydrophilic matrix [142, 157].

Polylactic acid (PLA) is a compostable polyester obtained from depolymerization of lactic acid previously obtained by fermentation of crops as sugar feedstock and corn. This thermoplastic can be extruded and thermoformed and has been widely used for food packaging in rigid applications [158]. Nevertheless, PLA shows some disadvantages as low heat stability, its production competes with food resources and demands new recycling systems [159]. Polyhydroxyalkanoates (PHAs) are microbial polyesters produced by fermentation of sugar or lipids [143]. Films of PHAs exhibit lower hydrophilicity than polysaccharides as cellulose or starch but nowadays its use is limited due to their high production costs [159].

The addition of chemical agents to the pulp slurry (furnish composition) and/or on sheet surface to improve the strength of the paper is a well-established technology in the papermaking. The most are additives are starch, polyacrylamides, alkyl ketene dimer (AKD), alkenyl succinic anhydride (ASA) and polyvinyl alcohol [160, 161]. The transition from petroleum-based additives to biomaterials to improve strength has become essential, both because of their impact on the environment and because of the uncertainty from a food safety point of view [27, 162].

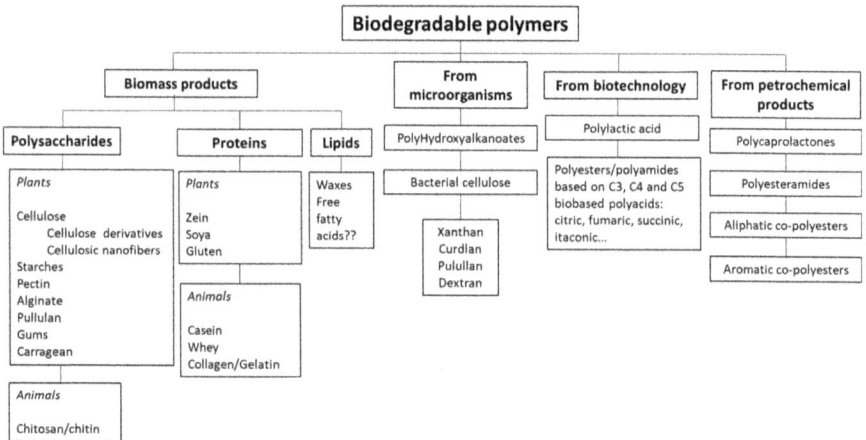

Figure 6. Origin of several exemplary biodegradable polymers.

Some innovative developments, include the addition of hemicelluloses (xylan and arabinogalactan and galactoglucomannans), obtained from the treatment of lignocellulosic material, to improve paper strength [26, 163, 164]. Additional steps to increase the amount of carboxyl groups in hemicelluloses, such as TEMPO-mediated oxidation or carboxymethylation [165, 166] or combining xylans with polyacrylamide [167], have been found to achieve strength improvements of up to 50 %.

Packaging could also contribute with other functionalities additional to its main purpose [59]. In these cases, it is possible to distinguish among intelligent or smart packaging, active packaging or functional packaging [115]. The first type includes possibilities as monitoring the conditions of food and include elements as labels or tags. Active packaging acts on the packaging environment, for instance by removing oxidants or contaminants. Functional packaging is designed to impart new functionalities not associated to the packaging materials (plastics or paper) as hydrophobicity or microbial resistance. These solutions contribute to enlarge the self-life of food and obviously reduce the impact of food losses, but also introduce complexity in the packaging design and could be an obstacle to packaging recyclability [115].

More complex solutions, as nanocomposites, could be difficult to justify in packaging for economic considerations and/or difficulties for implementation in the fabrication process. Bio-polymers can be isolated from biomass (xylans, cellulose…) synthesized from bio-derived monomers, as polylactic acid, or produced from microorganisms as the bacterial nanocellulose [155]. Quality of polymers can be measured in terms of physical, thermal, mechanical or barrier properties but there is few methodological information on their effect on recyclability, compostability or on the migration of substances to food-stuff. Migration of substances from packaging is an item of interest mainly with the recycling of paper/carboard packaging and could limit the global recycling rate or, in avoid the substitution of plastics by cardboard in packaging products.

Functional packaging is an interesting proposal to valorize the packaging and to reduce the impact of the food losses because of the increment in the shelf-life of food [155]. Antimicrobial properties of bio-materials used as packaging barriers has been previously reported with polymers as chitosan or carboxymethyl cellulose as promising barriers although the limited solubility of these substances requires chemical modifications. The impact of these modifications on the packaging surface has only been considered on the antimicrobial activity but no information is found on the impact on factors as recyclability or migration of substances caused by the chemical modifications. With respect to cellulose, it has few antimicrobial activity but could be improved with the inclusion of molecules as lignols on its surface. Both, cellulose and lignols, could be obtained from lignocellulosic fractionation or directly as a product and a by-product of the paper industry [5, 60, 61]. The addition of antimicrobial capacity to packaging is one of forms to preserve food but other capacities as oxygen scavengers, ethylene absorbers, antioxidants or moisture controllers has been also studied [168]. Antimicrobial in food packaging avoid the growth of microorganism as bacteria and, consequently extend the life of the food and reduce losses. This form of application is more effective than the direct addition of antimicrobials to foods because packaging surface continuously releases the agent to food. Moreover, the direct addition of antimicrobials to food may cause unwanted effects due to their reaction with proteins or lipids present in food. So, recommendation of multi-layer films (with the agent in the inner layer) is preferable to the direct addition. Organic substances, such as essential oils, proteins, peptides or polyphenols have demonstrated antimicrobial capacity. Also, metal oxides, as titanium dioxide and zinc oxide, and metals, as silver particles, present antimicrobial activity. Natural and renewable materials have been tested as film-forming for active packaging. Chitosan has excellent properties and even some antimicrobial properties, which must be improved with the incorporation of more active substances. Other natural polymers studied as film-forming to contain the antimicrobial agent, includes proteins, starch and, more recently, nanocelluloses, which also impart strength to the packaging. However, the cost of obtaining some of these products may condition their use in food packaging.

Besides antimicrobial activity, other functionalities can be added to the packaging as hydrophobicity, which great interest for the transport of fresh food as fruits and vegetables because humidity is responsible of the loss of strength in corrugated board packaging. As in the case of antimicrobial properties, lignin or polyphenols (isolated from lignocellulosic wastes) could be used in the modification of paper surfaces, directly on the cellulosic fibers surfaces or on the bio-material used as packaging barrier. The expected effect is the change of the packaging surface towards a more hydrophobic material able to protect the packaging from humidity [160, 161]. As natural candidates to impart this hydrophobicity to the surface of the packaging, there are different preparations of starch, a material traditionally used by the paper industry for this purpose. The paper industry has also developed various types of nanocelluloses, chemically identical to cellulose but structurally very different. With a very high surface-to-mass ratio, they can be effective coatings for packaging paper, to which they provide resistance and hydrophobicity, even when applied at very low doses. Finally, the greater energy efficiency of modern pulp mills makes available significant amounts of lignin (used as fuel in pulp production). Among the new uses sought for it, its hydrophobic nature provides resistance to humidity when suspensions of lignin nanoparticles are applied to the surface of packaging papers.

Considering the use of fractions obtained from agro-food wastes for the improvement of packaging from a circular economy approach, these improvements will be understood from various points of view, including not only resistance and functionalities but also sustainability according to partial and global life cycle assessments and techno-economic analyses to be performed all along the alternative packaging value-chain and their possible impact on food safety [106, 169-173].

In view of the diverse packaging materials, paper and board present several advantages due to their high recyclability and biodegradability. Paper and board are the most recycled packaging material in Europe. According to Eurostat [169] in 2019, recycling rates of main packaging materials waste were 82% for paper and board, 48.6 for plastic, 77.4 for metallic and 75.4 for glass. Paper and board also presents a good biodegradability. According to EN 13432 [170], the standard used to evaluate composting and biodegradation of packaging materials, unmodified packaging materials of natural origin, such us paper pulp shall be accepted as being biodegradable without testing. In addition, paper bags presents better anaerobic degradation than bioplastic ones [171]. As reported by Otto et al. [172], consumers judge packaging material by criteria of circular economy, natural looking material, and design. Paper and board packaging sustainability is well rated by consumers as by the scientific facts.

New developments to give paper/board more and better features must take into account packaging recyclability and biodegradability [173]. Some polymer films (such as polyethylene, poly(ethylene-co-vinyl alcohol), and polyethylene terephthalate-polyethylene) used to provide both moisture and grease resistance to food contact papers are neither compostable nor recyclable [174], On the contrary, nanofibrillar cellulose-based products are biodegradable and compostable [175].

In the food contact area, substances from packaging have the potential to migrate into foodstuffs. In Europe, following CEPI Food Contact Guidelines (CEPI) there is not a unified EU-wide list of substances authorized for use in the manufacturing of paper and board food contact materials. Several national legislation and best practice recommendations for paper and board listed substances but differ in the substances listed and/or in their limits and restrictions. Substances not appearing on these lists may be used provided that they are not excluded by any relevant industry commitment and one of the following circumstances applies:

a) There is no detectable migration into food or food simulants (detection limit of at least 10 µg/kg food) and the substance is not classified as 'mutagenic', 'carcinogenic' or 'toxic to reproduction' and the substance is not engineered as a nanomaterial
b) The substance is a direct food additive and/or is included in the positive list of other food contact materials and any restrictions are complied with.
c) A risk assessment (for example, based on the threshold of Toxicological Concern approach) has been carried out.

These substances, widely denominated intentionally added substances (IAS), have centered the evaluation of the safety of food contact materials until recently [176]. In addition to substances of known origin (IAS), the recycled paper and board may contain non-intentionally added substances (NIAS) [177]. The main origin of NIAS in recycled papers are the inks, coatings, adhesives and contaminants from previous uses and there are also lists of NIAS that must be tested. Recycled paper and board is not the clean and well controlled material ideal for food contact, but due to sustainability reasons, its use should be continued. Measures, such as functional barriers or functional sorbents, should be taken in order to reduce migration into food [178].

7.   Future perspectives

The first wave of innovation in biomaterials came in last century before the petroleum era, plant- or animal-based bio-based materials such as wood, paper, leather, textiles and numerous derivatives [179]. During the 19th and 20th centuries there has been an intense advance in industry, due to the scientific and technological knowledge related to the chemical nature of matter, reaction mechanisms and the role of physical and catalytic processes, achieving the transformation of crude oil into a wide variety of raw materials such as plastics, paints, detergents, textile products, etc., which changed the way of life of human beings.

The second wave of biomaterials was driven by the birth of biotechnology, especially due to recombinant DNA technology, which achieved great improvements in obtaining products from laundry detergents to animal feed [179]. Currently, the biological revolution due to the enormous progress in biology in the last fifty years, has positioned us at the beginning of a new stage of innovation in chemical production and in the third wave of innovation of biomaterials [179, 180].

Three types of biomaterials can be considered: drop-ins (a molecule traditionally obtained by petrochemical means, produces using biotransformations, i.e. biobased version of polyethylene), bio-replacements (product obtained from a biobased molecule, with the same technical characteristics and costs, but with an important reduction in environmental impact, i.e. polylactic acid) and bio-better (compounds obtained by unique biosynthetic pathways, which allow completely new material properties to be combined, i.e. novel biotech films that combine barrier properties with recyclability) [179].

Numerous tools are now available to address the needs of chemicals and materials, and challenges that arise from addressing them. This requires multidisciplinary teams, considering areas such as synthetic biology, metabolic engineering, molecular biology, microbiology, systems biology, synthetic chemistry, chemical engineering, bioinformatics, artificial intelligence, systems integration, metrology, chemical production, law, and bioethics [180].

To achieve industrialization of biology, biological routes and chemical routes must be considered as a whole, biological synthesis options have to be considered in the same way that individual chemical reactions are when developing a synthetic route [180]. Regarding the use of microorganisms in synthesis, it is important to advance in several aspects: host microorganism (or chassis), pathways and fermentation; and in their integration with raw materials and processing.

Efforts are needed to improve the design of enzymes with specific catalytic activities to reduce bioproduction costs. It is also necessary to continue developing basic science and technologies that allow the efficient development of organismal chassis and pathways, expanding the palette of "domesticated" microorganisms to increase the possibilities of using different raw materials. Finally, the design, creation and cultivation of robust strains that are genetically stable and maintain their yield over time will reduce costs both in the use and in the scale-up of microbial bioprocesses [179].

The use of cardboard packaging with improved features and new functionalities, can revolutionize the transport of goods and, in particular, of food. The consequences in the improvements of the packaging based on cellulose fibers would mean the displacement of materials derived from crude oil. The high recyclability of this type of packaging and its renewable nature means that its environmental impact is not so pronounced. Likewise, new packaging functionalities can lead to better food preservation and, consequently, reduce transport losses and improve logistics.

Therefore, a very promising alternative to reconnect human activity with biogeochemical cycles is the paradigm shift from a linear to a circular economy, as well as from resource depletion to renewable biomass exploitation, by means of bioeconomy [4, 13, 14, 20, 82]. The aim is to replace non-renewable food packaging by sustainable, compostable, recyclable biomaterials based on the biorefinery concept as a holistic, fully-responsible approach while maintaining or even enhancing the excellent properties now featured by the fossil-based materials used for food packaging [142, 162, 181].

References

[1] Energy Outlook 2022. Information on https://www.bp.com/en/global/corporate/energy-economics/energy-outlook.html

[2] United Nations Development (UNDP). Sustainable Developments Goals. Information on https://www.undp.org/sustainable-development-goals

[3] X.G. Zhu, S.P. Long, D.R. Long, What is the maximum efficiency with which photosynthesis can convert solar energy into biomass? Curr. Op. Biotecnol. 19(2) (2008) 153-159. https://doi.org/10.1016/j.copbio.2008.02.004

[4] P.R. Yaashikaa, P.S. Kumar, S. Varjani, Valorization of agro-industrial wastes for biorefinery process and circular bioeconomy: A critical review, Bioresour. Technol. 343 (2022) 126126. https://doi.org/10.1016/j.biortech.2021.126126

[5] F. Garcia-Ochoa, P. Vergara, M. Wojtusik, S. Gutierrez, V.E. Santos, M. Ladero, J.C. Villar, Multi-feedstock lignocellulosic biorefineries based on biological processes: An overview, Ind. Crops Prod. 172 (2021) 114062. https://doi.org/10.1016/j.indcrop.2021.114062

[6] T. Raj, K. Chandrasekhar, A.N. Kumar, S.H. Kim, Lignocellulosic biomass as renewable feedstock for biodegradable and recyclable plastics production: A sustainable approach, Renew. Sustain. Energy Rev. 158 (2022) 112130. https://doi.org/10.1016/j.rser.2022.112130

[7] J. Paini, V. Benedetti, S.S. Ail, M.J. Castaldi, M. Baratieri, F. Patuzzi, Valorization of wastes from the food production industry: A review towards an integrated agri-food processing biorefinery, Waste Biomass Valoriz. 13 (2021) 31-50. https://doi.org/10.1007/s12649-021-01467-1

[8] M.K. Awasthi, J.A. Ferreira, R. Sirohi, S. Sarsaiya, B. Khoshnevisan, S. Baladi, ... M.J. Taherzadeh, A critical review on the development stage of biorefinery systems towards the management of apple processing-derived waste, Renew. Sustain. Energy Rev. 143 (2021) 110972. https://doi.org/10.1016/j.rser.2021.110972

[9] M. del Mar Contreras, I. Romero, M. Moya, E. Castro, Olive-derived biomass as a renewable source of value-added products, Proc. Biochem. 97 (2020) 43-56. https://doi.org/10.1016/j.procbio.2020.06.013

[10] V. Yadav, A. Sarker, A. Yadav, A.O. Miftah, H.M. Iqbal, Integrated biorefinery approach to valorize citrus waste: A sustainable solution for resource recovery and environmental management, Chemosphere 293 (2022) 133459. https://doi.org/10.1016/j.chemosphere.2021.133459

[11] M. del Mar Contreras, J.M. Romero-García, J.C. López-Linares, I. Romero, E. Castro, Residues from grapevine and wine production as feedstock for a biorefinery, Food Bioprod. Proc. 134 (2022) 56-79. https://doi.org/10.1016/j.fbp.2022.05.005

[12] Information in the Global Timber Outlook 2020 - Gresham House. Webpage: chrome-extension://efaidnbmnnnibpcajpcglclefindmkaj/https://greshamhouse.com/wp-content/uploads/2020/07/GHGTO2020FINAL.pdf

[13] E.T. Kostas, J.M. Adams, H.A. Ruiz, G. Durán-Jiménez, G.J. Lye, Macroalgal biorefinery concepts for the circular bioeconomy: A review on biotechnological developments and future perspectives, Renew. Sustain. Energy Rev. 151 (2021) 111553. https://doi.org/10.1016/j.rser.2021.111553

[14] J. Ullmann, D. Grimm, D., Algae and their potential for a future bioeconomy, landless food production, and the socio-economic impact of an algae industry, Organic Agric. 11(2) (2021) 261-267. https://doi.org/10.1007/s13165-020-00337-9

[15] F. Licciardello, F., Packaging, blessing in disguise. Review on its diverse contribution to food sustainability, Trends Food Sci. Technol. 65 (2017) 32-39. https://doi.org/10.1016/j.tifs.2017.05.003

[16] Information in Eurostat: Packaging Waste Statistics. https://ec.europa.eu/eurostat/statistics-explained/ index. php/ Packaging_waste_statistics

[17] Y. Su, B. Yang, J. Liu, B. Sun, C. Cao, X. Zou, ... Z. He, Prospects for replacement of some plastics in packaging with lignocellulose materials: A brief review, BioRes. 13(2) (2018) 4550-4576. https://doi.org/10.15376/biores.13.2.Su

[18] N.M. Stark, L.M. Matuana, L. M., Trends in sustainable biobased packaging materials: A mini review, Mat. Today Sustain. 15 (2021) 100084. https://doi.org/10.1016/j.mtsust.2021.100084

[19] T. Efferth, N.W. Paul, Threats to human health by great ocean garbage patches, The Lancet Planetary Health.1 (8) (2017) 301-303. https://doi.org/10.1016/S2542-5196(17)30140-7

[20] M. Beltran, B. Tjahjono, A. Bogush, J. Julião, E.L.S. Teixeira. Food plastic packaging transition towards circular bioeconomy: A Systematic review of literature, Sustainability 13 (2021) 3896. https://doi.org/10.3390/su13073896

[21] M. Smith, D.C. Love, C.M. Rochman, R.A. Neff, Microplastics in seafood and the implications for human health, Curr. Environ Health Rep 5 (2018) 375-386. https://doi.org/10.1007/s40572-018-0206-z

[22] Eurostat: Packaging Waste Statistics. Information on: https://ec.europa.eu/eurostat/statisticsexplained/index.php?title=Packaging_waste_stati stics

[23] C. Matthews, F. Moran, A. K. Jaiswa, A review on European Union's strategy for plastics in a circular economy and its impact on food safety, J. Cleaner Prod. 283 (10) (2021) 125263. https://doi.org/10.1016/j.jclepro.2020.125263

[24] R. Rodrigues Fioritti. E. Revilla, J.C. Villar, M.L. D' Almeida, N. Gómez, Improving the strength of recycled liner for corrugated packaging by adding virgin fibres: Effect of refrigerated storage on paper properties, Packag. Technol, Sci. 34 (2021) 263-272. https://doi.org/10.1002/pts.2556

[25] W. Abdelmoez, I. Dahab, E.M. Ragab, O.A. Abdelsalam, A. Mustafa, A, Bio-and oxo-degradable plastics: Insights on facts and challenges, Polym. Adv. Technol. 32(5) (2021) 1981-1996. https://doi.org/10.1002/pat.5253

[26] K. Helanto, L. Matikainen, R. Talja, O.J. Rojas, Bio-based polymers for sustainable packaging and biobarriers: A critical review, BioRes. 14(2) (2019) 4902-4951. https://doi.org/10.15376/biores.14.2.Helanto

[27] Y. Liu, S. Ahmed, D.E. Sameen, Y. Wang, R. Lu, J. Dai, S. Li, W. Qin, A review of cellulose and its derivatives in biopolymer-based for food packaging application, Trends Food Sci. Technol. 112 (2021) 532-546. https://doi.org/10.1016/j.tifs.2021.04.016

[28] S. Sid, R.S. Mor, A. Kishore, V.S. Sharanagat, Bio-sourced polymers as alternatives to conventional food packaging materials: A review, Trends Food Sci. Technol. 115 (2021) 87-104. https://doi.org/10.1016/j.tifs.2021.06.026

[29] J. Liu, L. Zhang, W. Shun, J. Dai, J., Y. Peng, X. Liu, Recent development on bio-based thermosetting resins. J. Polym. Sci. 59(14) (2021) 1474-1490. https://doi.org/10.1002/pol.20210328

[30] E. Drago, R. Campardelli, M. Pettinato, P. Pereg, Innovations in Smart Packaging Concepts for Food: An Extensive Review. Foods. 9 (2020) 1628. https://doi.org/10.3390/foods9111628

[31] A. L. Woiciechowski, C. J. D. Neto, L. P. de Souza Vandenberghe, D. P. de Carvalho Neto, A. C. N. Sidney, L. A. J. Letti, S.G. Karp, L. A. Z. Torres, C. R. Soccol, Lignocellulosic biomass: Acid and alkaline pretreatments and their effects on biomass recalcitrance - Conventional processing and recent advances, Bioresour. Technol. 304 (2020) 122848. https://doi.org/10.1016/j.biortech.2020.122848

[32] C. G. Yoo, X. Meng, Y. Pu, A.J. Ragauskas, The critical role of lignin in lignocellulosic biomass conversion and recent pretreatment strategies: A comprehensive review, Bioresour. Technol. 301 (2020) 122784. https://doi.org/10.1016/j.biortech.2020.122784

[33] D. Sidiras, D. Politi, G. Giakoumakis, I. Salapa, Simulation and optimization of organosolv based lignocellulosic biomass refinery: A review, Bioresour. Technol. 343 (2022) 126158. https://doi.org/10.1016/j.biortech.2021.126158

[34] R. Roy, M.S. Rahman, D.E. Raynie, Recent advances of greener pretreatment technologies of lignocellulose, Curr. Res. Green Sustain. Chem. 3 (2020) 100035. https://doi.org/10.1016/j.crgsc.2020.100035

[35] A.R. Mankar, A. Pandey, A. Modak, K.K. Pant, Pretreatment of lignocellulosic biomass: A review on recent advances, Bioresour. Technol. 334 (2021) 125235. https://doi.org/10.1016/j.biortech.2021.125235

[36] L.V. Daza Serna, C.E. Orrego Alzate, C.A. Cardona Alzate, Supercritical fluids as a green technology for the pretreatment of lignocellulosic biomass, Bioresour. Technol. 199, (2016) 113-120. https://doi.org/10.1016/j.biortech.2015.09.078

[37] W. Den, V. K. Sharma, M. Lee, G. Nadadur, R. S. Varma, Lignocellulosic Biomass Transformations via Greener Oxidative Pretreatment Processes: Access to Energy and Value-Added Chemicals, Front. Chem., 6 (2018) 141. https://doi.org/10.3389/fchem.2018.00141

[38] A. Satlewal, R. Agrawal, S. Bhagia, J. Sangoro, A. J. Ragauskas, Natural deep eutectic solvents for lignocellulosic biomass pretreatment: Recent developments, challenges and novel opportunities, Biotechnol. Adv. 36 (2018) 2032-2050. https://doi.org/10.1016/j.biotechadv.2018.08.009

[39] Y.Liu, W. Chen, Q. Xia, B. Guo, Q.Wang, S.Liu, Y. Liu, J. Li, H. Yu, Efficient cleavage of lignin-carbohydrate complexes and ultrafast extraction of lignin oligomers from wood biomass by microwave-assisted treatment with deep eutectic solvent. ChemSusChem, 10 (2017) 1692-1700. https://doi.org/10.1002/cssc.201601795

[40] O.R. Alara, N.H. Abdurahman, C.I. Ukaegbu, Extraction of phenolic compounds: A review. Curr. Res. Food Sci. 4 (2021) 200-214. https://doi.org/10.1016/j.crfs.2021.03.011

[41] S. Perino, F. Chemat, Green process intensification techniques for bio-refinery. Curr. Opin. Food Sci. 25 (2019) 8-13. https://doi.org/10.1016/j.cofs.2018.12.004

[42] G. Grillo, L. Boffa, S. Talarico, R. Solarino, A. Binello, G. Cavaglià, S. Bensaid, G. Telysheva, G. Cravotto, Batch and flow ultrasound-assisted extraction of grape stalks: Process intensification design up to a multi-kilo scale. Antioxidants. 9 (2020) 1-30. https://doi.org/10.3390/antiox9080730

[43] M. González-Miquel, I. Díaz, Green solvent screening using modeling and simulation. Curr. Opin. Green Sustain. Chem. 29 (2021) 100469. https://doi.org/10.1016/j.cogsc.2021.100469

[44] D.P. Makris, S. Lalas, Glycerol and glycerol-based deep eutectic mixtures as emerging green solvents for polyphenol extraction: The evidence so far. Molecules. 25 (2020) 5842. https://doi.org/10.3390/molecules25245842

[45] L.S. Torres-Valenzuela, A. Ballesteros-Gómez, S. Rubio, Green Solvents for the Extraction of High Added-Value Compounds from Agri-food Waste. Food Eng. Rev., 12 (2020) 83-100. https://doi.org/10.1007/s12393-019-09206-y

[46] M. Ruesgas-Ramón, M.C. Figueroa-Espinoza, E. Durand, Application of Deep Eutectic Solvents (DES) for Phenolic Compounds Extraction: Overview, Challenges, and Opportunities. J. Agric. Food Chem. 65 (2017) 3591-3601. https://doi.org/10.1021/acs.jafc.7b01054

[47] R. Cai, Y. Yuan, L. Cui, Z. Wang, T. Yue, Cyclodextrin-assisted extraction of phenolic compounds: Current research and future prospects. Trends Food Sci. Technol. 79 (2018) 19-27. https://doi.org/10.1016/j.tifs.2018.06.015

[48] O. Gligor, A. Mocan, C. Moldovan, M. Locatelli, G. Crişan, I.C.F.R. Ferreira, Enzyme-assisted extractions of polyphenols - A comprehensive review. Trends Food Sci. Technol. 88 (2019) 302-315. https://doi.org/10.1016/j.tifs.2019.03.029

[49] I. Drevelegka, A.M. Goula, Recovery of grape pomace phenolic compounds through optimized extraction and adsorption processes. Chem. Eng. Process. - Process Intensif. 149 (2020) 107845. https://doi.org/10.1016/j.cep.2020.107845

[50] A. Sridhar, M. Ponnuchamy, P.S. Kumar, A. Kapoor, D.V.N. Vo, S. Prabhakar, Techniques and modeling of polyphenol extraction from food: a review; Springer International Publishing, Vol. 19 (2021) ISBN 0123456789. https://doi.org/10.1007/s10311-021-01217-8

[51] N.P. Kelly, A.L. Kelly, J.A. O'Mahony, Strategies for enrichment and purification of polyphenols from fruit-based materials. Trends Food Sci. Technol. 83 (2019) 248-258. https://doi.org/10.1016/j.tifs.2018.11.010

[52] S. Nanda, R. Azargohar, A.K. Dalai, J.A. Kozinski, An assessment on the sustainability of lignocellulosic biomass for biorefining. Renew. Sustain. Energy Rev. 50 (2015) 925-941. https://doi.org/10.1016/j.rser.2015.05.058

[53] R. Birner, Bioeconomy Concepts, in Bioeconomy, I. Lewandowski (Ed.), Springer International Publishing (2018) 17-38. https://doi.org/10.1007/978-3-319-68152-8_3

[54] Z.H. Liu, N. Hao, Y.Y. Wang, C. Dou, F. Lin, R. Shen,... J.S. Yuan, Transforming biorefinery designs with 'Plug-In Processes of Lignin'to enable economic waste valorization, Nature Comm. 12(1) (2021) 1-13. https://doi.org/10.1038/s41467-021-23920-4

[55] M. Galkin (2021), From stabilization strategies to tailor-made lignin macromolecules and oligomers for materials. Curr. Op. Green Sustain. Chem. 28 (2021) 100438. https://doi.org/10.1016/j.cogsc.2020.100438

[56] R.J. Kahn, C.Y. Lau, J. Guan, C.H. Lam, J. Zhao, Y. Ji,... S.Y. Leu, Recent advances of lignin valorization techniques toward sustainable aromatics and potential benchmarks to fossil refinery products. Bioresour. Technol. 346 (2022) 126419. https://doi.org/10.1016/j.biortech.2021.126419

[57] J.A. Poveda-Giraldo, J.C. Solarte-Toro, C.A.C. Alzate, The potential use of lignin as a platform product in biorefineries: A review, Ren. Sustain. Energy Rev. 138 (2021) 110688. https://doi.org/10.1016/j.rser.2020.110688

[58] M.V. Galkin, J.S. Samec, Lignin valorization through catalytic lignocellulose fractionation: a fundamental platform for the future biorefinery. ChemSusChem, 9(13) (2016) 1544-1558. https://doi.org/10.1002/cssc.201600237

[59] C.G. Yoo, X. Meng, Y. Pu, A.J. Ragauskas, The critical role of lignin in lignocellulosic biomass conversion and recent pretreatment strategies: A comprehensive review, Bioresour. Technol. 301 (2020) 122784. https://doi.org/10.1016/j.biortech.2020.122784

[60] M.M. Abu-Omar, K. Barta, G.T. Beckham, J.S. Luterbacher, J. Ralph, R. Rinaldi, ...F. Wang. Guidelines for performing lignin-first biorefining. Energy Environ. Sci. 14(1) (2012) 262-292. https://doi.org/10.1039/D0EE02870C

[61] W. Schutyser, T. Renders, S. Van den Bosch, S.-F. Koelewijn, G.T. Beckham, B.F. Sels, Chemicals from lignin: an interplay of lignocellulose fractionation, depolymerisation, and upgrading, Chem. Soc. Rev. 47 (2018) 852-908. https://doi.org/10.1039/C7CS00566K

[62] R. Ma, Y. Xu, X. Zhang, Catalytic oxidation of biorefinery lignin to value-added chemicals to support sustainable biofuel production, ChemSusChem. 8 (2015) 24-51. https://doi.org/10.1002/cssc.201402503

[63] J. Becker, C. Wittmann, A field of dreams: Lignin valorization into chemicals, materials, fuels, and health-care products, Biotechnol. Adv. 37 (2019) 107360. https://doi.org/10.1016/j.biotechadv.2019.02.016

[64] S. Roth, A.C. Spiess, Laccases for biorefinery applications: a critical review on challenges and perspectives, Bioprocess Biosyst. Eng. 38 (2015) 2285-2313. https://doi.org/10.1007/s00449-015-1475-7

[65] L. Munk, A.K. Sitarz, D.C. Kalyani, J.D. Mikkelsen, A.S. Meyer, Can laccases catalyze bond cleavage in lignin?, Biotechnol. Adv. 33 (2015) 13-24. https://doi.org/10.1016/j.biotechadv.2014.12.008

New Materials for a Circular Economy   Materials Research Forum LLC
Materials Research Foundations 149 (2023) 24-69   https://doi.org/10.21741/9781644902639-2

[66] A. Zerva, S. Simić, E. Topakas, J. Nikodinovic-Runic, Applications of microbial laccases: Patent review of the past decade (2009-2019), Catalysts. 9 (2019) 1023. https://doi.org/10.3390/catal9121023

[67] R.R. Singhania, A.K. Patel, T. Raj, C.-W. Chen, V.K. Ponnusamy, N. Tahir, S.-H. Kim, C.-D. Dong, Lignin valorisation via enzymes: A sustainable approach, Fuel. 311 (2022) 122608. https://doi.org/10.1016/j.fuel.2021.122608

[68] L. Martínková, B. Křístková, V. Křen, Laccases and tyrosinases in organic synthesis, Int. J. Mol. Sci. 23 (2022) 3462. https://doi.org/10.3390/ijms23073462. https://doi.org/10.3390/ijms23073462

[69] M. Singhvi, B.S. Kim, Lignin valorization using biological approach, Biotechnol. Appl. Biochem. 68 (2021) 459-468. https://doi.org/10.1002/bab.1995

[70] Y. Liu, G. Luo, H.H. Ngo, W. Guo, S. Zhang, Advances in thermostable laccase and its current application in lignin-first biorefinery: A review, Bioresour. Technol. 298 (2020) 122511. https://doi.org/10.1016/j.biortech.2019.122511

[71] M.B. Agustin, D.M. Carvalho, M.H. Lahtinen, K. Hilden, T. Lundell, K.S. Mikkonen, Laccase as a tool in building advanced lignin-based materials, ChemSusChem. 14 (2021) 4615-4635. https://doi.org/10.1002/cssc.202101169

[72] Y. Gu, L. Yuan, L. Jia, P. Xue, H. Yao, Recent developments of a co-immobilized laccase-mediator system: a review, RSC Adv. 11 (2021) 29498-29506. https://doi.org/10.1039/D1RA05104K

[73] A. Dong, K.M. Teklu, W. Wang, X. Fan, Q. Wang, M. Ardanuy, Z. Dong, New strategy for grafting hydrophobization of lignocellulosic fiber materials with octadecylamine using a laccase/TEMPO system, Int. J. Biol. Macromol. 160 (2020) 192-200. https://doi.org/10.1016/j.ijbiomac.2020.05.167

[74] R. Weiss, G.M. Guebitz, A. Pellis, G.S. Nyanhongo, Harnessing the power of enzymes for tailoring and valorizing lignin, Trends Biotechnol. 38 (2020) 1215-1231. https://doi.org/10.1016/j.tibtech.2020.03.010

[75] M.M. Cajnko, J. Oblak, M. Grilc, B. Likozar, Enzymatic bioconversion process of lignin: mechanisms, reactions and kinetics, Bioresour. Technol. 340 (2021) 125655. https://doi.org/10.1016/j.biortech.2021.125655

[76] R. Hellmayr, S. Bischof, J. Wühl, G.M. Guebitz, G.S. Nyanhongo, N. Schwaiger, F. Liebner, R. Wimmer, Enzymatic conversion of lignosulfonate into wood adhesives: A next step towards fully biobased composite materials, Polymers. 14 (2022) 259. https://doi.org/10.3390/polym14020259

[77] D. Huber, A. Ortner, A. Daxbacher, G.S. Nyanhongo, W. Bauer, G.M. Guebitz, Influence of oxygen and mediators on laccase-catalyzed polymerization of lignosulfonate, ACS Sustain. Chem. Eng. 4 (2016) 5303-5310. https://doi.org/10.1021/acssuschemeng.6b00692

[78] S. Slagman, H. Zuilhof, M.C.R. Franssen, Laccase-mediated grafting on biopolymers and synthetic polymers: A critical review, ChemBioChem. 19 (2018) 288-311. https://doi.org/10.1002/cbic.201700518

[79] P. Tang, T. Zheng, C. Yang, G. Li, Enhanced physicochemical and functional properties of collagen films cross-linked with laccase oxidized phenolic acids for active edible food packaging, Food Chem. 393 (2022) 133353. https://doi.org/10.1016/j.foodchem.2022.133353

[80] M. Jimenez Bartolome, S. Bischof, A. Pellis, J. Konnerth, R. Wimmer, H. Weber, N. Schwaiger, G.M. Guebitz, G.S. Nyanhongo, Enzymatic synthesis and tailoring lignin properties: A systematic study on the effects of plasticizers, Polymer. 202 (2020) 122725. https://doi.org/10.1016/j.polymer.2020.122725

[81] S. Winestrand, L. Järnström, L.J. Jönsson, Fractionated lignosulfonates for laccase-catalyzed oxygen-scavenging films and coatings, Molecules. 26 (2021) 6322. https://doi.org/10.3390/molecules26206322

[82] R.D. Di Lorenzo, I. Serra, D. Porro, P. Braunduardi. State of the art on the microbial production of industrially relevant organic acids. Catalysts 12 (2022) 234-247. https://doi.org/10.3390/catal12020234

[83] Y. Wan, J.M. Lee. Toward value-added dicarboxylic acids from biomass derivatives via thermocatalytic conversion, ACS Catal. 11(5) (2021) 2524-2560. https://doi.org/10.1021/acscatal.0c05419

[84] T. Werpy, G. Petersen, Top Value-Added Chemicals from Biomass Volume I-Results of screening for potential candidates from sugars and synthesis gas energy efficiency and renewable energy, National Renewable Energy Lab., Golden, CO (US) (2004). https://doi.org/10.2172/15008859

[85] T.J. Farmer, R.L. Castle, J.H. Clark, D.J. Macquarrie, Synthesis of unsaturated polyester resins from various bio-derived platform molecules. Int J Mol Sci. 16(7) (2015) 14912-32. https://doi.org/10.3390/ijms160714912

[86] A.M. Diez-Pascual, Tissue engineering bionanocomposites based on poly(propylene fumarate). Polymers 9(7) (2017) 1-19. https://doi.org/10.3390/polym9070260

[87] V. Martin-Dominguez, J. Estevez, F. De Borja Ojembarrena, V.E. Santos, M. Ladero, Fumaric acid production: A biorefinery perspective. Fermentation 4(2) (2018). https://doi.org/10.3390/fermentation4020033

[88] M. Sano, T. Tanaka, H. Ohara, Y. Aso, Itaconic acid derivatives: structure, function, biosynthesis, and perspectives. Appl Microbiol Biotechnol. 104(21) (2020) 9041-51. https://doi.org/10.1007/s00253-020-10908-1

[89] R.K. Das, S.K. Brar, M. Verma, Fumaric Acid. Platform Chemical Biorefinery. Elsevier Inc. (2016) 133-157. https://doi.org/10.1016/B978-0-12-802980-0.00008-0

[90] A. Jiménez-Quero, E. Pollet, M. Zhao, E. Marchioni, L. Avérous, V. Phalip, Itaconic and fumaric acid production from biomass hydrolysates by Aspergillus strains. J Microbiol Biotechnol. 26(9) (2016) 1557-65. https://doi.org/10.4014/jmb.1603.03073

[91] A. Kuenz, S. Krull, Biotechnological production of itaconic acid-things you have to know. Appl Microbiol Biotechnol. 102(9) (2018) 3901-14. https://doi.org/10.1007/s00253-018-8895-7

[92] Fumaric Acid Market | 2022 - 27 | Industry Share, Size, Growth - Mordor Intelligence. Information on: https://www.mordorintelligence.com/industry-reports/fumaric-acid-market#faqs

[93] Q. Xu, S. Li, H. Huang, J. Wen, Key technologies for the industrial production of fumaric acid by fermentation. Biotechnol. Adv. 30(6) (2012) 1685-96. https://doi.org/10.1016/j.biotechadv.2012.08.007

[94] V. Martin-Dominguez, P.I. Aleman-Cabrera, L. Eidt, U. Pruesse, A. Kuenz, M. Ladero, V.E. Santos, Production of Fumaric Acid by Rhizopus arrhizus NRRL 1526: A Simple Production Medium and the Kinetic Modelling of the Bioprocess. Fermentation 8 (2022) https://doi.org/10.3390/fermentation8020064

[95] Z. Zhou, G. Du, Z. Hua, J. Zhou, J. Chen, Optimization of fumaric acid production by Rhizopus delemar based on the morphology formation. Bioresour. Technol. 102(20) (2011) 9345-9. https://doi.org/10.1016/j.biortech.2011.07.120

[96] B.E. Teleky, D.C. Vodnar, Biomass-derived production of itaconic acid as a building block in specialty polymers. Polymers 11(6) (2019). https://doi.org/10.3390/polym11061035

[97] M. Okabe, D. Lies, S. Kanamasa, E.Y. Park, Biotechnological production of itaconic acid and its biosynthesis in Aspergillus terreus. Appl Microbiol Biotechnol. 84(4) (2009) 597-606. https://doi.org/10.1007/s00253-009-2132-3

[98] H. Hosseinpour Tehrani, J. Becker, I. Bator, K. Saur, S. Meyer, A.C. Rodrigues Lóia, Integrated strain- And process design enable production of 220 g·L-1 itaconic acid with Ustilago maydis. Biotechnol Biofuels. 12(1) (2019) 1-11. https://doi.org/10.1186/s13068-019-1605-6

[99] D. Gopaliya, V. Kumar, S.K. Khare, Recent advances in itaconic acid production from microbial cell factories. Biocatal. Agric. Biotechnol. 36 (2021) https://doi.org/10.1016/j.bcab.2021.102130

[100] J.R. Elmore, G.N. Dexter, D. Salvachúa, J. Martinez-Baird, E.A. Hatmaker, J.D. Huenemann, Production of itaconic acid from alkali pretreated lignin by dynamic two stage bioconversion. Nat. Commun. 12(1) (2021) https://doi.org/10.1038/s41467-021-22556-8

Materials Research Forum LLC
https://doi.org/10.21741/9781644902639-2

[101] A. Jiménez-Quero, E. Pollet, M. Zhao, E. Marchioni, L. Avérous, V. Phalip, Fungal fermentation of lignocellulosic biomass for itaconic and fumaric acid production. J Microbiol. Biotechnol. 27(1) (2017) 1-8. https://doi.org/10.4014/jmb.1607.07057

[102] L. Shen, H. Xu, L. Kong, Y.Yang, Non-toxic crosslinking of starch using polycarboxylic acids: kinetic study and quantitative correlation of mechanical properties and crosslinking degrees, J. Pol. Environ. 23(4) (2015) 588-594. https://doi.org/10.1007/s10924-015-0738-3

[103] J. Lu, J.Li, H.Gao, D. Zhou, H. Xu, Y. Cong, W. Zhang, F. Xin, M. Jiang, Recent progress on bio-succinic acid production from lignocellulosic biomass, World J. Microbiol. Biotechnol. 37(1) (2021) 1-8. https://doi.org/10.1007/s11274-020-02979-z

[104] G. Priya, U. Narendrakumar, I. Manjubala, Thermal behavior of carboxymethyl cellulose in the presence of polycarboxylic acid crosslinkers, J. Thermal Anal. Calor. 138(1) (2019) 89-95. https://doi.org/10.1007/s10973-019-08171-2

[105] C.J. Chiang, R.C. Hu, Z.C. Huang, Y.P. Chao, Production of Succinic Acid from Amino Acids in Escherichia coli, J. Agric. Food Chem. 69(29) (2021) 8172-8178. https://doi.org/10.1021/acs.jafc.1c02958

[106] M. Ladero, J. Esteban, J.M. Bolívar, V.E. Santos, V. Martín-Domínguez, A. García-Martín, A. Lorente, I.A. Escanciano, Food waste biorefinery for bioenergy and value-added products, in: A. Sinharoy, P.N.L. Lens (Eds.), Renewable Energy Technologies for Energy Efficient Sustainable Development, Springer Chan, New York, 2022, pp. 185-224. https://doi.org/10.1007/978-3-030-87633-3_8

[107] R. Dickson, E. Mancini, N. Garg, J.M. Woodley, K.V. Gernaey, M. Pinelo, J. Liu, S.S. Mansouri, Sustainable bio-succinic acid production: Superstructure optimization, techno-economic, and lifecycle assessment, Energy Environ. Sci. 14(6) (2021) 3542-3558. https://doi.org/10.1039/D0EE03545A

[108] E.O. Jokodola, V. Narisetty, E. Castro, S. Durgapal, F. Coulon, R. Sindhu, P. Binod, J.R. Banu, G. Kumar, V. Kumar, Process optimisation for production and recovery of succinic acid using xylose-rich hydrolysates by Actinobacillus succinogenes, Bioresour. Technol. 344 (2022) 126224. https://doi.org/10.1016/j.biortech.2021.126224

[109] M. Ferone, A. Ercole, F. Raganati, G. Olivieri, P. Salatino, A. Marzocchella, Efficient succinic acid production from high-sugar-content beverages by Actinobacillus succinogenes, Biotechnol. Prog. 35(5) (2019) e2863. https://doi.org/10.1002/btpr.2863

[110] K. Filippi, N. Georgaka, M. Alexandri, H. Papapostolou, A. Koutinas, Valorisation of grape stalks and pomace for the production of bio-based succinic acid by Actinobacillus succinogenes, Ind. Crops Prod. 168 (2021) 113578. https://doi.org/10.1016/j.indcrop.2021.113578

[111] I.A. Escanciano, M. Ladero, V.E. Santos, On the succinic acid production from xylose by growing and resting cells of Actinobacillus succinogenes: a comparison, Biomass Conv. Bioref. (2022) 1-14. https://doi.org/10.1007/s13399-022-02943-x

[112] S. Mores, L.P. de Souza Vandenberghe, A.I.M. Júnior, J.C. de Carvalho, A.F.M. de Mello, A. Pandey, C.R. Soccol, Citric acid bioproduction and downstream processing: Status, opportunities, and challenges, Bioresour. Technol. 320 (2021) 124426. https://doi.org/10.1016/j.biortech.2020.124426

[113] S. Chavan, B. Yadav, A. Atmakuri, R.D. Tyagi, J.W.C. Wong, P. Drogui, Bioconversion of organic wastes into value-added products: A review, Bioresour. Technol. 344 (2022) 126398. https://doi.org/10.1016/j.biortech.2021.126398

[114] O. Sawant, S. Mahale, V. Ramchandran, G. Nagaraj, A. Bankar, Fungal citric acid production using waste materials: a mini-review, J. Microbiol. Biotechnol. Food Sci. (2021) 821-828. https://doi.org/10.15414/jmbfs.2018.8.2.821-828

[115] L. Wen, Y. Liang, Z. Lin, D. Xie, Z. Zheng, C. Xu, B. Lin, Design of multifunctional food packaging films based on carboxymethyl chitosan/polyvinyl alcohol crosslinked network by using citric acid as crosslinker, Polymer 230 (2021) 124048. https://doi.org/10.1016/j.polymer.2021.124048

[116] M.M. Hassan, N. Tucker, M.J. Le Guen, Thermal, mechanical and viscoelastic properties of citric acid-crosslinked starch/cellulose composite foams, Carbohyd. Polym. 230 (2020) 115675. https://doi.org/10.1016/j.carbpol.2019.115675

[117] P.G. Ponnusamy, J. Sundaram, S. Mani, Preparation and characterization of citric acid crosslinked chitosan-cellulose nanofibrils composite films for packaging applications, J. Appl. Pol. Sci. 139(17) (2022) 52017. https://doi.org/10.1002/app.52017

[118] K.A. Uyanga, O.P. Okpozo, O.S. Onyekwere, W.A. Daoud,). Citric acid crosslinked natural bi-polymer-based composite hydrogels: effect of polymer ratio and betacyclodextrin on hydrogel microstructure. React. Funct. Polym. 154 (2020) 104682. https://doi.org/10.1016/j.reactfunctpolym.2020.104682

[119] J. Shang, Z. Shao, X. Chen, Electrical behavior of a natural polyelectrolyte hydrogel: chitosan/Carboxymethylcellulose hydrogel. Biomacromolecules 9 (2008) 1208-1213. https://doi.org/10.1021/bm701204j

[120] L. Bao, X. Chen, B. Yang, Y. Tao, Y. Kong, Construction of electrochemical chiral interfaces with integrated polysaccharides via Amidation. ACS Appl. Mater. Interfaces 8 (2016) 21710-21720. https://doi.org/10.1021/acsami.6b07620

[121] P. Chen, F. Xie, F. Tang, T. McNally, Glycerol plasticisation of chitosan/carboxymethyl cellulose composites: role of interactions in determining structure and properties. Int. J. Biol. Macromol. 163 (2020) 683-693. https://doi.org/10.1016/j.ijbiomac.2020.07.004

Materials Research Forum LLC
https://doi.org/10.21741/9781644902639-2

[122] J.C. Roy, A. Ferri, S. Giraud, G. Jinping, F. Salauen, Chitosan-carboxymethylcellulosebased polyelectrolyte complexation and microcapsule shell formulation. Int. J. Mol. Sci. 19 (2018) 2521/2521-2521/2519. https://doi.org/10.3390/ijms19092521

[123] L. Wang, X. Yang, W.A. Daoud, High power-output mechanical energy harvester based on flexible and transparent Au nanoparticle-embedded polymer matrix. Nano Energy, 55 (2019) 433-440. https://doi.org/10.1016/j.nanoen.2018.10.030

[124] E. Akar, A. Altınışık, Y. Seki, Preparation of pH- and ionic-strength responsive biodegradable fumaric acid crosslinked carboxymethyl cellulose. Carbohydr. Polym. 90 (2012) 1634-1641. https://doi.org/10.1016/j.carbpol.2012.07.043

[125] M. Dilaver, K. Yurdakoc, Fumaric acid cross-linked carboxymethylcellulose/poly(vinyl alcohol) hydrogels. Polym. Bull. 73 (2016) 2661-2675. https://doi.org/10.1007/s00289-016-1613-7

[126] K.A. Uyanga, W.A. Daoud, Green and sustainable carboxymethyl cellulose-chitosan composite hydrogels: Effect of crosslinker on microstructure. Cellulose 28 (2021) 5493-5512. https://doi.org/10.1007/s10570-021-03870-2

[127] Y. Zhanga, S. Zhoua, X. Fangb, X. Zhoua, J. Wanga, F. Baib, S. Pengb, Renewable and flexible UV-blocking film from poly(butylene succinate) and lignin, Eur. Pol. J. 116 (2019) 265-274. https://doi.org/10.1016/j.eurpolymj.2019.04.003

[128] Mazière, P. Prinsen and A. García, R. Luque , C. Len, A review of progress in (bio)catalytic routes from/to renewable succinic acid, Biofuels, Bioprod. Bioref. 11 (2017) 908-931. https://doi.org/10.1002/bbb.1785

[129] M.B. Coltelli, I. Della Maggiore, M. Bertold, F. Signori, S. Bronco, F. Ciardelli, Poly(lactic acid) properties as a consequence of poly(butylene adipate-co-terephthalate) blending and acetyl tributyl citrate plasticization, J. Appl. Pol. Sci. (2008) 110, 1250-1262. https://doi.org/10.1002/app.28512

[130] A. Pellis, J.W. Comerforda, A.J. Maneffaa, M.H. Sipponen, J. H. Clark, T. J. Farmer, Elucidating enzymatic polymerisations: Chain-length selectivity of Candida antarctica lipase B towards various aliphatic diols and dicarboxylic acid diesters, Eur. Pol. J. 106 (2018) 79-84. https://doi.org/10.1016/j.eurpolymj.2018.07.009

[131] L. Daviot, T. Len, C. Sze Ki Lin and C. Len, Microwave-assisted homogeneous acid catalysis and chemoenzymatic synthesis of dialkyl succinate in a flow reactor, Catalysts 9 (2019) 272. https://doi.org/10.3390/catal9030272

[132] F. L. Aguzín, M.L. Martínez, A.R. Beltramone, C.L. Padró, N.B. Okulik, Esterification of succinic acid using sulfated zirconia supported on SBA-15, Chem. Eng. Technol. 44 (2021), 1185-1194. https://doi.org/10.1002/ceat.202000333

Materials Research Forum LLC
https://doi.org/10.21741/9781644902639-2

[133] D. Pithadia , A. Patel, Conversion of bioplatform molecule, succinic acid to value-added products via esterification over 12-tungstosilicic acid anchored to MCM-22, Biomass Bioen. 151 (2021) 106178. https://doi.org/10.1016/j.biombioe.2021.106178

[134] V.R. Umrigar, M. Chakraborty,P.A. Parikh, Optimization of microwave assisted esterification of succinic acid using Box-Behnken design approach. Information on: https://doi.org/10.21203/rs.3.rs-1206807/v1 https://doi.org/10.21203/rs.rs-1206807/v1

[135] S. Dinh Le, S. Nishimura, K. Ebitani, Direct esterification of succinic acid with phenol using zeolite beta catalyst, Catal. Comm. 122 (2019) 20-23. https://doi.org/10.1016/j.catcom.2019.01.006

[136] K. Sonker, N. Tiwari,1 R. K. Nagarale, V. Verma, Synergistic effect of cellulose nanowhiskers reinforcement and dicarboxylic acids crosslinking towards polyvinyl alcohol properties, J. Pol. Sci. part A: Pol. Chem. 54 (2016), 2515-2525. https://doi.org/10.1002/pola.28129

[137] M. Asghera, S. A. Qamara, M. Bilalb, H. M.N. Iqbalc, Bio-based active food packaging materials: Sustainable alternative to conventional petrochemical-based packaging materials. Food Res. Int. 137 (2020) 109625. https://doi.org/10.1016/j.foodres.2020.109625

[138] A. Cimini, M. Moresi, Carbon footprint of a pale lager packed in different formats: Assessment and sensitivity analysis based on transparent data. J. Cleaner Prod. 112 (2016) 4196-4213. https://doi.org/10.1016/j.jclepro.2015.06.063

[139] P. Nechita. Review on polysaccharides used in coatings for food packaging papers. Coatings 10(6) (2020) 566. https://doi.org/10.3390/coatings10060566

[140] T. M. Brouwer, E. U. Thoden van Velzen, K. Ragaert, R. Klooster. Technical limits in circularity for plastic packages. Sustainability 12(23) (2020) 10021. https://doi.org/10.3390/su122310021

[141] M. Hoque, S. Gupta, R. Santhosh, I. Syed, P. Sarkar. Chapter 3. Biopolymer-based edible films and coatings for food applications, in: K. Pal, I. Banerjee, P. Sarkar, A. Bit, D. Kim, A. Anis, S. Maji (Eds.), Food, Medical, and Environmental Applications of Polysaccharides. Elsevier, 2021, pp. 81-107. https://doi.org/10.1016/B978-0-12-819239-9.00013-0

[142] S. Mangaraj, A. Yadav, L.M. Bal, S. K. Dash, N.K. Mahanti, Application of biodegradable polymers in food packaging industry: A comprehensive review, J Package Technol. Res. 3 (2019) 77-96. https://doi.org/10.1007/s41783-018-0049-y

[143] V. Kumar Rastogi, P. Samyn, Bio-Based coatings for paper applications, Coatings 5 (2015) 887-930 https://doi.org/10.3390/coatings5040887

[144] K. Khwaldia, E. Arab-Tehrany, S. Desobry, Biopolymer coatings on paper packaging materials, Compr. Rev. Food Sci. Food Saf. 9 (2010) 82-91. https://doi.org/10.1111/j.1541-4337.2009.00095.x

[145] H. W. Maurer, Starch in the paper industry, Starch, (2009) 657-713. https://doi.org/10.1016/B978-0-12-746275-2.00018-5

[146] H. Li, Y. Qi, Y.Zhao, J. Chi, S. Cheng, Starch and its derivatives for paper coatings: A review, Prog. Org. Coat. 135 (2019) 213-227. https://doi.org/10.1016/j.porgcoat.2019.05.015

[147] S.H. Osong, Mechanical pulp-based nanocellulose processing and applications relating to paper and paperboard, composite films, and foams (2016) Mid Sweden University Doctoral Thesis 245. Information on: http://www.diva-portal.org/smash/record.jsf?pid=diva2%3A1033818&dswid=-7058

[148] A.F.Turbak, F.W. Snyder, K.R. and Sandberg, Micro-fibrillated cellulose and process for producing it, U.S. Patent CH 648071 (A5). (1985)

[149] T. Isogai, T. Saito, A. Isogai, Wood cellulose nanofibrils prepared by TEMPO electro-mediated oxidation, Cellulose 18(2) (2011) 421-431. https://doi.org/10.1007/s10570-010-9484-9

[150] U. Fillat, B. Wicklein, R. Martín-Sampedro, D. Ibarra, E. Ruiz-Hitzky, C. Valencia, A. Sarrión, E. Castro, M.E. Eugenio, Assessing cellulose nanofiber production from olive tree pruning residue, Carbohydr. Polym. 179 (2018) 252-261. https://doi.org/10.1016/j.carbpol.2017.09.072

[151] N. Lin, J. Tang, A. Dufresne, M. K. C. Tam, Advanced functional materials from nanopolysaccharides, in: N. Lin, J. Tang, A. Dufresne, M.K.C. Tam (Eds.), Springer Series in Biomaterials Science and Engineering, Springer Nature Singapore, 2019, 15 pp. 1-54. https://doi.org/10.1007/978-981-15-0913-1

[152] A. Balea, E. Fuente, M.C. Monte, N. Merayo, C. Campano, C. Negro, A. Blanco, Industrial application of nanocelluloses in papermaking: a review of challenges, technical solutions, and market perspectives. Molecules 25 (3) 526 (2020) 526 1-30. https://doi.org/10.3390/molecules25030526

[153] A.F. Lourenço, J.A.F. Gamelas, P. Sarmento, P.J. Ferreira, Cellulose micro and nanofibrils as coating agent for improved printability in office papers, Cellulose 27 (2020) 6001-6010. https://doi.org/10.1007/s10570-020-03184-9

[154] C. Salas, M. Hubbe, O.J. Rojas, Nanocellulose applications in papermaking, in: Production of materials from sustainable biomass resources, Biofuels and biorefineries, Springer, Singapore, 2019, pp 61-69. https://doi.org/10.1007/978-981-13-3768-0_3

[155] U. Fillat, P. Vergara, N. Gómez. J.C. Villar. Effect of different nanofibers obtained from lignocellulose as barriers for paper packaging. March 2021. XXV TECNICELPA - International Forest, Pulp and Paper Conference, XI CIADICYP - 2021

[156] M. Nasrollahzadeh, Z. Nezafat, Z. Nezafat, N. Shafiei, N.S.S. Bidgoli, Food packaging applications of biopolymer-based (nano)materials, in: M. Nasrollahzadeh (Ed.), Biopolymer-Based Metal Nanoparticle Chemistry for Sustainable Applications, 2021, pp. 137-186. https://doi.org/10.1016/B978-0-323-89970-3.00004-4

[157] S. Galus, J. Kadzinska, Food applications of emulsion-based edible films and coatings, Trends Food Sci. Technol. 45 (2015) 273-283. https://doi.org/10.1016/j.tifs.2015.07.011

[158] K. J. Jem, B. Tan, The development and challenges of poly (lactic acid) and poly(glycolic acid). Adv. Ind. Eng. Polym. Res. 3 (2020) 60-70. https://doi.org/10.1016/j.aiepr.2020.01.002

[159] A. Z. Naser, I. Deiaba, B. M. Darras, Poly(lactic acid) (PLA) and polyhydroxyalkanoates (PHAs), green alternatives to petroleum-based plastics: a review, RSC Adv. 11 (2021) 17151-17196. https://doi.org/10.1039/D1RA02390J

[160] M.A. Hubbe. Prospects for maintaining strength of paper and paperboard products while using less forest resources: A review. BioRes 9(1) (2014)1634-1763. https://doi.org/10.15376/biores.9.1.1634-1763

[161] T. Fadiji, T. Berry, C.J. Coetzee, L. Opara Investigating the mechanical properties of paperboard packaging material for handling fresh produce under different environmental conditions: experimental analysis and finite element modelling. J. Appl. Pack. Res. 9(2) (2017) 20-34.

[162] S. Sid, R.S. Mor, A. Kishore, V.S. Sharanagat, V. S. Bio-sourced polymers as alternatives to conventional food packaging materials: A review. Trends Food Sci. Technol. 115 (2021) 87-104. https://doi.org/10.1016/j.tifs.2021.06.026

[163] Y. Hamzeh, A. Ashori, Z. Khorasani, A. Abdulkhani, A. Abyaz, Pre-extraction of hemicelluloses from bagasse fibers: Effects of dry-strength additives on paper properties. Ind. Crops Prod. 43 (2013) 365-371. https://doi.org/10.1016/j.indcrop.2012.07.047

[164] H. Lindqvist, J. Homback, A. Rosling, K. Salminen, B. Holmbom, M. Auer, A. Sundberg, Galactoglucomannan derivatives and their application in papermaking. Biores 8(1) (2013) 994-1010. https://doi.org/10.15376/biores.8.1.994-1010

[165] X. Song, M.A: Hubbe, M. A. Enhancement of paper dry strength by carboxymethylated β-D-glucan from oat as additive. Holzforschung, 68(3) (2013) 257-263. https://doi.org/10.1515/hf-2013-0108

[166] X. Song, M.A: Hubbe, TEMPO-mediated oxidation of oat β-D-glucan and its influences on paper properties. Carbohyd. Pol. 99 (2014) 617-623. https://doi.org/10.1016/j.carbpol.2013.08.070

[167] G.B. Xu, W.Q. Kong, C.F. Liu, R.C. Sun, J.L. Ren, Synthesis and characteristic of xylan-grafted-polyacrylamide and application for improving pulp properties. Materials 10(8) (2017) 971. https://doi.org/10.3390/ma10080971

[168] M.S. Firouz, K. Mohi-Alden, M. Omid, A critical review on intelligent and active packaging in the food industry: Research and development. Food Res. Intern. 141 (2021) 110113. https://doi.org/10.1016/j.foodres.2021.110113

[169] Eurostat Recycling rates for packaging waste. Information on: https://ec.europa.eu/eurostat/databrowser/view/ten00063/default/table

[170] EN 13432: 2001 Packaging - Requirements for Packaging Recoverable Through Composting and Biodegradation - Test Scheme and Evaluation Criteria for the Final Acceptance of Packaging

[171] G. Dolci, V. Venturelli, A. Catenacci, R. Ciapponi, F. Malpei, Romano S.E. Turri, M. Grosso, Evaluation of the anaerobic degradation of food waste collection bags made of paper or bioplastic, J. Environ. Manage. 305 (2022) 114331. https://doi.org/10.1016/j.jenvman.2021.114331

[172] S. Otto, M. Strenger, A. Maier-Nöth, M. Schmid, Food packaging and sustainability - Consumer perception vs. correlated scientific facts: A review, J. Cleaner Prod. 298, (2021) 126733. https://doi.org/10.1016/j.jclepro.2021.126733

[173] C. Andersson, New ways to enhance the functionality of paperboard by surface treatment - a review, Packag. Technol. and Sci. 21 (2008). https://doi.org/10.1002/pts.823

[174] G. Glenn, R. Shogren, X. Jin, W. Orts, W. Hart-Cooper, L. Olson, Per- and polyfluoroalkyl substances and their alternatives in paper food packaging, Compr Rev Food Sci Food Saf. 20 (2021) 2596-2625. https://doi.org/10.1111/1541-4337.12726

[175] M. Vikman, J. Vartiainen, I. Tsitko, P. Korhonen, biodegradability and compostability of nanofibrillar cellulose-based products, J. Polym. Environ. 23 (2015) 206-215. https://doi.org/10.1007/s10924-014-0694-3

[176] CEPI (Confereration of European Paper Infustries) Food Contact Guidelines for the compilance of paper & board materials and articles. Information on: https://www.cepi.org/wp-content/uploads/2020/09/Food-Contact-Guidelines_2019.pdf

[177] C. Nerín, S. Bourdoux, B. Faust, T. Gude, C. Lesueur, T. Simat, A. Stoermer, E. Van Hoek, P. Oldring, Guidance in selecting analytical techniques for identification and quantification of non-intentionally added substances (NIAS) in food contact materials (FCMS). Food Addit. Contam. Part A: Chem. Anal. Control Expo. Risk Assess. 39 (2022) 620-643. https://doi.org/10.1080/19440049.2021.2012599

Materials Research Forum LLC
https://doi.org/10.21741/9781644902639-2

[178] K. Grob, How to make the use of recycled paperboard fit for food contact? A contribution to the discussion, Food Addit. Contam. Part A: Chem. Anal. Control Expo. Risk Assess. 39 (2022) 198-213. https://doi.org/10.1080/19440049.2021.1977853

[179] T. Brennan, M. Chui, W. Chyan, A. Spamann. McKinsey & Company, 2021. The third wave of biomaterials: When Innovation meets demand. Information on: https://www.mckinsey.com/industries/chemicals/our-insights/the-third-wave-of-biomaterials-when-innovation-meets-demand.

[180] National Research Council, 2015. Industrialization of Biology: A Roadmap to Accelerate the Advanced Manufacturing of Chemicals. Washington D.C. The National Academies Press. Information on: https://doi.org/10.17226/19001. https://doi.org/10.17226/19001

[181] M.K. Verma, S. Shakya, P. Kumar, J. Madhavi, J. Murugaiyan, M.V.R. Rao. Trends in packaging material for food products: Historical background, current scenario, and future prospects. J. Food. Sci. Technol. 58(11) (2021) 4069-4082. https://doi.org/10.1007/s13197-021-04964-2

New Materials for a Circular Economy
Materials Research Foundations 149 (2023) 70-104

Materials Research Forum LLC
https://doi.org/10.21741/9781644902639-3

Chapter 3

Solar Energy and Reused Materials in the Recycling of Plastics

C. Puente-Rueda[1], M.A. Saénz-Nuño[2], J. Luis-Zamora[2], C. Puente-Agueda[1]

[1] ICAI School of Engineering, Comillas Pontifical University, 28015 Madrid, Spain

[2] Institute for Research in Technology (IIT), ICAI School of Engineering, Comillas Pontifical University, 28015 Madrid, Spain

Abstract

The scarcity of technology in developing countries is a reality, while resources such as, plastic caps or aluminum cans are relatively easy to obtain anywhere in the world. Moreover, these materials are accumulating both in the oceans and in landfills and finding a solution to contain this phenomenon is of the outmost importance. This project aims to reincorporate such materials, normally treated as waste, to the production cycle. With the help of a solar oven, manufactured with recycled materials, plastic can be molded and used to create simple pieces, thus contributing to its reuse as part of sustainable development.

Keywords

Plastic, Recycle, Solar Oven, Solar Energy, Circular Economy, Prosthesis

Contents

1. Introduction

At temperatures around 160°C, easily reached in a household or industrial oven, some common plastics such as polyethylene become mouldable, opening endless possibilities for the reuse of these plastics to form simple parts, thus contributing to circular economy and sustainable development.

Polyethylene is one of the most common plastics, due to its low price and simplicity of manufacture, which leads to a production of approximately 80 million tonnes per year worldwide. This massive output also results in its accumulation inland and in the ocean as a waste product, making it accessible anywhere in the world and nearly costless. Furthermore, basic resources such as wood, aluminium cans and glass are available in all regions, even those in developing countries where other means of technology are scarce.

These materials, normally treated as waste, have a fundamental role to play in circular economy in developing countries. Whereas a small solar thermal plant or another type of generator that can heat up a chamber to reach plastic-moulding temperatures takes immense resources of time and money, the design and construction of a solar oven with recycled materials, is a possibility that makes reusing plastic more accessible in these places.

For reaching such high temperatures, using only renewable energy and with the materials described before, a solar oven prototype has been designed, tested, and finally developed. The oven's design has been focused on reusing materials that are accessible anywhere in the world and are able to create the perfect environment to collect direct solar radiation and reach high temperatures, also retaining heat. Its fabrication and assembly processes have been conceived so that they can be reproduced using simple tools and with no special technical knowledge required. As a result, the prototype is made mostly out of wooden crates, cardboard, snack bags and glass, and can rise its temperature up to more than 120°C during a 6-hour solar exposure period. [1]

In addition to the research and design of this solar oven, several applications of moulded plastic pieces have been studied. One being the fabrication and testing of a fully functioning hand prosthesis made entirely of high-density polyethylene obtained from bottle caps and using moulds from aluminium cans. [2,3]

2. Plastics

Plastics are synthetic polymers: long chains of atoms artificially bonded together. The different properties of these materials vary according to the length and shape of these chains, giving plastics an innumerable range of uses and versatility.

2.1 Origin and evolution

The first semi-synthetic polymer developed in 1869 was the forerunner of a true revolution. Derived from cellulose and camphor, this new material replaced other naturally obtained ones such as ivory, leather or turtle shells. These natural materials were previously used for everything: piano keys, knife handles, eyeglass frames, combs and brush handles, to name the most common. By the mid-18th century, the hunting of hawksbill turtles for their shells had escalated to the point where it almost caused their extinction, similarly to what happened with ivory, where demand far outstripped supply. With the development of these early types of plastics, the demand for these natural materials was severely cut back, slowing down the hunt for turtles and elephants sufficiently so that these animals are still around today. Therefore, plastic, which has historically been attributed as the destroyer of marine life, actually saved it in the first place. [4]

After years of development, a fully synthetic material was finally produced in 1907: Bakelite. Its invention meant that man was no longer limited to materials from nature, but could create them in a laboratory. Plastic was no longer derived from cellulose, now it was produced from oil. This new substance was produced in the desired shape, which was then able to be preserved in various conditions of temperature and humidity.

Plastic reached every home and industry in the 1960s and 1970s, replacing steel in cars, wood in furniture and even paper and glass in packaging. After several decades of boom, the problem of plastic accumulation in the environment began to emerge and efforts were made to raise public awareness with several worldwide campaigns to promote recycling. Although much progress has been made in recycling, it is far from perfect. Plastic accumulation is an increasing problem and solutions related to the circular economy must take care of this problem, since plastic is essential for development.

Plastics have been a historic breakthrough and are present throughout our business network; if their use were to be drastically reduced, it would make economic development impossible.

On one hand, in Spain there are 93,000 jobs directly related to the plastics sector. This market is valued at some 30,000 million euros and involves more than 3,000 companies, most of which are SMEs. It accounts for 21% of the world's manufacturing industry and 2.7% of Spain's GDP. [5]

Moreover, as plastics are lighter and stronger materials, the weight savings from using them to replace other materials leads to savings in raw materials and emissions. In transportation, they make vehicles lighter and consume less fuel. Plastic food packaging allows food to be preserved for longer periods of time, making it easier for transport, thus reducing food waste. In addition, 60% of plastics are long-lasting: more than 5 years of guaranteed and operational life and can have a useful life of up to 50 years.

On the other hand, the Spanish economy generated last year 2021, around 1,6 million tons of plastic waste, of which more than 650 thousand tons were recycled. The rest either ended up in landfills or were dumped directly into the ocean. The United Nations Environment Program has calculated that more than 126 tons of plastics are thrown into the sea daily from Spain. That is, almost 46 thousand tons per year. [6]

In conclusion, the future of plastics does not lie in their elimination, but in their correct recycling and circular solutions.

2.2   Plastic types

According to [2], [7] and [8], plastics are classified by their internal structure and behavior, all of them being grouped into two main groups: thermoplastics and thermosets. Both groups have long chains of atoms linked together. The main difference between them is the ability of these chains to return to their original position under stress or drastic temperature changes.

Thermoplastics are those that become deformable when exposed to a heat source. Their chains are not permanently joined, so they can slide and acquire new positions that are maintained once the material has cooled. This type of plastic is the easiest to recycle.

By contrast, the chains of thermosetting plastics are held together by bonds previously formed by degradation or setting. These bonds make it impossible for the chains to slip, thus maintaining their shape permanently. When these plastics melt, their properties change completely, making recycling this type of plastic meaningless. Within this family are elastomers. They have chains that are joined at different points, but also folded. This causes these plastics to deform under stress and then return to their original shape once the stress has ceased.

In addition, there are an infinite number of plastics in these categories. To facilitate their identification and recycling in 1988, the Plastics Industry Society developed a system for their classification according to the resin from which the product was manufactured. In this system, called the Resin Identification Code, abbreviated as RIC, plastics fall into seven different categories, the acronyms of which are: PET, HDPE, PVC, LDPE, PP, PS and Others. In the first version of this classification, the numbers were surrounded by a triangle with arrows, commonly known as the recycling symbol. This first version caused confusion among consumers, as many thought that the symbol meant that all these plastics were recyclable. Finally, in 2013, it was decided to change the arrows to a solid triangle to prevent this confusion. [9]

PET (Polyethylene terephthalate)

It is a petroleum-derived thermoplastic, easily moldable and adaptable to any shape and design. Its properties include its flexibility and strength. It is resistant to wear, chemicals and corrosion and behaves well in the presence of humidity. It can be used as a substitute for PVC. It has a high degree of transparency, which is why it is used in the lighting sector. Its mechanical qualities make it possible to manufacture yarns from this plastic, forming the well-known synthetic fabrics, such as polyester. Among its most common uses, it is also used to make containers for the food industry (water and soft drink bottles), due to its low degree of toxicity. The main disadvantage is that there is a risk of antimony and phthalate gases being released during use. Objects made from this material can be formed by blow molding, injection molding and extrusion. It is 100% recyclable and is currently the easiest plastic to recycle.

HDPE (High Density Polyethylene)

High density polyethylene is chemical resistant, flexible and has a high durability, it is also translucent, less transparent than LDPE. These properties, together with its low cost, result in its use in the manufacture of containers, lids, closures, pipes, ducts and packaging. Its forming process is mainly by extrusion blow molding or injection molding. This plastic is also recyclable.

PVC (Polyvinyl Chloride)

This thermoplastic is a chemical combination of carbon, hydrogen and chlorine and is manufactured from abundant resources such as common salt and petroleum. Because it contains chlorine, it does not burn easily, so it is used as an insulator in cables. It is also used to manufacture packaging, except in the food industry, where its use is prohibited due to its toxicity. It is used to make pipes, drains, hoses and medical devices such as catheters and blood bags. It is very difficult to recycle and this can be a problem, since this plastic is particularly long-lasting: it can last more than 70 years in good condition.

LDPE (Low Density Polyethylene)

Low-density polyethylene is one of the safest plastics and can be directly reused. It is used in bags for waste and agricultural uses, as well as in cling film. It can be recycled but is more difficult to recycle than type 1 and 2 plastics.

PP (Polypropylene)

Polypropylene is a low density, but high stiffness thermoplastic used to manufacture lightweight products. It is resistant to chemicals but weak against ultraviolet radiation. It is used to manufacture food packaging, household goods, pipes and films. It is widely used in the agricultural and automotive industries. It is relatively easy to recycle.

PS (Polystyrene)

Polystyrene has a low temperature resistance, softening between 85 and 105 degrees. This makes it easy to mold it by heating. It has the lowest thermal conductivity of all thermoplastics and is therefore mainly used as a thermal insulator, for fillings and frozen food packaging. In addition, thanks to its low electrical conductivity, it is the ideal insulator in electrical installations. The main disadvantage is that it is difficult to recycle.

Others (Resins and Polyurethane)

This category includes materials that are a mixture of various plastics or resins. Adhesive elements, container handles, foams, fillers and other elements of the wood and carpentry industry are included in this category. Plastics in this category are not recycled, as the mixture of several plastics makes it impossible to separate them for recycling.

2.3   Plastics chosen for melting

For this research work, as it is stated in [2], any thermoplastic whose melting temperature is reachable within the solar oven can be used. The plastic chosen is high-density polyethylene, classified as type 2. This material can be found in plastic bottle caps, and it is accessible worldwide. If needed, it can be bought at a very low price, although the point of using high-density polyethylene is to recycle already existing plastic that would be otherwise treated as waste. This thermoplastic can be melted by applying heat and then molded without losing its properties at around 135°C, making it the ideal material use it as raw material in the oven and give it a second life.

The theoretical characteristics of high-density polyethylene are as follows: [10, 11]

Table 1. Mechanical characteristics of HDPE. Sources: [2, 10, 11]

Parameters	Values
Density	$0.952 - 0.965$ g/cm^3
Price	$1.26 - 1.39$ €/kg
Elastic modulus	$1.07 - 1.09$ GPa
Poisson's ratio	$0.41 - 0.427$
Mechanical resistance to compression	$18.6 - 24.8$ MPa
Mechanical resistance to bending	$30.9 - 43.4$ MPa
Tensile strength	$22.1 - 31$ MPa
Fracture toughness	$1.52 - 1.82$ MPa m$^{1/2}$
Hardness (ambient temperature)	$7.9 - 9.9$ HV
Melting temperature	$130 - 137$ °C

2.4   Mechanical tests

To justify the use of this plastic to create new pieces, the mechanical properties shown above must remain intact after remolding. For this purpose, test probes from molded polyethylene were made and tested for bending and traction. The process to manufacture these pieces was carried out in a conventional furnace, since this part of the research was developed before the construction of the solar oven prototype.

To make the test tubes, the following steps were carried out, and were repeated twice. This allowed to create two layers that could be perfectly joined together, allowing the sample to have the desired thickness.

1. Forming the can mold.

With gloves, scissors and pliers, the bases of the cans are cut off and the central cylinder is slit to obtain a rectangular sheet of aluminum. This sheet is shaped into a rectangular box to insert the plastic pieces afterwards.

Figure 1. Test tube mold. Source: [2]

2. Cutting of the bottle caps into 6 pieces.

The previously washed bottle caps are cut into six pieces to facilitate their casting and adhesion to each other. The smaller the size of the cuts, the more homogeneous the piece is. Caps of different colors can be used, as long as they are made with the same plastic, but it is more aesthetic and gives a greater sensation of robustness if they are of the same color.

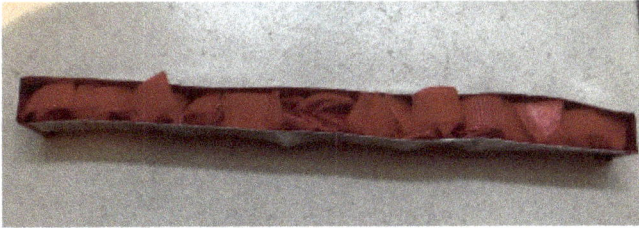

Figure 2. Mold with caps. Source: [2]

3.Baking and resting stages.

In this step, heating time in the oven is alternated with cooling time. The plastic parts need the minutes of rest to release the vapors that have been generated inside the plastic. At the same time, the molten plastic is pressed with a tool so that no air bubbles are trapped inside the part that can weaken its properties.

These times are as follows:

i. 15 minutes of baking at 160°.

ii. 5 minutes of resting time.

iii. 15 minutes of baking at 180°.

iv. 5 minutes resting time.

v. 15 minutes of baking at 180°.

4. Final resting and separation of the parts from the mold.

After letting the test probe to cool, it is separated from the mold.

Finally, having repeated steps 1 to 4 twice, the two layers must be introduced together again in the molds and heated for 15 minutes at 180 degrees.

New Materials for a Circular Economy   Materials Research Forum LLC
Materials Research Foundations 149 (2023) 70-104   https://doi.org/10.21741/9781644902639-3

Figure 3. Finished test tube. Source: [2]

The procedure for testing the strength of the material can be conducted in a laboratory by means of a tensile test and a bending test. Both tests must be done using a specific testing machine, which in this research work was HOYTOM L.T machine. This machine allows to exert force manually on the material through a hydraulic system and has a meter to read the breakage limit. [12,13]

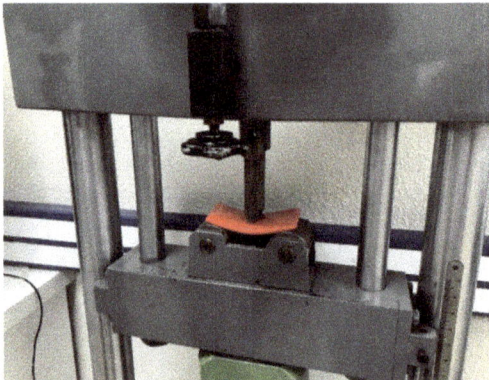

Figure 4. Test probe during bending test. Source: [2]

The mechanical properties of the plastic can be also tested in an artisanal way, without the need of specific machinery. In developing countries or places where this kind of machinery is not available, this becomes an interesting option. The artisanal bending test can be performed applying force in the middle of the probe by hanging objects of a known weight and then measuring how many of them can the probe withstand before breaking.

In the work shown in [2], the rustic test was carried out refilling half-liter water bottles for exerting the middle force. They are filled with water for its easy mass-volume rate, where water's density is approximately equivalent to 1kg/L.

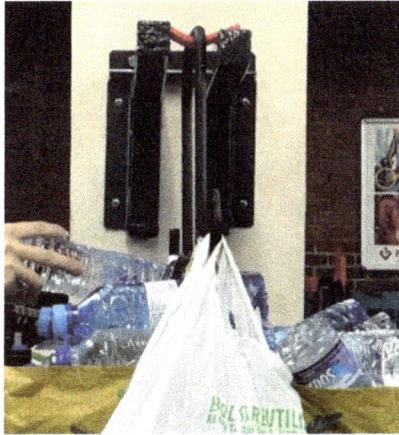

Figure 5. Probe in rustic bending test. Source: [2]

The results of both professional and rustic tests yielded mechanical flexural strength results of around 32.66 MPa, within the theoretical range of high-density polyethylene. In other words, this proves that a part made of recycled plastic can withstand a weight of up to 20 kilograms without bending and up to 50 kilograms before breaking. This opens up endless possibilities in the use of parts manufactured with this material and this process.

Once it has been proven that the recycled material is strong enough, research has been focused on a way of melting the plastic that is sustainable and feasible anywhere in the world. To this end, the solar oven prototype presented below has been created, which, in addition to being an affordable way to use existing resources, contributes to circular economy, giving cans and bottle cups the chance to be treated as resources themselves, instead of waste.

3.   Solar oven prototype

For the design of the oven, studies on the different types of handmade solar ovens manufactured so far were developed, paying special attention to the materials used, the parts they have and how the evolution of their interior temperature is measured.

In this project, the handmade oven is presented as a tool capable of playing a fundamental role in many aspects of the daily life of people in developing countries. The main advantage of the oven made with recycled materials is precisely that its manufacturing cost is practically non-existent, since the raw material can be easily found in these areas. In addition, it is low maintenance and the energy from which it draws, photovoltaic solar energy, is an unlimited and abundant resource.

3.1 Solar ovens to date

As it is stated in [1], existing artisanal solar ovens are used for low temperature cooking in ecological environments or as simple experiments. These ovens can be manufactured professionally by companies that specialize in this field or they can be built at home. The most common materials used in their construction are wood, steel, glass and aluminum.

Both home-made and professional ovens work as follows: the inside of the oven is laminated with reflective material, which mirror the sun's rays that penetrate through the translucent lid, causing them to bounce off the walls until they are absorbed by the material in the main compartment, which heats up.

3.1.1 Parts

Most solar oven designs consist of the following parts: two drawers, a reflective interior, a translucent cover, a rear hatch and adhesive to keep everything in position.

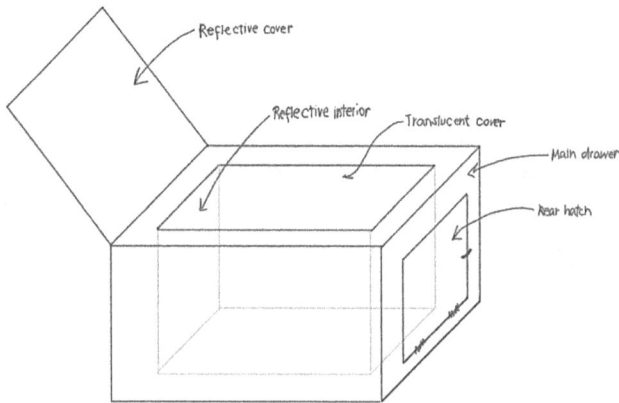

Figure 6. Parts of a solar oven. Own source.

- **Main drawer**

The main compartment of the oven has the function of housing the container that will be heated. Its walls are covered with reflective materials to reflect the sun's rays. Their sizes vary considerably among the different popular designs.

Materials used to date for this part of the oven are usually wood or cardboard, since they can be easily obtained and manipulated to obtain the desired dimensions. Both can withstand high temperatures without burning (cardboard 250°C and wood 400°C) and are good thermal insulators.

There is the possibility to place two boxes one inside the other, so that the air between them acts as an insulator and the heat loss from the inner box is considerably reduced. In addition to air, there is the option of using other types of insulation such as sawdust, wood shavings or newspaper scraps between the two drawers. Other insulation commonly used for heaters and furnaces are rock wool or mineral wool and polystyrene sheets.

The crate's shape is also a key aspect to consider. The wood can be cut so that it is an inclined box, this will allow the oven to be positioned so that the sun's rays enter directly and reach the element to be heated without bouncing off any walls. If the drawer is straight, it usually has reflective elements that concentrate the sun's rays inside it.

- **Reflective covers**

The use of external reflective covers is quite common, as it allows to collect solar rays coming from different directions. For its correct operation, it must be positioned at a certain angle. This angle varies depending on the position of the sun and the dimensions of the base and the reflectors.

- **Reflective interior**

The inside of the oven must be coated with a reflective material, which allows the sun's rays to bounce off it and end up on the element to be heated. The materials used for this application are usually aluminum foil or reflective paint. This reflective material is normally used for the inner base of the oven, as well as black paint or black cardboard. The color black absorbs all the wavelengths emitted by the sun and this makes it easier for the base to heat up, transmitting this heat to the inner vessel.

- **Translucent cover**

A fundamental element to allow the sun's rays to enter the box is the cover. To be able to do this, the cover must be made of a transparent material, such as glass or plastic. Glass covers are usually thicker and therefore insulate the heat better, preventing it from escaping through the lid.

- **Adhesive**

To assemble all the parts of the oven, an adhesive or similar is needed. The most common are white glue, duct tape and silicone. There are different options suitable for choosing adhesives, but the most important thing to consider is that it must withstand high temperatures, since it will go both inside and outside the oven.

- **Rear hatch**

As part of the outer drawer, a rear door can be incorporated to keep the top cover sealed and to use this door as access to the oven interior. Without specific tools, the installation of this door becomes difficult, as it requires cutting the outer drawer and then installing the door's opening and closing mechanism. In addition, it must be securely fastened to the rest of the drawer to prevent heat leakage.

3.1.2 Temperature measurement

The temperature measurement of the handmade solar ovens manufactured to date is practically non-existent. In those that do incorporate a temperature meter, the device used for this purpose is usually similar to an oven thermometer, which allows the temperature to be measured instantaneously and with very low precision. As for the gathering of these temperatures over time, the only extended way is the manual writing of the temperature measured in the oven every few minutes, a process that is far from being automatic and simple.

3.2   Materials

The criteria followed for the design of the prototype follow two lines: the first is that they must be recycled materials, accessible in underprivileged areas, and the second is that they must be capable of providing the ideal environment to collect the sun's rays and achieve adequate heat retention.

The materials chosen to build the oven are: two wooden crates, cardboard, snack bags, a refrigerator glass, black paint and joining elements such as nails, duct tape and white glue.

3.3   Design

Before construction could begin, a model of the oven with its different parts was designed using SolidWorks CAD software. This permitted to see how the prototype would look and get a sense of its dimensions.

Figure 7. Solar oven model designed with SOLIDWORKS. Source: [1]

- **Main drawer and inner drawer**

For the outer drawer, a wooden box of dimensions 500x330x175mm was chosen. This box had a thickness of between 7mm at the bottom and 9mm at the sides. The initial function of this box was to transport bottled wine and it was easy to obtain. Being made of wood, it facilitates its later painting with black paint and the incorporation of nails to center the flaps, explained later.

The inner drawer also consists of a reused wooden box of dimensions 320x227x133mm and with a thickness of 10mm at the bottom and 3mm on the sides.

Wood does not need any special physical or chemical treatment, so its recycling is cheap and economical. In addition, its porosity makes it a good thermal insulator, being able to retain heat inside the box and minimize heat leakage. These properties of wood make the election of this material for the main box and the interior an appropriate choice.

Figure 8. Main drawer and inner drawer. Source: [1]

- **Interior and exterior coating**

To ensure that the sun's rays are reflected and trapped inside the oven, two types of coatings are required. One will be reflective and the other black.

The reflective coating will be placed on the inside of the outer drawer, the outside of the small drawer and on the flaps. To fulfill its function, it must be made of a material with a high reflectivity. Aluminum reflects more than 90% of visible radiation, making it an ideal material for lining the inside of the solar oven. In addition, aluminum is the most abundant metal in the earth's crust and can be found in a great number of everyday products. For this purpose, the aluminum found in the inner layer of snack bags, previously emptied and washed, will be used.

These types of bags are complex to recycle and in nature take between 100 and 200 years to degrade on their own. This is because they have up to seven layers, in which aluminum and different types of plastic, such as polypropylene (PP) or polyethylene terephthalate (PET), are interspersed. Its use in the construction of the oven lining gives a new life to this product, which would otherwise be incinerated or disposed of in a landfill and could reach all types of ecosystems with the consequences that this entails.

Figure 9. Snack bags with aluminium layers. Source: [1]

The other type of coating corresponds to one made with matte black paint. This type of paint makes it possible to absorb all the radiation that reaches the surface, causing it to heat up. This ensures that the outer face of the outer box is heated and minimizes heat loss on these surfaces. Moreover, at the base of the inner drawer, the black paint makes it easier for that surface to reach very high temperatures and heat is transmitted to the mold by conduction. For this oven prototype, it has been possible to try an anti-heat black paint. This paint is a type of topcoat made with special resins that resist high temperatures. It can withstand maximum temperatures between 350°C and 600°C. It is used to decorate and protect objects that reach those high temperatures, such as: fireplaces, barbecues, stoves, boilers or heating pipes, so it is perfect for our application.

- **Cover**

The piece used as cover must be thick enough to retain heat, but at the same time transparent enough to allow the sun's rays to penetrate. A 3mm thick glass will be used for this purpose. This glass used to belong to the shelf of a refrigerator and has been reconditioned for this purpose by removing the plastic edges and then putting a wooden frame to make it safer to manipulate the glass.

- **Adhesive**

Six different types of adhesives were used in the construction process, chosen to match the characteristics required for each joint and according to their availability in countries in disadvantaged areas.

White glue was used on the glass frame, allowing an effective adhesion of the wood to each other and also to the glass. Nails were used for the union of the wooden stop strip to the box and also used to hold the exterior reflective cardboard. Duct tape was used to line the inside of the drawers with the snack bags. To fasten the aluminum used in the reflectors to the cardboard reflector base, staples were utilized. Clamps and rope made a simple, adjustable mechanism for adjusting the angle of the reflectors in reference to the horizontal plane. A rubber band was required to hold de panels against the outer crate. This form of temporary fastening allows the panels to be easily removed and transported separately from the main crate.

- **Reflective flaps**

Consists of four panels cut from cardboard and lined with aluminum, stapled to the cardboard. The panels are fastened to each other by the clamp and rope system mentioned above and to the drawer by means of a rubber band.

There are two trapezoidal panels of dimensions 94.5x49.5mm with a rectangular end of 49.5x11mm that is attached to the outer drawer and two panels of 83x33mm, with a rectangular end of 33x10.5mm. These sizes were chosen based on the available cardboard. They could be manufactured in different sizes as long as they allow the reflection of the sun's rays inside the oven.

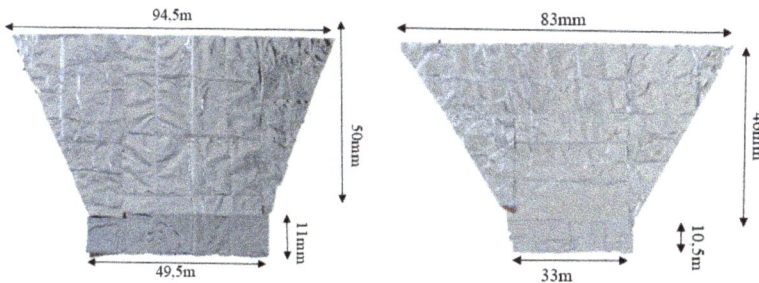

Figure 10. Big and small flaps. Source: [1]

- **Rear door**

For this prototype, it was decided not to incorporate a back door, since the framed glass serves the function of a door. In addition, in order to cut the wood, specific tools would be necessary, which are not available everywhere.

3.4 Construction process

This section details the steps followed in the construction of the furnace, in sequential order.

1. Cleaning of the wooden boxes.

Before proceeding to paint or stick the aluminum layers to the wooden drawers, both the wooden boxes and the bags were cleaned. Both boxes were cleaned with soap and water with the help of a brush.

2. Cutting the snack bags.

For the aluminum of the bags to be ready to line the crates, first they were cut on the side of each bag and then all the bags were spread out to avoid wrinkles in the material.

3. Painting the bottom of the small drawer and the outer sides of the large drawer.

The paint used is matte black, spread with a brush over the previously cleaned surface. It is then left to dry for a few hours.

4. Lining of both drawers with aluminum

Several layers of lining were applied so that there were no visible gaps in the wood. To stick together the aluminum to the box, duct tape was used.

Figure 11. Inner box. Source: [1]

Figure 12. Outer box. Source: [1]

5. Cutting out the cardboard and lining with aluminum to create the reflectors.

Two pairs of flaps were made, which would be placed with their base on the side of the main box.

Figure 13. Finished flaps. Source: [1]

6. Cleaning and disassembling of the glass cover.

Pliers were used to remove the plastic coating from the glass. The frame came off easily and cleaning was done carefully so as not to damage the glass.

7. Framing the glass

To create the glass frame, a wooden strip with a square profile was purchased. Because of this shape, it was easy to create a stop to hold the glass in place. Four pieces were cut with a hacksaw, and then glued together with white wood glue.

Figure 14. Close up of the framed glass. Source: [1]

8. Trimming and adding the wood strip to the outer drawer.

Since all the materials were reused and none were purchased, they didn't fit perfectly. Between the outer drawer and the framed glass (lid) there was a gap of about two centimeters. In order to make the lock as airtight as possible, a wooden strip was nailed to the inside of the oven with nails.

Nails were also nailed to the outside of the main drawer. The reflectors are supported on these nails.

Figure 15. Side view of the solar oven prototype. Source: [1]

8. Introducing the temperature sensors.

The sensors were placed one on the inside of the oven and the other on the outside, identified by a sign reading "oven" or "environment".

Figure 16. Inside temperature sensor. Source: [1]

Figure 17. Outside temperature sensor. Source: [1]

Final solar oven prototype result:

Figure 18. Finished solar oven prototype. Source: [1]

3.5   Temperature measurement hardware

To perform the temperature measurements inside the oven, SPI sensors of type Max6675 have been used, together with type K thermocouples. The sensors are connected to a breadboard that reaches an Arduino UNO board. See detailed software in [1].

With this software, it is possible to represent in a real time graph the temperature measurement inside and outside the oven provided by the sensors and save these measurements automatically.

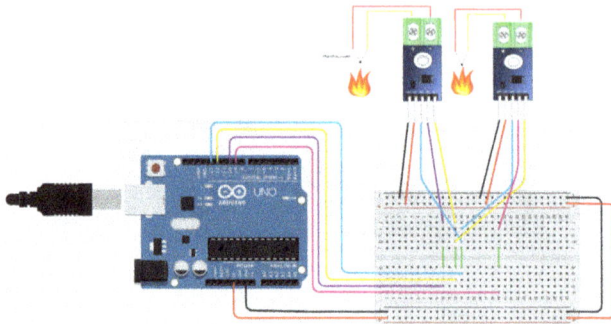

Figure 19. Temperature measurement system. Source: [1]

3.6 Collected data

To see the evolution of the inside temperature in relation to the outside temperature, both measurements were collected with the sensors and the previously mentioned program for a period of 7 days. As for the outside temperature records, the first data was obtained from the AEMET and eltiempo.es pages and the last four days, the ambient temperature data was also obtained through the Arduino program.

A joint plot of the temperature variation over the sampling time is presented below. As it can be seen, the temperature inside the oven varies practically in the same way the different days. First, there is a zone of very pronounced growth, then it stabilizes and, depending on the cloudy intervals, this stabilization lasts more or less time. Finally, as the solar incidence decreases in the afternoon, the temperature begins to fall slowly until it is equal to the outside temperature.

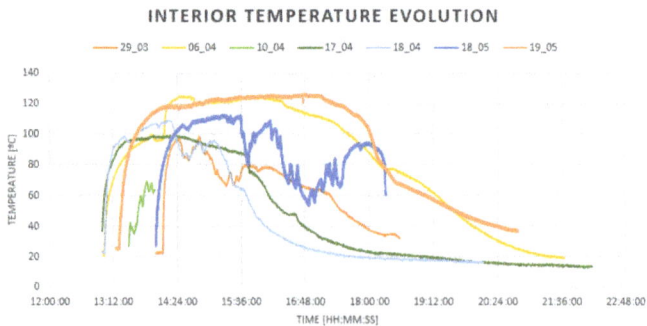

Figure 20. Registered temperature data. Source: [1]

To explain the operation of the furnace, the data log for March 29, 2021, is presented below. On that day the outside temperature had a maximum of 20 degrees Celsius and a minimum of 10 degrees Celsius (see figure 22). The next graph shows the cloudiness on the same day. There was a peak around 3pm where the sky was fully covered. When this cloudiness passed, the sky remained clear until nightfall (see figure 23).

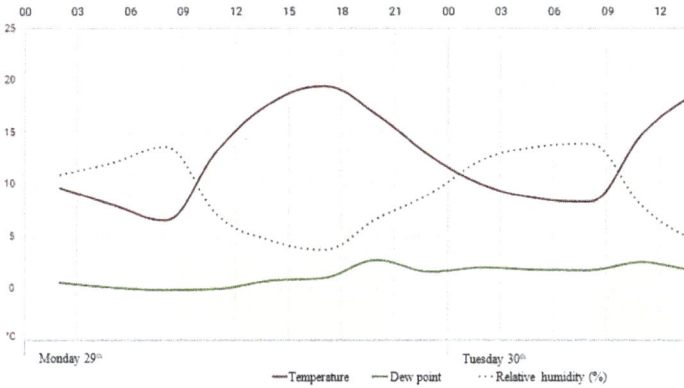

Figure 21. Temperature data from March 29. Source: AEMET

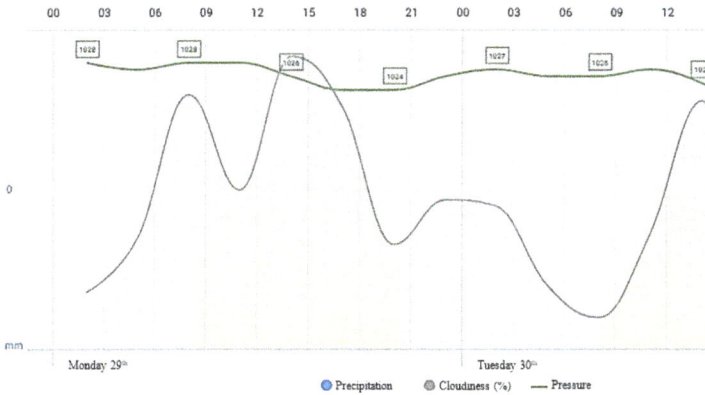

Figure 22. Cloudiness data from March 29. Source: AEMET

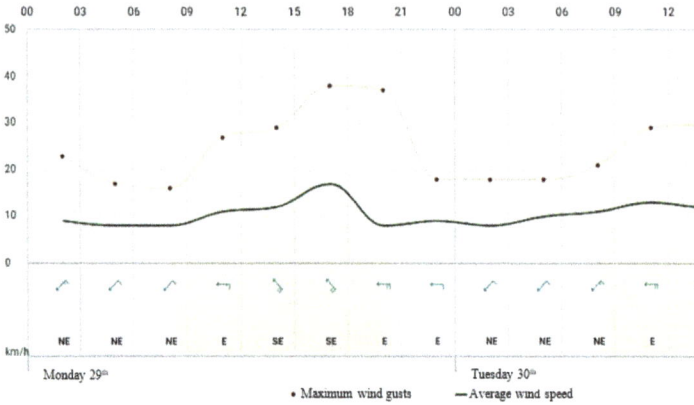

Figure 23. Wind data from March 29. Source: AEMET

INTERIOR TEMPERATURE EVOLUTION

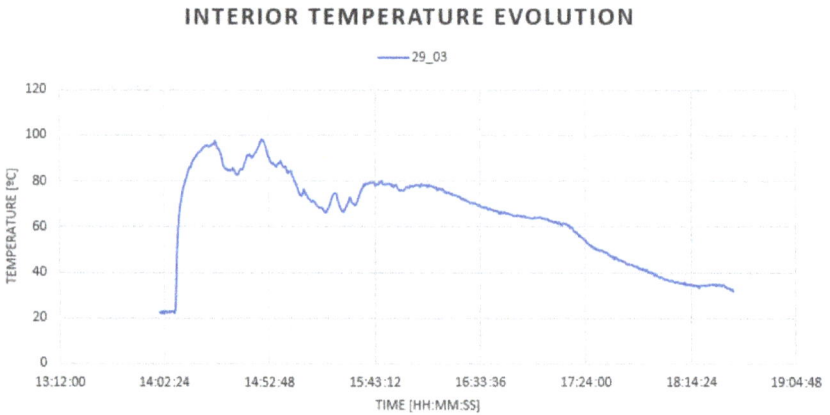

Figure 24. Temperature recorded data from March 29. Source: [1]

Contrasting the cloudiness plot [Figure 23] with that of the recorded temperature [Figure 25], we find a peak drop in temperature around 15:30h, which coincides with a peak in recorded cloudiness. This was to be expected since the furnace is heated by the sun's rays. It should be noted that the temperature does not drop abruptly during the cloudy hours, but instead, the oven cools down slowly. This is because the oven has an effective insulation,

Materials Research Forum LLC
https://doi.org/10.21741/9781644902639-3

making it able to retain some heat in the less bright hours. Also, the 29th of May was windy, this made the reflectors unable to maintain their position well enough to concentrate the heat. Consequently, the average temperature reached was lower than that of the following days. At the same time, the maximum outdoor temperature reached was 20°C at 16 pm, which does not coincide with the peak indoor temperature of 98°C at 14:50. From this it can be deduced that if it had not been for the cloudiness, the indoor temperature would have continued to rise, since it had not reached the maximum.

For another specific day, May 19, the following temperature graph was obtained, showing the measurement of the programmed indoor and outdoor temperature sensors, as well as the temperature provided by AEMET.

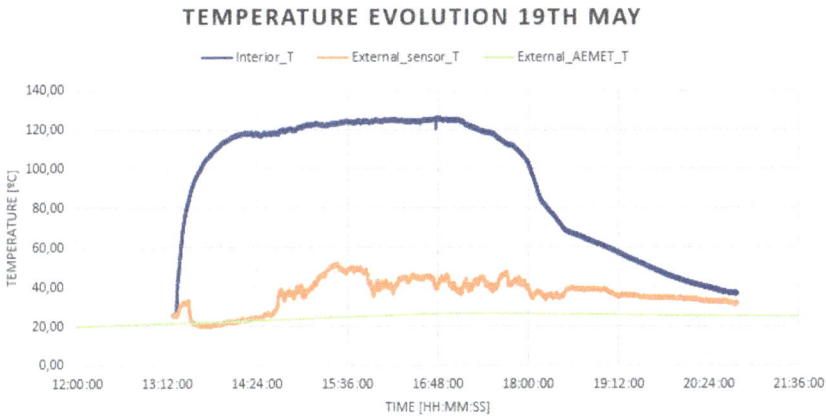

Figure 25. Graph of temperature variation for 19th of May. Source: [1]

This perfectly reflects what would happen in the oven on a completely sunny day. First, there is the steep upward slope, then about 4 hours pass where the temperature remains constant in between 110°C and 130°C, and finally, the temperature begins to drop towards an asymptote that coincides with the outside temperature. Such a prolonged plateau in time was observed because the sky was completely clear that day and there was not a cloud that could cover the incident radiation in the oven.

This is the ideal type of temperature graph for our plastic melting application, because that 4-hour plateau where the temperature remains high and constant, gives enough time for the plastic pieces to be melted and remolded.

New Materials for a Circular Economy   Materials Research Forum LLC
Materials Research Foundations 149 (2023) 70-104   https://doi.org/10.21741/9781644902639-3

With this study, it has been proved that it is possible to use recycled materials for the construction of a solar oven that has exceeded 120 degrees Celsius. After analyzing the recorded temperature data, the following conclusions were reached:

First: By means of recycled materials and an imperfect insulation system, elevated temperatures can be reached, the maximum recorded is 126.25°C. Although recycled materials are not ideal because they must be modified to adapt to the dimensions and shape of the other available elements, it has been demonstrated that a good result can be achieved.

Second: The radiation that causes the main heating in the furnace is direct radiation; diffuse radiation plays an almost negligible role. This conclusion is reached because during cloudy intervals when there is no direct radiation incident on the furnace, there is a sharp drop in temperature. On days when there is no cloud cover, the temperature rises steadily and then is maintained for a prolonged period of time.

Third: Other parameters such as humidity or wind have their own effect on the heating of the oven. Humidity is not determinant in the heating of the furnace. Relative humidity data was initially collected because it was thought that the relative humidity percentage was directly proportional to cloudiness, but this has been shown not to be the case. Wind makes it difficult for the reflectors to focus the sun's rays on the oven's inner box, causing the oven to heat up more slowly.

A possible application of this recycling method of plastic is a functional prosthetic hand. This prosthetic hand is presented once the strength of the plastic has been validated and it has been demonstrated that parts can be fabricated inside the solar oven. A prosthetic hand made from this material is strong enough to withstand all the daily tasks and stresses involved and, at the same time, it is tough enough to tolerate impacts.

4.   Functional hand prosthesis

To begin with the prosthesis design, a first prototype was made from scratch in SolidWorks, based on the designs and mechanisms studied previously [14,15]. This design tries to avoid complex geometries and is focused on maximum simplicity since the manufacturing process of the parts greatly limit the precision that can be obtained. To this end, it was decided to avoid curved areas and to try to use quite simple geometries.

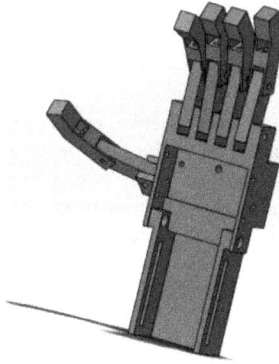

Figure 26. Hand prostheses design. Source: [3]

4.1 Design of the prosthesis parts

The aim of the design is to use as few parts as possible, so they can be made inside the solar oven reusing the molds, allowing to minimize manufacturing times. The prosthesis design consists of the following parts:

Table 2. Pieces of the hand prostheses. Source: [3]

Part	Number of pieces
Upper finger	5
Upper finger side	10
Lower finger	5
Finger half	5
Hand side	2
Knuckle	5
Knuckle thumb	2
Base knuckle	1
Palm hand	1
Palm support	1
Arm	1
Arm rope anchor	1
Arm side	2
Arm anchor	2

New Materials for a Circular Economy   Materials Research Forum LLC
Materials Research Foundations 149 (2023) 70-104   https://doi.org/10.21741/9781644902639-3

As it is explained in [3], these different parts have to be glued together applying heat, forming three main functional pieces: the fingers, the hand and the arm. They will be later assembled with rolled pieces of the can sheets, simulating the hand joints.

- **Fingers**

To simplify the process, all the fingers were designed to be the same size. All fingers except the thumb have two parts. The upper part of the finger is tilted to avoid having to include an extra phalanx in the mechanism. As part of the extension mechanism of the fingers, an elastic cord will be attached to the hole at the back of the finger. This will hold the finger in a stretched position and will automatically return the finger to this position after flexing it. Similarly, there will be a rigid cord in the front part of the finger that when it is tightened will allow the finger's flexion to be able to grab objects with the prosthesis.

The lower part of the finger is formed by a simple piece which has two holes that cross it longitudinally. The rigid rope will pass through one and the rubber band through the other. The two large holes will be used to attach this middle piece to the rest of the finger and to the knuckles of the hand.

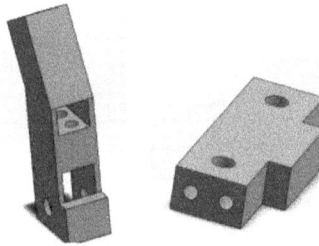

Figure 27. Upper (left) and lower(right) parts of the finger. Source: [3]

- **Hand**

In this principal piece all the fingers will be anchored. Again, there are a number of holes for the strings and rubber bands. All the rubber bands of the fingers will be joined together and tied to a single hole located in the center of this hand piece. The rigid ropes will each have a hole through which they will be inserted, as they will have to be tied individually to the piece on the side of the arm where they can be tensioned. There are also holes on the sides of the hand so that ropes can be attached, and the stump can be held in place.

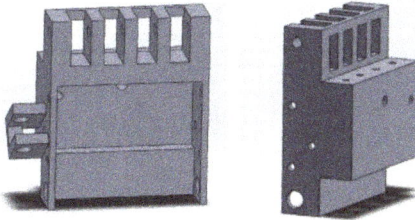

Figure 28. Hand piece view of the front and back. Source: [3]

- **Arm**

The arm part contains the holes in which the ropes are attached and where their tension can be adjusted individually. The lateral slots of the piece are designed to braid a rubber band to hold the prosthesis to the user's arm.

4.2   Creation of the pieces

The process to create the various parts is nearly the same as the one used to fabricate the test probes in 2.4. The difference is that in this process, the mold for each piece is different and must be drawn first on the aluminum sheet, as they have complex shapes.

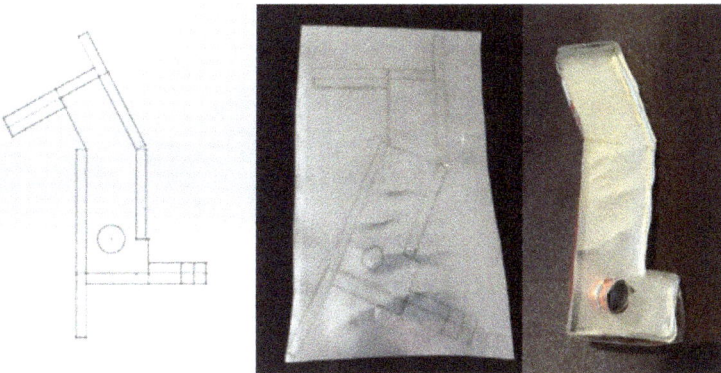

Figure 29. Upper finger side mold. Source: [3]

New Materials for a Circular Economy   Materials Research Forum LLC
Materials Research Foundations 149 (2023) 70-104   https://doi.org/10.21741/9781644902639-3

Figure 30. Manufacturing of the lower finger. Source: [3]

4.3   Assembly of the prosthesis

4.3.1 Elastic cords

A bungee cord is tied to the back of every fingertip. This string goes down through the back hole in the middle of the fingertip and through the knuckle hole. The string is passed through the hole in the middle of the palm towards the back of the palm. At this hole, all the strings are tied together in a knot. These strings must keep the fingers in a stretched position. In the case of the big finger the string ends in a separate hole as shown in the picture below.

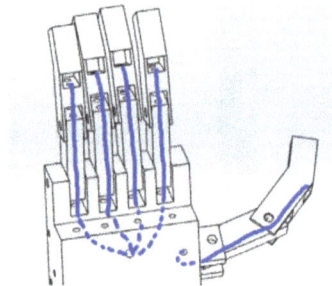

Figure 31. Elastic cord routing. Source: [3]

4.3.2 Rigid strings

Stiff strings are tied to the front of the fingertip. They go down through the front hole in the middle of the fingertip and pass directly through the holes in the back of the palm. Finally, they go down to the holes in the arm anchor, where they are tightened. These ropes cause the fingers to flex when the wrist rotates. In the case of the thumb, the rope passes through the same hole as in the case of the bungee cord.

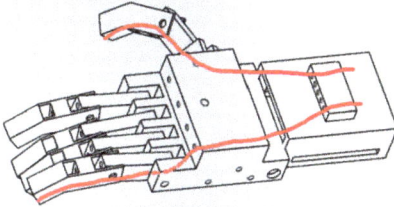

Figure 32.Rigid strings routing. Source: [3]

4.3.3 Attachment to the hand

To support the user's arm, a rope is passed through the sides of the arm piece and tied like shoelaces (green rope). Similarly, the residual limb is sustained with a rigid rope through the holes in the sides of the hand (yellow rope).

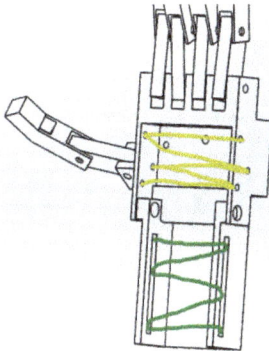

Figure 33. Strings for arm support. Source: [3]

Figure 34. Assembled prosthesis in user's arm simulating a stump. Source: [3]

When the prosthesis is assembled, the only thing left to do is to adjust the tension of all the strings, checking their functioning in the flexed and stretched hand positions.

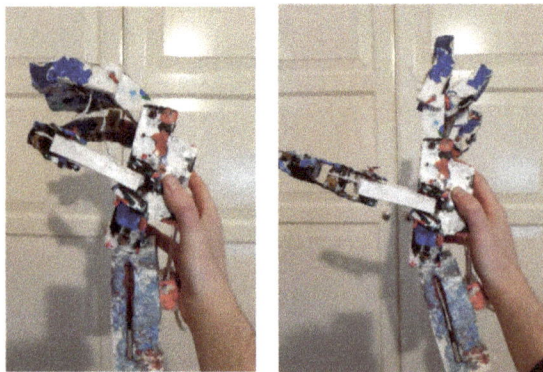

Figure 35. Flexed and stretched positions. Source: [3]

4.4   Results and conclusions

The prosthesis manufactured in this project is characterized by great robustness, as well as being proven resistant to impacts, abrasion and corrosion. The parts are hard and resistant and do not break easily. The prosthesis adapts very well to the hand because it allows the ropes to be properly tensioned to the arm and fist. The prosthesis is functional and performs the flexion and extension movement, although not very effectively. The movement is neither fluid nor precise, and this is mainly due to the heavy weight of the prosthesis. It has been possible to grab an object with it, but not in a random placement, so its usefulness for this task is quite low.

An average of 300 high-density polyethylene caps were used to make the prosthesis that is the subject of the following report, resulting in a prototype weighing 800g. Likewise, the elaboration of all the molds has required 20 aluminum cans of 355ml. Regardless of the possibility of obtaining the raw material from recycling points, the approximate cost of making the prosthesis through the direct acquisition of the caps and cans based on the market value of the materials would be about 2 euros, which reflects the suitability of the process for supplying areas that are immersed in social or military conflicts. However, it is emphasized that the project can be carried out at zero cost, and this has been the mentality during its development.

5.   SGDs alignment and circular economy

The Sustainable Development Goals (SDGs) are a universal call to action to end poverty, protect the planet and improve the lives and prospects of people around the world. The Agenda for Sustainable Development, drafted in 2015, aims to achieve these 17 goals by 2030. [16]

This project, through its alignment with the goals, aims to serve as an impulse to generate the necessary transformations in the environment for which it is intended, while seeking to serve as an example for other engineering projects in the sustainability field. The objectives that most closely align with this project are the following:

(#6) Clean Water and Sanitation.

At temperatures above 100°C, where water starts to boil and evaporate, bacteria and other living organisms die. Heating water inside the solar oven and then condensing it can be a way of cleaning the water from bacteria where there is no access to other heating methods and clean water.

(#7) Affordable and Clean Energy.

Solar radiation is a clean and renewable energy source, accessible anywhere on the planet. Moreover, its use for this purpose is free, making it affordable for anyone. This energy is also clean, since its use does not generate any waste.

New Materials for a Circular Economy   Materials Research Forum LLC
Materials Research Foundations 149 (2023) 70-104   https://doi.org/10.21741/9781644902639-3

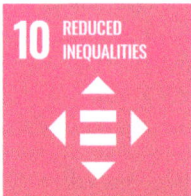

(#10) Reduced Inequalities.

The oven can be used to manufacture hand prostheses, which greatly facilitate the daily tasks of those who need one, helping disable people to have equal opportunities with others of the same age, gender or birthplace.

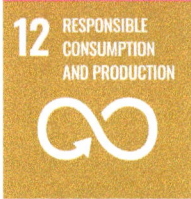

(#12) Responsible Consumption and Production.

The alignment with this SDG is clear as it is related to recycling plastic, aluminum and other materials. The solar oven is made from 100% recycled materials and, in addition, the raw material needed to manufacture the pieces is plastic taken from plastic bottle caps and the molds are made from soda cans.

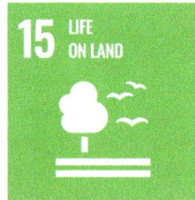

(#13) Climate Action, (#14) Life Below Water and (#15) Life on Land.

The accumulation of plastic in oceans, landfills and even in urban areas is considered a global emergency. In order to alleviate the effect this has on ecosystems and the natural development of life, it is necessary to become aware of it and start acting individually, with special emphasis on recycling and reusing these plastics.

For this reason, the pieces will be made from discarded plastic, contributing to the cleanliness of the environment while raising awareness among the population that these materials are a resource in themselves, offering an infinite number of possibilities.

6. Future lines of work

Recycled materials and plastic reuse are issues of growing interest. To continue with this research work, the insulation of the oven could be improved and further tested until it reaches 160 degrees Celsius. To improve heat retention, different insulation types could be studied and tested between the two drawers. Possible recycled and accessible materials to fulfill the insulation function could be: cork, sawdust, paper or natural fibers.

In addition, aspects of the prosthesis to be further developed are the design of the final prosthesis and the crushing process of the caps, that is not fully optimal to this date. The next prosthesis design should be focused on material reduction, since the prosthesis has a weight of about 800 grams and it is quite uncomfortable to support that load. Once the weight issue is solved, the hand mechanism should be improved to make the movement more precise and effective when grabbing objects. Another aspect to analyze is the durability and mechanical resistance of the prosthesis, both with simulations and with long-term wear and fatigue tests.

A fundamental aspect in terms of future lines will be to address the process of cutting and preparation of the caps. As previously mentioned, this aspect is the most time consuming in the manufacturing process of the plastic parts. It would be interesting to develop a tool that allows this task to be carried out and that does not require electrical energy in order to maintain a manufacturing process without costs and adaptable to territories without large resources.

Continuing with the development and optimization of the prosthesis, a fundamental aspect to work on is the appearance. To begin with, it would be very useful to develop a method of sanding or polishing the prosthesis to improve its surface finish. Next, it would be necessary to incorporate the possibility of making personalized designs on it, which would especially help children to normalize a physical disability and make it more attractive to wear. In addition, this process could be transferred to a foot prosthesis, always taking into account the cushioning when walking, which could be good with the use of rubber.

Acknowledgments

The acknowledgement of this work goes to two students who became interested in the circular economy and the way in which a waste product could be used as a resource, with the purpose of helping people: Ing. Beatriz Quiralte Moreno and Ing. Álvaro Valdés Gómez. Thank you for your excellent work, which has boosted research in these fields. Special thanks also to Dr. María Ana Saénz Nuño, without your guidance and advice none of this would have been possible.

References

[1]   Puente Rueda, Cristina. Final thesis to qualify for the degree in industrial engineering at ICAI, Universidad Pontificia de Comillas. Design, construction and characterization of a handmade solar oven for social purposes. Sáenz Nuño, María Ana; Zamora Macho, Juan Luis. 2021, Madrid.

[2]   Quiralte Moreno, Beatriz. Final thesis to qualify for the degree in industrial engineering at ICAI, Universidad Pontificia de Comillas. Design and development of low-cost prosthesis. Sáenz Nuño, María Ana; Díez Sánchez, Noelia. 2019, Madrid.

[3]   Valdés Gómez, Álvaro. Master's thesis to qualify for the master's degree in industrial engineering at ICAI, Universidad Pontificia de Comillas. Replication process of the joints of a human hand for the development of a low-cost prosthesis using recycled plastic. Sáenz Nuño, María Ana. 2021, Madrid.

[4]   UCL. When Plastics Saved Turtles. 2019. Retrieved from:
https://blogs.ucl.ac.uk/researchers-in-museums/2019/05/25/when-plastics-saved-turtles/

[5]   Information on https://esplasticos.es

[6]   UNEP. Marine Litter Assessment in the Mediterranean. 2015. Retrieved from:
https://wedocs.unep.org/handle/20.500.11822/7098

[7]   Information on https://www.plasticsindustry.org

[8]   Information on https://ecoembesdudasreciclaje.es

[9]   Information on https://newsroom.astm.org/astm-plastics-committee-releases-major-revisions-resin-identification-code-ric-standard

[10]   Adrew J. Peacock. Handbook of Polyethylene. CRC Press. 2000, New York.
https://doi.org/10.1201/9781482295467

[11]   Information on: https://www.quiminet.com

[12]   Material Mecanics. F.P. Beer, E. R. Johnston, J.T. Dewolf and D. F. Mazurek.
McGraw-Hill, 2010

[13]   William D. Callister, Introduction to Science and Materials Engineering. Limusa
Wiley, 2013.

[14]   Frank H. Netter. Atlas of human anatomy. Elsevier. 2018.

[15]   Bruno Sospedra Griñó. Final thesis to qualify for the degree in mecanical
engineering at Jaime I University. Mechanical design of an anthropomorphic
multifingered infractated hand prosthesis. De la Esperanza, Javier Andrés. 2015,
Castellón.

[16]   United Nations. Sustainable Development Goals. 2015. Retrieved from:
https://www.un.org/sustainabledevelopment/es/development-agenda/

New Materials for a Circular Economy
Materials Research Foundations 149 (2023) 105-141

Materials Research Forum LLC
https://doi.org/10.21741/9781644902639-4

Chapter 4

Biopolymers and their Application-Oriented Composites into a Circular Economy Context

Ruben Teijido[1,2], Leire Ruiz Rubio[1,2]*, Qi Zhang[2,3], Senentxu Lanceros[2,3],
José Luis Vilas Vilela[1,2]

[1]Macromolecular Chemistry Group (LQM), Physical Chemistry Department, Faculty of Science and Technology, University of the Basque Country (UPV/EHU), 48940 Leioa, Spain

[2]BCMaterials, Basque Center for Materials, Applications and Nanostructures, UPV/EHU Science Park, 48940 Leioa, Spain

[3]IKERBASQUE, Basque Foundation for Science, Plaza Euskadi, 5, Bilbao 48009, Spain

leire.ruiz@ehu.eus

Abstract

Society is increasingly concerned about environmental problems derived from the inappropriate use of polymeric materials and their inefficient recycle. Furthermore, considering that petroleum is not an endless raw material, the search for new sustainable materials has been gaining great interest in recent years. These materials should not only be sustainable but also fulfil the requirements requested by the industry. In this context, biopolymers stand out for their natural origin and their potential to substitute some of the synthetic polymers. On the other hand, the design of new composites capable of adapting and responding to the increasingly sophisticated applications presented by this new era of additive manufacturing or the internet of things is crucial. Moreover, the new business horizons related to the implementation of industry 4.0 require the development of complex sustainable materials for their development but considering a Circular Economy and sustainability perspectives.

Keywords

Biopolymers, Sustainability, Polysaccharides, Biodegradation

Contents

1.   Introduction

Modern societies are, now more than ever, concerned about the problems derived from the non-renewability of most of the key energy sources and materials required to guarantee fair living conditions. Governments have already realized the importance in controlling these scarce strategic resources. Therefore, a paradigm shift is needed in order not to slow down the development of our society and continue to meet the ever-increasing demands for energy and material resources. Thus, it is increasingly necessary to turn towards both energy and material sustainability. Ubiquitous plastics are among the most important materials present in daily life basic commodities and components as well as in most advanced applications. With an estimated global production of plastics of 367 million tons in 2020 [1], of which only 10% is properly disposed and recycled [2], the dependency of plastics industry on fossil resources needs to be reduced. A new way for obtaining these materials from more renewable sources must be explored. At this regard, nature itself could be the source of a plethora of promising biopolymeric materials from which substituents for synthetic petroleum derived plastics could be developed for multiple and diverse advanced applications. Unfortunately, conventional synthetic polymers are thoughtfully designed to accomplish their specific functions, thus, making them difficult to replace. Furthermore, actual biopolymers and biopolymer-based composites usually lack of suitable mechanical or thermal properties, in some cases having difficult processability and applicability at high temperatures [3]. However, one of the greatest advantages of biopolymers is their abundance. They are evenly distributed throughout the world and, in many cases, are cheap and easy to obtain. However, it is still necessary to increase and deepen the knowledge about the chemical and physical properties of biopolymers so that they can represent a true alternative to non-renewable ones [4]. As part of new bio-based tailored composite materials, in which either the matrixes [5], the fillers (biofibres [6], bionanoparticles [7]…), or even both [8] are made out of biopolymers, a precise control of the interfaces is required to provide them with specific properties including self-healing, stimulus responsiveness, stiffness, crosslinking density or hydrophobicity, among others.

Although the term biopolymer is easily associated with natural origin biodegradable polymers, it is worth mentioning that this extensive group of polymeric materials includes not only those obtained from nature. Thus, some synthetic polymers fabricated in a laboratory, thanks to the field of biotechnology, as well as some non-biodegradable polymers with natural origins (those derived from the petrochemistry industry) may be included in this classification [9]. Biodegradability is often assumed to be a property of every natural origin biopolymer, but these natural plastics are not always biodegradable or compostable. In fact, many physicochemical processes with multiple variables are involved in the biodegradation of a polymer. Thus, an early classification of the biodegradation processes can be made based on whether only microorganisms are needed (biodegradation) or if the presence of water (hydro-biodegradation), sunlight (photo-biodegradation) or ambient erosion (bioerodable) are required to assist with the delinking of complex macromolecules, to be afterwards processed by microorganisms. Another form of biodegradability is also compostability, in which the degradation products have beneficial effects improving soils conditions [2]. Furthermore, biodegradation rate is intrinsically related to the chemical structure of each system components, with their specific chemical bonding defining the time for complete biodegradation [10]. Thus, as it will be described in this chapter, there are also synthetically obtained bioplastics that could, as well, fit into the biopolymer classification and used in the building of bio-based composite materials, due to their biodegradability, such as polyesters. Fig. 1 includes some of the most interesting biopolymers classified by their origins and biodegradability, which will be described in the following pages.

Thus, this chapter begins with a review of the most important biopolymers, including their origin, obtaining methodologies, and abundancy, as well as their main chemical and thermomechanical characteristics. Then, the main applications of biocomposites are reviewed, including biocomposites used for biomedical applications, flexible electronics and packaging.

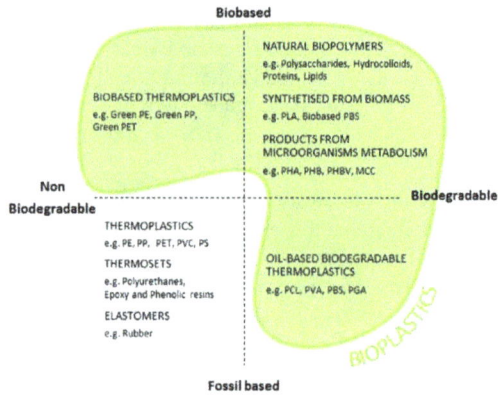

Figure 1. Biopolymers classification according to their origins and biodegradability. Reproduced with permission from [11]. Copyright 2021 Springer Nature.

2.   Biopolymers from natural origin

Natural polymeric materials are worldwide ubiquitous as they form the basic templates that allow life to exist on earth[9] by accomplishing one or more of the key functions essential for any living organism. These materials may serve as energetic reservoirs (polysaccharides, lipids), structural supports (polysaccharides, phospholipids, and proteins) or develop specific functionalities (polynucleotides and most polypeptides/proteins), such as preserving the genetic information, facilitating biotic/abiotic factors recognition or allowing movement and/or relationship with surrounding media. This wide spectrum of functionalities is possible thanks to each of their unique chemical and structural identities. Thus, when employed to develop new bio-based composites, it is of key importance to choose the adequate biopolymer, always considering the aimed final properties and compatibilities with other elements in each system.

With the actual concerns over materials sustainability, biopolymers from natural origin, either extracted from biomass or produced by microorganisms (yeast, algae, bacteria), have gathered a great amount of interest. Studies on these materials have already given results with their utilization in food packaging, agricultural films, membrane applications and sustainable clothing, among others. Despite these advances, their employment as alternatives for synthetic non-biodegradable polymers in advanced application instruments or devices is, in most cases, in a laboratory research phase, working towards their optimization for industrial-scale processes.

New Materials for a Circular Economy   Materials Research Forum LLC
Materials Research Foundations 149 (2023) 105-141   https://doi.org/10.21741/9781644902639-4

2.1   Biopolymers from biomass

Waste biomass is the main feedstock for obtaining some of the most interesting bio-polymers. These materials, naturally present on all living beings are highly abundant, accessible and evenly shared worldwide, thus, in general having a fairly high sustainability.

Polysaccharides

Polysaccharides are the most abundant biopolymers on earth due to their main biological function as energy reservoirs in most living organisms. Polysaccharides are non-toxic, highly biodegradable, thermostable to some extent and shear-stable [12,13]. The structure of the main polysaccharides is depicted in Fig 2.

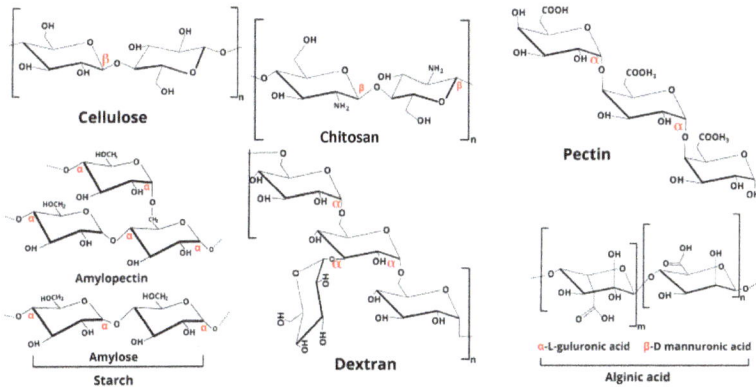

Figure 2. Chemical structure of the main polysaccharides obtained from waste biomass. Reproduced with permission from [14]. Copyright 2022 Elsevier.

The basic chemical structure of polysaccharides consists of a backbone of monosaccharides rings linked together through covalent glycoside bonds. However, despite their similarity, their rich and diverse chemical functionalities provide them with some differential properties. Furthermore, the presence of these hydroxyl, carboxyl and amide groups opens the possibility for chemical modifications aimed at optimizing their properties or providing with new interesting ones, regarding the chemical system in which are introduced or their future application.

Alginate: Alginate is the main polysaccharide naturally found in the cell wall of algae, providing enough flexibility and structural strength to withstand constant waves and tidal forces [15]. It can also be found in some bacteria capsules (*Azotobacter* sp., *Pseudomonas* sp.) where it provides adherence to surfaces by promoting the formation of biofilms, thus,

New Materials for a Circular Economy   Materials Research Forum LLC
Materials Research Foundations 149 (2023) 105-141   https://doi.org/10.21741/9781644902639-4

facilitating bacterial colonization [16]. Structurally, alginate is a linear random copolymer of D-mannuronic (M units) and L-guluronic (G units) (Fig.2). These units are arranged irregularly in G-G, G-M and M-M blocks with different overall G/M ratio depending on its source of origin [17], heavily influencing its final physicochemical properties [18]. The main characteristic of alginate is its ability for ionic gelation through complexation of some metallic cations, specially Ca^{2+} when its G-G blocks adopt an egg-shell fashioned structure [19]. Alginate with higher M/G ratios have a more flexible structure and higher biodegradability, but lower tendency for gelation with cations, while lower M/G ratios provide stiffness and a higher tendency and selectivity for Ca^{2+} complexation [20]. Moreover, its functionality richness in hydroxyl and carboxyl groups opens the possibilities for chemical functionalization and/or crosslinking treatments [21], in order to optimize its properties for each application.

Alginate extraction from seaweed biomass (predominantly from *laminaria, macrocystis* and *ascophyllum* genera) is a multi-stage process, in which the feedstock is milled into powder, then rehydrated in 2% w/v formaldehyde, followed by acid (HCl 0.2 - 2% (w/v), 40 - 60 °C, 2 - 4 h.) or alkaline treatment to break the plant cell walls. The solid residue is extracted with Na_2CO_3 at 2 - 4% (w/v), 40 - 60 °C, 2 - 3 h to obtain water soluble alginate, which is finally precipitated with ethanol and dried (overnight, 50 – 60 °C) [22]. Although there are novel alternative methods such as ultrasound, microwaves or enzymes by which they usually improve the extraction yield, their utilization at industrial scale is not yet economically justified.

Cellulose: As the main component of green plant cell walls, being also present in some algae, fungi and bacterial species, cellulose is the most abundant biopolymer on earth. Aside from its cost effectiveness, when compared with most synthetic polymers, it has many other properties of interest for designing biocomposites. Among them, its biodegradability, biocompatibility, low density, non-abrasiveness and simple processability are the most remarkable. Furthermore, cellulose is naturally arranged in long linear chains hierarchically organized from nanofibrils to microfibrils through intra and interchain hydrogen-bond type interactions, thus, showing excellent mechanical properties as well as a high specific surface area [23,24]. This natural self-arrangement into fibrils, together with its high crystallinity degree ($\approx$ 66%, Sigma-Aldrich) has made cellulose highly employed as reinforcing material in the form of fibres and nanoparticles (nanocrystals, nanowhiskers,...) [25]. As reinforcement, it also contributes with good mechanical properties (high stiffness-strength), thermal behavior ($T_g \approx$ 220 °C) and stability (decomposition over 250 °C) [26], low thermal conductivity (0.04 - 0.05 W·m^{-1}·K^{-1}[27]), low thermal expansion and high dielectric properties [28].

The extraction of cellulose from different biomass sources has been deeply studied over the years by the paper industries that resulted in physical [29], chemical [30] and enzymatic available methodologies [31]. The most significant drawback when using cellulose is its high water uptake capability. As for most biopolymers, both its water content and its molecular weight (highly dependent on the precise extraction methodology) exert a great influence in all its properties [32]. However, considering all cellulose pros and cons, the remarkable interest for the applications specially in the biomedical field are focusing in drug controlled release carriers, implantology and prosthetics, wound healing, bone tissue engineering and 3D-printing [24].

Chitin and Chitosan: After cellulose, chitin is the second most abundant biopolymer. It is one of the main building materials of crustaceans and insect exoskeletons, being also present also the cell walls of some fungi species. Several studies for chitin applications are being developed [33]. However, despite its abundancy and due to its chemical intractability and insolubility, produced by the presence of N-acetylated units in its chemical structure [34] (Fig.2), chitin use has been relegated after its most well-known derivative, chitosan. Chitosan, which can be obtained from chitin by its chemical [35] or enzymatic deacetylation [36] is the most abundant amino-polysaccharide, and is characterized by unique features that provides it with some remarkable qualities not shared by other polysaccharides as antioxidant [37] and strong antimicrobial activities [38,39], and the intrinsic ability for chelating metal ions [40,41]. As copolymers of D-glucosamine (deacetylated units) and N-acetyl-D-glucosamine (acetylated units) (Fig.1), chitin and chitosan are characterized by their deacetylation degree (DD). DD and molecular weight are the two most important structural parameters determining their processability, final properties and performances in many applications [42].

The main feedstocks for chitin are shrimp and crab shells, abundant by-products in food-processing industries. A common industrial chemical methodology for chitin extraction is the deproteinization of waste shells biomass through its treatment with inorganic and organic common acids (HCl, HNO_3, H_2SO_4, CH_3COOH and HCOOH) and bases (NaOH) [43]. Although the acid-alkaline treatment is predominant for industrial applications, it can induce depolymerization and uncontrolled deacetylation problems, with temperature and exposure time being critical factors [44]. Thus, alternative chemical methods [45], for example, some biological methods involving the use of proteolytic and demineralizing enzymes [46], as well as mixed methods [47], are being studied, and most of them are still under optimization for their industrial uses, due to a considerable increase of the process complexity and costs.

Hyaluronic acid: Hyaluronic acid is by far the less available polysaccharide, as it is only found on cells and tissues of humans (and some other vertebrates) anatomy, especially in the synovial fluid, the eye vitreous humor, and loose connective tissues [48]. Traditionally hyaluronic acid has been isolated from rooster combs and bovine vitreous humor through complex and expensive chemical methods, *i.e.* employing lysis buffers of 4M guanidinium hydrochloride. Major disadvantages for traditional methods are the difficulty to obtain high molecular weight products, due to its complexation tendency with proteoglycans [49]. However, as it was also found in some *streptococcus* sp. and *staphylococcus* sp. bacteria this has come to be its main source for industrial applications, thus it could be classified under microorganism origin polysaccharides.

Hyaluronic acid is an alternating copolymer comprised of glucuronic acid (carboxylated unit) and N-acetyl-glycosamine (acetylated unit) monomers (Fig.2), with unbranched and linear structure, usually in a random coil configuration and polyanionic at physiological pH values [48].

Despite the initial drawbacks, this biopolymer displays unique properties, such as high hygroscopicity, acting as a wetting agent in living tissues and viscoelasticity at higher molecular weights, which allows it to maintain extra cellular matrix properties [50]. Furthermore, hyaluronic acids high biocompatibility, long-term safety, proven ability to reduce bacterial adhesion and biofilm formation [51], non-immunogenicity and not generation of toxic products upon degradation [52], are qualities of great interest. The mentioned properties grant hyaluronic acid a deserved spot among the most useful polysaccharides for developing biocomposites aimed at biomedical applications [53–55].

Starch: Starch is the main energy-storage carbohydrate present in all higher plants as the end-product of photosynthesis, second only to cellulose in abundance. It serves as a main energy source for animals that feed from plants, representing no less than 70% of the human caloric intake worldwide. It is mainly found in leaves as a D-glucose storage material, but also in fruits, tubers or seeds for long-term storing [56]. Structurally, it is composed of two homopolymers of D-glucose residue monomers, amylose (15 - 30%), linear and unbranched, and amylopectin (70 - 85%), with $\approx$ 5% of branching linkages (1→6 instead of 1→4 glycosidic bonds) (Fig.1) [57]. With molecular weights ranging from 500 to 5 x 10^8 Da and a crystallinity degree of 15 - 40%, starch presents itself as a granulate material insoluble in water, mixed with proteins, lipids or minerals, with different sizes, morphologies and compositions depending on its original species [58]. Although the main sources of commercially used starch are cereals, such as corn (70%) or wheat (8%) and some tubers such as potatoes (12%), alternative and more sustainable sources with specific extraction methodologies are being developed [59]. Among them are the residuals of all kinds of wasted fruits and vegetables, due to a large amount generated every year (14.8% of all food residues worldwide, 50% in households as average) [58]. Meanwhile, specific extraction methodologies are usually chosen regarding the original source and parameters such as purity, total yielding or final structural integrity. Some of the most employed methods include both chemical and/or physical processes as dry/wet extractions, wet milling, sonication, freezing-thawing followed by saline extraction and enzymatic methods, among others [60].

Starch properties are highly dependent on the amylose/amylopectin proportion and their structural conformation [61]. Among starch properties of interest for bio-based composite materials are:

- Temperature modulated viscosity (pasting), water swelling and viscoelastic behavior, due to gelatinization and retrogradation processes, during heating/cooling through glass transition and gelatinization values ($T_g \approx 50 - 60$ °C, $T_{gel} \approx 60 - 80$ °C)[61]
- Amylose hydrophobic molecules complexation capability (cyclohexane, chloroform, benzene derivatives …) by forming helix-type coils around them, with the internal side having a more hydrophobic character [62].

- Excellent biocompatibility and biodegradability, as it is the only polysaccharide directly edible and naturally metabolized by humans, ideal for the preparation of edible films and coatings [63] or for drug delivery applications [64].
- The possibility for a large number of physical treatments (heat-moisture, annealing, microwave, ultrahigh pressure, ultrasonic, milling ...) [61] and chemical modifications (oxidations, formation of ether/ester derivatives ...) [56] in order to modulate its thermo-mechanical properties and compatibilities for specific chemical systems and applications.

2.2   Polypeptides/proteins

Structurally, polypeptides and proteins are more complicated than polysaccharides. Instead of a well-defined structure of similar repeating units they are composed of specific peptide-bonded amino acids sequences, forming their primary structure. Furthermore, their 3D-conformation is unique for each protein type, organized in 3 additional structural levels: secondary, tertiary and quaternary, due to bending and folding of their primary structure and inter-association between proteins [9]. This structuring is originated from the different physicochemical interactions between neighboring amino acid residues or backbone moieties and is key to endow proteins many biological functions in living organisms. Modification of this native structure has been extensively studied and produced by means of heat denaturation, irradiation, mechanical treatments, pressure, acidic or alkali treatments, hydrolytic enzymes, addition of inorganic salts or chemical crosslinking [4]. The growing acceptance of the possibility for their structural modification by the materials science community, which can, among others, provide proteins with excellent films or fiber forming properties, has increased the interest for their use in bio-based composite materials [65].

Unique characteristics of proteins, not found in other biopolymers, are their electrostatic charges, amphiphilic nature, conformational denaturation [4] and ability to offer sites at the side chains for attachment that could include drugs, crosslinking agents, or pendant groups to add functionalities or manipulate *ad hoc* their mechanical and chemical properties [66]. Furthermore, protein-based materials have already demonstrated the capability for smart acting in response to diverse physicochemical stimuli such as temperature, magnetic/electric field exposure or enzymatic stimuli [67]. However, the use of most complicated protein manipulation techniques is limited due to the high cost needed for industrial fabrication.

Proteins from vegetable biomass
Obtainable from vegetable biomass, these biopolymers share the huge abundancy and widespread availability of some polysaccharides. These extracted proteins belong to three main groups: leguminous proteins, cereal proteins and oil seed proteins, with some distinctions[68]. Among its wide diversity, in this chapter one of the main representatives of each group will be covered, namely, soy protein (leguminous), gluten (cereal) and peanut protein (oil seed).

New Materials for a Circular Economy   Materials Research Forum LLC
Materials Research Foundations 149 (2023) 105-141   https://doi.org/10.21741/9781644902639-4

Soy Protein: Processed soybeans obtained from soya plants (*glycine* max) are the world's largest source of animal protein feed and the second largest source of vegetable oil [69], with a widespread presence in human daily life. Soy protein is usually found in soybeans on discrete protein bodies acting as amino acids reservoirs for the future plant [9].

Soybean proteins are extracted during oil processing in the agricultural field. The most common methodology consists on the alkali-acid solution precipitation based on the principle that plant proteins are soluble at basic pH values and precipitate at acidic pH under isoelectric condition [70]. This method guarantees high extraction rate, especially when combined with thermal treatments to assist solubilizing the insoluble fractions (*okara*) [71] and high purity, with a low cost and an easily scalable operation. However, its pollutant potential is high, producing great amounts of alkaline and acidic water residues. Thus, important efforts are being made to develop new extraction methodologies [72,73], aiming at obtaining soy protein isolates (SPI) which can contain up to 90% protein or concentrates with up to 70% protein mixed with water soluble-polysaccharides [9].

Remarkable soy protein properties include a great compatibility with other biopolymers for advanced fully bio-based composites [74], SPI photoluminescent activity [75] and due to a strong amphiphilic nature, a great emulsifying property [76] and hydrophobic inner cores, ideal for containment and controlled release applications [77]. However, soy protein's strong interchain interactions and crystallinity induces stiffness and brittleness [78]. It is this lack of a high mechanical strength, together with low water resistance properties [79], that leads the researchers to renew efforts for their improvement through diverse methods, *i.e.* chemical crosslinking [80,81].

Gluten: Gluten is the cohesive and viscoelastic proteic by-product left behind after the extraction of starch granules and other soluble components from cereal processed seeds, such as wheat flour, its main feedstock, through wet extraction processes [82]. Mainly composed of 2 proteins, gliadin (M_w = 30 - 75 kDa) (Fig.2B) and glutenin (M_w = 30 - 90 kDa), gluten is an amorphous polymeric material with a density of 0.69 g/mL. Contrary to other proteins it is a heat stable material with a specific heat capacity of 0.41 $J \cdot g^{-1} \cdot °C^{-1}$ and a T_g ranging from 173 to 175 °C [83]. Wheat flour contains up to 15% proteins from which 90% is gluten with different proteic composition in different wheat varieties. Its main function is as amino acids reservoir and a binding and extending agent for other biopolymers providing improved textures and higher moisture retention [84].

Gluten displays excellent film forming properties, efficient oxygen barrier properties [85] and high cohesion. Its mechanical properties are directly derived from the total content in cysteine residues and the disulphide bonds between such amino acids. Various methods as heat treatments or pH adjustments can promote or hinder disulphide bonds density thus allowing to tune its mechanical performance or even transforming it into a thermoset material [86].

New Materials for a Circular Economy
Materials Research Foundations 149 (2023) 105-141

Materials Research Forum LLC
https://doi.org/10.21741/9781644902639-4

Proteins obtained from animal biomass

Before presenting some of the most important biopolymers extracted from animal biomass that display useful qualities for being used in future bio-based composites, it is important to consider that their availability, costs and overall sustainability would never be as optimal as for those obtained from vegetable biomass. Animal husbandry activities are mainly aimed at feeding humans or other animals. For the use of these biopolymers not to enter into the competition with the food industry, increasing the costs and hindering their sustainability, it is necessary to use extraction methods that primarily have these industries leftovers as raw materials.

Collagen/Gelatin: Collagens are a family of glycoproteins present in the extracellular matrix of animal organs and tissues, acting as a load distributing element and providing optimal environment for highly active cells, regulating their morphology and containing receptors for cell interactions. The structure of all collagens is based on repeating triple helix complexes of amino acids chains (collagen monomers), with a predominant presence of glycine, up to 57% accompanied by residues from 18 other amino acids [87]. A typical collagen molecule is 280 nm long and weighs 360 kDa[88].

Collagen is much more permeable than other proteinic biopolymers. However, its structure is doubly crosslinked intra and inter-triplehelix units, through enzymatic and glucose degradation associated processes. This natural crosslinking provides collagen with superior and tunable mechanical properties [89] and a great stability against chemical or enzymatic degradation [90].

Meat is the main raw material for collagen extraction derived from the slaughter of animals. However, industrial aliments waste by-products as bones, tendons, skin, fatty tissues, horns, hooves, blood and internal organs should be preferred to increase its sustainability [88]. The most common industrial extraction process involves chemical and enzymatic hydrolysis, but several new techniques (ultrasound assisted, biological processes) are also under development [91].

Gelatin can be obtained by the hydrolytic degradation of collagen, either thermally or enzymatically assisted, as the known colorless material widely used as gelling agent in food production [87]. Depending on collagen pre-treatment, two types of gelatin (types A and B) may be obtained. Type A is obtained from acid treatments, it has an isoelectric point between pH 6 and 9 and a more loosened structure, while type B gelatin is produced by alkali treatments, shows an isoelectric point at pH 5 and its crosslinking density is higher [92]. Gelatin has a strong emulsifying power and antioxidant properties [93] and its specific structure of amino acids provides several medical benefits, such as, reduced inflammatory response and long-term neuroprotective effects [94].

Elastin: Elastin is a 70 kDa elastomeric, insoluble, fibrous protein, forming an crosslinked fiber system, which is key in the extracellular matrix providing resilience and elasticity to tissues and organs [95]. Being three orders of magnitude more flexible than collagen, elastin is the main component in organs in movement and high deformation adaptability demand such as lungs, skin and the arteries. In its native form elastin is capable of maintaining its original shape without memory effect after hundreds of millions of cycles [96].

Tropoelastin, the elastin precursor, could be considered as its monomerical unit, bears a high proportion of hydrophobic amino acids (proline, glycine, desmosine and isodesmosine) and a high crosslinking density. Thus, it is highly insoluble in water and easily extractable by exposing its raw materials (animal skin and connective tissues) to high temperatures and extreme pH conditions [97].

The most interesting properties of elastin for its use in advanced biomaterials are its self-assembly capability, long-term stability, biological activity and compatibility with collagen. High densely crosslinked elastin-collagen networks have been developed employing the 1-Ethyl-3-(3-dimethylaminopropyl)carbodiimide/N-hydroxysuccinimide (EDC/NHS) route [98]. Elastin Young's modulus has been determined from bovine ligaments in the range 0.4 - 1.2 MPa [99]. However, all these properties fall behind its most outstanding feature, its linear elastic extension in the range of 103 - 105%, larger than that of any other known biological materials [100].

Fibrin: Fibrinogens circulating in human blood are globular-chain 45 nm elongated proteins, which upon cleavage produce fibrin that undergoes crosslinking to produce 3D networks usually involved in wound healing [101]. This process can be modulated trough different pH values; calcium salts concentration and the presence of other plasma proteins (thrombocins). Covalently crosslinked fibrin networks display rheological properties and viscoelastic behavior, of great use when mimicking the properties of some native tissues [102].

Fibrin principal drawback is its low availability compared with other biopolymers, being also characterized by complicated extraction methods involving plasma fractionation and cryoprecipitation. Some alternative precipitation routes employing ammonium sulphate, ethanol, or polyethylene-glycol have been explored [103]. Thus, huge efforts are still needed to develop optimized techniques that guarantee non-toxicity and absence of immunogenic reactions, as the main researched applications are in the biomedical field [104].

Silks: In general terms, silks are spun fibrous proteinic biopolymers excreted by some biological systems. The main source of silk is *Bombyx* mori silkworms' cocoons, through a labor-intensive agriculture-based operation called sericulture. *B.* mori silk is made of two structural polypeptide fibroin materials, highly composed of glycine, alanine, and serine amino acids[105], and many others. These two polypeptides are different by their molecular weight with a heavy chain of 390 kDa and a light chain of just 26 kDa and are joined together through disulphide bonds. Usually both are present in a 1:1 molar ratio, and in an amphiphilic organization of alternate hydrophilic and hydrophobic blocks. Including a coating of glue-like silk sericins, silk filaments have diameters between 10 to 25 µm [106]. Silk fibroin properties are defined by multiple variables including the crystallinity

New Materials for a Circular Economy   Materials Research Forum LLC
Materials Research Foundations 149 (2023) 105-141   https://doi.org/10.21741/9781644902639-4

degree, the disulphide crosslinking density, the organization of crystalline domains and the amphiphilicity of unfolded segments [107].

Silk fibroins show an excellent combination of thermal and chemical stability, strength and toughness, having shown interesting functional properties as shear piezoelectricity and a facile processing to achieve fibres, films, particles, non-woven materials or 3D porous structures [105,106]. Other important physical characteristics of silks are silk filaments total length (400 - 1600 m), fibre fineness (2 - 12 deniers equivalent to 2 - 12 g / 9 km of fibers) and their density (1.30 - 1.34 g·cm⁻³) [108].

3. Biopolymers from microorganisms

Thanks to the recent advances in bioengineering and synthetic biology, bacteria represent reliable sources for some innovative biopolymers of interest to help the substitution of petrochemistry derived polymers used to develop composites for advanced applications. Bacteria are well known for producing several biopolymer species, *i.e.* bacterial polysaccharides such as alginate, chitin, cellulose or hyaluronic acid, but also gellan dextran or xanthan [109]. In general, these polysaccharides are able to bestow to their composites the same properties as those derived from vegetal biomasses but have the main advantage of a highly obtainable purity [110]. Aside from polysaccharides, different bacteria species may synthetize amino acids composed of polyamides, polyphosphates from inorganic phosphate sources or polyesters, among which the most interesting ones are polyhydroxyalkanoates [111].

3.1 Polyhydroxyalkanoates

Polyhydroxyalkanoates (PHAs) are environmentally friendly and sustainable polyesters produced by some bacteria species from sugars and fatty acids. The main function of these granulated bioproducts is to serve as water insoluble carbon reservoirs, produced and stored under certain circumstances of environmental stress, with carbon substrates abundancy and limitation of other nutrients (N, P, K, Mg, O…) [112], which force microorganisms to deviate from the tricarboxylic acid route to secondary metabolic pathways [113].

Structurally, PHAs are linear polyesters formed by monomers with hydroxyl and carboxyl functionalities linked together through ester bonds, formed between these chemically complementary groups in neighbouring monomers [114]. Polymerization degree of PHAs is always higher than 150 with molecular weights ranging from 5×10^4 to 2×10^7 Da, depending on the number of carbon atoms in their monomers. Thus, these polymers are classified in short chain length PHAs with 3- 5 C atoms per monomer and produced by a wide range of bacteria, such as *Cupriavidus* necator, or medium chain length PHAs with 6 - 14 C atoms per monomer produced mostly by *Pseudomonas* sp. [115].

Most important products of polyhydroxyalkanoates on an industrially scaled production are poly(3-hydroxybutyrate) (PHB) and poly(3-hydroxybutyrate-co-3-hydroxyvalerate) (PHBV). One of the big advantages of PHAs is the possibility of adjusting the environmental conditions and variating/alternating nutrients availability to synthetize them

New Materials for a Circular Economy   Materials Research Forum LLC
Materials Research Foundations 149 (2023) 105-141   https://doi.org/10.21741/9781644902639-4

into homopolymers, random or block copolymers or highly functionalized polymers including triple bonds, halogen groups or epoxy, carbonyl, cyano or phenyl functionalities in their final structure[115]. These synthetic opportunities open a wide range of obtainable biopolymeric materials with the properties adjusted and the natural inherent defects compensated, regarding their final function in composite systems [116].

Among PHAs disadvantages, their high production costs, due to the current-state of industrial biotechnology production processes and poor thermo-mechanical properties are worth considering [117]. Nevertheless, PHAs degradation rates and physicochemical properties can be modulated by combining them in composite systems. Some of the most interesting properties displayed by composite systems in which PHAs are combined with other materials include: higher biocompatibility, improved mechanical [118] and barrier performances [119], electrical properties, enhanced thermal properties [120], antimicrobial properties, water absorption and diffusion properties and surface morphologies [116].

4.   Biopolymers from synthetic origin

In addition the biopolymers obtained by biosynthesis, different chemical synthetic methodologies (addition, condensation, metathesis polymerizations… [121]) are able to produce some biopolymers with similar ecologically benign characteristics. Among these synthetic biomaterials poly(α-hydroxy esters), obtained through biotechnology methods and polycaprolactone (PCL), which despite being derived from the petrochemistry industry maintains a high biodegradability, are among the most commonly employed in composite materials.

4.1   Biopolymers from biotechnology

Synthetic polyesters or poly(α-hydroxy esters) are the most studied class of synthetic biopolymers. The different obtention methodologies contemplate the use of various initial monomers and condensation or ring opening polymerizations. While the first is indicated for obtaining low molecular weight polymers (< 30 kDa) by means of direct esterification of hydroxy acids or diacids and diols, the latter is preferred for high molecular weight polymers with shorter reaction times and the absence of undesired byproducts. In both cases, monomers purity has to be high to avoid hindering the final molecular weight due to early undesired termination reactions and the use of non-solvents polymerizations is advised, in order to obtain highly biodegradable materials.

Most well-known representatives within the poly(α-hydroxy esters) family are: poly(glycolic acid) (PGA), poly(lactic acid) (PLA) and poly(lactide-co-glycolide acid) (PLGA) (Fig.3).

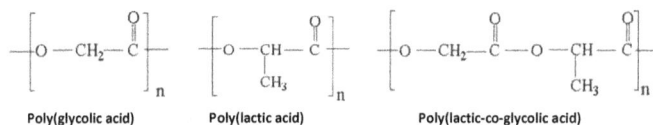

Poly(glycolic acid)   Poly(lactic acid)   Poly(lactic-co-glycolic acid)

Figure 3. Repeating polymeric units in poly(α-hydroxy esters).

PGA is the simplest linear aliphatic polyester, synthetized from cyclic lactone monomers (glycolides) through ring opening polymerization. Its most interesting properties include a fairly high tensile modulus and low solubility in organic solvents due to its high crystallinity degree ($\approx$ 45 - 55%) and good thermal properties with melting temperatures over 220 °C and a T_g between 35 and 40 °C [122]. Despite its poor solubility, several processing methodologies, such as extrusion, solvent casting, particulate leaching, injection and compression molding, PGA has been obtained with a wide variety of forms and internal structures for diverse applications [123]. With an excellent fiber-forming capability, PGA has been investigated for biomedical applications since 1969 when the first resorbable suture was approved by the FDA. However, PGA homopolymer losses its strength and mass in periods variating from 1 - 2 to 6 - 12 months which may limit some of their applications in making the use of copolymers and blends with other poly(α-hydroxy esters) that is required to overcome this biodegradation processes [124].

PLA is commonly obtained through polycondensation of either L-lactic acid (the natural isomer) or D-lactic acid (the synthetic isomer), or ring opening polymerization of cyclic lactones. Commercial PLAs have different D/L lactide ratios which allow the control of their final properties. When synthetized exclusively from L-lactide acids, poly(L-lactide acid) (PLLA) is obtained with a melting temperature above 175 °C, a T_g between 60 and 65 °C and a crystallinity degree of approximately 37%, which provides with good tensile strength, low extension and high modulus ($\approx$ 4.8 GPa). Meanwhile, when both D- and L-monomers are employed, PDLLA is obtained with a highly amorphous structure, resulting in a lower modulus, a lower T_g (55 - 60 °C) and faster biodegradation rates [125].

PLGAs are extensively studied synthetic copolymers with specific formulations based on its constituent homopolymers PGA and PLLA and commonly synthetized by catalyzed ring opening polymerizations. Different PLGAs inherit the properties of their constituent monomers in agreement with the specific LA/GA ratios. Higher LA proportions result in a highly hydrophobic material with lower stiffness and degradation rates, while a predominance of GA induces stronger, more hydrophilic materials. This parameter is of key importance as it can have significant effects for applications such as drug encapsulation and controlled delivery rates or in the formations (size and morphology) of PLGA nanoparticles for anticancer therapies[126].

4.2 Biopolymers derived from petrochemistry

Although in principle, polymers obtained from the petrochemistry industry are often considered non-biodegradable, there exist some examples of synthetic biopolymers which despite their initial monomers being petroleum-derived may result in biodegradable and eco-friendly final materials that attract a great interest for fully bio-based composites. A prominent example is poly(ε-caprolactone) (PCL), a semi-crystalline polyester synthetized by catalyzed ring opening polymerization of fairly cheap monomers, lactones (Fig.4).

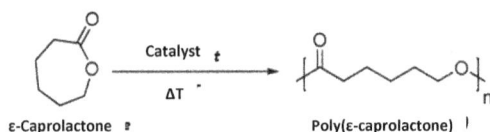

Figure 4. PCL initial monomer and its polymeric repeating unit.

PCL thermal and mechanical properties have been extensively characterized, and it has a T_g of -60 °C with melting temperature between 60 and 65 °C, low tensile strength (23 MPa) and high elongation at break (from 7 to 47-fold). Main disadvantage of PCL is its low degradation rate (2 - 3 years). However, to overcome this disadvantage, its solubility has been probed with a broad range of solvents as well as a high compatibility with many other biopolymers, thus, increasing its processability and possibilities for combination in bio-based systems to modulate its inherent characteristics. A good example is the copolymerization of PCL with PDLLA which allows obtaining less stiff and rapidly biodegradable biopolymers, making them suitable for biomedical applications, among others [121,122,125].

5. Applications of biopolymer composites

5.1 Biomedicine

The excellent biological properties that biopolymers commonly present make them excellent candidates for the development of composite materials for biomedical applications such as tissue engineering or implants. One of the most interesting applications in this area is tissue engineering, since these composites can provide a suitable three-dimensional environment for cell growth and differentiation.

Wound dressing: Dressings must allow complete restoration of the wound tissue through an integrated response of different cell types in the injury to promote wound healing and reconstitution (Fig. 5) [127]. Among the possible materials that have been studied for the development of dressings, composites formed by metallic nanoparticles and biopolymers are of great interest due to the excellent results. Thus, for example, silver nanoparticles (AgNP) have been widely used to prevent the proliferation of bacteria. In addition, calcium

New Materials for a Circular Economy
Materials Research Foundations 149 (2023) 105-141

Materials Research Forum LLC
https://doi.org/10.21741/9781644902639-4

is essential for the hemostasis of mammalian skin and modulates the proliferation and differentiation of keratinocytes. On the other hand, the presence of zinc is essential for protein synthesis and T cell development, although it can also act as an antioxidant agent [128].

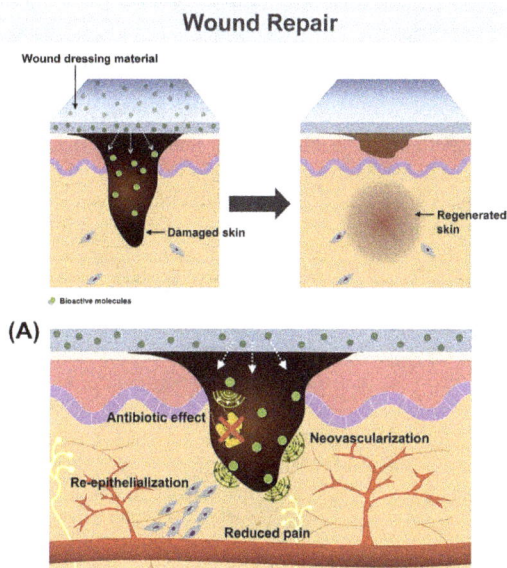

Figure 5. Scheme of healing process carried out in an injury covered with a wound dressing. Reproduced with permission from [129]. Copyright 2017 Elsevier.

There are various methods to generate composites to be used in wound dressings applications [130]. Electrospinning represents an efficient method to obtain porous films with a high surface area. These types of materials have been used for the treatment of wounds and burns. Morsy *et al*. [131] developed an antibacterial biocomposite as a wound dressing made of electrospun gelatin and silver nanoparticles. The results confirmed that AgNPs prepared in situ enhanced the antibacterial activity of electrospun biocomposites against gram-positive and gram-negative bacteria. Other biocomposites based on polysaccharides and silver nanoparticles have also been described in the literature. Thus, biopolymers such as chitosan [132], starch [133] or alginate [134], among others, have also been used as matrix in biocomposites filled with AgNP, obtaining good antibacterial properties.

New Materials for a Circular Economy   Materials Research Forum LLC
Materials Research Foundations 149 (2023) 105-141   https://doi.org/10.21741/9781644902639-4

Tissue engineering: Although there are examples of biocomposites used in vascular grafts or the regeneration of other organs such as the liver [135,136], the main developments in tissue engineering with these materials have focused on bone and cartilage.

Bone is an organic and inorganic composite material with high mechanical resistance, so when biomaterials are used as scaffolds for the regeneration of bone tissue, the ones that are having the greatest success are those formed by composites of porous bioactive/inorganic composite polymers. Specifically, the most widespread fillers are those made of bioactive ceramics since they exhibit good bioactivity, flexibility and capacity to adapt to bone defects [137,138]. There are currently several commercial formulations based on biopolymer and ceramic compounds. The vast majority of them take type I collagen and mineral calcium phosphate as their matrix material, such as: Collagraft®, Collapat II®, FormaGraft® or Vitoss®, among other examples [139]. Although collagen is the most widely used biopolymer as a matrix for this type of composite, both commercially and experimentally, there are examples of other types of biopolymers with this use. Among the most promising biopolymers for the formation of scaffolds for bone regeneration, it is worthy to highlight gelatin, chitosan, PHA, PHBV or silk fibroin [140–142].

As an example, a silk fibroin/hydroxyapatite composite-based coating was fabricated to improve osseointegration in porous titanium implants (Fig.6). In addition, it should be noted that this study was carried out under diabetic conditions through the activation of the phosphatidylinositol 3-kinase (PI3K)/protein kinase B (AKT) signaling pathway [143]. This work aimed to develop materials that facilitate osseointegration in those cases where patients suffer from diabetes mellitus, since this disease makes it difficult to integrate the implant, which usually results in its rejection. One of the main hypothesis is that the overproduction of reactive oxygen species (ROS) in the bone-implant interface can generate this problem. Several studies have shown that the overproduction of ROS was related to the regulation of PI3K/Akt signaling, this being the key to regulating the biological behaviors (adhesion, migration, proliferation and differentiation) of osteoblasts [144]. The authors evaluated the osseointegration of their materials by implanting them in diabetic rabbits. Thus, in the case of titanium substrates, the overproduction of ROS derived from diabetes produced poor osseointegration. On the other hand, it should be noted that the developed biocomposites significantly suppressed the overproduction of ROS in diabetic conditions, improving the functional recovery of osteoblasts, their adhesion and morphology, which resulted in improved osseointegration.

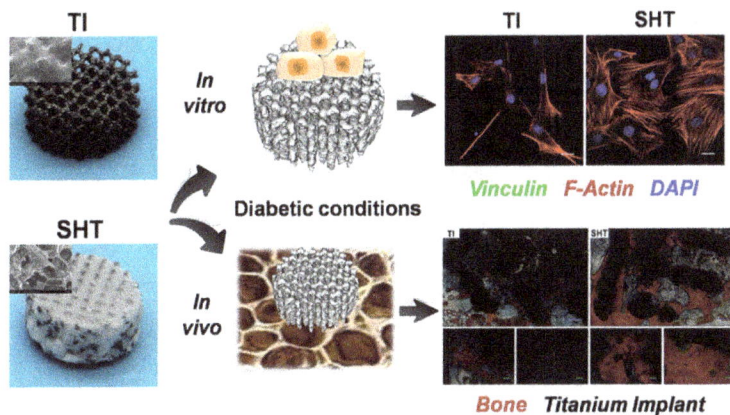

Figure 6. Scheme of the study developing silk fibroin/hydroxyapatite composite-based coatings to improve osteointegration in porous titanium implants [143]. *Copyright 2022 American Chemical Society.*

On the other hand, cartilage is a tissue with highly demanding functions, so the biocomposites used for the development of scaffolds must be able to be flexible and able to support while providing a medium for the growth of chondrocytes [145]. The most used biopolymers for the development of these scaffolds for cartilage regeneration are collagen and hyaluronic acid. Despite the predominance of hyaluronic acid and collagen, new biocomposites with interesting properties for cartilage regeneration are being developed. Thus, bicompounds based on a mixture of two gellan gums and lignin have been obtained [146]. It should be noted that the presence of lignin gives the materials anti-oxidant properties, eliminating free radicals, and bacteriostatic properties, reducing the proliferation of *S. aureus* and *S. epidermidis*. In addition, the human mesenchymal stem cell (hMSC) viability in these materials was greater than 70%, thus allowing adequate cell regeneration.

Biopolymer composites in implants: Biocomposites are being used in the manufacture of permanent implants due to their biocompatibility. Nevertheless, biodegradable biocomposites may be of special interest in the development of temporal implants, such as cardiac stents. These implants are used to widen coronary arteries that have become blocked or narrowed due to the deposition of cholesterol, fats and other blood components in them (Fig. 7). Until now, the most used biopolymers in this type of implants have been PLA and PHB [147]. Some of the developments currently marketed include pharmaceuticals. For example, ABSORB II developed by Abbott Vascula is a bioresorbable everolimus-eluting stent composed of PLA. Studies conducted with

ABSORB II showed results, after 1 year, similar to those of the everolimus-eluting metallic stent (Xience, Abbott Vascular) [148].

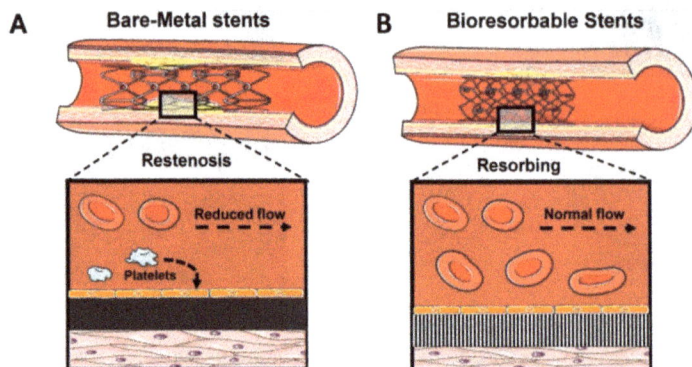

*Figure 7. Scheme of the vas*cular response of A) bare metal stent and B) Bioresorbable stent. Reproduced with permission from [149]. Copyright 2022 Elsevier.

The latest developments reported in the area are directed towards the personalized stents by manufacturing them on demand through 3D printing, although some research is also focusing on adding new biological properties to their developments (anti-inflammatory...). In the search for more efficient personalized medicine, the combination of magnetic resonance angiography (MRA) and additive manufacturing processes may facilitate the development of patient-specific coronary stents. Thus, stents have been printed on demand by printing PCL-based materials [150]. The developed material was based on heparinized PCL to promote endothelial cell adhesion, propagation and proliferation and inhibit excessive proliferation of other cells. The 3D-printed structures were functionalized with heparin by covalent grafting. 3D-printed stents were implanted in the abdominal aorta of rabbits to assess implantation feasibility and biocompatibility. The 3D printed stents shown good mechanical properties and in vitro and in vivo (rabbit) data described good biological properties in terms of biocompatibility, cell viability, etc. Yang *et al.* [151] developed a new PLA-based stent combined with a zwitterionic poly(carboxybetaine acrylate-dopamine co-methacrylate) copolymer (PCBDA) in order to obtain stents with antiproliferative and anti-inflammatory properties. These added properties are of particular importance in combating restenosis after stent implantation.

Another example of the use of biocomposites is the development of nerve guide conduits used for the repair of peripheral nerve defects. Thus, the biocompatibility of biopolymers is of special interest [152]. Zhao *et al.* developed chitosan/graphene oxide (GO) films, with 0 – 1 wt.% of GO, as the optimal material for the development of this type of device. The researchers evaluated the effects of this biocomposite on Schwann cell proliferation and adhesion. Cells cultured on these materials exhibited elongated spindle shapes and increased expression of factors related to nerve regeneration and for Schwann cell differentiation and nerve repair, among other favorable characteristics.

5.2 Applications in batteries, capacitors and flexible electronics

The search for new eco-sustainable materials has also been reflected in areas such as optics and electronics, where there is a growing interest in new bionanocomposites [153]. Specifically, the combination of carbon-based nanomaterials, mainly 1D (nanodiamonds, carbon quantum dots) and 2D (graphene, graphene oxide, carbon nanotubes), are widely investigated as nanofillers with biopolymers in order to obtain tailored mechanical, optical and electronic properties while being environmentally friendly [154,155].

Among the fillers used in electronic applications, two-dimensional (2D) metal carbides and nitrides, also known as MXenes, stand out as promising materials. MXenes have been successfully used in supercapacitors due to their high electrical conductivity similar to that of metal and their pseudocapacitance [156,157]. However, the poor mechanical properties limit their use in flexible electronics. The incorporation of these materials in polymeric supports can improve their mechanical properties and thus their applicability. Although alginate is mostly used in biomedicine-related applications, more studies are being reported in other areas. Specifically, Shahzad et al. [158] developed nanocomposites of alginate and Ti_3C_2Tx MXene for shielding against electromagnetic interference (EMI). Composites with percentages by weight of Ti_3C_2Tx between 90 and 10% were developed. The films of the different formulations were manufactured by vacuum-assisted filtration. The formulations exhibited an EMI shielding effectiveness of up to 92 dB, for a 45 μm thick film. This performance is related to the excellent electrical conductivity of the Ti_3C_2Tx films reaching up to 4600 S·cm^{-1}. In another example of biocomposites with MXenes, Tian et al. fabricated biocomposites using MXenes and cellulose nanofibers (CNF) as matrix. The presence of CNF and the interactions between CNF and MXenes allow these nanocomposites to improve mechanical resistance, up to 29 MPa, without sacrificing electrochemical performance, 2.95×10^4 S·m^{-1} and gravimetric capacitance ≈ 298 F·g^{-1}.

In addition, taking CNF as matrix, different biocomposites for flexible electronics have also been explored. An interesting example is silver nanowire (AgNW)/CNF transparent conductive paper for the development of foldable solar cells [159]. The films were obtained by means of hot pressing and exhibited high electrical conductivity and transparency. The low weight of the material makes it very suitable for lightweight and portable electronic devices. Importantly, repeated folding tests showed high durability. In addition, the authors developed solar cells with these biocomposites, achieving an energy conversion of 3.2%.

On the other hand, in the area of sensors, composites made from biopolymers are also being developed. Thus, examples of polysaccharide-based nanocomposites for sensor development can be found in the literature [160]. For example, Bouvree et al. [161] developed nanocomposite ducts using chitosan as a matrix and carbon nanoparticles. These nanocomposites were used to sense vapors of polar solvents (water and methanol). The developed transducers gave a quantitative response when exposed to water vapor and presented good selectivity, allowing vapors to be classified, based on the relative response amplitude (Ar), in the following order: water > methanol > toluene.

5.3   Biocomposites for packaging applications

The growing concern about environmental problems and single-use packaging, banned in the coming years by the European Union, has boosted the research on environmentally sustainable packaging. In this context, biocomposites are crucial for the development of this industry. The materials used up to now have a series of limitations such as the management of the wastes, the recycling and depletion of resources. The biocomposites used in packaging must meet a series of requirements. These materials must be able to protect and preserve the products, while being sustainable and biodegradable [11].

One of the most used bioplastics for the development of packaging is PLA due to its mechanical and thermal properties comparable to conventional plastics such as PS and PET. However, they also present a series of weaknesses so that it is necessary to improve them, such as their fragility, low impact resistance or little toughness, among others. There are numerous studies each year reporting improved properties of PLA from its composites. For example, mixtures of PLA with other polymers have been developed to improve their crystallinity [162] or their mechanical properties [163,164], and their resistance against visible ultraviolet has also been improved by embedment of ZnO nanoparticles [165].

Currently there is also a strong interest in the development of new materials from starch-based biocomposites [166], especially, and to a lesser extent on other biopolymers such as chitosan [167] or carboxymethylcellulose [168]. As in PLA, the development of composites allows improving some of the properties of biopolymers that bar their uses (low mechanical properties or low resistance to water).

Conclusion

In this chapter, the characteristics of biopolymers, which could play an important role in the upcoming circular economy context, as well as some of the applications of their biocomposites have been presented. The diverse nature, cheapness and widespread availability of biopolymers make them accessible for many research groups all over the globe, allowing them to focus their efforts on these emerging materials. Furthermore, their wide range of compatibility with other materials will allow, in the following years, given enough investments and research efforts are put into them, to give form to cheaper and more efficient advanced bio-based composites. These new materials are expected to show a much lower environmental impact than those with traditional petrol-derived plastics, greatly contributing to approaching their end-of-life cycles to the statements of a circular economy production. Many of the biocomposites developed have shown similar and in some cases superior characteristics to synthetic polymers, a good starting point aiming at the replacement of classic polymers in a new era of sustainable polymeric materials. This will allow helping reduce plastic waste, with strong environmental impact. In addition, environmental and research policies throughout the world in this field are leading to increased efforts on biopolymer-based compounds development. The areas of application of these materials are very diverse, ranging from biomedicine or flexible electronics to packaging, among others. Over time, it is expected that the areas in which these materials can be used will increase and take hold.

Aknowledgement

R.T. wants to thank the Basque Government for funding under an FPI grant. Authors thanks Basque Government for Grupos Consolidados and ELKARTEK program.

References

[1]   Information on https://www.statista.com

[2]   A. Vinod, M.R. Sanjay, S. Suchart, P. Jyotishkumar, Renewable and sustainable biobased materials: An assessment on biofibers, biofilms, biopolymers and biocomposites, J. Clean. Prod. 258 (2020) 120978. https://doi.org/10.1016/j.jclepro.2020.120978

[3]   A.G. Facca, M.T. Kortschot, N. Yan, Predicting the tensile strength of natural fibre reinforced thermoplastics, Compos. Sci. Technol. 67 (2007) 2454–2466. https://doi.org/10.1016/j.compscitech.2006.12.018

[4]   N.P. Sukumaran, S. Gopi, Overview of biopolymers, Elsevier Inc., 2021. https://doi.org/10.1016/b978-0-12-819240-5.00001-8

[5]   S. Edebali, Methods of engineering of biopolymers and biocomposites, Elsevier Ltd., 2020. https://doi.org/10.1016/B978-0-12-819988-6.00015-X

[6]   Q. Tarrés, F. Vilaseca, P.J. Herrera-Franco, F.X. Espinach, M. Delgado-Aguilar, P. Mutjé, Interface and micromechanical characterization of tensile strength of bio-based composites from polypropylene and henequen strands, Ind. Crops Prod. 132 (2019) 319–326. https://doi.org/10.1016/j.indcrop.2019.02.010

[7]   G.J. Tudryn, L.C. Smith, J. Freitag, R. Bucinell, L.S. Schadler, Processing and Morphology Impacts on Mechanical Properties of Fungal Based Biopolymer Composites, J. Polym. Environ. 26 (2018) 1473–1483. https://doi.org/10.1007/s10924-017-1047-9

[8]   H. Ye, Y. Wang, Q. Yu, S. Ge, W. Fan, M. Zhang, Z. Huang, M. Manzo, L. Cai, L. Wang, C. Xia, Bio-based composites fabricated from wood fibers through self-bonding technology, Chemosphere. 287 (2022) 132436. https://doi.org/10.1016/j.chemosphere.2021.132436

[9]   R. Francis, S. Sasikumar, G.P. Gopalan, Synthesis, structure, and properties of biopolymers (natural and synthetic), Polym. Compos. Biocomposites. 3 (2013) 11–107. https://doi.org/10.1002/9783527674220.ch2

[10]   S. Lambert, M. Wagner, Environmental performance of bio-based and biodegradable plastics: The road ahead, Chem. Soc. Rev. 46 (2017) 6855–6871. https://doi.org/10.1039/c7cs00149e

[11]   F. Versino, O.V. López, M.A. García, Green Biocomposites for Packaging Applications, in: 2021: pp. 1–30. https://doi.org/10.1007/978-981-33-4091-6_1

[12]   T. Lou, X. Wang, G. Song, G. Cui, Synthesis and flocculation performance of a chitosan-acrylamide-fulvic acid ternary copolymer, Carbohydr. Polym. 170 (2017) 182–189. https://doi.org/10.1016/j.carbpol.2017.04.069

[13]   R. Yang, H. Li, M. Huang, H. Yang, A. Li, A review on chitosan-based flocculants and their applications in water treatment, Water Res. 95 (2016) 59–89. https://doi.org/10.1016/j.watres.2016.02.068

[14]   A. Ahmad, Y. Gulraiz, S. Ilyas, S. Bashir, Polysaccharide based nano materials: Health implications, Food Hydrocoll. Heal. 2 (2022) 100075. https://doi.org/10.1016/j.fhfh.2022.100075

[15]   J. Venkatesan, R. Nithya, P.N. Sudha, S.K. Kim, Role of alginate in bone tissue engineering, 1st ed., Elsevier Inc., 2014. https://doi.org/10.1016/B978-0-12-800268-1.00004-4

[16]   V. Urtuvia, N. Maturana, F. Acevedo, C. Peña, A. Díaz-Barrera, Bacterial alginate production: an overview of its biosynthesis and potential industrial production, World J. Microbiol. Biotechnol. 33 (2017) 0. https://doi.org/10.1007/s11274-017-2363-x

[17]   J. Minghou, W. Yujun, X. Zuhong, G. Yucai, Studies on the M:G ratios in alginate, Hydrobiologia. 116–117 (1984) 554–556. https://doi.org/10.1007/BF00027745.

[18]   X. Zhang, X. Wang, W. Fan, Y. Liu, Q. Wang, L. Weng, Fabrication , Property and Application of Calcium Alginate Fiber : A Review, (2022) 1–18.

[19]   A. Dodero, M. Alloisio, S. Vicini, M. Castellano, Preparation of composite alginate-based electrospun membranes loaded with ZnO nanoparticles, Carbohydr. Polym. 227 (2020) 115371. https://doi.org/10.1016/j.carbpol.2019.115371

[20]   H. Hecht, S. Srebnik, Structural Characterization of Sodium Alginate and Calcium Alginate, Biomacromolecules. 17 (2016) 2160–2167. https://doi.org/10.1021/acs.biomac.6b00378

[21]   S. Sellimi, I. Younes, H. Ben Ayed, H. Maalej, V. Montero, M. Rinaudo, M. Dahia, T. Mechichi, M. Hajji, M. Nasri, Structural, physicochemical and antioxidant properties of sodium alginate isolated from a Tunisian brown seaweed, Int. J. Biol. Macromol. 72 (2015) 1358–1367. https://doi.org/10.1016/j.ijbiomac.2014.10.016

[22]   S. Saji, A. Hebden, P. Goswami, C. Du, A Brief Review on the Development of Alginate Extraction Process and Its Sustainability, Sustain. 14 (2022) 1–20. https://doi.org/10.3390/su14095181

[23]   A. Sharma, M. Thakur, M. Bhattacharya, T. Mandal, S. Goswami, Commercial application of cellulose nano-composites – A review, Biotechnol. Reports. 21 (2019) e00316. https://doi.org/10.1016/j.btre.2019.e00316

[24]   B. Joseph, V.K. Sagarika, C. Sabu, N. Kalarikkal, S. Thomas, Cellulose nanocomposites: Fabrication and biomedical applications, J. Bioresour. Bioprod. 5 (2020) 223–237. https://doi.org/10.1016/j.jobab.2020.10.001

[25]   T. Li, C. Chen, A.H. Brozena, J.Y. Zhu, L. Xu, C. Driemeier, J. Dai, O.J. Rojas, A. Isogai, L. Wågberg, L. Hu, Developing fibrillated cellulose as a sustainable technological material, Nature. 590 (2021) 47–56. https://doi.org/10.1038/s41586-020-03167-7

[26]   L. Szcześniak, A. Rachocki, J. Tritt-Goc, Glass transition temperature and thermal decomposition of cellulose powder, Cellulose. 15 (2008) 445–451. https://doi.org/10.1007/s10570-007-9192-2

[27]   B. Petter Jelle, Nano-based thermal insulation for energy-efficient buildings, Elsevier Ltd, 2016. https://doi.org/10.1016/B978-0-08-100546-0.00008-X

[28]   K. Daicho, T. Saito, S. Fujisawa, A. Isogai, The Crystallinity of Nanocellulose: Dispersion-Induced Disordering of the Grain Boundary in Biologically Structured Cellulose, ACS Appl. Nano Mater. 1 (2018) 5774–5785. https://doi.org/10.1021/acsanm.8b01438

[29]   N.M. Park, S. Choi, J.E. Oh, D.Y. Hwang, Facile extraction of cellulose nanocrystals, Carbohydr. Polym. 223 (2019) 115114. https://doi.org/10.1016/j.carbpol.2019.115114

[30]   A.Y. Melikoğlu, S.E. Bilek, S. Cesur, Optimum alkaline treatment parameters for the extraction of cellulose and production of cellulose nanocrystals from apple pomace, Carbohydr. Polym. 215 (2019) 330–337. https://doi.org/10.1016/j.carbpol.2019.03.103

[31]   M.P. Menon, R. Selvakumar, Derived From Biomass for Environmental, (2017) 42750–42773. https://doi.org/10.1039/C7RA06713E

[32]   Q. Lin, Y. Huang, W. Yu, Effects of extraction methods on morphology, structure and properties of bamboo cellulose, Ind. Crops Prod. 169 (2021) 113640. https://doi.org/10.1016/j.indcrop.2021.113640

[33]   N. Berezina, Production and application of chitin, Phys. Sci. Rev. 1 (2016) 1–8. https://doi.org/10.1515/psr-2016-0048

[34] C.K.S. Pillai, W. Paul, C.P. Sharma, Chitin and chitosan polymers: Chemistry, solubility and fiber formation, Prog. Polym. Sci. 34 (2009) 641–678. https://doi.org/10.1016/j.progpolymsci.2009.04.001

[35] I. Younes, M. Rinaudo, Chitin and chitosan preparation from marine sources. Structure, properties and applications, Mar. Drugs. 13 (2015) 1133–1174. https://doi.org/10.3390/md13031133

[36] I. Tsigos, A. Martinou, D. Kafetzopoulos, V. Bouriotis, Chitin deacetylases : new , versatile tools in, TibTech. 18 (2000) 129–135.

[37] F. Luan, L. Wei, J. Zhang, W. Tan, Y. Chen, F. Dong, Q. Li, Z. Guo, Preparation and characterization of quaternized chitosan derivatives and assessment of their antioxidant activity, Molecules. 23 (2018). https://doi.org/10.3390/molecules23030516

[38] C.L. Ke, F.S. Deng, C.Y. Chuang, C.H. Lin, Antimicrobial actions and applications of Chitosan, Polymers (Basel). 13 (2021). https://doi.org/10.3390/polym13060904

[39] S. Ahmed, M. Ahmad, S. Ikram, Advanced materials chitosan : A natural antimicrobial agent : A review, J. Appl. Chem. 2 (2014) 493–503.

[40] D. Liu, Z. Li, Y. Zhu, Z. Li, R. Kumar, Recycled chitosan nanofibril as an effective Cu(II), Pb(II) and Cd(II) ionic chelating agent: Adsorption and desorption performance, Carbohydr. Polym. 111 (2014) 469–476. https://doi.org/10.1016/j.carbpol.2014.04.018

[41] A. del Carpio-Perochena, C.M. Bramante, M.A.H. Duarte, M.R. de Moura, F.A. Aouada, A. Kishen, Chelating and antibacterial properties of chitosan nanoparticles on dentin, Restor. Dent. Endod. 40 (2015) 195. https://doi.org/10.5395/rde.2015.40.3.195

[42] H. El Knidri, R. Belaabed, A. Addaou, A. Laajeb, A. Lahsini, Extraction, chemical modification and characterization of chitin and chitosan, Int. J. Biol. Macromol. 120 (2018) 1181–1189. https://doi.org/10.1016/j.ijbiomac.2018.08.139

[43] M. Pakizeh, A. Moradi, T. Ghassemi, Chemical extraction and modification of chitin and chitosan from shrimp shells, Eur. Polym. J. 159 (2021) 110709. https://doi.org/10.1016/j.eurpolymj.2021.110709

[44] K.K. Gadgey, K. Kumar Gadgey Head, A. Bahekar Head, A. Professor, K. Kumar Gadgey, A. Bahekar, Studies on Extraction Methods of Chitin from Crab Shell and Investigation of its Mechanical Properties, IJMET_08_02_027 Int. J. Mech. Eng. Technol. 8 (2017) 220–231.

[45] D. Zhao, W.C. Huang, N. Guo, S. Zhang, C. Xue, X. Mao, Two-step separation of chitin from shrimp shells using citric acid and deep eutectic solvents with the assistance of microwave, Polymers (Basel). 11 (2019). https://doi.org/10.3390/polym11030409

[46]   S. Kaur, G.S. Dhillon, Recent trends in biological extraction of chitin from marine shell wastes: A review, Crit. Rev. Biotechnol. 35 (2015) 44–61. https://doi.org/10.3109/07388551.2013.798256

[47]   V.L. Pachapur, K. Guemiza, T. Rouissi, S.J. Sarma, S.K. Brar, Novel biological and chemical methods of chitin extraction from crustacean waste using saline water, J. Chem. Technol. Biotechnol. 91 (2016) 2331–2339. https://doi.org/10.1002/jctb.4821

[48]   P. Saranraj, M.A. Naidu, Hyaluronic Acid Production and its Applications-A Review, Int. J. Pharm. Biol. Arch. 4 (2013) 853–859.

[49]   T.E. Hardingham, H. Muir, Hyaluronic acid in cartilage and proteoglycan aggregation, Biochem. J. 139 (1974) 565–581. https://doi.org/10.1042/bj1390565

[50]   P. Zhai, X. Peng, B. Li, Y. Liu, H. Sun, X. Li, The application of hyaluronic acid in bone regeneration, Int. J. Biol. Macromol. 151 (2020) 1224–1239. https://doi.org/10.1016/j.ijbiomac.2019.10.169

[51]   F. Zamboni, C. Okoroafor, M.P. Ryan, J.T. Pembroke, M. Strozyk, M. Culebras, M.N. Collins, On the bacteriostatic activity of hyaluronic acid composite films, Carbohydr. Polym. 260 (2021) 117803. https://doi.org/10.1016/j.carbpol.2021.117803

[52]   K.M. Lopez, S. Ravula, R.L. Pérez, C.E. Ayala, J.N. Losso, M.E. Janes, I.M. Warner, Hyaluronic Acid-Cellulose Composites as Patches for Minimizing Bacterial Infections, ACS Omega. 5 (2020) 4125–4132. https://doi.org/10.1021/acsomega.9b03852

[53]   Y. Wang, G. Liu, L. Wu, H. Qu, D. Song, H. Huang, C. Wu, M. Xu, Rational design of porous starch/hyaluronic acid composites for hemostasis, Int. J. Biol. Macromol. 158 (2020) 1319–1329. https://doi.org/10.1016/j.ijbiomac.2020.05.018

[54]   S. Maiz-Fernández, L. Pérez-Álvarez, L. Ruiz-Rubio, J.L. Vilas-Vilela, S. Lanceros-Mendez, Polysaccharide-Based In Situ Self-Healing Hydrogels for Tissue Engineering Applications, Polymers (Basel). 12 (2020) 2261. https://doi.org/10.3390/polym12102261

[55]   M. Dovedytis, Z.J. Liu, S. Bartlett, Hyaluronic acid and its biomedical applications: A review, Eng. Regen. 1 (2020) 102–113. https://doi.org/10.1016/j.engreg.2020.10.001

[56]   J.F. Robyt, Starch: Structure, Properties, Chemistry, and Enzymology, in: B.O. Fraser-Reid, T. Kuniaki, J. Thiem (Eds.), Glycoscience, Springer International Publishing, Heidelberg, 2008, pp. 1438–1468. https://doi.org/10.1201/9781315371399-3

[57]   S. Pérez, P.M. Baldwin, D.J. Gallant, Structural Features of Starch Granules I, In J. B.Miller and R. Whistler, Starch. Chemistry and Technology, Third Edit, Elsevier Inc., Amsterdam, 2009. https://doi.org/10.1016/B978-0-12-746275-2.00005-7

[58]   D.H. Kringel, S.L.M. El Halal, E. da Rosa, A.R. Guerra, Methods for the extraction of roots, tubers, pulses, pseudocereals, and other unconventional starches sources: A review, Methods Starch Extr. (2020). https://doi.org/10.1002/star.201900234

[59]   B.L. Tagliapietra, M.H.F. Felisberto, E.A. Sanches, P.H. Campelo, M.T.P.S. Clerici, Non-conventional starch sources, Curr. Opin. Food Sci. 39 (2021) 93–102. https://doi.org/10.1016/j.cofs.2020.11.011

[60]   S.L.M. El Halal, D.H. Kringel, E. da R. Zavareze, A.R.G. Dias, Methods for Extracting Cereal Starches from Different Sources: A Review, Starch/Staerke. 71 (2019) 1–14. https://doi.org/10.1002/star.201900128

[61]   Z. Sui, X. Kong, Physical modifications of starch, Springer Nature, Singapore, 2018. https://doi.org/10.1007/978-981-13-0725-6

[62]   T. Kuge, K. Takeo, Complexes of starchy materials with organic compounds, Agric. Biol. Chem. 32 (1968) 1232–1238. https://doi.org/10.1080/00021369.1968.10859210

[63]   E. Basiak, A. Lenart, F. Debeaufort, Effect of starch type on the physico-chemical properties of edible films, Int. J. Biol. Macromol. 98 (2017) 348–356. https://doi.org/10.1016/j.ijbiomac.2017.01.122

[64]   O.P. Troncoso, F.G. Torres, Non-conventional starch nanoparticles for drug delivery applications, Med. Devices Sensors. 3 (2020) 1–16. https://doi.org/10.1002/mds3.10111

[65]   E.J. Bealer, S. Onissema-karimu, A. Rivera-galletti, M. Francis, J. Wilkowski, X. Hu, Protein – Polysaccharide Composite Materials :, Polymers (Basel). (2020) 1–28.

[66]   Z. Zhang, O. Ortiz, R. Goyal, J. Kohn, Biodegradable Polymers, in K. Modjarrad, S. Ebnesajjad (Eds.), Handbook of polymer applications in medicine and medical devices, Elsevier Inc., Chadds Ford, 2014, pp. 303-335. https://doi.org/10.1016/B978-0-323-22805-3.00013-X

[67]   L. Wang, X. Huang, New protein-based smart materials, In G. Wei, S. G. Kumbar, Artificial Protein and Peptide Nanofibers, Woodhead Publishing, Sawston, 2020, pp. 415-436. https://doi.org/10.1016/B978-0-08-102850-6.00017-6

[68]   D. Lin, W. Lu, A.L. Kelly, L. Zhang, B. Zheng, S. Miao, Interactions of vegetable proteins with other polymers: Structure-function relationships and applications in the food industry, Trends Food Sci. Technol. 68 (2017) 130–144. https://doi.org/10.1016/j.tifs.2017.08.006

[69]   Information on https://www.geographynotes.com/vegetable-oil/sources-of-vegetable-oil-world-economic-geography/5532

[70]   Q. Chen, L. Chaihu, X. Yao, X. Cao, W. Bi, J. Lin, D.D.Y. Chen, Molecular Property-Tailored Soy Protein Extraction Process Using a Deep Eutectic Solvent, ACS

Sustain. Chem. Eng. 9 (2021) 10083–10092.
https://doi.org/10.1021/acssuschemeng.1c01848

[71]   K.E. Preece, N. Hooshyar, A. Krijgsman, P.J. Fryer, N.J. Zuidam, Intensified soy protein extraction by ultrasound, Chem. Eng. Process. - Process Intensif. 113 (2017) 94–101. https://doi.org/10.1016/j.cep.2016.09.003

[72]   K.E. Preece, N. Hooshyar, N.J. Zuidam, Whole soybean protein extraction processes: A review, Innov. Food Sci. Emerg. Technol. 43 (2017) 163–172. https://doi.org/10.1016/j.ifset.2017.07.024

[73]   M.M. Rahman, S. Dutta, B.P. Lamsal, High-power sonication-assisted extraction of soy protein from defatted soy meals: Influence of important process parameters, J. Food Process Eng. 44 (2021) 1–11. https://doi.org/10.1111/jfpe.13720

[74]   J.R. Kim, A.N. Netravali, Self-healing green composites based on soy protein and microfibrillated cellulose, Compos. Sci. Technol. 143 (2017) 22–30. https://doi.org/10.1016/j.compscitech.2017.02.030

[75]   L. Xu, S. Zhong, Y. Gao, X. Cui, Seeking brightness from natural resources: Soy protein isolate and its multifunctional applications, Dye. Pigment. 196 (2021) 109768. https://doi.org/10.1016/j.dyepig.2021.109768

[76]   R. Mozafarpour, A. Koocheki, E. Milani, M. Varidi, Extruded soy protein as a novel emulsifier: Structure, interfacial activity and emulsifying property, Food Hydrocoll. 93 (2019) 361–373. https://doi.org/10.1016/j.foodhyd.2019.02.036

[77]   D. Yuan, F. Zhou, P. Shen, Y. Zhang, L. Lin, M. Zhao, Self-assembled soy protein nanoparticles by partial enzymatic hydrolysis for pH-Driven Encapsulation and Delivery of Hydrophobic Cargo Curcumin, Food Hydrocoll. 120 (2021) 106759. https://doi.org/10.1016/j.foodhyd.2021.106759

[78]   F. Song, D.L. Tang, X.L. Wang, Y.Z. Wang, Biodegradable soy protein isolate-based materials: A review, Biomacromolecules. 12 (2011) 3369–3380. https://doi.org/10.1021/bm200904x

[79]   R. Kumar, L. Wang, L. Zhang, Structure and Mechanical Properties of Soy Protein Materials Plasticized by Thiodiglycol, J. Appl. Polym. Sci. 111 (2008) 970–977. https://doi.org/10.1002/app

[80]   Z. Yue-Hong, Z. Wu-Quan, G. Zhen-Hua, G. Ji-You, Effects of crosslinking on the mechanical properties and biodegradability of soybean protein-based composites, J. Appl. Polym. Sci. 132 (2015) 1–9. https://doi.org/10.1002/app.41387

[81]   F. Li, T. Liu, W. Gu, Q. Gao, J. Li, S.Q. Shi, Bioinspired super-tough and multifunctional soy protein-based material via a facile approach, Chem. Eng. J. 405 (2021) 126700. https://doi.org/10.1016/j.cej.2020.126700

[82]   L. Day, Wheat gluten: production, properties and application, In G.O. Phillips, P.A. Williamsn(Eds.), Handbook of food proteins, Woodhead Publishing Limited, Sawston, 2011, pp. 267-288. https://doi.org/10.1533/9780857093639.267

[83]   Z. Wang, S. Ma, B. Sun, F. Wang, J. Huang, X. Wang, Q. Bao, Effects of thermal properties and behavior of wheat starch and gluten on their interaction: A review, Int. J. Biol. Macromol. 177 (2021) 474–484. https://doi.org/10.1016/j.ijbiomac.2021.02.175

[84]   J.R. Biesiekierski, What is gluten?, J. Gastroenterol. Hepatol. 32 (2017) 78–81. https://doi.org/10.1111/jgh.13703

[85]   T. Sartori, G. Feltre, P.J. do Amaral Sobral, R. Lopes da Cunha, F.C. Menegalli, Properties of films produced from blends of pectin and gluten, Food Packag. Shelf Life. 18 (2018) 221–229. https://doi.org/10.1016/j.fpsl.2018.11.007

[86]   N.K. Kim, F.G. Bruna, O. Das, M.S. Hedenqvist, D. Bhattacharyya, Fire-retardancy and mechanical performance of protein-based natural fibre-biopolymer composites, Compos. Part C Open Access. 1 (2020) 100011. https://doi.org/10.1016/j.jcomc.2020.100011

[87]   J. Alipal, N.A.S. Mohd Pu'ad, T.C. Lee, N.H.M. Nayan, N. Sahari, H. Basri, M.I. Idris, H.Z. Abdullah, A review of gelatin: Properties, sources, process, applications, and commercialisation, Mater. Today Proc. 42 (2019) 240–250. https://doi.org/10.1016/j.matpr.2020.12.922

[88]   T.F. Da Silva, A.L.B. Penna, Colágeno: Características químicas e propriedades funcionais, Rev Inst Adolfo Lutz. 71 (2012) 530–539

[89]   B. Gurumurthy, A. V. Janorkar, Improvements in mechanical properties of collagen-based scaffolds for tissue engineering, Curr. Opin. Biomed. Eng. 17 (2021) 100253. https://doi.org/10.1016/j.cobme.2020.100253

[90]   M. Meyer, Processing of collagen based biomaterials and the resulting materials properties, Biomed. Eng. Online. 18 (2019) 1–74. https://doi.org/10.1186/s12938-019-0647-0

[91]   M.M. Schmidt, R.C.P. Dornelles, R.O. Mello, E.H. Kubota, M.A. Mazutti, A.P. Kempka, I.M. Demiate, Collagen extraction process, Int. Food Res. J. 23 (2016) 913–922.

[92]   S. Cao, Y. Wang, L. Xing, W. Zhang, G. Zhou, Structure and physical properties of gelatin from bovine bone collagen influenced by acid pretreatment and pepsin, Food Bioprod. Process. 121 (2020) 213–223. https://doi.org/10.1016/j.fbp.2020.03.001

[93]   C. Liu, Y. Xia, M. Hua, Z. Li, L. Zhang, S. Li, R. Gong, S. Liu, Z. Wang, Y. Sun, Functional properties and antioxidant activity of gelatine and hydrolysate from deer antler base, Food Sci. Nutr. 8 (2020) 3402–3412. https://doi.org/10.1002/fsn3.1621

[94] L.S. Kumosa, V. Zetterberg, J. Schouenborg, Gelatin promotes rapid restoration of the blood brain barrier after acute brain injury, Acta Biomater. 65 (2018) 137–149. https://doi.org/10.1016/j.actbio.2017.10.020

[95] A. Shuttleworth, Extracellular Matrix, in: Encycl. Immunol., 2nd Editio, 1998: pp. 861–866. https://doi.org/10.1006/rwei.1999.0226

[96] J.H. Kristensen, M.A. Karsdal, Chapter 30 - Elastin, In M. A. Karsdal Biochem. Collagens, Laminins Elastin Struct. Funct. Biomarkers, 2016, pp. 197–201. https://doi.org/https://doi.org/10.1016/C2015-0-05547-2

[97] M. Nadalian, S.M. Yusop, W.A.W. Mustapha, M.A. Azman, A.S. Babji, Extraction and characterization of elastin from poultry skin, AIP Conf. Proc. 1571 (2013) 692–695. https://doi.org/10.1063/1.4858735

[98] D. V. Bax, H.E. Smalley, R.W. Farndale, S.M. Best, R.E. Cameron, Cellular response to collagen-elastin composite materials, Acta Biomater. 86 (2019) 158–170. https://doi.org/10.1016/j.actbio.2018.12.033

[99] M.L. Del Prado-Audelo, N. Mendoza-Muñoz, L. Escutia-Guadarrama, D.M. Giraldo-Gomez, M. González-Torres, B. Florán, H. Cortes, G. Leyva-Gómez, Recent advances in elastin-based biomaterials, J. Pharm. Pharm. Sci. 23 (2020) 314–332. https://doi.org/10.18433/JPPS31254

[100] L.D. Muiznieks, F.W. Keeley, Molecular assembly and mechanical properties of the extracellular matrix: A fibrous protein perspective, Biochim. Biophys. Acta - Mol. Basis Dis. 1832 (2013) 866–875. https://doi.org/10.1016/j.bbadis.2012.11.022

[101] B. Bujoli, J.C. Scimeca, E. Verron, Fibrin as a multipurpose physiological platform for bone tissue engineering and targeted delivery of bioactive compounds, Pharmaceutics. 11 (2019) 1–15. https://doi.org/10.3390/pharmaceutics11110556

[102] N. Atiqah Maaruf, N. Jusoh, J. Bahru, Angiogenic and Osteogenic Properties of Fibrin in Bone Tissue Engineering, Malaysian J. Med. Heal. Sci. 18 (2022) 85–94.

[103] S.M. Alston, K.A. Solen, A.H. Broderick, S. Sukavaneshvar, S.F. Mohammad, New method to prepare autologous fibrin glue on demand, Transl. Res. 149 (2007) 187–195. https://doi.org/10.1016/j.trsl.2006.08.004

[104] K. Froelich, R.C. Pueschel, M. Birner, J. Kindermann, S. Hackenberg, N.H. Kleinsasser, R. Hagen, R. Staudenmaier, Optimization of fibrinogen isolation for manufacturing autologous fibrin glue for use as scaffold in tissue engineering, Artif. Cells, Blood Substitutes, Biotechnol. 38 (2010) 143–149. https://doi.org/10.3109/10731191003680748

[105] A. Reizabal, D.M. Correia, C.M. Costa, L. Perez-Alvarez, J.L. Vilas-Vilela, S. Lanceros-Méndez, Silk Fibroin Bending Actuators as an Approach Toward Natural Polymer Based Active Materials, ACS Appl. Mater. Interfaces. 11 (2019) 30197–30206. https://doi.org/10.1021/acsami.9b07533

[106] T. Asakura, S. Kametani, Y. Suzuki, Silk, In Encyclopedia of Polymer Science and Technology, (Ed.), Wiley, New York, 2018, pp. 1–19. https://doi.org/10.1002/0471440264.pst339.pub2

[107] A. Reizabal, C.M. Costa, P.G. Saiz, B. Gonzalez, L. Pérez-Álvarez, R. Fernández de Luis, A. Garcia, J.L. Vilas-Vilela, S. Lanceros-Méndez, Processing Strategies to Obtain Highly Porous Silk Fibroin Structures with Tailored Microstructure and Molecular Characteristics and Their Applicability in Water Remediation, J. Hazard. Mater. 403 (2021) 123675. https://doi.org/10.1016/j.jhazmat.2020.123675

[108] N. V. Padaki, B. Das, A. Basu, Advances in understanding the properties of silk, Elsevier Ltd., 2015. https://doi.org/10.1016/B978-1-78242-311-9.00001-X

[109] R.R. McCarthy, M.W. Ullah, P. Booth, E. Pei, G. Yang, The use of bacterial polysaccharides in bioprinting, Biotechnol. Adv. 37 (2019) 107448. https://doi.org/10.1016/j.biotechadv.2019.107448

[110] M.F. Moradali, B.H.A. Rehm, Bacterial biopolymers: from pathogenesis to advanced materials, Nat. Rev. Microbiol. 18 (2020) 195–210. https://doi.org/10.1038/s41579-019-0313-3

[111] D. Tan, Y. Wang, Y. Tong, G.Q. Chen, Grand Challenges for Industrializing Polyhydroxyalkanoates (PHAs), Trends Biotechnol. 39 (2021) 953–963. https://doi.org/10.1016/j.tibtech.2020.11.010

[112] S. Ray, V.C. Kalia, Biomedical Applications of Polyhydroxyalkanoates, Indian J. Microbiol. 57 (2017) 261–269. https://doi.org/10.1007/s12088-017-0651-7

[113] D. Tan, J. Yin, G.Q. Chen, Production of Polyhydroxyalkanoates, Elsevier B.V., 2016. https://doi.org/10.1016/B978-0-444-63662-1.00029-4

[114] T.R. Ahammad, S. Z.; Gomes, J.; Sreekrishnan, Wastewater treatment forproductionofH2S-free biogas, J. Chem. Technol. Biotechnol. 83 (2008) 1163–1169. https://doi.org/10.1002/jctb

[115] J. Możejko-Ciesielska, R. Kiewisz, Bacterial polyhydroxyalkanoates: Still fabulous?, Microbiol. Res. 192 (2016) 271–282. https://doi.org/10.1016/j.micres.2016.07.010

[116] A.J. Emaimo, A.A. Olkhov, A.L. Iordanskii, A.A. Vetcher, Polyhydroxyalkanoates Composites and Blends: Improved Properties and New Applications, J. Compos. Sci. 6 (2022). https://doi.org/10.3390/jcs6070206

[117] Y. Zheng, J.C. Chen, Y.M. Ma, G.Q. Chen, Engineering biosynthesis of polyhydroxyalkanoates (PHA) for diversity and cost reduction, Metab. Eng. 58 (2020) 82–93. https://doi.org/10.1016/j.ymben.2019.07.004

[118] J. Jo, H. Kim, S.Y. Jeong, C. Park, H.S. Hwang, B. Koo, Changes in mechanical properties of polyhydroxyalkanoate with double silanized cellulose nanocrystals using

different organosiloxanes, Nanomaterials. 11 (2021). https://doi.org/10.3390/nano11061542

[119] M. Liu, I.A. Kinloch, R.J. Young, D.G. Papageorgiou, Modelling mechanical percolation in graphene-reinforced elastomer nanocomposites, Compos. Part B Eng. 178 (2019) 1–28. https://doi.org/10.1016/j.compositesb.2019.107506

[120] P. Cataldi, P. Steiner, T. Raine, K. Lin, C. Kocabas, R.J. Young, M. Bissett, I.A. Kinloch, D.G. Papageorgiou, Multifunctional Biocomposites Based on Polyhydroxyalkanoate and Graphene/Carbon Nanofiber Hybrids for Electrical and Thermal Applications, ACS Appl. Polym. Mater. 2 (2020) 3525–3534. https://doi.org/10.1021/acsapm.0c00539

[121] C.P. Kubicek, Synthetic biopolymers, In A. Glieder, C.P. Kubicek, D. Mattanovich, B. Wiltschi, M. Sauer (Eds.), Synthetic Biology, Springer, Cham, 2016, p. 307-335. https://doi.org/10.1007/978-3-319-22708-5_9

[122] M. Rahman, M.R. Hasan, Synthetic biopolynmers, In M.A.J. Mazumder, A. Al-Ahmed, H. Sheardown (Eds.), Funct. Biopolym., Springer Nature, Oxford, 2017, pp. 1–43. https://doi.org/10.1007/978-3-319-95990-0

[123] K.J. Jem, B. Tan, The development and challenges of poly (lactic acid) and poly (glycolic acid), Adv. Ind. Eng. Polym. Res. 3 (2020) 60–70. https://doi.org/10.1016/j.aiepr.2020.01.002

[124] S.A. Benner, A.M. Sismour, Synthetic biology, Nat. Rev. Genet. 6 (2005) 533–543. https://doi.org/10.1038/nrg1637

[125] P.A. Gunatillake, R. Adhikari, N. Gadegaard, Biodegradable synthetic polymers for tissue engineering, Eur. Cells Mater. 5 (2003) 1–16. https://doi.org/10.22203/eCM.v005a01

[126] S. Rezvantalab, N.I. Drude, M.K. Moraveji, N. Güvener, E.K. Koons, Y. Shi, T. Lammers, F. Kiessling, PLGA-based nanoparticles in cancer treatment, Front. Pharmacol. 9 (2018) 1–19. https://doi.org/10.3389/fphar.2018.01260

[127] A. Stejskalová, B.D. Almquist, Using biomaterials to rewire the process of wound repair, Biomater. Sci. 5 (2017) 1421–1434. https://doi.org/10.1039/C7BM00295E

[128] H. Wang, Z. Xu, Q. Li, J. Wu, Application of metal-based biomaterials in wound repair, Eng. Regen. 2 (2021) 137–153. https://doi.org/10.1016/j.engreg.2021.09.005

[129] S.-B. Park, E. Lih, K.-S. Park, Y.K. Joung, D.K. Han, Biopolymer-based functional composites for medical applications, Prog. Polym. Sci. 68 (2017) 77–105. https://doi.org/10.1016/j.progpolymsci.2016.12.003

[130] S. Ahmadian, M. Ghorbani, F. Mahmoodzadeh, Silver sulfadiazine-loaded electrospun ethyl cellulose/polylactic acid/collagen nanofibrous mats with antibacterial properties for wound healing, Int. J. Biol. Macromol. 162 (2020) 1555–1565. https://doi.org/10.1016/j.ijbiomac.2020.08.059

[131] R. Morsy, M. Hosny, F. Reicha, T. Elnimr, Developing a potential antibacterial long-term degradable electrospun gelatin-based composites mats for wound dressing applications, React. Funct. Polym. 114 (2017) 8–12. https://doi.org/10.1016/j.reactfunctpolym.2017.03.001

[132] P.T.S. Kumar, S. Abhilash, K. Manzoor, S.V. Nair, H. Tamura, R. Jayakumar, Preparation and characterization of novel β-chitin/nanosilver composite scaffolds for wound dressing applications, Carbohydr. Polym. 80 (2010) 761–767. https://doi.org/10.1016/j.carbpol.2009.12.024

[133] S. Sethi, Saruchi, Medha, S. Thakur, B.S. Kaith, N. Sharma, S. Ansar, S. Pandey, V. Kuma, Biopolymer starch-gelatin embedded with silver nanoparticle–based hydrogel composites for antibacterial application, Biomass Convers. Biorefinery, (2022) 1–22. https://doi.org/10.1007/s13399-022-02437-w

[134] S.Y. Seo, G.H. Lee, S.G. Lee, S.Y. Jung, J.O. Lim, J.H. Choi, Alginate-based composite sponge containing silver nanoparticles synthesized in situ, Carbohydr. Polym. 90 (2012) 109–115. https://doi.org/10.1016/j.carbpol.2012.05.002

[135] B. Lee, E.-J. Choi, E.-J. Lee, S.-M. Han, D.-H. Hahm, H.-J. Lee, I. Shim, The Neuroprotective Effect of Methanol Extract of Gagamjungjihwan and Fructus Euodiae on Ischemia-Induced Neuronal and Cognitive Impairment in the Rat, Evidence-Based Complement. Altern. Med. 2011 (2011) 1–9. https://doi.org/10.1093/ecam/nep028

[136] B. Seal, Polymeric biomaterials for tissue and organ regeneration, Mater. Sci. Eng. R Reports. 34 (2001) 147–230. https://doi.org/10.1016/S0927-796X(01)00035-3

[137] M. Mousa, N.D. Evans, R.O.C. Oreffo, J.I. Dawson, Clay nanoparticles for regenerative medicine and biomaterial design: A review of clay bioactivity, Biomaterials. 159 (2018) 204–214. https://doi.org/10.1016/j.biomaterials.2017.12.024

[138] N. Ramesh, S.C. Moratti, G.J. Dias, Hydroxyapatite-polymer biocomposites for bone regeneration: A review of current trends, J. Biomed. Mater. Res. Part B Appl. Biomater. 106 (2018) 2046–2057. https://doi.org/10.1002/jbm.b.33950

[139] R. Yunus Basha, S.K. T.S., M. Doble, Design of biocomposite materials for bone tissue regeneration, Mater. Sci. Eng. C. 57 (2015) 452–463. https://doi.org/10.1016/j.msec.2015.07.016

[140] S.K. Misra, S.P. Valappil, I. Roy, A.R. Boccaccini, Polyhydroxyalkanoate (PHA)/Inorganic Phase Composites for Tissue Engineering Applications, Biomacromolecules. 7 (2006) 2249–2258. https://doi.org/10.1021/bm060317c

[141] M. Saleem, S. Rasheed, C. Yougen, Silk fibroin/hydroxyapatite scaffold: a highly compatible material for bone regeneration, Sci. Technol. Adv. Mater. 21 (2020) 242–266. https://doi.org/10.1080/14686996.2020.1748520

[142] F. Sun, H. Zhou, J. Lee, Various preparation methods of highly porous hydroxyapatite/polymer nanoscale biocomposites for bone regeneration, Acta Biomater. 7 (2011) 3813–3828. https://doi.org/10.1016/j.actbio.2011.07.002

[143] X.-Y. Ma, D. Cui, Z. Wang, B. Liu, H.-L. Yu, H. Yuan, L.-B. Xiang, D.-P. Zhou, Silk Fibroin/Hydroxyapatite Coating Improved Osseointegration of Porous Titanium Implants under Diabetic Conditions via Activation of the PI3K/Akt Signaling Pathway, ACS Biomater. Sci. Eng. 8 (2022) 2908–2919. https://doi.org/10.1021/acsbiomaterials.2c00023

[144] L. Wang, X. Zhao, B. Wei, Y. Liu, X. Ma, J. Wang, P. Cao, Y. Zhang, Y. Yan, W. Lei, Y. Feng, Insulin improves osteogenesis of titanium implants under diabetic conditions by inhibiting reactive oxygen species overproduction via the PI3K-Akt pathway, Biochimie. 108 (2015) 85–93. https://doi.org/10.1016/j.biochi.2014.10.004

[145] D. Qin, N. Wang, X.-G. You, A.-D. Zhang, X.-G. Chen, Y. Liu, Collagen-based biocomposites inspired by bone hierarchical structures for advanced bone regeneration: ongoing research and perspectives, Biomater. Sci. 10 (2022) 318–353. https://doi.org/10.1039/D1BM01294K

[146] M.A. Bonifacio, S. Cometa, A. Cochis, A. Scalzone, P. Gentile, A.C. Scalia, L. Rimondini, P. Mastrorilli, E. De Giglio, A bioprintable gellan gum/lignin hydrogel: a smart and sustainable route for cartilage regeneration, Int. J. Biol. Macromol. 216 (2022) 336–346. https://doi.org/10.1016/j.ijbiomac.2022.07.002

[147] Y. Shen, X. Yu, J. Cui, F. Yu, M. Liu, Y. Chen, J. Wu, B. Sun, X. Mo, Development of Biodegradable Polymeric Stents for the Treatment of Cardiovascular Diseases, Biomolecules. 12 (2022) 1245. https://doi.org/10.3390/biom12091245

[148] J.A. Ormiston, P.W. Serruys, E. Regar, D. Dudek, L. Thuesen, M.W. Webster, Y. Onuma, H.M. Garcia-Garcia, R. McGreevy, S. Veldhof, A bioabsorbable everolimus-eluting coronary stent system for patients with single de-novo coronary artery lesions (ABSORB): a prospective open-label trial, Lancet. 371 (2008) 899–907. https://doi.org/10.1016/S0140-6736(08)60415-8

[149] J. Zong, Q. He, Y. Liu, M. Qiu, J. Wu, B. Hu, Advances in the development of biodegradable coronary stents: A translational perspective, Mater. Today Bio. 16 (2022) 100368. https://doi.org/10.1016/j.mtbio.2022.100368

[150] Y. Shen, C. Tang, B. Sun, Y. Zhang, X. Sun, M. EL-Newehy, H. EL-Hamshary, Y. Morsi, H. Gu, W. Wang, X. Mo, 3D printed personalized, heparinized and biodegradable coronary artery stents for rabbit abdominal aorta implantation, Chem. Eng. J. 450 (2022) 138202. https://doi.org/10.1016/j.cej.2022.138202

[151] L. Yang, H. Wu, Y. Liu, Q. Xia, Y. Yang, N. Chen, M. Yang, R. Luo, G. Liu, Y. Wang, A robust mussel-inspired zwitterionic coating on biodegradable poly(L-lactide) stent with enhanced anticoagulant, anti-inflammatory, and anti-hyperplasia properties, Chem. Eng. J. 427 (2022) 130910. https://doi.org/10.1016/j.cej.2021.130910

[152] Y.-N. Zhao, P. Wu, Z.-Y. Zhao, F.-X. Chen, A. Xiao, Z.-Y. Yue, X.-W. Han, Y. Zheng, Y. Chen, Electrodeposition of chitosan/graphene oxide conduit to enhance peripheral nerve regeneration, Neural Regen. Res. 18 (2023) 207. https://doi.org/10.4103/1673-5374.344836

[153] M. Darder, P. Aranda, E. Ruiz-Hitzky, Bionanocomposites: A New Concept of Ecological, Bioinspired, and Functional Hybrid Materials, Adv. Mater. 19 (2007) 1309–1319. https://doi.org/10.1002/adma.200602328

[154] E. Colusso, A. Martucci, An overview of biopolymer-based nanocomposites for optics and electronics, J. Mater. Chem. C. 9 (2021) 5578–5593. https://doi.org/10.1039/D1TC00607J

[155] M.C. Demirel, M. Vural, M. Terrones, Composites of Proteins and 2D Nanomaterials, Adv. Funct. Mater. 28 (2018) 1704990. https://doi.org/10.1002/adfm.201704990

[156] M. Khazaei, A. Mishra, N.S. Venkataramanan, A.K. Singh, S. Yunoki, Recent advances in MXenes: From fundamentals to applications, Curr. Opin. Solid State Mater. Sci. 23 (2019) 164–178. https://doi.org/10.1016/j.cossms.2019.01.002

[157] Y. Xia, T.S. Mathis, M.Q. Zhao, B. Anasori, A. Dang, Z. Zhou, H. Cho, Y. Gogotsi, S. Yang, Thickness-independent capacitance of vertically aligned liquid-crystalline MXenes, Nature. 557 (2018) 409–412. https://doi.org/10.1038/s41586-018-0109-z

[158] F. Shahzad, M. Alhabeb, C.B. Hatter, B. Anasori, S. Man Hong, C.M. Koo, Y. Gogotsi, Electromagnetic interference shielding with 2D transition metal carbides (MXenes), Science (80-.). 353 (2016) 1137–1140. https://doi.org/10.1126/science.aag2421

[159] M. Nogi, M. Karakawa, N. Komoda, H. Yagyu, T.T. Nge, Transparent Conductive Nanofiber Paper for Foldable Solar Cells, Sci. Rep. 5 (2015) 17254. https://doi.org/10.1038/srep17254

[160] Y. Liu, S. Yang, W. Niu, Simple, rapid and green one-step strategy to synthesis of graphene/carbon nanotubes/chitosan hybrid as solid-phase extraction for square-wave voltammetric detection of methyl parathion, Colloids Surfaces B Biointerfaces. 108 (2013) 266–270. https://doi.org/10.1016/j.colsurfb.2013.03.003

[161] A. Bouvree, J.-F. Feller, M. Castro, Y. Grohens, M. Rinaudo, Conductive Polymer nano-bioComposites (CPC): Chitosan-carbon nanoparticle a good candidate to design polar vapour sensors, Sensors Actuators B Chem. 138 (2009) 138–147. https://doi.org/10.1016/j.snb.2009.02.022

[162] T. Ghosh, S.M. Bhasney, V. Katiyar, Blown films fabrication of poly lactic acid based biocomposites: Thermomechanical and migration studies, Mater. Today Commun. 22 (2020) 100737. https://doi.org/10.1016/j.mtcomm.2019.100737

Materials Research Forum LLC
https://doi.org/10.21741/9781644902639-4

[163] S. Agustin-Salazar, P. Cerruti, L.Á. Medina-Juárez, G. Scarinzi, M. Malinconico, H. Soto-Valdez, N. Gamez-Meza, Lignin and holocellulose from pecan nutshell as reinforcing fillers in poly (lactic acid) biocomposites, Int. J. Biol. Macromol. 115 (2018) 727–736. https://doi.org/10.1016/j.ijbiomac.2018.04.120

[164] E. Fortunati, F. Luzi, D. Puglia, R. Petrucci, J.M. Kenny, L. Torre, Processing of PLA nanocomposites with cellulose nanocrystals extracted from Posidonia oceanica waste: Innovative reuse of coastal plant, Ind. Crops Prod. 67 (2015) 439–447. https://doi.org/10.1016/j.indcrop.2015.01.075

[165] E. Lizundia, L. Ruiz-Rubio, J.L. Vilas, L.M. León, Poly(L -lactide)/zno nanocomposites as efficient UV-shielding coatings for packaging applications, J. Appl. Polym. Sci. 133 (2016) n/a-n/a. https://doi.org/10.1002/app.42426

[166] B. Khan, M. Bilal Khan Niazi, G. Samin, Z. Jahan, Thermoplastic Starch: A Possible Biodegradable Food Packaging Material-A Review, J. Food Process Eng. 40 (2017) e12447. https://doi.org/10.1111/jfpe.12447

[167] H. Wang, J. Qian, F. Ding, Emerging Chitosan-Based Films for Food Packaging Applications, J. Agric. Food Chem. 66 (2018) 395–413. https://doi.org/10.1021/acs.jafc.7b04528

[168] F. Zhu, Polysaccharide based films and coatings for food packaging: Effect of added polyphenols, Food Chem. 359 (2021) 129871. https://doi.org/10.1016/j.foodchem.2021.129871

New Materials for a Circular Economy
Materials Research Foundations 149 (2023) 142-172

Materials Research Forum LLC
https://doi.org/10.21741/9781644902639-5

Chapter 5

Recent Advances in Polymeric Systems for CO_2 Capture: A Small Catalogue

Raquel López-Gallego[1], Gaurav Sharma[2*], Amit Kumar[2], María Moral-Zamorano[1],
Alberto García-Peñas[1*]

[1] Departamento de Ciencia e Ingeniería de Materiales e Ingeniería Química, IAAB, Universidad Carlos III de Madrid, Avda. de la Universidad, 30, 28911, Leganés, Madrid, Spain

[2] International Research Centre of Nanotechnology for Himalayan Sustainability (IRCNHS), Shoolini University, Solan 173212, Himachal Pradesh, India

[3] College of Materials Science and Engineering, Shenzhen Key Laboratory of Polymer Science and Technology, Guangdong Research Center for Interfacial Engineering of Functional Materials, Nanshan District Key Lab. for Biopolymers and Safety Evaluation, Shenzhen University, Shenzhen 518060, PR China

[4] School of Science and Technology, Glocal University, Saharanpur, India

[5] Department of Chemistry, Shahr-e-Qods Branch, Islamic Azad University, Tehran, Iran

gaurav8777@gmail.com; alberto.garcia.penas@uc3m.es

Abstract

The advances in terms of new materials used for CO_2 collectors are growing due to the necessity to reduce the greenhouse gases emission and to respond to the environmental regulations. There are multiple systems which have shown remarkable properties for removal of specific pollutants. Nevertheless, the necessity of developing devices more effective, specific and with higher yield is necessary to reach net zero emissions by 2050. The use of polymeric matrices for CO_2 capture purpose is growing due to the versatility of these materials together with their low cost, regarding in comparison to other systems. There are multiple advances in terms of CO_2 collectors, and specifically for membranes due to the possibility to modulate these polymers and their properties, but also because these can be combined with other materials thus increasing the possibilities of their applications. This chapter shows some of these recent advances attending to the permeability of these materials, as one of the most important materials with unique characteristics studied for CO_2 capture.

Keywords

CO_2 Capture, Polymers, Permeability, Green Chemistry, Circular Economy

New Materials for a Circular Economy   Materials Research Forum LLC
Materials Research Foundations 149 (2023) 142-172   https://doi.org/10.21741/9781644902639-5

1. Introduction

One of the most important causes associated with climate change is the increase of carbon dioxide (CO_2) in the atmosphere. This gas absorbs a part of the sun radiation (infrared radiation), and subsequently the radiation is emitted producing the greenhouse effect [1].

CO_2 is mainly produced by energy generation, industrial activities, transportation, agriculture and waste burning. A great part of these emissions is clearly associated with the combustion of fuels (petroleum, natural gas or coal) which are clearly identified with energy generation, but there are also other contributions related to industry and transport.

The CO_2 emissions continue globally increasing despite the diverse crisis experienced these years. For instance, the CO_2 emissions were slightly reduced during the pandemic associated with lockdowns of COVID-19 but regrettably the emissions of 2021 show a new rebound from 2020 levels [2]. A continuous tracing is carried out by The Carbon Monitor Program, measuring the CO_2 emissions from energy production, industry, transportation, and residential areas [3, 4]. These measurements are necessary to control the emissions and specifically to reach the goal of the European Union to become climate-neutral by 2050, and also to respond to the Paris Agreement (COP21).

The accumulation of CO_2 in the atmosphere could be responsible for many environmental problems associated with the greenhouse effect promoting the thawing of glacial masses, ocean acidification, rising sea levels, desertification, alteration in ecosystems or the decrease in harvest yields (Figure 1) [5]. Consequently, urgent and important actions should be taken in order to reduce these CO_2 emissions. In this sense, specific technologies and devices are developed to reduce these emissions, which are mainly based on capture and storage systems of carbon dioxide. Carbon capture and storage (CCS) is defined as the process for capturing carbon dioxide before this one is released in the atmosphere, and its subsequent transportation and storage in underground geological formation.

Contents

New Materials for a Circular Economy   Materials Research Forum LLC
Materials Research Foundations 149 (2023) 142-172   https://doi.org/10.21741/9781644902639-5

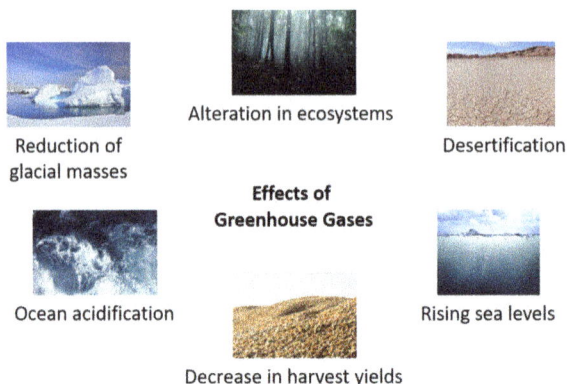

Figure 1. Effects of greenhouse gases.

Traditionally, the capture of CO_2 is possible through mainly three technologies which allow reducing the number of compounds or pollutants, increasing the concentration of CO_2 and reducing the number of steps of the process: post-combustion, pre-combustion and oxy-fuel combustion. Subsequently, the separation can be carried out by different methods, some of them are still under developing but numerous techniques started the commercialization. These are mainly based on the use of absorption, adsorption, cryogenic distillation or membranes [6]. The use of one method will response to the conditions of the gas stream sources where the CO_2 needs to be extracted [1].

There are many factors to consider before selecting a capture method, and applying the design parameters. The operating conditions will define the parameters associated with the higher yield, efficiency and minimum cost, and consequently the technology applied. For example, nominal CO_2 removal, lean solution feed temperature, pressure, flow rate and gas composition will define the operating conditions for an absorption column [7].

Nevertheless, a correct selection of the technology will be essential for getting the best results. Generally, simulations supported with experimental data will assure the success of a system for CO_2 capture. Nowadays, new advances are developing in terms of efficiency, yield, energy consumption or costs, among others.

In this work, we are focused on the efficiency, permeability and selectivity associated with some of the most prominent polymeric systems recently studied in diverse CO_2 collectors. The interest of composites and hybrid polymers is still rising due to the versatility of the polymeric matrices which are able to improve the properties of other inorganic and/or organic materials. Also, the use of hybrid materials allows getting other functionalities and features which could enhance the current systems in terms of sustainability.

1.1   The use of polymers

The plastic consumption is still increasing despite the new regulations associated with sustainability and the circular economy, among others. The reason is the enormous range of applications where polymers can be used regarding other materials. The modularity of the molecular features of polymers allows getting tailored materials with specific properties.

The global production of plastics is still increasing showing the importance of these materials [8]. Nevertheless, the environmental problems associated with the plastics, and specifically with microplastics, have provided them a bad reputation that interestingly seems not affected to its production.

Plastics can also be an interesting tool for sustainable development due to their properties and the possibility of its recycling (Figure 2), and can be used for air purification or water remediation. A special relevance of the use of polymers is observed for technologies associated with CO_2 capture [9]. Specifically, tunable porous organic polymers (POPs) show interesting properties for gas separation and storage due to their versatility. Some of their most important properties are associated with their low density, large surface area, tailorable surface properties, thermal stability, and chemically inert [5, 10, 11]. All these properties seem to guarantee the use of POPs as excellent materials for selective CO_2 capture [5].

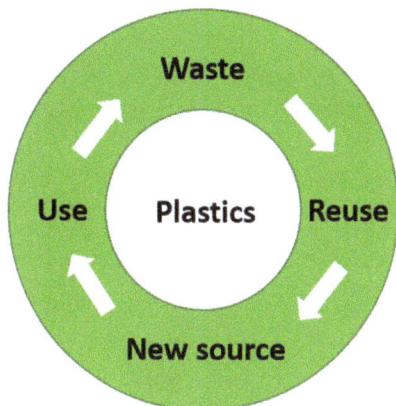

Figure 2. Recyclability of plastics.

The control over the POPs through the control of porous size and distributions, together with the degree of functionalization allows improving the selectivity, affinity and efficiency of the CO_2 adsorbents [10]. The use of other structures or materials regarding polymers shows interesting results associated with the adsorption capacity but there are some shortcomings that the use of polymers or polymeric matrices can solve. For example, a higher CO_2 adsorption functionality can be obtained through the use of amine groups into the diverse inorganic porous structures, but some problems can appear associated with amine leaching. On the other hand, a high density is observed for many ionic liquids used for CO_2 capture. The new progresses in POPs can solve part of these problems incorporating also other features such as high flexibility, low cost or other functionalities [12].

Selectivity will play another important role for CO_2 capture over other gases, considering that these flows are composed also of nitrogen, oxygen and water, among other compounds (SOx, NOx, particulate matter…). The effectiveness of the process will be clearly associated with CO_2 selectivity over other gases, but this one will also be affected by the regeneration cycles [12]. Successive regeneration cycles could affect the effectiveness of the adsorbent and consequently the yield of the process, and also the energy consumption of this process.

The facilitated transport membranes (FTMs) allow getting a reversible reaction between CO_2 and amine groups regarding other polymeric membranes. In addition, lower partial pressure induces a higher CO_2 permeance and selectivity [13]. The relationship between pressure, temperature, permeability and selectivity will define the properties of these matrices for CO_2 capture. Another important point will be associated with the CO_2 concentration and composition of the flow to be treated. In many cases, these membranes are working perfectly for dilute CO_2 sources [14]. Also, FTMs show high yields for temperatures higher than 100 °C [14].

There are many possibilities for CO_2 capture using polymeric matrices due to their versatility where many parameters can affect the mentioned parameters. For instance, the crosslinking in polymer will modify the CO_2 selectivity [5].

New properties or functionalities can also be developed if mixtures, composites or hybrid materials are prepared. The preparation of these structures response to the necessity to fix some shortcoming associated with other inorganic or organic materials. These hybrid polymers can improve some properties and also allow getting new functionalities depending on the final requirements.

1.2   Applications: Membranes

The use of polymers for CO_2 capture mainly focuses on the development of membranes and processes of adsorption and absorption, covering a great part of the systems designed for CO_2 collection (Figure 3). The necessities associated with these technologies change depending on the conditions and the requirements, consequently the materials involved in

New Materials for a Circular Economy   Materials Research Forum LLC
Materials Research Foundations 149 (2023) 142-172   https://doi.org/10.21741/9781644902639-5

these systems will determine their final physicochemical properties, the energy consumption and the regeneration cycles.

CO$_2$ capture systems

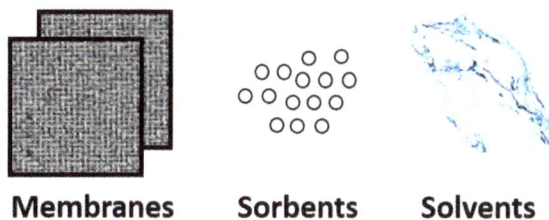

| Membranes | Sorbents | Solvents |

Figure 3. CO$_2$ capture systems.

The versatility and tunable properties of polymeric systems explains that the major part of membranes is based on polymers. The permeability will be an essential parameter for the development of new membranes, i.e., a high permeability will define their efficiency [1]. In a similar way, CO$_2$ selectivity will affect the efficiency of the process and must be studied especially when multiple gases are involved in the flow to be treated.

In general, the membranes can be composed of polymeric, inorganic and hybrid materials. Normally, these hybrid materials are based on polymers with embedded inorganic compounds. The development of polymeric structures can also be understood from their exceptional processability regarding other materials, and in many cases, low cost. Also, the use of membranes collects other benefits such as simplicity or higher energy efficiency.

The membranes can be divided depending on the type of flow treated: flue gases, syngas processing and gas or biogas. Each feed will be also treated depending on the composition, concentration, pressure, and necessities [14].

1.3   A catalogue of polymeric systems for CO$_2$ capture

This chapter tries to collect some of the most interesting recent advances in terms of promising polymeric structures for CO$_2$ collectors. The classification was prepared considering the complexity of the final material, and was divided into general polymers, mixtures of polymers and hybrid materials and composites.

There are many polymer-based materials recently published, and consequently this work only collects some of them. This information could be interesting for industry and researchers due to a general prospective of polymers used for CO$_2$ capture can be obtained with a naked eye. Nevertheless, these pages only show some examples about the huge work carried out from many research institutions and universities.

2. New polymeric structures

2.1 Polymeric systems

Table 1 exhibits the recent advances in pure polymers and blends used for CO_2 collectors, showing the value of CO_2 permeability together with the used specific conditions, and the reference to the research work. In addition, the values obtained from the scientific papers are exhibited using the same units for a better comparison.

The polymers exhibit permeability values higher than 100 Barrer, excluding the pure homopolymers due to the combination of numerous monomers for getting copolymers, or the blend between different polymeric structures may provide better properties. The preparation of a block copolymer can release values of permeability up to 20400 Barrer like in the case of PEO-b-PES structures [15].

A high permeability value is observed for polyvinylamine (PVAm), and can also increase if ethylenediamine (EDA), methylcellulose (MC) or monoethanolamine (MEA) are part of the membranes [16-19].

The method of preparation of polymeric systems will affect the properties of the membrane, specifically if the morphology of the surface area and porosity are modified. The use of different experimental conditions and treatments can promote these changes. In this sense, interesting values are observed for triptycene-contained polymers of intrinsic microporosity and polyimides of intrinsic microporosity [20]. Also, the use of the different solvents can be responsible for these changes, but also the modifiers, crosslinkers or additives used during the process. Similarly, the processing of the resulting materials will change the values of permeability, for example, the preparation of fibers through electrospinning. For example, poly(1-trimethylsilyl-1-propyne) (PTMSP) prepared by electrospinning shows a permeability of 39919 Barrer. [21].

The thermal treatment applied will also play an important role over the CO_2 permeability, such as it can be deduced from some polymeric structures of the Table 1. A good example is the thermally rearranged polybenzoxazole (TR-PBO500) whose permeability values reached 2326 Barrer.

Table 1. General polymers and polymeric blends used for CO_2 collectors.

Polymer	Description	Δp (atm)	T (K)	CO_2 permeability (Barrer)	Reference
PVC-g-POEM	Graft copolymer based on poly(vinyl chloride) (PVC) (main chains) and poly(oxyethylene methacrylate) (POEM) (side chains).	1	308	107	[22]

PAMAM/PEG DMA/4GMAP	Membrane of poly(amidoamine) (PAMAM) dendrimers withpoly(ethylene glycol) dimethacrylates (PEGDMAs).	1	313	131.6	[23]
POEM-g-PVC	Graft copolymer based on poly(vinyl chloride) (PVC) (main chains) and poly(oxyethylene methacrylate) (POEM) (side chains) (after treatment with acetone vapor).	1	308	145	[22]
PVA/PAA-C4H9	Membrane of poly(vinyl alcohol) (PVA) crosslinked with poly-N-isobutylallylamine (PAA-C4H9).	2	383	154.6	[24]
DGBAmE/TMC	Membrane of diethylene glycol bis(3-aminopropyl) ether (DGBAmE) and trimesoyl Chloride (TMC).	5	295	183.4	[25]
TR Ac-450	Thermally rearranged (TR) polymers derived from polyimides based on 3,3'-dihydroxy-4,4'-diamino-biphenyl and 2,2'-bis-(3,4-dicarboxyphenyl) hexafluoropropane dianhydride (HAB-6FDA) prepared with acetate.	10	308	320	[26]
TEGMVE/VEEM(14/1)	Membrane formed of copolymer of vinyl ether monomers having a triethylene glycol segment [2-(2-(2-methoxyethoxy)ethoxy)ethyl vinyl ether (TEGMVE)] with vinyl ether monomer having a methacrylate unit [2-(2-vinyloxyethoxy)ethyl methacrylate: VEEM] with a feed ratio of 14:1.	1	298	410	[27]
PEGDA/PEGMEA99	Membrane formed of 1% poly(ethylene glycol) diacrylate and 99% poly(ethylene glycol) methyl ether acrylate (PEGDA-co-PEGMEA99).	4.4	308	570	[28]
SpiroTR-PBO-6F	Membrane formed by thermally rearranged spiro-polybenzoxazole based on 4,4'-hexafluoroisopropylidene .	1	308	675	[29]

HSBI-4-CF3	Membrane formed by the C_3F functionalized polymer of intrinsic microporosity (PIM) synthetized with 3,3,3',3'-tetramethyl spirobisindane-6,6'-diol and 3(4)-trifluoromethylbenzaldehyde.	2	308	686	[30]
GPP1000-60	3-Glycidyloxypropyltrimethoxysilane (GOTMS), poly(ethylene oxide) (PEO) and poly(ethylene glycol) (PEG) blend (1000 is the molecular weight of blended PEG and 60 Is the wt% of the blended PEG).	2	308	696	[31]
Pebax/CC-PEINTs	Membrane formed with Pebax (block copolymers made up of rigid polyamide blocks and soft polyether blocks) and acid etched polyethyleneimine nanotubes.	2	303	710	[32]
PEGDA/TRIS-A	Membrane formed of [tris-(trimethylsiloxy)silyl] propyl acrylate (TRIS-A) and poly(ethylene glycol) diacrylate (PEGDA).	15	308	716	[33]
PEO-PBT/PEG-DBE	Membrane of polyethylene oxide, polybutylene terephthalate (PEO/PBT) multiblock copolymer, andpolyethylene glycol dibutyl ether as smart additive.	10	303	750	[34]
COF(6-9.6)-based MMM	Mixed-matrix membranes of poly(ethylene glycol) methyl ether acrylate and poly(ethylene glycol) diacrylate with covalent-organic frameworks (COF) integrated formed with 1,3,5-tris(4-aminophenyl)benzene (TAPB) (6 mg) and 1,1'-bis(2,4-dinitrophenyl)-[4,4'-bipyridine]-1,1'-diium dichloride (BDB) (9.6 mg).	5	308	803.9	[35]
P(DADMACA-co-PVAm)	Composite membrane of poly(diallyldimethylammonium carbonate-co-vinylamine) (P(DAD-MACA-co-VAm)) and polysulfone (PSf).	1	298	921	[36]

Pebax 2533/PEG-*b*-PPFPA	Membrane formed of polyether block amide *Pebax® 2533* and diblock polymers of poly(ethy-lene glycol)-b-poly(pentauoropropyl actylate) (PEG-bPPFPA).	3,5	308	924	[37]
AO-PIM-1	Amidoxime-functionalized prototypical solution-processable polymer of intrinsic microporosity, PIM-1 (AO-PIM-1) synthetized from of1,4-dicyanotetra- fluorobenzene (DCTB), 5,5',6,6'-tetrahydroxy-3,3,3',3'-tetramethyl-1,1'-spirobisindane (TTSBI), and dimethylacetamide (DMAc).	2	308	1153	[38]
TBDA2-SBI-PI	Polyimide (PI) of intrinsic microporosity derived of spirobisindane-based dianhydride (SBI) and 3,9-diamino-4,10-dimethyl-6H,12H-5,11-methanodibenzo[1,5]diazocine (TBDA2).	1	308	1213	[39]
MEDA-TMC	Membrane formed of the polymerization from *N*-methyldiethanolamine (MEDA) and trimesoyl chloride (TMC) on crosslinked polydimethylsiloxane (PDMS) coating polysulfone (PS) support membrane.	1	298	1275	[40]
SPDA-BSBF	Polyimides of intrinsic microporosity based on 3,3'-dibromo-9,9'-spirobifluorene-2,2'-diamine (BSBF) and dianhydride pyromellitic dianhydride (SPDA).	2	308	1340	[41]
PDXLA-co-PDXLEA75	Membrane based on copolymer of poly(1,3 dioxolane) acrylate (PDXLA) and poly(1,3 dioxolane) acrylate with ethoxy ω terminal groups (PDXLEA) (75 wt% of PDXLEA)	6.8	343	1400	[42]

DMAEMA-AA	Polymer membrane form with copolymer of DMAEMA (2-*N*,*N*-dimethyl aminoethyl methacrylate) and AA (acrylic acid) and polysulfone (PS) ultra-filtration membrane as support.	0.011	299	1512	[43]
CPBOc(90:10)	Cardo-polybenzoxazole polymer membrane (CPBO) (90:10 mole ratio of of 3,3_-dihydroxybenzidine and 9,9-bis(3-amino-4-hydroxyphenyl)fluorene used in the polymerization).	3.5	308	1539	[44]
DPPD-TMPD	Polymer membrane formed by 3,8-diphenylpyrene-1,2,6,7-tetracarboxylic dianhydride (DPPD) and 2,4,6-tri- methyl-m-phenylenediamine (TMPD).	3	303	1600	[45]
PIM-PI-1	Polymer with intrinsic microporosity (PIM) based on the monomer formed by the synthesis of 5,5',6,6'-tetrahydroxy-3,3,3',3'-tetramethyl-1,10-spirobisindane, dianhydride and 1,4-phenyldiamine.	1	303	2156.5	[46]
TR-PBO500	Thermally rearranged (at 500 °C) membrane of polybenzoxazole based on 4,4'-hexafluoroisopropylidene diphthalic anhydride (6FDA) and 2,2-bis(3-amino-4-hydroxyphenyl)hexafluoropropane (bisAPAF).	1.0-6.0	Room temperature	2326	[47]
DC-PIM1	Carboxylated polymers of intrinsic microporosity synthetized with 5,5',6,6'-tetrahydroxy-3,3,3',3'-tetramethylspirobisindane with tetrafluoroterephthalonitrile and heptafluoro-*p*-tolylphenylsulfone with degree of hydrolysis of ~22%.	3.4	298	2345	[48]
KAUST-PI-1	Triptycene-contained polyimides membrane of intrinsic microporosity (PIM-PIs) synthetized from 9,10-diisopropyl-2,3,6,7-tetramethoxyanthracene, benzene diazonium chloride, 4,5-dichlorophthalonitrile, acetic	2	308	2389	[20]

	anhydride and 2,3,5,6-tetramethyl-1,4-phenylene diamine.				
PDMS	Membrane of polydimethylsiloxane.	-	294	2652.4	[49]
CS/Pebax® 1657	Hydrogel membrane composed by chitosan (CS) and polyether-block-amide (Pebax).	2	358	2884	[50]
6FDA-DAM + 10 wt% PBP-menm	Matrix membranes (MMMs) based on polyimide (synthesized from 4,4'-(hexafluoroisopropylidene) diphthalic anhydride.(6FDA) and2,4,6-trimethyl-m-phenylenediamine (DAM)) and porous organic polymers of 1,2-methylethylenediamine (10 wt%)(menm = 1,2-methylethylenediamine).	1	308	2988	[51]
CA/PM-4 (1:3 % w/w)	Membrane of cellulose acetate (CA) blended with ortho-linked thiazole-based polyimine (PM-4) (1:3 % w/w).	3	308	3000	[52]
Teflon AF2400®	Membrane of polytetrafluoroethylene.	-	294	3127.1	[49]
KAUST-PI-7	Triptycene-based polymer of intrinsic microporosity-polyimides prepared from the dianhydride i-C 3 triptycene propyl-based dianhydride monomer (TPDA) and dimethylnaphthidine (DMN).	2	308	4391	[53]
K-PPO/BnOH	Membrane of polyphenylene oxide (PPO) polymer and benzyl alcohol (BnOH).	2	303	4651.2	[54]
TFTPN-TPE-25	Intrinsically microporous polymer (PIM) synthesized by polycondensation reaction of 1,1',2,2',-tetrahydroxy-tetraphenylethylene (TPE) (at 25%) with 2,3,5,6-tetrafluoroterephthalonitrile (TFTPN).	2	308	5203	[55]

PVA/PAA-C3H7/Poly(silo xane)	Membrane of poly-*N*-isopropylallylamine (PAA-C3 H7) into cross-linked polyvinylalcohol-poly(siloxane).	2	383	6500	[56]
PIM-EA-TB	Polymers of intrinsic microporosity based on Tröger's base (TB, 2,8-dimethyl-6H,12H-5,11-methanodibenzo[b,f][1,5]diazocine) and 2,6(7)-diamino-9,10-dimethylethanoanthracene (EA).	6.9	298	7140	[57]
PVAm	Membrane of polyvinyl amine	1.2	318	10222	[16]
cPIM-1/PPN2-3%	Porus polymer network (PPNs) membrane formed by PIM-1 (5,5',6,6'-tetrahydroxy-3,3,3',3'-tetramethyl-1,1'-spirobisindane and 2,3,5,6-tetrafluoroterephthalonitrile) with 3% of 4,4'-diacetylbiphenyl.	2	303	11511	[58]
PEO-b-PES (T=343 K)	Membrane based on block copolymer of polyethersulfone (PES) andpolyethylene oxide (PEO).	0.2	343	20400	[15]
PVAm-EDA	Polyvinylamin and ethylenediamine (*EDA*) membrane.	1.1 – 16	room temperature	30350	[17]
PTMSP	Poly(1-trimethylsilyl-1-propyne) (PTMSP) electrospun fiber.	3	298	39919	[21]
PVAm-MC	Polyvinylamin and methylcellulose (*MC*) membrane.	3	295	59000	[18]
PVAm-MEA	Polyvinylamin and monoethanolamine (*MEA*) membrane.	0.4	323	79800	[19]
PVAm-PIP/PS	Composite membrane of polyvinylamine (PVAm) crosslinged with piperazine (PIP) and polysulfone (PS).	1	295	195000	[59]

2.2 Hybrid materials and composites

The preparation of hybrid materials can improve the properties of the polymeric matrix and also develop new characteristics. Table 2 shows some of the most recent hetero structures whose CO_2 permeability values are higher than corresponding polymers due to mainly two reasons. The main one is associated with the improvement of the final properties associated with the polymeric matrix, and the second one is clearly explained through the contribution of the filler (Figure 4).

Constituent types

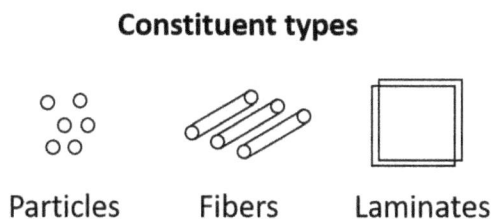

Particles   Fibers   Laminates

Figure 4. Morphologies of fillers used for different purposes.

In general, there are multiple advances for polymers of intrinsic microporosity (PIMs) such as it can be observed in the literature, where really high CO_2 permeability values are observed. Among all these structures, the case of DC-PIM1, which is a carboxylated polymer of intrinsic microporosity, achieves values of 2345 Barrer [48].

The preparation of composites or blends based on Pebax seems to increase the permeability such as it can be observed for Pebax/TCP 80 or CS/Pebax® whose values are 1673.2 and 2884 Barrer, respectively [50, 60].

The high value of CO_2 permeability obtained poly(1-trimethylsilyl-1-propyne) (PTMSP) can also rise if a hybrid material is prepared, such as it can be deduced from the addition of porous aromatic frameworks (PAFs) into polymeric matrix [61]. In this case, the permeability can increase from 39919 to 50600 Barrer.

In a similar way, polyvinyl amine (PVAm) is another polymer used for the preparation of membranes. There are multiple research works where the PVAm is used as polymeric matrix providing high results in comparison with the pure polymer [62].

There are also many works where poly(ether ether ketone) (SPEEK) and poly(4-methyl-2-pentyne) (PMP) as polymeric matrices, among others [63, 64].

Table 2. Hybrid materials based on polymers used for CO_2 collectors.

Polymer	Filler	Description	Δp (atm)	T (K)	CO_2 permeability (Barrer)	Reference
Im-PI	Zn 10%	Composite membrane formed of 4,4'-(hexafluoroisopropylidene)diphthalic anhydride (6FDA) and 2-(4-aminophenyl)-5-aminobenzimidazole (PABZ) polyimide (Im-PI) with coordinated zinc (10% mass ratio).	3	298	9.6	[65]
Im-PI	Co 5%	Composite membrane formed of 4,4'-(hexafluoroisopropylidene)diphthalic anhydride (6FDA) and 2-(4-aminophenyl)-5-aminobenzimidazole (PABZ) polyimide (Im-PI) with coordinated cobalt (5% mass ratio).	3	298	12.7	[65]
PIL	SiO_2	1-(4-vinylbenzyl)-3-butyl imidazolium bis(trifluoromethane) sulfonimides (polyionic liquid, PIL) grafted on silicon dioxide nanoparticles.	0.35	298	53.5	[66]
Polyether-diamine	Silica(insitu)	Polyether-diamine membrane with silica.	4	303	125	[67]
PEPG3300	Silica(insitu)	Poly (propylene-glycol)-block-poly(ethyleneglycol)-block-poly(prop-yleneglycol) (MW = 3300g/mol) membrane with silica.	-	303	132,3	[68]

Matrimid5218	Nanohydrogels	Matrix of *Matrimid® 5218* (a polyimide) with poly(N-isopropylacrylamide) nanohydrogels (NHs) incorporated.	2	303	278	[69]
Pebax MMMs	CuBTC	Matrix membranes (MMMs) based on poly(ether-block-amide) with copper-1,3,5-benzenetricarboxylate (CuBTC) (15 wt.%).	1	308	286	[70]
[EMIM][OAc]/ Pebax MMMs	CuBTC	Matrix membranes (MMMs) based on 1-ethyl-3-methylimidazolium acetate [EMIM][OAc], copper-1,3,5-benzenetricarboxylate (CuBTC) ,and poly(ether-block-amide) (Pebax).	1	308	335	[70]
PEA	POSS	Composite of poly(ethylene adipate) or PEA and polyhedral oligomeric silsesquioxanes (*POSS*).	1	308	380	[71]
PEI	$Cu^{2+}(CH_3COO^-)_2$	Metal-induced ordered microporous polymers (MMPs) of polyethylenimine (PEI) with $Cu^{2+}(CH_3COO^-)_2$.	2	298	400	[72]
BPPO	Fumed silica	Membrane of bromomethylated poly(phenylene oxide) (*BPPO*) with silica nanoparticles.	0,68	room temperature	436	[73]
PEO	66 wt% $[C_2mim][C(CN)_3]$	Ionic liquid (IL)-based semi-interpenetrating polymer network (sIPN) membranes based on poly(ethylene oxide) (PEO) network and linear nitrile butadiene rubber, incorporating up to 66 wt% of 1-ethyl-3-methylimidazolium tricyanomethanide ($[C_2mim][C(CN)_3]$).	1	308	633	[74]

1,2-polybutadiene	MgO	Membrane of 1,2-polybutadiene with dispersed magnesium oxide (MgO).	12	308	650	[75]
PEG	Silica(insitu)	Composite of polyethylene glycol (PEG) matrix with silica.	4	318	845	[31]
PVA/PAA-C4H9	AIBA-K	Poly(vinyl alcohol) (PVA) and 2-bromobutane poly(allylamine) matrix with 2-aminoisobutyric acid-potassium salt (AIBA-K).	3	379	938	[76]
PAN-PU/PIM(20 wt%)	UIO66-NH$_2$	Composite membrane of three layers. A first layer of PAN (polyacrylonitrile), a second one formed of amine functionalized metal organic framework made up of [Zr$_6$O$_4$(OH)$_4$] clusters with 1,4-benzodicarboxylic acid struts (UIO66) and a third layer of polymer of intrinsic microporosity (PIM) and polyurethane (20 wt% PIM in PU) blend.	1	308	999.2	[77]
P[vbim][NTf$_2$]/[emim][B(CN)$_4$]	ZIF-8	Polymer of 1-vinyl-3-butyl imidazolium-bis (trifluoromethylsulfonyl) imidate ([vbim][NTf$_2$]) and 1-ethyl-3-methylimidazolium tetracyanoborate with integred zeolite imidazolate framework-8 (ZIF-8).	3.5	308	1062.4	[78]
PVC-g-POEM	ZIF-8	Mixed matrix membrane of zeolite imidazole frameworks (H_ZIF-8) dispersed in poly(vinyl chloride)-g-poly(oxyethylene methacrylate) (PVC-g-POEM).	-	308	1195.4	[79]

PIM	MTZ	Carboxylated polymer of intrinsic microporosity (PIM) of 5,5′,6,6′-tetrahydroxy-3,3,3′,3′-tetramethylspirobisindane with tetrafluoroterephthalonitrile and heptafluoro-p-tolylphenylsulfone functionalized with metiltetrazole functional group.	3.4	298	1391	[80]
PNVF-co-Vam	PZEA-Sar	Selective layer contained poly(N-vinylformamide-co-vinylamine) (PNVF-co-VAm) as fixed-site carrier and 2-(1-piperazinyl) ethylamine sarcosinate (PZEA-Sar) as mobile carrier.	4	340	1449	[81]
Pebax®1657	PEI-MCM-41	Membrane of polyethylenimine functionalized ordered mesoporous silica (PEI-MCM-41) with poly(ether-block-amide) (Pebax® MH 1657) as matrix.	2	298	1521	[82]
Pebax	Pebax/TCP 80 gel membrane	Gel membrane of poly(ether-block-amide) (Pebax) matrix with tricaproin (80 wt.%).	0.06	308	1673.2	[60]
PEO	Silica (in situ)	Composite of polyethylene oxide membrane (PEO) with silica.	2	318	1840	[83]
PPO-PEO-PPO	Silica(insitu)	Membrane formed by poly(propylene oxide)–poly(ethylene oxide)–poly(propylene oxide) (PPO–PEO–PPO) triblock copolymers and silica nanoparticles.	3,5	308	2000	[84]

SPEEK	SO3-MIL-101(Cr)	Sulfonic acid groups functionalized MIL-101(Cr) (metal-organic framework of high thermal and chemical stability) incorporated to sulfonated poly(ether ether ketone) (SPEEK) mixed matrix membrane.	1.5	303	2064	[64]
SPBI-1-EDA/PEI	AIBA-K	Membrane based on sulfonated polybenzimidazole (SPBI), ethylenediamine (EDA), polyethylenimine (PEI), 2-aminoisobutyric acid-potassium salt solution (AIBA-K).	2	373	2539	[85]
PVA/PAA	Silica	Silica mixed matrix membranes of poly(vinyl alcohol) (PVA) and poly(allyamine) (PAA) containing amine carriers and fumed silica (FS).	4,76	380	2901	[86]
PVA/poly(siloxane)	CNTs	Carbon nanotubes (CNTs) dispersed in poly(vinyl alcohol) (PVA) and poly(siloxane) layer.	2	380	>3290	[87]
Teflon AF 1600	Silicalite-1 (MFI)	Surface modification calcined zeolites dispersed homogeneously in solutions of Teflon AF 1600.	1	298	3720	[88]
PIM	(UiO-66-NH$_2$	Membrane of metal organic framework cluster (UiO-66-NH$_2$, [Zr$_6$O$_4$(OH)$_4$] functionalized at 10 wt% with a polymer with intrinsic microporosity (PIM) based on the monomer formed by the synthesis of 5,5',6,6'-tetrahydroxy-3,3,3',3'-tetramethyl-1,10-spirobisindane, dianhydride and 1,4-phenyldiamine.	1	303	3827.3	[46]

PAMAM	EDA	Starburst polyamidoamine (PAMAM) generation 0 dendrimers having ethylenediamine (EDA) as the core.	1.2	298	4100	[89]
PDMS	Zeolite	An amine-terminated polydimethylsiloxane (PDMS) matrix with zeolite.	-	-	~4500	[90]
PTFE	[P$_{4444}$][Gly]	Polytetrafluoroethylene (PTFE) microporous membrane with tetrabutylphosphonium glycinate ([P$_{4444}$][Gly]).	10	373	5000	[91]
PVA/PAA	AIBA-K	Membrane of poly(vinyl alcohol) (PVA) with poly(allylamine) and 2-aminoisobutyric acid-potassium salt (AIBA-K).	2	393	8200	[92]
PTFE	[P$_{2225}$][Gly]	Polytetrafluoroethylene (PTFE) microporous membrane with triethyl(pentyl)phosphonium glycinate ([P$_{2225}$][Gly]).	10	373	~11200	[93]
PMP	Fumed silica	Poly(4-methyl-2-pentyne) (PMP) crosslinked using 4,4'-(hexafluoroisopropylidene) diphenyl azide (HFBAA) matrix with fumed silica.	2	308	11250	[63]
PIM	SBF	Membrane of 2,2',3,3'-tetrahydroxy-9,9'-spirobifluorene based polymer (polymer of intrinsic microporosity, PIM) with spirobifluorene (SBF) as the main structural unit.	1	298	13900	[94]

Materials Research Forum LLC
https://doi.org/10.21741/9781644902639-5

PIM-1	SWCNTs	Polymer of intrinsic microporosity (PIM-1) matrix with single-walled carbon nanotubes (*SWCNT*).	2	303	15721	[95]
PTMSP	PAFs	Membranes composed on poly(1-trimethylsilyl-1-propyne) (PTMSP) and porous aromatic frameworks (PAFs) (based on tetrakis(4-bromophenyl) methane and nanoparticles of Li_6C_{60}).	1	298	50600	[61]
PVAm	HT	Hydrotalcite (HT) channel in a polyvinylamine (PVAm) fixed carrier membrane.	1.1	298	159350	[62]
Cellulose acetate	MWCNTs	Membrane formed by cellulose acetate with multi-walled carbon nanotubes (*MWCNTs*).	3	298	185417	[96]
PEIE	HT	Membrane of hydrotalcite (HT) and copolymer of polyethyleneimine (PEI) and epichlorohydrin.	1.1	298	284650	[62]
PS-TMC	DAmBS-DGBAmE	Composite membrane with sodium 3,5-diaminobenzoate (DAmBS) and diethylene glycol bis(3-aminopropyl) ether (DGBAmE) in an aqueous phase, trimesoyl chloride (TMC) as the monomer of the organic phase and polysulfone (PS) as the support membrane.	1	298	583100	[97]

3. Conclusions and future perspectives

The use of polymeric systems in CO_2 collectors is increasing for several reasons, where it stands up its versatility and low cost in comparison with other materials. Particularly, polymeric systems based on polyvinylamin (PVAm) shows the highest values of CO_2 permeability such as it was observed in Table 1 for PVAm-MC, PVAm-MEA and PVAm-PIP/PS whose values reached 59000, 79800 and 195000 Barrer, respectively.

The preparation of composites can increase the spectra of properties or provide other functionalities. In this sense, it was observed that other polymeric matrices can develops higher permeability values such as it was observed in Table 2 for systems based on cellulose acetate, polyethyleneimine (PEI) and polysulfone reaching values up to 185417, 284650 and 583100 Barrer, respectively.

Consequently, the new developments in terms of materials for CO_2 collectors probably will be focused on polymeric composites due to those systems can substantially change their properties depending on the constituent added and the type of polymer used. In addition, the ratio between polymer and filler, the type of filler used and the type of processing will define the final properties.

Finally, the development of these systems will be focused on the use of green chemistry for its preparation, and taking into account the recyclability of the structures after use.

References

[1] S. Wang, X. Li, H. Wu, Z. Tian, Q. Xin, G. He, D. Peng, S. Chen, Y. Yin, Z. Jiang, M.D. Guiver, Advances in high permeability polymer-based membrane materials for CO2 separations, Energy Environ. Sci. 9 (2016) 1863-1890. https://doi.org/10.1039/C6EE00811A

[2] Z. Liu, Z. Deng, S.J. Davis, C. Giron, P. Ciais, Monitoring global carbon emissions in 2021, Nat. Rev. Earth Environ. 3 (2022) 217-219. https://doi.org/10.1038/s43017-022-00285-w

[3] Z. Liu, P. Ciais, Z. Deng, R. Lei, S.J. Davis, S. Feng, B. Zheng, D. Cui, X. Dou, B. Zhu, R. Guo, P. Ke, T. Sun, C. Lu, P. He, Y. Wang, X. Yue, Y. Wang, Y. Lei, H. Zhou, Z. Cai, Y. Wu, R. Guo, T. Han, J. Xue, O. Boucher, E. Boucher, F. Chevallier, K. Tanaka, Y. Wei, H. Zhong, C. Kang, N. Zhang, B. Chen, F. Xi, M. Liu, F.M. Breon, Y. Lu, Q. Zhang, D. Guan, P. Gong, D.M. Kammen, K. He, H.J. Schellnhuber, Near-real-time monitoring of global CO2 emissions reveals the effects of the COVID-19 pandemic, Nat Commun. 11 (2020) 5172. https://doi.org/10.1038/s41467-020-20254-5

[4] Z. Liu, P. Ciais, Z. Deng, S.J. Davis, B. Zheng, Y. Wang, D. Cui, B. Zhu, X. Dou, P. Ke, T. Sun, R. Guo, H. Zhong, O. Boucher, F.M. Breon, C. Lu, R. Guo, J. Xue, E. Boucher, K. Tanaka, F. Chevallier, Carbon Monitor, a near-real-time daily dataset of global CO2 emission from fossil fuel and cement production, Sci. Data 7 (2020) 392. https://doi.org/10.1038/s41597-020-00708-7

[5] S. Mane, Z.-Y. Gao, Y.-X. Li, D.-M. Xue, X.-Q. Liu, L.-B. Sun, Fabrication of microporous polymers for selective CO2 capture: the significant role of crosslinking and crosslinker length, J. Mater. Chem. A 5 (2017) 23310-23318. https://doi.org/10.1039/C7TA07188D

[6] P. Styring, Carbon Dioxide Capture Agents and Processes, in: P. Styring, E. A. Quadrelli, K. Armstrong, editors. Carbon Dioxide Utilisation: Elsevier; 2015, pp. 19-32. https://doi.org/10.1016/B978-0-444-62746-9.00002-5

[7] T. Nagy, K. Koczka, E. Haáz, A.J. Tóth, L. Rácz, P. Mizsey, Efficiency Improvement of CO2 Capture, Period. Polytech. Chem. Eng. (2016).

[8] Information on https://www.plasticseurope.org/en/resources/publications/4312-plastics-facts-2020

[9] L. Pires da Mata Costa, D. Micheline Vaz de Miranda, A.C. Couto de Oliveira, L. Falcon, M. Stella Silva Pimenta, I. Guilherme Bessa, S. Juarez Wouters, M.H.S. Andrade, J.C. Pinto, Capture and Reuse of Carbon Dioxide (CO2) for a Plastics Circular Economy: A Review, Processes 9 (2021). https://doi.org/10.3390/pr9050759

[10] W. Wang, M. Zhou, D. Yuan, Carbon dioxide capture in amorphous porous organic polymers, J. Mater. Chem. A 5 (2017) 1334-1347. https://doi.org/10.1039/C6TA09234A

[11] H. Gao, Q. Li, S. Ren, Progress on CO2 capture by porous organic polymers, Curr. Opin. Green Sustain. Chem. 16 (2019) 33-38. https://doi.org/10.1016/j.cogsc.2018.11.015

[12] N. Huang, G. Day, X. Yang, H. Drake, H.-C. Zhou, Engineering porous organic polymers for carbon dioxide capture, Sci. China Chem. 60 (2017) 1007-1014. https://doi.org/10.1007/s11426-017-9084-7

[13] Y. Han, W.S.W. Ho, Mitigated carrier saturation of facilitated transport membranes for decarbonizing dilute CO2 sources: An experimental and techno-economic study, JMS Letters 2 (2022). https://doi.org/10.1016/j.memlet.2022.100014

[14] Y. Han, W.S.W. Ho, Polymeric membranes for CO2 separation and capture, J. Membr. Sci. 628 (2021). https://doi.org/10.1016/j.memsci.2021.119244

[15] B. Xue, X. Li, L. Gao, M. Gao, Y. Wang, L. Jiang, CO2-selective free-standing membrane by self-assembly of a UV-crosslinkable diblock copolymer, J. Mater. Chem. 22 (2012). https://doi.org/10.1039/c2jm31037f

[16] M. Sandru, S.H. Haukebø, M.-B. Hägg, Composite hollow fiber membranes for CO2 capture, J. Membr. Sci. 346 (2010) 172-186. https://doi.org/10.1016/j.memsci.2009.09.039

[17] S. Yuan, Z. Wang, Z. Qiao, M. Wang, J. Wang, S. Wang, Improvement of CO2/N2 separation characteristics of polyvinylamine by modifying with ethylenediamine, J. Membr. Sci. 378 (2011) 425-437. https://doi.org/10.1016/j.memsci.2011.05.023

[18] Z. Ma, Z. Qiao, Z. Wang, X. Cao, Y. He, J. Wang, S. Wang, CO2 separation enhancement of the membrane by modifying the polymer with a small molecule containing amine and ester groups, RSC Adv. 4 (2014). https://doi.org/10.1039/c4ra01107d

[19] Z. Qiao, Z. Wang, S. Yuan, J. Wang, S. Wang, Preparation and characterization of small molecular amine modified PVAm membranes for CO2/H2 separation, J. Membr. Sci. 475 (2015) 290-302. https://doi.org/10.1016/j.memsci.2014.10.034

[20] B.S. Ghanem, R. Swaidan, E. Litwiller, I. Pinnau, Ultra-microporous triptycene-based polyimide membranes for high-performance gas separation, Adv. Mater. 26 (2014) 3688-3692. https://doi.org/10.1002/adma.201306229

[21] W. Zheng, Z. Liu, R. Ding, Y. Dai, X. Li, X. Ruan, G. He, Constructing continuous and fast transport pathway by highly permeable polymer electrospun fibers in composite membrane to improve CO2 capture, Sep. Purif. Technol. 285 (2022). https://doi.org/10.1016/j.seppur.2021.120332

[22] S.H. Ahn, S.J. Kim, D.K. Roh, H.-K. Lee, B. Jung, J.H. Kim, Controlling gas permeability of a graft copolymer membrane using solvent vapor treatment, Macromol. Res. 22 (2013) 160-164. https://doi.org/10.1007/s13233-014-2014-0

[23] I. Taniguchi, T. Kai, S. Duan, S. Kazama, H. Jinnai, A compatible crosslinker for enhancement of CO2 capture of poly(amidoamine) dendrimer-containing polymeric membranes, J. Membr. Sci. 475 (2015) 175-183. https://doi.org/10.1016/j.memsci.2014.10.015

[24] Y. Zhao, W.S. Winston Ho, Steric hindrance effect on amine demonstrated in solid polymer membranes for CO2 transport, J. Membr. Sci. 415-416 (2012) 132-138. https://doi.org/10.1016/j.memsci.2012.04.044

[25] S.C. Li, Z. Wang, C.X. Zhang, M.M. Wang, F. Yuan, J.X. Wang, S.C. Wang, Interfacially polymerized thin film composite membranes containing ethylene oxide groups for CO2 separation, J. Membr. Sci. 436 (2013) 121-131. https://doi.org/10.1016/j.memsci.2013.02.038

[26] D.F. Sanders, R. Guo, Z.P. Smith, K.A. Stevens, Q. Liu, J.E. McGrath, D.R. Paul, B.D. Freeman, Influence of polyimide precursor synthesis route and ortho-position functional group on thermally rearranged (TR) polymer properties: Pure gas permeability and selectivity, J. Membr. Sci. 463 (2014) 73-81. https://doi.org/10.1016/j.memsci.2014.03.032

[27] T. Sakaguchi, F. Katsura, A. Iwase, T. Hashimoto, CO2-permselective membranes of crosslinked poly(vinyl ether)s bearing oxyethylene chains, Polymer 55 (2014) 1459-1466. https://doi.org/10.1016/j.polymer.2014.02.012

[28] H. Lin, E. Wagner, J. Swinnea, B. Freeman, S. Pas, A. Hill, S. Kalakkunnath, D. Kalika, Transport and structural characteristics of crosslinked poly(ethylene oxide) rubbers, J. Membr. Sci. 276 (2006) 145-161. https://doi.org/10.1016/j.memsci.2005.09.040

[29] S. Li, H.J. Jo, S.H. Han, C.H. Park, S. Kim, P.M. Budd, Y.M. Lee, Mechanically robust thermally rearranged (TR) polymer membranes with spirobisindane for gas

separation, J. Membr. Sci. 434 (2013) 137-147.
https://doi.org/10.1016/j.memsci.2013.01.011

[30] Y. Weng, W. Ji, C. Ye, H. Dong, Z. Gao, J. Li, C. Luo, X. Ma, Simultaneously enhanced CO2 permeability and CO2/N2 selectivity at sub-ambient temperature from two novel functionalized intrinsic microporous polymers, J. Membr. Sci. 644 (2022). https://doi.org/10.1016/j.memsci.2021.120086

[31] J. Xia, S. Liu, C.H. Lau, T.-S. Chung, Liquidlike poly(ethylene glycol) supported in the organic-inorganic matrix for CO2 removal, Macromolecules 44 (2011) 5268-5280. https://doi.org/10.1021/ma200885k

[32] Y. Sun, M. Gou, Highly efficient of CO2/CH4 separation performance via the pebax membranes with multi-functional polymer nanotubes, Microporous Mesoporous Mater. 342 (2022). https://doi.org/10.1016/j.micromeso.2022.112120

[33] V.A. Kusuma, B.D. Freeman, S.L. Smith, A.L. Heilman, D.S. Kalika, Influence of TRIS-based co-monomer on structure and gas transport properties of cross-linked poly(ethylene oxide), J. Membr. Sci. 359 (2010) 25-36. https://doi.org/10.1016/j.memsci.2010.01.049

[34] W. Yave, A. Car, S.S. Funari, S.P. Nunes, K.-V. Peinemann, CO2-philic polymer membrane with extremely high separation performance, Macromolecules 43 (2009) 326-333. https://doi.org/10.1021/ma901950u

[35] Y. Zhang, L. Ma, Y. Lv, T. Tan, Facile manufacture of COF-based mixed matrix membranes for efficient CO2 separation, Chem. Eng. J. 430 (2022). https://doi.org/10.1016/j.cej.2021.133001

[36] P. Li, Z. Wang, Y. Liu, S. Zhao, J. Wang, S. Wang, A synergistic strategy via the combination of multiple functional groups into membranes towards superior CO2 separation performances, J. Membr. Sci. 476 (2015) 243-255. https://doi.org/10.1016/j.memsci.2014.11.050

[37] J.M.P. Scofield, P.A. Gurr, J. Kim, Q. Fu, S.E. Kentish, G.G. Qiao, Development of novel fluorinated additives for high performance CO2 separation thin-film composite membranes, J. Membr. Sci. 499 (2016) 191-200. https://doi.org/10.1016/j.memsci.2015.10.035

[38] R. Swaidan, B.S. Ghanem, E. Litwiller, I. Pinnau, Pure- and mixed-gas CO2/CH4 separation properties of PIM-1 and an amidoxime-functionalized PIM-1, J. Membr. Sci. 457 (2014) 95-102. https://doi.org/10.1016/j.memsci.2014.01.055

[39] Z. Wang, D. Wang, J. Jin, Microporous polyimides with rationally designed chain structure achieving high performance for gas separation, Macromolecules 47 (2014) 7477-7483. https://doi.org/10.1021/ma5017506

[40] F. Yuan, Z. Wang, S. Li, J. Wang, S. Wang, Formation-structure-performance correlation of thin film composite membranes prepared by interfacial polymerization

for gas separation, J. Membr. Sci. 421-422 (2012) 327-341.
https://doi.org/10.1016/j.memsci.2012.07.035

[41] X. Ma, O. Salinas, E. Litwiller, I. Pinnau, Novel spirobifluorene- and dibromospirobifluorene-based polyimides of intrinsic microporosity for gas separation applications, Macromolecules 46 (2013) 9618-9624. https://doi.org/10.1021/ma402033z

[42] J. Liu, S. Zhang, D.-e. Jiang, C.M. Doherty, A.J. Hill, C. Cheng, H.B. Park, H. Lin, Highly polar but amorphous polymers with robust membrane CO2/N2 separation performance, Joule 3 (2019) 1881-1894. https://doi.org/10.1016/j.joule.2019.07.003

[43] J. Shen, J. Qiu, L. Wu, C. Gao, Facilitated transport of carbon dioxide through poly(2-N,N-dimethyl aminoethyl methacrylate-co-acrylic acid sodium) membrane, Sep. Purif. Technol. 51 (2006) 345-351. https://doi.org/10.1016/j.seppur.2006.02.015

[44] Y.F. Yeong, H. Wang, K. Pallathadka Pramoda, T.-S. Chung, Thermal induced structural rearrangement of cardo-copolybenzoxazole membranes for enhanced gas transport properties, J. Membr. Sci. 397-398 (2012) 51-65. https://doi.org/10.1016/j.memsci.2012.01.010

[45] J.L. Santiago-García, C. Álvarez, F. Sánchez, J.G. de la Campa, Gas transport properties of new aromatic polyimides based on 3,8-diphenylpyrene-1,2,6,7-tetracarboxylic dianhydride, J. Membr. Sci. 476 (2015) 442-448. https://doi.org/10.1016/j.memsci.2014.12.007

[46] A. Husna, I. Hossain, O. Choi, S.M. Lee, T.H. Kim, Efficient CO2 separation using a PIM-PI-functionalized UiO-66 MOF incorporated mixed matrix membrane in a PIM-PI-1 polymer, Macromol. Mater. Eng. 306 (2021). https://doi.org/10.1002/mame.202100298

[47] S. Kim, S.H. Han, Y.M. Lee, Thermally rearranged (TR) polybenzoxazole hollow fiber membranes for CO2 capture, J. Membr. Sci. 403-404 (2012) 169-178. https://doi.org/10.1016/j.memsci.2012.02.041

[48] N. Du, M.M. Dal-Cin, G.P. Robertson, M.D. Guiver, Decarboxylation-induced cross-linking of polymers of intrinsic microporosity (PIMs) for membrane gas separation, Macromolecules 45 (2012) 5134-5139. https://doi.org/10.1021/ma300751s

[49] C. Makhloufi, D. Roizard, E. Favre, Reverse selective NH3/CO2 permeation in fluorinated polymers using membrane gas separation, J. Membr. Sci. 441 (2013) 63-72. https://doi.org/10.1016/j.memsci.2013.03.048

[50] Y. Liu, S. Yu, H. Wu, Y. Li, S. Wang, Z. Tian, Z. Jiang, High permeability hydrogel membranes of chitosan/poly ether-block-amide blends for CO2 separation, J. Membr. Sci. 469 (2014) 198-208. https://doi.org/10.1016/j.memsci.2014.06.050

[51] Y. Lee, C.Y. Chuah, J. Lee, T.-H. Bae, Effective functionalization of porous polymer fillers to enhance CO2/N2 separation performance of mixed-matrix membranes, J. Membr. Sci. 647 (2022). https://doi.org/10.1016/j.memsci.2022.120309

[52] E. Akbarzadeh, A. Shockravi, V. Vatanpour, High performance compatible thiazole-based polymeric blend cellulose acetate membrane as selective CO2 absorbent and molecular sieve, Carbohydr. Polym. 252 (2021) 117215. https://doi.org/10.1016/j.carbpol.2020.117215

[53] R. Swaidan, M. Al-Saeedi, B. Ghanem, E. Litwiller, I. Pinnau, Rational design of intrinsically ultramicroporous polyimides containing bridgehead-substituted triptycene for highly selective and permeable gas separation membranes, Macromolecules 47 (2014) 5104-5114. https://doi.org/10.1021/ma5009226

[54] Y. Wu, N. Xing, S. Li, L. Yang, Y. Ren, Y. Liu, X. Liang, Z. Guo, H. Wang, H. Wu, Z. Jiang, In situ knitted microporous polymer membranes for efficient CO2 capture, J. Mater. Chem. A 9 (2021) 2126-2134. https://doi.org/10.1039/D0TA08453K

[55] X. Ma, I. Pinnau, A novel intrinsically microporous ladder polymer and copolymers derived from 1,1′,2,2′-tetrahydroxy-tetraphenylethylene for membrane-based gas separation, Polym. Chem. 7 (2016) 1244-1248. https://doi.org/10.1039/C5PY01796C

[56] Y. Zhao, W.S.W. Ho, CO2-selective membranes containing sterically hindered amines for CO2/H2 separation, Ind. Eng. Chem. Res. 52 (2012) 8774-8782. https://doi.org/10.1021/ie301397m

[57] M. Carta, R. Malpass-Evans, M. Croad, Y. Rogan, J.C. Jansen, P. Bernardo, F. Bazzarelli, N.B. McKeown, An efficient polymer molecular sieve for membrane gas separations, Science 339 (2013) 303-307. https://doi.org/10.1126/science.1228032

[58] W. Han, C. Zhang, M. Zhao, F. Yang, Y. Yang, Y. Weng, Post-modification of PIM-1 and simultaneously in situ synthesis of porous polymer networks into PIM-1 matrix to enhance CO2 separation performance, J. Membr. Sci. 636 (2021). https://doi.org/10.1016/j.memsci.2021.119544

[59] Z. Qiao, Z. Wang, C. Zhang, S. Yuan, Y. Zhu, J. Wang, S. Wang, PVAm-PIP/PS composite membrane with high performance for CO2/N2 separation, AIChE J. 59 (2013) 215-228. https://doi.org/10.1002/aic.13781

[60] Y. Wu, D. Zhao, S. Chen, J. Ren, K. Hua, H. Li, M. Deng, The effect of structure change from polymeric membrane to gel membrane on CO2 separation performance, Sep. Purif. Technol. 261 (2021). https://doi.org/10.1016/j.seppur.2020.118243

[61] C.H. Lau, K. Konstas, C.M. Doherty, S. Kanehashi, B. Ozcelik, S.E. Kentish, A.J. Hill, M.R. Hill, Tailoring physical aging in super glassy polymers with functionalized porous aromatic frameworks for CO2 capture, Chem. Mater. 27 (2015) 4756-4762. https://doi.org/10.1021/acs.chemmater.5b01537

[62] J. Liao, Z. Wang, C. Gao, S. Li, Z. Qiao, M. Wang, S. Zhao, X. Xie, J. Wang, S. Wang, Fabrication of high-performance facilitated transport membranes for CO2 separation, Chem. Sci. 5 (2014) 2843-2849. https://doi.org/10.1039/C3SC53334D

[63] L. Shao, J. Samseth, M.-B. Hägg, Crosslinking and stabilization of nanoparticle filled PMP nanocomposite membranes for gas separations, J. Membr. Sci. 326 (2009) 285-292. https://doi.org/10.1016/j.memsci.2008.09.053

[64] Q. Xin, T. Liu, Z. Li, S. Wang, Y. Li, Z. Li, J. Ouyang, Z. Jiang, H. Wu, Mixed matrix membranes composed of sulfonated poly(ether ether ketone) and a sulfonated metal-organic framework for gas separation, J. Membr. Sci. 488 (2015) 67-78. https://doi.org/10.1016/j.memsci.2015.03.060

[65] Y. Shi, J. Liang, B. Babu Shrestha, Z. Wang, Y. Zhang, J. Jin, Enhancing the CO2 plasticization resistance of thin polymeric membranes by designing Metal-polymer complexes, Sep. Purif. Technol. 289 (2022). https://doi.org/10.1016/j.seppur.2022.120699

[66] Z. Wang, H. Chen, Y. Wang, J. Chen, M.A. Arnould, B. Hu, I. Popovs, S.M. Mahurin, S. Dai, Polymer-grafted porous silica nanoparticles with enhanced CO2 permeability and mechanical performance, ACS Appl. Mater. Interfaces 13 (2021) 27411-27418. https://doi.org/10.1021/acsami.1c04342

[67] M.L. Sforça, I.V.P. Yoshida, S.P. Nunes, Organic-inorganic membranes prepared from polyether diamine and epoxy silane, J. Membr. Sci. 159 (1999) 197-207. https://doi.org/10.1016/S0376-7388(99)00059-9

[68] H. Kim, C. Lim, S.-I. Hong, Gas permeation properties of organic-inorganic hybrid membranes prepared from hydroxyl-terminated polyether and 3-isocyanatopropyltriethoxysilane, J. Sol-Gel Sci. Technol. 36 (2005) 213-221. https://doi.org/10.1007/s10971-005-3782-y

[69] X. Li, M. Wang, S. Wang, Y. Li, Z. Jiang, R. Guo, H. Wu, X. Cao, J. Yang, B. Wang, Constructing CO2 transport passageways in Matrimid® membranes using nanohydrogels for efficient carbon capture, J. Membr. Sci. 474 (2015) 156-166. https://doi.org/10.1016/j.memsci.2014.10.003

[70] N. Habib, O. Durak, M. Zeeshan, A. Uzun, S. Keskin, A novel IL/MOF/polymer mixed matrix membrane having superior CO2/N2 selectivity, J. Membr. Sci. 658 (2022). https://doi.org/10.1016/j.memsci.2022.120712

[71] M.L. Chua, L. Shao, B.T. Low, Y. Xiao, T.-S. Chung, Polyetheramine-polyhedral oligomeric silsesquioxane organic-inorganic hybrid membranes for CO2/H2 and CO2/N2 separation, J. Membr. Sci. 385-386 (2011) 40-48. https://doi.org/10.1016/j.memsci.2011.09.008

[72] Z. Qiao, S. Zhao, M. Sheng, J. Wang, S. Wang, Z. Wang, C. Zhong, M.D. Guiver, Metal-induced ordered microporous polymers for fabricating large-area gas separation

membranes, Nat. Mater. 18 (2019) 163-168. https://doi.org/10.1038/s41563-018-0221-3

[73] H. Cong, X. Hu, M. Radosz, Y. Shen, Brominated poly(2,6-diphenyl-1,4-phenylene oxide) and its silica nanocomposite membranes for gas separation, Ind. Eng. Chem. Res. 46 (2007) 2567-2575. https://doi.org/10.1021/ie061494x

[74] A.S.L. Gouveia, E. Bumenn, K. Rohtlaid, A. Michaud, T.M. Vieira, V.D. Alves, L.C. Tomé, C. Plesse, I.M. Marrucho, Ionic liquid-based semi-interpenetrating polymer network (sIPN) membranes for CO2 separation, Sep. Purif. Technol. 274 (2021). https://doi.org/10.1016/j.seppur.2021.118437

[75] S. Matteucci, R.D. Raharjo, V.A. Kusuma, S. Swinnea, B.D. Freeman, Gas permeability, solubility, and diffusion coefficients in 1,2-polybutadiene containing magnesium oxide, Macromolecules 41 (2008) 2144-2156. https://doi.org/10.1021/ma702459k

[76] H. Bai, W.S.W. Ho, Carbon dioxide-selective membranes for high-pressure synthesis gas purification, Ind. Eng. Chem. Res. 50 (2011) 12152-12161. https://doi.org/10.1021/ie2007592

[77] S.-T. Fan, M. Tan, W.-T. Liu, B.-J. Li, S. Zhang, MOF-layer composite polyurethane membrane increasing both selectivity and permeability: Pushing commercial rubbery polymer membranes to be attractive for CO2 separation, Sep. Purif. Technol. 297 (2022). https://doi.org/10.1016/j.seppur.2022.121452

[78] L. Hao, P. Li, T. Yang, T.-S. Chung, Room temperature ionic liquid/ZIF-8 mixed-matrix membranes for natural gas sweetening and post-combustion CO2 capture, J. Membr. Sci. 436 (2013) 221-231. https://doi.org/10.1016/j.memsci.2013.02.034

[79] W.S. Chi, S.J. Kim, S.J. Lee, Y.S. Bae, J.H. Kim, Enhanced performance of mixed-matrix membranes through a graft copolymer-directed interface and interaction tuning approach, ChemSusChem 8 (2015) 650-658. https://doi.org/10.1002/cssc.201402677

[80] N. Du, G.P. Robertson, M.M. Dal-Cin, L. Scoles, M.D. Guiver, Polymers of intrinsic microporosity (PIMs) substituted with methyl tetrazole, Polymer 53 (2012) 4367-4372. https://doi.org/10.1016/j.polymer.2012.07.055

[81] Y. Han, W. Salim, K.K. Chen, D. Wu, W.S.W. Ho, Field trial of spiral-wound facilitated transport membrane module for CO2 capture from flue gas, J. Membr. Sci. 575 (2019) 242-251. https://doi.org/10.1016/j.memsci.2019.01.024

[82] H. Wu, X. Li, Y. Li, S. Wang, R. Guo, Z. Jiang, C. Wu, Q. Xin, X. Lu, Facilitated transport mixed matrix membranes incorporated with amine functionalized MCM-41 for enhanced gas separation properties, J. Membr. Sci. 465 (2014) 78-90. https://doi.org/10.1016/j.memsci.2014.04.023

[83] J. Xia, S. Liu, T.-S. Chung, Effect of end groups and grafting on the CO_2 separation performance of poly(ethylene glycol) based membranes, Macromolecules 44 (2011) 7727-7736. https://doi.org/10.1021/ma201844y

[84] C.H. Lau, D.R. Paul, T.S. Chung, Molecular design of nanohybrid gas separation membranes for optimal CO_2 separation, Polymer 53 (2012) 454-465. https://doi.org/10.1016/j.polymer.2011.12.011

[85] H. Bai, W.S.W. Ho, New carbon dioxide-selective membranes based on sulfonated polybenzimidazole (SPBI) copolymer matrix for fuel cell applications, Ind. Eng. Chem. Res. 48 (2008) 2344-2354. https://doi.org/10.1021/ie800507r

[86] R. Xing, W.S.W. Ho, Crosslinked polyvinylalcohol-polysiloxane/fumed silica mixed matrix membranes containing amines for CO_2/H_2 separation, J. Membr. Sci. 367 (2011) 91-102. https://doi.org/10.1016/j.memsci.2010.10.039

[87] Y. Zhao, B.T. Jung, L. Ansaloni, W.S.W. Ho, Multiwalled carbon nanotube mixed matrix membranes containing amines for high pressure CO_2/H_2 separation, J. Membr. Sci. 459 (2014) 233-243. https://doi.org/10.1016/j.memsci.2014.02.022

[88] G. Golemme, A. Bruno, R. Manes, D. Muoio, Preparation and properties of superglassy polymers - zeolite mixed matrix membranes, Desalination 200 (2006) 440-442. https://doi.org/10.1016/j.desal.2006.03.396

[89] A.S. Kovvali, H. Chen, K.K. Sirkar, Dendrimer membranes: A CO_2-selective molecular gate, J. Am. Chem. Soc. 122 (2000) 7594-7595. https://doi.org/10.1021/ja0013071

[90] Ş.B. Tantekin-Ersolmaz, Ç. Atalay-Oral, M. Tatlıer, A. Erdem-Şenatalar, B. Schoeman, J. Sterte, Effect of zeolite particle size on the performance of polymer-zeolite mixed matrix membranes, J. Membr. Sci. 175 (2000) 285-288. https://doi.org/10.1016/S0376-7388(00)00423-3

[91] S. Kasahara, E. Kamio, T. Ishigami, H. Matsuyama, Amino acid ionic liquid-based facilitated transport membranes for CO_2 separation, Chem. Commun. (Camb) 48 (2012) 6903-6905. https://doi.org/10.1039/c2cc17380h

[92] J. Zou, W.S.W. Ho, CO_2-selective polymeric membranes containing amines in crosslinked poly(vinyl alcohol), J. Membr. Sci. 286 (2006) 310-321. https://doi.org/10.1016/j.memsci.2006.10.013

[93] S. Kasahara, E. Kamio, H. Matsuyama, Improvements in the CO_2 permeation selectivities of amino acid ionic liquid-based facilitated transport membranes by controlling their gas absorption properties, J. Membr. Sci. 454 (2014) 155-162. https://doi.org/10.1016/j.memsci.2013.12.009

[94] C.G. Bezzu, M. Carta, A. Tonkins, J.C. Jansen, P. Bernardo, F. Bazzarelli, N.B. McKeown, A spirobifluorene-based polymer of intrinsic microporosity with improved

performance for gas separation, Adv. Mater. 24 (2012) 5930-5933.
https://doi.org/10.1002/adma.201202393

[95] M.M. Khan, V. Filiz, G. Bengtson, S. Shishatskiy, M.M. Rahman, J. Lillepaerg, V. Abetz, Enhanced gas permeability by fabricating mixed matrix membranes of functionalized multiwalled carbon nanotubes and polymers of intrinsic microporosity (PIM), J. Membr. Sci. 436 (2013) 109-120.
https://doi.org/10.1016/j.memsci.2013.02.032

[96] A.L. Ahmad, Z.A. Jawad, S.C. Low, S.H.S. Zein, A cellulose acetate/multi-walled carbon nanotube mixed matrix membrane for CO2/N2 separation, J. Membr. Sci. 451 (2014) 55-66. https://doi.org/10.1016/j.memsci.2013.09.043

[97] M. Wang, Z. Wang, S. Li, C. Zhang, J. Wang, S. Wang, A high performance antioxidative and acid resistant membrane prepared by interfacial polymerization for CO2 separation from flue gas, Energy Environ. Sci. 6 (2013) 539-551.
https://doi.org/10.1039/C2EE23080A

New Materials for a Circular Economy   Materials Research Forum LLC
Materials Research Foundations 149 (2023) 173-201   https://doi.org/10.21741/9781644902639-6

Chapter 6

Smart Anticorrosive Polymeric Coatings for Prolonging the Lifetime of Metallic Structures

Reza Naderi*, Najmeh Asadi

School of Metallurgy and Materials Engineering, College of Engineering, University of Tehran, Tehran, Iran

rezanaderi@ut.ac.ir

Abstract

Organic coatings as a physical barrier are applied on metallic substrates to suppress the metal deterioration against corrosive media. When the coatings are scratched or damaged, no active corrosion protection is provided. In other words, when the barrier properties fail, protection performance decreases and corrosion reactions occur on the metallic substrate surface. To overcome the problem, corrosion inhibitors have been proposed to be included directly or indirectly. One of the most effective strategies to develop smart coatings is to incorporate inhibitor-container into the formulation. Layers by layers (LbLs), layered double hydroxides (LDHs) with anion exchange capability, halloysite nanotubes (HNTs), particles with cation exchange capability such as clay montmorillonite, zeolite, etc. are some of the reservoirs that will discuss in this chapter. The containers carry inhibitor and liberate it on-demand. So, the protection performance of the coatings is guaranteed for the longer times.

Keywords

Smart Coatings, Inhibitor, Organic Coating, Active Corrosion Protection, Nano Reservoir

Contents

1.   Corrosion Protection by Polymeric Coatings

A corrosion phenomenon is a serious problem in industrial environment making enormous cost for maintenance and repair of the metallic structures. Some solutions for reducing the corrosion rate of the metallic structures have been suggested such as selection of corrosion resistive metal/alloy [1], cathodic protection [2–4], anodic protection for the metal with active/passive behavior [5-7], utilizing corrosion inhibitors [8-10] and applying protective coating on the metallic structures [11-15]. For many years, organic coatings have been used as a reliable protection method for metal structures, especially steel structures [16-18]. Polymeric coatings have been served to protect metals from corrosion deterioration by making a physical barrier against aggressive agents [19]. Protection performance of the coatings strongly depends on crosslinking density of the polymeric network [20]. Another parameter which could be effective on the barrier property of the coatings is the film's thickness as the higher thickness provides modified corrosion protection. Generally, the protective performance of the organic coatings is based on two main properties: barrier (for water, oxygen), and blocking of the ionic paths between cathodic and anodic sites along the polymer/metal interface. The barrier properties refer to the passage of aggressive species through the film, toward the metallic substrate [19]. As a result, the polymer layers act as a barrier between the metallic substrate and water and ion species, preventing direct contact between them. However, the coating may degrade due to external factors such as UV radiation, temperature, and mechanical actions (scratches or cracks). Due to the penetration of water and aggressive species into the polymeric network, pores and micro-cracks propagate through the network, resulting in the initiation of corrosion [21].

Polymeric coatings have the disadvantage of permeability to corrosive species such as oxygen, water, and ions [19, 21]. On the other hand, when a coating is damaged, conductive pathways is developed and barrier property of the coating is reduced.

In fact, organic coatings postpone the reach of aggressive agents to the metal/coating interface where the electrochemical reactions take place. In the other words, organic coatings act only as a physical barrier that prolongs the conductive ways to metal / coating interface, thus delaying the development of electrochemical reactions on the metal surface. Anyway, the coating may be damaged by light, heat, mechanical impacts, the penetration of chemical agents, etc., and such conductive pathways toward metal / coating interface may create. So, it is needed to prolong the protective properties of the coating and to provide active protection.

New Materials for a Circular Economy   Materials Research Forum LLC
Materials Research Foundations 149 (2023) 173-201   https://doi.org/10.21741/9781644902639-6

Anticorrosion pigments can help overcome this problem by being incorporated into coating formulation. So, when water penetrates to the coating films, partial dissolution of pigment and releasing inhibitors causes the formation of a protective layer on metal substrate. For example, Naderi et.al. [22-29], in order to improve the anti-corrosion performance of epoxy coating, introduced phosphate-based pigments. They investigated the corrosion inhibitive effect of zinc phosphate (ZP) [25, 26, 28], zinc aluminum phosphate (ZAP) [29] and zinc aluminum polyphosphate (ZAPP) [25, 26, 30] through various electrochemical tests and surface analysis methods. According to the results, ZAPP and ZP pigments showed respectively the superior and inferior inhibition for the mild steel immersed in 3.5 wt. % NaCl which was due to the development of a protective layer on the surface of metal in the presence of modified pigment particles. Interestingly, ZAPP caused an approximately 95 % improvement in the polarization resistance of the mild steel. They incorporated ZAPP into a solvent born epoxy coating applied on mild steel. The acquired data of coating and charge transfer showed that during 175-day exposure to NaCl solution the best protection was provided by the epoxy coating which was pigmented at lambda 0.6 [25]. Besides, the coating revealed an enhanced cathodic disbanding resistance [24]. Satisfactory inhibitive property of zinc aluminum polyphosphate have been also confirmed by other researchers [32]. Mahdavian and his co-worker [33], assessed the effectiveness of zinc gluconate (ZG), zinc acetylacetonate (ZAA), zinc acetate (ZA) as anticorrosion pigments for protective coatings. According to the results, the inhibitive effect of the used pigments in 3.5 wt. % NaCl followed the order of: ZG > ZAA > ZA. In the case of ZG and ZAA, the increasing in the values of polarization resistance (sum of coating and charge transfer resistances) with the immersion time suggested that probably a protective layer gradually was formed on the mild steel surface. Zubielewicz and his co-worker [17], claimed that introducing the non-toxic passivating pigments of calcium zinc phosphate and zinc ferrite into waterborne resins of urethane and styrene acrylic could successfully passivate the steel surface.

Green hybrid pigments based on plant extracts composed of organic and inorganic compounds have also been reported [34-43]. For boosting the anti-corrosion features of epoxy coating, Salehi and his co-workers [34] made a pigment by urtica dioica (nettle leaves) extract and zinc acetate. As a result of incorporation of the pigment into an epoxy ester coating applied on mild steel, the coating was able to improve its protective performance by providing smart corrosion protection [34]. They stated that the formation of a film which was mainly $Zn(OH)_2$ precipitation in cathodic sites as well as Zn^{2+}-utrica dioica and Fe^{2+}-utrica dioica complexes in anodic sites was responsible for the behavior. The production of green pigments based on the combination of nettle with zinc nitrate [38] and cerium nitrate [37] has been also reported.

The addition of corrosion inhibitors into the coating's formulation can be considered as another strategy for enhancing the anti-corrosion properties of organic coatings. The addition of corrosion inhibitors into the coatings limits the corrosion reaction rate on a metal surface. This goal can be achieved in two ways; direct and indirect addition of inhibiting species into the coating's formulations. The latter one is more attractive for researchers. Such containers are employed to carry inhibitors and suppress the direct

New Materials for a Circular Economy   Materials Research Forum LLC
Materials Research Foundations 149 (2023) 173-201   https://doi.org/10.21741/9781644902639-6

contact of inhibitors with the polymeric network. This is so vital especially when the coating's resin and inhibitor are not compatible [44]. In the direct approach, the presence of inhibitor molecules can create some defects in the coating's network and decrease barrier properties. The creation of defects may result in the inhibitors leach out. For this reason, the employing of containers can help to reserve inhibitors for longer time and release it on-demand based on a trigger. Direct addition of corrosion inhibitors can result in some problems like deteriorated interactions between the corrosion inhibitor molecules and coating compounds, disruption of coating curing, and uncontrollable inhibitor leakage [45, 46]. The direct and indirect addition of inhibitor has been compared by Snihirova et. al. [44]. They introduced cerium nitrate and 2,5-dimercapto-1,3,4-thiadiazolate (DMTA) with and without $CaCO_3$ as container to a water based epoxy coating deposited on an Al substrate. Upon adding cerium nitrate directly to the formulation, the electrochemical impedance spectroscopy (EIS) results indicated a drop in the impedance module. Other report is the work of Zheludkevich et. al. [47] . In order to reduce the negative effects of direct inhibitor addition on the barrier properties of the sol-gel coating applied to an aluminum alloy (2024 AA), they immobilized Ce cations in amorphous ZrO_2 nanoparticles. Therefore, when cerium cations were directly added to the sol-gel matrix, inhibitor release was fast, while it was slower for Ce-doped ZrO_2.

As a consequence, the critical strategy frequently suggested in the literature is to reserve the inhibitor molecules and to release them on demand. Some containers which are commonly used for this purpose are studied in the next section.

2.   Corrosion Inhibitor Reservoir

2.1   Layer by Layer (LbL)

Layer by layers (LbLs) are constructed from an inorganic core coated by opposite charge of polyelectrolyte and entrapping an inhibitor between the layers. There are two types of polyelectrolytes: polyanions and polycations. Polyanions possess numerous ionizable acidic groups including carboxylic acid. Alkaline and neutral solutions ionize these acidic groups. A polyelectrolyte can also contain polycations that contain basic functional groups. Acidic and neutral solutions can ionize polycations. Several layers of polyelectrolytes can be used in LbL method, including polycations and polyanions [45, 48]. When the polyelectrolyte layer becomes uncharged, the attraction between the layer and the loading molecules decreases. As a result of pH-dependent action, some LbL anticorrosion nano-reservoirs release corrosion inhibitors (those loaded on polycation layers).

Ananda Jaysing Jadhav et. al. [49], utilized the inorganic material of zinc phosphate as the core and then deposited the two oppositely charge of polyelectrolytes (polyaniline and polyacrylic acid). This LbL used as container for entrapping of immidazole corrosion inhibitor. Immidazole immobilized between the two layers of polyelectrolyte. The schematic of LbL preparation is presented in Figure 1. They investigated the corrosion inhibition function of the synthesized nanocontainers incorporated in epoxy coating applied on mild steel.

From the release studies in different pHs, it was concluded that, at pH $\geq$ 8.0, a hydrophobic swollen state is produced by deprotonation of carboxylate groups (forming an "open" nano-shell which is penetrable to molecule passage) and at lower pHs (pH < 8.0), the protonation of carboxylate groups of the polymer makes shrunken state. So, in this case, pH is the trigger for releasing of the inhibitor. The schematic of inhibitor release in closed and open state respectively at low and high pH are illustrated in Figure 2. Regarding these interpretations, it can be expected that when these carriers are used in organic coatings, they are able to release the inhibitor based on demand. When the corrosion reactions take place on the surface, pH rises due to the production of hydroxyl groups resulting from the reduction of oxygen, so the inhibitor leaves the carrier and acts on the surface.

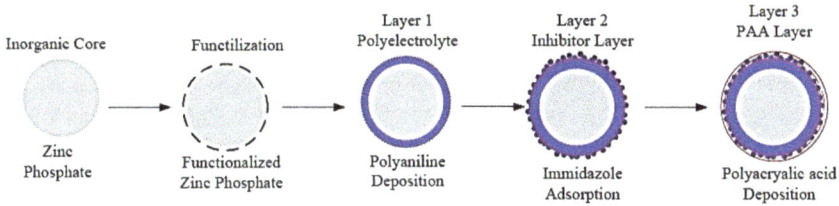

Figure 1. The schematic of formation of LbL with zinc phosphate as in-organic core [49].

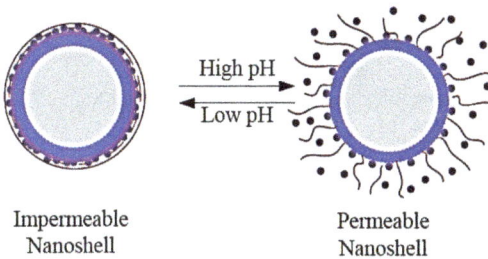

Figure 2. The schematic of liberation of inhibitor from the LbL with zinc phosphate as in-organic core [49].

In another report, Shchukin et. al. [50], utilized SiO_2 as inorganic core then deposited two layers of poly(ethylene imine) (PEI) and poly(styrene sulfonate) (PSS) with oppositely charge. They used benzotriazole (BTA) as corrosion inhibitor. For fabricating this LbL, the following procedure was done; the SiO_2 nanoparticles have negative charge (negative zeta potential), so those can adsorb PEI with positive charge. At the next step, the adsorption of the second negative layer of PSS is possible and at the end, benzotriazole as corrosion inhibitor were adsorbed on the particles. Since BTA is partially soluble in neutral pH water, the third inhibitor layer was adsorption in an acidic solution. To more loading of the inhibitor, the third step was repeated for several times. The electrochemical impedance spectroscopy data clearly showed the negative impact of direct addition of inhibitors to the sol-gel coating. In fact, because of the inhibitor incompatibility with the matrix the barrier properties of the coating have been dropped. This approved when the inhibitor-container incorporated to the coating, a better protection was provided. They stated that due to the increase in pH, surface of the LbL nanoparticles opens and the inhibitor is released.

Bhanvase *et al.* [51], synthesized LbL construction of cerium zinc molybdate as core, polyaniline and polyacrylic acid as polyelectrolyte and imidazole molecules as corrosion inhibitor (CZM/PANI/inhibitor/PAA) and the nanoreservoirs were loaded into an alkyd resin coating. As a result of the electrochemical analysis, the corrosion protection properties of the alkyd coating were considerably improved.

Sonawane and his co-workers have studied the LbL assemblies on ZnO nanoparticles [52]. PANI and PAA were used as polyelectrolyte monolayers, and the corrosion inhibitor (benzotriazole) was loaded between them on the positively charged PANI monolayers. Based on the Tafel plots and corrosion rate analyses, the nanoreservoirs of ZnO/PANI/PAA inhibitor displayed a good anticorrosion efficiency.

Izadi *et. al.* made LbL by Fe_3O_4 as core and nettle leaves extract as green corrosion inhibitor [45]. To prepare the particles, at the first step, Fe_3O_4 nanoparticles was functionalized by myristic acid (MA) negatively charge and then the functionalized Fe_3O_4 were covered by polyaniline (PANI) positive charge. Finally, the inhibitor loading step was done. To facilitate loading of nettle aqueous extract (NE) molecules on the nanoparticles, PANI was utilized as a weak polycation. The layer-by-layer assembled Fe_3O_4/polyaniline nano-carrier could release inhibitor molecules in alkaline and neutral solutions.

2.2   Double Layered Hydroxide (LDH)

The use of double layered hydroxide (LDH) structures is another way for storage of corrosion inhibitors. Their activity is based on the controlled release of anions such as vanadium ions or nitrates intercalated with hydroxide layers [53 – 56].

There is a general formula for LDH, $[M^{2+}_{1-x} M^{3+}_X (OH)_2]^{X+} + [A^{n-}_{X/2}]$ .mH_2O, where the cations M^{2+} and M^{3+} are located in the octahedral holes in a brucite-like layer, and the A^{n-} anion occupies is located in the hydrated interlayer galleries [57].

According to Zheludkevich *et al.* [53], the anions of (divanadate) located between Al/Zn hydroxide layers can be exchanged by chloride anions in the corrosive NaCl medium.

So, LDHs have two major advantages: storage and sustain release of inhibitors providing active corrosion protection and entrapping of the corrosive agents like Cl^- and SO_4^{2-} [58].

New Materials for a Circular Economy   Materials Research Forum LLC
Materials Research Foundations 149 (2023) 173-201   https://doi.org/10.21741/9781644902639-6

Performing EIS test on the aluminum substrates immersed in 0.05 M NaCl solution in the presence and absence of LDH particles, Zheludkevich *et al.* [53] showed that at the beginning of the experiment, only one time constant for the samples immersed in a solution containing LDH nanoparticles was observed in the spectrum, which was correspond to a dense passive film formed on aluminum while in the case of the samples immersed in an electrolyte solution without LDH nanoparticles, the spectrum was more complicated. After one-hour, two time constants were observed. The time constant at high frequencies was associated with the passive film and that at low frequencies was related to the electrochemical phenomena in the interface. After 7 days of immersion, the impedance values of samples immersed in the solution without LDH nanoparticles dropped so rapidly, indicating rapid corrosion. With further immersion, the impedance increased slightly due to the corrosion products forming on the metal surface. The impedance spectra of the samples immersed in the solution containing LDH nanoparticles showed greater stability.

Poznyak *et. al.* [58], produced nanocrystalline plate-like of Zn-Al and Mg-Al LDHs intercalated with quinaldate (QA) and 2-mercaptobenzothiazolate (MBT) organic anions.

The UV-visible and EIS results confirmed the inhibitors release from the LDHs, enhancing the corrosion protection of bare AA2024. Therefore, LDHs was suggested be introduced into organic coatings as nanocontainers for active protection.

The influence of the ions such as Cl^-, CO_3^{2-}, NO^{3-}, SO_4^{2-}, and pH on the release mechanisms and kinetics of the intercalated MoO_4^{2-} from $Zn_2Al/-LDH$ powders during the immersion in aqueous solutions was investigated by Shkirskiy *et. al.* [59]. They stated that all curves of release of inhibitor showed high rate during the initial 10s because of two reasons. First, the fast exchange with species adsorbed on the LDH plates from the air such as carbonates or by the release of MoO_4^{2-} adsorbed on the external surface of LDH. Additionally, in 0.005, 0.05, and 0.5 M NaCl without carbonate (6<pH<10), release was incomplete (less than 40%). By adding carbonates to the NaCl solutions, the complete release (100%) was achieved after 60 min. CO_3^{2-} was released by anion exchange, despite the fact that its concentration in carbonate solution at neutral pH was less than 0.5%. One hour of immersion at an alkaline pH (pH 12) resulted in 100% release. As a result of these conditions, $Zn_2Al/-LDH$ framework partial dissolution occurred along with the ion exchange mechanism. They also studied the effect of inhibitor release from the LDHs incorporated into an organic primer which was applied on hot deep galvanized steel (HDG). EIS and leaching measurements were conducted on the coated HDG with and without $Zn_2Al/- MoO_4^{2-}$ LDH fillers during immersion. Results were evaluated in order to determine the influence of the environment on the release rate of MoO_4^{2-}. According to these studies, Zn dissolution and polymer detachment were lower from hybrid coatings due to the release of MoO_4^{2-} from $Zn_2Al/- MoO_4^{2-}$ controlled by the anion composition of the corrosive solution. The release of MoO_4^{2-} in 0.5 M NaCl solution under air was suppressed for 24 h due to a lower affinity of Cl^- to $Zn_2Al/-LDH$ than MoO_4^{2-}. It was found that 0.1 M $NaHCO_3$ in 0.5 M NaCl provided an immediate release of MoO_4^{2-} and measurable inhibiting action, acting as a potential control tool for LDH hybrid coatings.

New Materials for a Circular Economy   Materials Research Forum LLC
Materials Research Foundations 149 (2023) 173-201   https://doi.org/10.21741/9781644902639-6

Examining the effect of Zn-Al-PO$_4{}^{3-}$ and Zn-Al-NO$_3{}^-$ incorporated into a silane sol-gel film as primer applied on mild steel substrate, Alibakhshi and his co-workers [60], indicated that phosphate loaded nanocontainers provided better inhibition effect than nitrate one. They also applied epoxy-polyamide top coat on the hybrid silane primer containing doped LHDs and concluded that the silane film loaded with Zn-Al-PO$_4{}^{3-}$ nanocontainer significantly reduced the cathodic delamination of epoxy coating on steel substrate. The worst delamination resistance was found for the blank system which was not incorporated with LDHs nanoparticles. In contrary, the best delamination resistance was observed for the coating system with the primer incorporated with Zn-Al-PO$_4{}^{3-}$. Zinc cations as an effective cathodic inhibitor forms an insoluble protective layer on micro-cathodic sites. Here, the zinc cations bring from the solution of LDHs in cathodic area owning high pHs. LDHs are solved at alkaline pHs and liberated zinc cations. On the other hand, phosphate anions released from the Zn-Al-PO$_4{}^{3-}$ LDHs combined with the zinc cations and precipitated in the delamination front. This was the reason why Zn-Al-PO$_4{}^{3-}$ caused the lowest cathodic delamination among the test samples.

Some other studies on LDHs are summarized in Table 1.

Table 1. The summary of the studies on LDHs.

LDH type	Inhibitor	Substrate	Reference
Zn-Al, Mg-Al	Quinaldate, 2-mercaptobenzothiazolate	AA2024	[58]
Zn-Al, Mg-Al	vanadate ions	AA2024	[53]
Mg–Al	tungstate anions	Mg alloy	[61]
Mg-Fe	CO$_3{}^{2-}$	pure Mg	[62]
Zn-Al, Mg-Al	2-benzothiazolylthio-succinic acid	carbon steel	[63]
Zn–Al	Vanadate, 2-mercapto benzothiazole, molybdate, phytic acid, 8-hydroxyquinoline	AA 2024	[64]
Zn-Al	molybdate	carbon steel	[65]
Mg–Al	sodium benzoate, 3-aminopropyltriethoxysilane, 8-hydroxyquinoline	Mg alloy	[66]

According to the literature, LDHs can also be produced in situ as protective films on metallic substrates [54, 55, 57, 67 – 72].

New Materials for a Circular Economy   Materials Research Forum LLC
Materials Research Foundations 149 (2023) 173-201   https://doi.org/10.21741/9781644902639-6

2.3   Mesoporous Materials

The use of mesoporous particles is also one of the ways to reserve the inhibitor in the coating and to control its release [73 – 77]. Among the mesoporous materials, inorganic mesoporous silica materials have amorphous structure consisting of oxygen and silicon atoms. Pore sizes range from 2 to 50 nm. Precursors, surfactants, temperature, solvent, and concentration are all factors that influence mesoporous material formation [78]. It is possible to have either an ordered mesopore structure or a disordered mesopore structure in the mesoporous materials. On an atomic scale, silica structures are disordered, but pore correlations are well organized. Based on the conditions during synthesizing, pores can be cylindrical or spherical [78]. The preparation of mesoporous silica involves the hydrolysis and condensation of silica precursors such as tetraethoxysilane (TEOS), followed by calcination or solvent extraction [78]. It is possible to alter the size, shape, and intermolecular interactions of the mesoporous material to produce hexagonal (MCM-41, SBA-15), cubic (MCM-48), lamellar (MCM-50), and other random arrangements of mesoporous structures [78]. Different synthesis conditions like silica precursor, type of surfactant, ionic strength, pH, temperature, reaction composition, and time of synthesis can impact on the surfactant micellar conformation, the silica-surfactant interactions, and the degree of silica polycondensation. The type of mesostructure material depends on these parameters [78]. The size of MSMs ranges from nano to microscales, with high surface areas (700×1000 m^2/g) and pore volumes (0.6×1 cm^2/g) [78]. The particles can be hydrophilic or hydrophobic based on the surface silanol groups. Therefore, it may be possible to provide various applications with functionalization or loading capabilities [78].

Sulfamethazine drug has been executed as an organic inhibitor which have been loaded into mesopore silica to achieve a smart coating [76]. The loaded particles with the morphology of hallow cylindrical and average pore diameter of 2.9 nm which was reduced to 1.1 nm after doping of the inhibitor could provide acceptable active corrosion protection for epoxy primer applied on mild steel.

For loading of inorganic inhibitor such as molybdate ions, Saremi and Yeganeh [73, 74] functionalized mesoporous silica nanoparticles with 1-(2-aminoethyl)- 3-aminopropyltrimethoxysilane then the intermediate cations of Al^{3+} ($AlCl_3$) were used for absorption of anions of the inhibitor. Finally, they incorporated the prepared loaded meso-porous silica into polypyrrole coating applied on carbon steel. By measuring the zeta potential of the surface of silica particles, they concluded that in acidic pHs, the particles have positive and negative charge in acidic and alkaline pHs, respectively. Hence, in alkaline media such as cathodic sites of corrosion reactions negatively charged molybdate ions is repelled. In fact, when electrochemical reactions occur on the steel surface, the pH rises due to the increase of hydroxyl ions, so the inhibitor is released and the inhibiting film is formed on the metal surface to prevent further metal lost.

They were also, incorporated the particles in alkyd and epoxy primer coatings applied on mild steel [78, 79].

New Materials for a Circular Economy   Materials Research Forum LLC
Materials Research Foundations 149 (2023) 173-201   https://doi.org/10.21741/9781644902639-6

A modification was done on MCM-41 particles to gain a novel nanocontainer for active corrosion protection goals [81]. At first, the MCM-41 particles were modified with carboxyl groups (-COOH) and then doped with benzotriazole (BTA). At the final step, they were wrapped by the highly branched polyethylenimine (PEI). The pH release behavior of the prepared particles revealed that liberating of the inhibitor was accelerated at high pHs or alkaline condition. The proposed mechanism for the behavior was corresponded to decrease of charge density and consequently shrinkage and desorption of PEI polymeric chains at alkaline pHs. The desorption of PEI from the carrier's surface caused the accelerated release of the loaded BTA. Moreover, the anionic characteristic of the nanocontainers networks would readily expand at high pHs, which help the release of BTA molecules. On the other hands, in comparison between neutral and highly acidic conditions, the deionization degree of carboxyl groups on the surface of modified-MCM increased at the lower pHs leading to decrease of the electrostatic attraction force between the MCM and PEI and so slightly more release of BTA. To assess the active action effect, they were embedded as the fabricated containers in epoxy primer coating on mild steel. According to the scratched coatings and salt spray results, when local corrosion attack appeared the containers released the inhibitor to diminish the corrosion damage by adsorption on the metallic surface.

Santa Barbara amorphous (SBA) is another type of mesoporous silica employed as container of inhibitors in organic coatings. Effect of mesoporous silica type SBA-15, functionalized with piperazine (MSBA) then loaded with 2-mercaptobenzothiazoleas (MBT@MSBA) and lanthanum cations (La@MSBA) as organic and inorganic corrosion inhibitors, on the protection performance of epoxy-silicone coating applied on mild steel has been investigated by Amini and his co-workers [77, 81]. For EIS assessment of the active corrosion protection of the epoxy-silicone coatings incorporated with the two-inhibitor loaded modified-SBA containers (C(MBT@MSBA) and C(La@MSBA)), the films applied on mild steel were artificially scratched. During 24 h exposure to sodium chloride solution, the incorporated coating experienced a negligible drop compared to the neat one. The impedance drops during the immersion time for the C(MBT@MSBA), C(La@MSBA) and neat samples were 20, 37 and 74%, respectively, meaning the inhibitors liberated from the modified-SBA containers could help the polymer coating to keep its corrosion protection properties within the exposure period.

In another research [83], the effect of dodecylamine loaded into a highly ordered hexagonal mesoporous silica (SBA-15) with internal diameter of 6-7 nm on the corrosion protection properties of an alkyd primer coating was studied. To examine the coating film active corrosion properties, scanning vibrating electrode technique (SVET) had been utilized. The SVET results showed that in the case of blank coating (without nanoparticles), an anodic peak appeared while no anodic peak was detected for the sample with loaded mesoporous even after 23 h of immersion in 0.05 mol/L NaCl.

Synthesizing superhydrophobic benzotriazole-loaded mesoporous silica film electrodeposited on 2024 Al alloy, some researchers [83, 84] reported that the release of inhibitor in the superhydrophobic film is responsive to pH value. When corrosion reactions occur, inhibitor would be released on-demand, protecting the metal substrates.

2.4 Ion Exchange Inorganic Materials

The cations such as Zn^{2+} and Ce^{3+} loaded in the minerals with ion exchange characteristic can be liberated on-demand. This behavior has many attracted many attentions in the field of smart anti-corrosion coatings [44, 85–87].

Sodium montmorillonite (Na+ MMT), zeolite and talc are some minerals with cation exchange properties that can be introduced to organic coating for smart protection purposes. Beside the barrier effect of the particles, they can reserve the inhibitive cations such as $Zn2+$ and liberate them on-demand.

Na^+ MMT with a lamellar morphology is one of the types of clay aluminosilicate with the general chemical formula of $(Ca,Na,H)(Al,Mg,Fe,Zn)2(Si,Al)4O10(OH)2.xH2O$. In the structure, the central octahedral sheet of alumina sited between the two tetrahedral sheets of silica.

A hexagonal network is formed by silicon-oxygen tetrahedra linked to neighboring tetrahedra by sharing three corners. The octahedral sheet is usually composed of aluminum or magnesium in six-fold coordination with oxygen from the tetrahedral sheet and with hydroxyl. The MMT particles are formed by electrostatic and van der Waals forces between adjacent layers [88-90].

To neutralize the charge, some cations such as Na^+ exist in the interlayer spaces which are able to exchange with inhibitive cations like Zn^{2+} and Ce^{3+}. This feature makes it possible to use montmorillonite as an inhibitor reservoir.

Motte et. al. [86], applied a hybrid silane coating consisting of tetraethylorthosilicate, methyltriethoxysilane, γ -glycidoxypropyltrimethoxysilane on galvanized steel. To improve active corrosion protection of the coating, they incorporated cerium montmorillonite (Ce^{3+} MMT) into the coating. The results of electrochemical impedance spectroscopy of the silanized galvanized steels in 0.1 M sodium chloride solution showed that incorporation of nanoparticles in to the coating's formulation can increase corrosion resistance of the substrate. The charge transfer resistance of the samples coated with the silane + Na^+ MMT composite film was approximately 10 times higher than that for the neat silane film without nanoparticles, and in the case of silane + Ce^{3+} MMT nanoparticles the increase in charge transfer resistance could be increased up to about 100 times. Two time constants were seen in the AC impedance spectra of the coating containing Na^+ MMT. The time constant in high frequencies was related to the silane coating and the time constant at medium frequencies was linked to the interface oxide layer. The resistance of the interface oxide layer was more affected by the nanoparticle than the resistance of the silane film. However, as time spent, the resistance of interface layer decreased. The spectrum related to the samples coated by silane incorporated Ce^{3+}-MMT particles indicated that the coating resistance decreased with increasing immersion time, which was related to the formation of conductive pathways [92-94] while the resistance of the interface layer increased. They claimed that this phenomenon is associated with the release of cerium ions in this area. As mentioned earlier, cerium ions increased the resistance of this layer by forming stable deposits.

The exchange of cerium ions with Na^+ and Zn^{2+} ions in the interface resulted in the release of cerium ions and formation of a stable film. Sodium ions access the interface through the electrolyte diffusion pathways and zinc ions are present in the interface due to corrosion of the galvanized steel surface. Moreover, they showed that the effect of Zn^{2+} ions is greater than Na^+ on the release rate of cerium ion.

2.5   Zeolite

Zeolite porous materials are categorized in the family of aluminosilicate clays with the chemical formula of $M_{2/n}O.Al_2O_3.xSiO_2.yH_2O$. M can be substitute by elements such as sodium (Na^+), lithium (Li^+), potassium (k^+), calcium (Ca^{2+}) and magnesium (Mg^{2+}) [95].

Anticorrosion performance of silane-zeolite composite coatings (bilayer ~ 10 micron) applied on aluminum 6061 was assessed in 3.5% NaCl solution [96]. In this regard, different concentrations of zeolite microparticles from 500 to 8000 ppm were tested and the optimum concentration of 4000 ppm was found leading to the best anticorrosion performance. In the presence of the particles, the permeability of the composite coating decreased due to higher crosslinking density confirmed by FTIR test. Besides, the enhanced corrosion properties can be related to the increase of the film thickness [97].

A similar investigation was done on AA2024 where the zeolite micro-particles served as the reservoir of Ce^{3+} cations as corrosion inhibitor to improve the anti-corrosion performance of silica-zirconia sol-gel films (~1.5 μm thickness) [98]. The corrosion problem of AA 2024 is the existence of cu rich intermetallic which is potentially cathodic relative to the aluminum matrix. So, sever anodic dissolution of the matrix around the particles happens [99-102]. Zeolite composite film enhanced the barrier properties of the coating compared with the neat coating. But in the case of cerium doped zeolite composite film, active corrosion protection was also presented. Due to hydrolyses of metal cations produced from anodic site, pH reduces locally to the lower pHs. Then, cerium cations can be liberated from the micro-particles. The released cerium cations precipitate in the local cathodic sites (intermetallic surfaces) where the local pH increases due to oxygen reduction reaction [98].

Rassouli et. al. [103, 104], synthesized zeolite particles and doped Zn^{2+} cations into the particles to promote the active corrosion protection of epoxy-ester coating. They claimed that when the water and other corrosive species penetrate into the organic coating, the zinc cations liberated from the zeolite structure may form a stable hydroxide film on micro cathodic sites suppressing cathodic reaction of oxygen reduction. Therefore, the iron dissolution would suppress sequentially. In fact, Zn^{2+} is exchanged with Na^+ ions from the corrosive NaCl media, liberating and reacting with the hydroxyl (OH^-) anions from oxygen reduction reaction (Eq. 1):

New Materials for a Circular Economy
Materials Research Foundations 149 (2023) 173-201

Materials Research Forum LLC
https://doi.org/10.21741/9781644902639-6

$$O_2 + 2\ H_2O + 4\ e^- \rightarrow 4\ OH^- \tag{1}$$

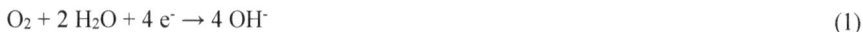

The results of electrochemical impedance spectroscopy test were indicative of the positive effect of incorporation of Zn^{2+} loaded zeolite into the epoxy-ester coatings applied on mild steel. After 21 days of immersion in 3.5 wt. % NaCl, the impedance module of the coated samples followed the order: Zn^{2+} doped zeolite composite epoxy ester coating > zeolite composite epoxy ester coating > neat epoxy ester coating. After 21 days of immersion, the impedance module of Zn^{2+} doped zeolite composite epoxy ester was 3 orders of magnitude higher than that for the neat coating. Beside the role of Zn^{2+} as corrosion inhibitor, the barrier effect of zeolite particles caused the impedance module of the incorporated epoxy ester coated plates was higher than that for the neat epoxy ester. They also investigated the effect of the particles on the active protection function of barrier coatings when they loaded with organic-inorganic corrosion inhibitors [105, 106].

2.6   Halloysite

Chemically, halloysite nanotubes (HNTs) are very similar to kaolinite [107, 108]. Both are composed of aluminosilicates with the chemical formula $[Al_2Si_2O_5(OH)_4.nH_2O]$.n indicates the number of water molecules, which is zero in kaolinite. The interlayer distance is 7.2 and 10 Å for kaolinite and halloysite, respectively. In fact, both kaolinite and halloysite are composed of a layer of Si-O tetrahedral plates and a layer of Al-O octahedral plates. The interlayer distance of halloysite is larger than that of kaolinite and water molecules exist between its layer [109].

The morphology of kaolinite particles is plate-shaped and the morphology of halloysite is tubular [107, 108]. The inner and outer surfaces of halloysite nanotubes have different chemical compositions. Those are respectively composed of Al-OH and Si-O and water molecules are between the layers [109, 110]. The distance between the hydrated halloysite is 10 Å. If the water is removed, the distance reduced to ~7 Å, which is equal to the interlayer distance of the kaolinite plates [108, 109]. Halloysite nanotubes have an inner and outer lumen of 15 and 50 nm, respectively and a length of 300 nm up to 1 μm [110-113].

The zeta potential values of halloysite particles is negative in the wide range of pH, which is approximately between the values of alumina and silica. So, at pH less than 8.5, the inner surface of halloysite nanotubes has a positive potential while at pH > 8.5 it has the negative potential, which is due to the protonation or deprotonation of the internal hydroxyl groups at acidic or alkaline condition. The outer surface also has a negative potential over a wide range of pH [114,115].

Existence of two internal and external surfaces with different chemical compositions as well as relatively large lumen diameters (~ 15 nm) provides halloysite an appropriate choice for loading of many materials such as drugs or corrosion inhibitors. As mentioned previously, positive inner surface and negative outer surface of halloysite make a potential

New Materials for a Circular Economy   Materials Research Forum LLC
Materials Research Foundations 149 (2023) 173-201   https://doi.org/10.21741/9781644902639-6

for loading of negative molecules into the lumen and preventing their adsorption on the outer surface [117].

To gain a smart anti-corrosion coating, Lvov and his co-workers [113] loaded benzotriazole (BTA) as a corrosion inhibitor in halloysite and then incorporated the BTA-loaded halloysite in SiO_x-ZrO_x hybrid coating applied on 2024 aluminum. The SVET results confirmed that active corrosion protection property was added to the sol-gel hybrid coating properties.

The corrosion current density around the scratched area of the ZrOx-SiOx sol-gel coating with and without BTA-loaded halloysite nanotubes was monitored and the results were indicative of further corrosion progression during the immersion time in sodium chloride solution for the halloysite-free coating. In contrary, the film containing BTA-loaded halloysite nanotubes didn't show any signs of corrosion current and corrosion progression during immersion in the aggressive solution.

Imidazole dicarboxylic acid as an organic inhibitor was also successfully loaded in the lumen of halloysite nanotubes to provide epoxy-ester coating with smart corrosion protection [118]. Halloysite nanotubes have a weak ion exchange characteristic, so it is so difficult to load cations as inorganic corrosion inhibitors. In this regard, Asadi *et. al.* [119], functionalized halloysite nanotubes with 3-aminopropyltriethoxysilane (APTES) to enhance loading of zinc cations. The inner and edge aluminol groups of HNTs were grafted by hydrolyzed APTES [117]. The existence of N atoms with the lone pair electrons in the chemical structure of APTES could raise the tendency of cations to adsorb in HNTs.

Functionalization of halloysite nanoparticles has been also reported for other purposes such as dispersion enhancement of halloysite nanotubes in polymeric composites to improve mechanical properties [120]. In this way, Asadi and her co-workers [119], investigated the release behavior of Zn^{2+} which was loaded into HNTs and functionalized HNTs (MHNTs) at different NaCl concentrations and pHs. The ICP-OES studies indicted higher concentration of Zn^{2+} when the HNTs were modified with APTES. Besides, as ionic strength of NaCl increased the concentration of released Zn^{2+} increased. This means more cations of Na^+ from the solution could be exchanged with Zn^{2+} cations of loaded MHNTs. Moreover, as pH rose the Zn^{2+} release decreased. In the other words, the concentration of Zn^{2+} released from the both loaded HNTs and modified halloysite nanotubes MHNTs was higher in acidic conditions. As pH increased, the zeta potential of HNTs particles got more negative charge so release of Zn^{2+} got more difficult. The effect of modified halloysite nanotubes (MHNTs) and Zn^{2+} loaded MHNTs on the anti-corrosion properties of epoxy-ester coating was also compared. The impedance results were indicative of the enhanced function of the coatings as after 50-days exposure in 3.5 wt. % NaCl solution, phase diagram of the both mild steel plates coated with neat epoxy ester and the one which was contain MHNTs showed two separate time constants. The time constant at high frequencies region was related to degrade coating and the second one at low frequencies corresponded to the corrosion reactions on the metallic surface. At the same immersion time, phase diagram of the mild steel plate coated with epoxy ester incorporated with Zn^{2+} loaded

New Materials for a Circular Economy   Materials Research Forum LLC
Materials Research Foundations 149 (2023) 173-201   https://doi.org/10.21741/9781644902639-6

MHNTs showed one broad time constant related to the coating. The impedance module of the mild steel plate coated with epoxy ester incorporated with Zn^{2+} loaded MHNTs was so much higher than that for the two other plates. It was 3.75×10^4, 6.08×10^5 and 1.18×10^{10} $\Omega.cm^2$ for the steels coated with neat epoxy ester, epoxy ester incorporated with MHNTs and epoxy ester incorporated with Zn^{2+} loaded MHNTs, respectively. Generally, according to the reported coating's resistances, MHNTs led to rising of the barrier properties of the epoxy ester. In the case of epoxy ester incorporated with Zn^{2+} loaded MHNTs in addition to barrier effect of MHNTs, the presence of the zinc cations diminished the corrosion reactions on the metallic surface and boosted the ionic resistance of the coating.

They also, simultaneously doped zinc cations and 4,5-imidazole dicarboxylic acid (IDC) as respectively inorganic and organic corrosion inhibitors into MHNTs [121]. The synergistic effect of the two inhibitors was proved by Tafel polarization and EIS experiments of the bare mild steel in 3.5 wt. % NaCl solution. So, the superior function of the epoxy ester coating incorporated by the double doped nanotubes may arises from the more inhibition efficiency of the released inhibitors in corrosion attacked zone resulting in the suppression of the diffusion of Fe^{2+} from the substrate to the electrolyte. The proposed mechanism of the inhibition was the precipitation of zinc hydroxide in cathode site and formation of Fe^{2+}-IDC chelate in anodic sites.

On the other hand, the significant barrier effect of MHNTs postponed the access of corrosive species to the metal/coating interface where the corrosion reactions take place.Double dope of 4,5-imidazole dicarboxylic acid (IDC) and zinc cations into talc nanoparticles which was dispersed into epoxy ester coating has been reported elsewhere [122]. Natural talc is a magnesium silicate with the morphology of flake like. It is composed of magnesium–oxygen/hydroxyl octahedral which are stacked between the two layers of silicon–oxygen tetrahedral [122, 123]. Its cation-exchange capability makes an opportunity to dope cations as corrosion inhibitor [124–127]. According to the ICP-OES and TOC results, the released concentration of the both Zn^{2+} and IDC inhibitors in the extracts in 3.5 wt.% NaCl solution was higher than that in the extracts in distilled water (DI water), meaning talc particles have partial cation and anion exchange capacity. This capability facilitates dope of zinc cations and IDC. The impedance results of mild steel coated with epoxy ester incorporated with the loaded talc particles confirmed the superiority of the coating double doped with the organic and in-organic corrosion inhibitors.

Recently, to increase the lumen diameter of HNTS and higher loading efficiency, some researchers [129], modified the inner lumen of HNTs (10-15 nm) using H_2SO_4. Also, to prevent the release of inhibitor, the loaded particles were encapsulated using urea-formaldehyde. The encapsulation using urea-formaldehyde was formerly suggested by Brown et. al. [130]. Finally, they utilized the particle in a silane coating formulation applied on AZ31 magnesium alloy.

Acid etching (H_2SO_4 of 1 M) was also reported to enlarge the lumen diameter of HNTs and subsequently to increase the loading of corrosion inhibitors such as Ce^{3+}-Zr^{4+}, 2-8-

New Materials for a Circular Economy   Materials Research Forum LLC
Materials Research Foundations 149 (2023) 173-201   https://doi.org/10.21741/9781644902639-6

hydroxyquinoline and mercaptobenzothiazole [131]. The loaded nanotubes were incorporated in silane hybrid coating applied of AZ91 magnesium alloy.

Xing et al. [132], loaded Na_2MoO_4 into HNTs through applying a vacuum condition and in order to control the release rate of Na_2MoO_4 from the lumen, they were used Ca^{2+} to form insoluble $CaMoO_4$ end stopper. They examined the anti-corrosion performance of the epoxy coatings with the dry thickness of $\sim 30 \pm 5$ μm containing different concentrations of the prepared particles on Q235 steel in 3.5 wt. % NaCl solution. As time elapsed the coating resistance of the specimens declined which might be attributed to the degradation of the coatings [93]. However, the degradation rate of the neat epoxy was faster than that of the composite coatings. There was an optimum concentration of nanoparticles that can be ensured more protective protection for the steel substrate. During the immersion times, the higher coating resistance along with the lower coating's degradation were attributed to the epoxy coating incorporated with 10 wt. % of Ca^{2+} treated Na_2MoO_4 loaded halloysite. Probably incorporation of more particles made more pore in the coating and decreased the coating barrier effect. Formation of stoppers at the end of the nanotubes has been also reported in the literature [132–134].

There are some reports in the literature concerning the loading of green corrosion inhibitors into the halloysite's lumen [132, 135].

The basil aqueous extract was loaded into the HNT's lumen by the cycle of vacuum – atmosphere pressure [136]. The loaded containers of HNTs were introduced into an alkyd coating applied on mild steel. The dry thickness of the coatings was ~ 25 μm. From the EIS studies, the optimum concentration of the loaded HNTs for improving the coating resistance of the alkyd coatings was ~ 0.5 wt % as in the case of the higher concentration the barrier effect of the coatings declined. The proposed mechanism for the inhibition of the green corrosion inhibitor was formation of chelates between Fe^{2+} cations resulting from corrosion of the substrate and the basil molecules of the green organic inhibitor. In fact, in the case of un-inhibited solution (3.5 wt. % NaCl), the hydroxyl (OH^-) anions and Fe^{2+} cations are produced on the cathodic and anodic area of the steel surface, respectively. Consequently, the iron hydroxide/oxide products (black/brownish rust) are formed on the metal surface. In the presence of the green organic inhibitor, Fe^{2+} cations interacted with the inhibitor molecules released from the loaded HNT's. The formed chelates can be responsible for the reduction of the iron dissolution rate.

In another report [133], two green corrosion inhibitors of vanillin and thyme oil were doped into HNTs and copper ions were used for end stopper formation. The inhibitor loading process was done in two steps. At the first step, vanillin solved in deionized water and HNTs were added to the solution. The obtained suspension was under the vacuum until the water was evaporated. The halloysites doped vanillin was added to the solution of thyme oil in ethanol. Again, a vacuum condition was applied until the ethanol was evaporation. Finally, the product was centrifuged, washed and dried to ready for anti-corrosion purposes.

Conclusion

The chapter deals with the strategies introduced for providing smart corrosion protection to barrier organic coatings. Application of smart coatings providing long-term protection can prolong the metallic structures lifetime which will reduce costs in various industries. At the same time, reducing the need for maintenance and repainting may result in significant economic savings. Incorporating new generations of anticorrosion pigments into the coatings could be considered as an appropriate way to raise the protection properties of the polymeric films by partial dissolution of the pigment particles coming in contact with the diffused water. As another strategy, encapsulation of corrosion inhibitors causes the release of inhibiting moieties on-demand and prevents the direct contact of inhibitor with the polymeric network. layer by layers (LbLs), layered double hydroxides (LDHs), some in-organic particles with cation exchange capability such as talc and montmorillonite, mesoporous materials such zeolite and mesoporous silica and halloysite nanotubes are the most widely used particles which could be utilized to host the corrosion inhibitors.

References

[1] J. Akpoborie, O. S. I. Fayomi, O. Agboola, O. D. Samuel, B. U. Oreko, A. A. Ayoola, Electrochemical Corrosion Phenomenon and Prospect of Materials Selection in Curtailing the Challenges, in IOP Conference series: Materials Science and Engineering, 1107 (2021) 012072. https://doi.org/10.1088/1757-899X/1107/1/012072

[2] H. Marchebois, M. Keddam, C. Savall, J. Bernard, S. Touzain, Zinc-rich powder coatings characterisation in artificial sea water EIS analysis of the galvanic action, Electrochim. Acta 49 (2004) 1719-1729. https://doi.org/10.1016/j.electacta.2003.11.031

[3] S. Shen, Y. Zuo, X. Zhao, The effects of 8-hydroxyquinoline on corrosion performance of a Mg-rich coating on AZ91D magnesium alloy, Corros. Sci. 76 (2013) 275-283. https://doi.org/10.1016/j.corsci.2013.06.050

[4] U. M. Angst, A Critical Review of the Science and Engineering of Cathodic Protection of Steel in Soil and Concrete, corrosion 75 (2019) 1420-1433. https://doi.org/10.5006/3355

[5] Y. Zuo, P. Zhao, J. Zhao, The influences of sealing methods on corrosion behavior of anodized aluminum alloys in NaCl solutions, Surf. coatings Technol. 166 (2003) 237-242. https://doi.org/10.1016/S0257-8972(02)00779-X

[6] B. C. Blawert, W. Dietzel, E. Ghali, G. Song, Anodizing Treatments for MagnesiumAlloysandTheirEffecton Corrosion Resistance in Various Environments, Adv. Eng. Mater. 8 (2006) 511-533. https://doi.org/10.1002/adem.200500257

[7] A. K. Sharma, Anodizing titanium for space applications, Thin Solid Films 208 (1992) 48-54. https://doi.org/10.1016/0040-6090(92)90946-9

[8] A. K. Dubey, G. Singh, Corrosion Inhibition of Mild Steel in Sulphuric Acid Solution by Using Polyethylene Glycol Methyl Ether (PEGME), Port. Electrochim. Acta 25 (2007) 221-235. https://doi.org/10.4152/pea.200702221

[9] N. Asadi, M. Ramezanzadeh, G. Bahlakeh, B. Ramezanzadeh, Utilizing Lemon Balm extract as an effective green corrosion inhibitor for mild steel in 1M HCl solution: A detailed experimental, molecular dynamics, Monte Carlo and quantum mechanics study, J. Taiwan Inst. Chem. Eng. 95 (2019) 252-272. https://doi.org/10.1016/j.jtice.2018.07.011

[10] S. S. Abd, E. Rehim, H. H. Hassan, M. A. Amin, Corrosion inhibition study of pure Al and some of its alloys in 1 . 0 M HCl solution by impedance technique, Corros. Sci. 46 (2004) 5-25. https://doi.org/10.1016/S0010-938X(03)00133-1

[11] N. Parhizkar, T. Shahrabi, B. Ramezanzadeh, Synthesis and characterization of a unique isocyanate silane re duce d graphene oxide nanosheets; Screening the role of multifunctional nanosheets on the adhesion and corrosion protection performance, J. Taiwan Inst. Chem. Eng. 82 (2018) 281-299. https://doi.org/10.1016/j.jtice.2017.10.033

[12] A. Kalendová, Alkalizing and neutralizing effects of anticorrosive pigments containing Zn, Mg, Ca, and Sr cations, Prog. Org. Coatings 38 (2000) 199-206. https://doi.org/10.1016/S0300-9440(00)00103-X

[13] M. Behzadnasab, S. M. Mirabedini, K. Kabiri, S. Jamali, Corrosion performance of epoxy coatings containing silane treated ZrO 2 nanoparticles on mild steel in 3 . 5 % NaCl solution, Corros. Sci. 53 (2011) 89-98. https://doi.org/10.1016/j.corsci.2010.09.026

[14] S. Nabavian, R. Naderi, N. Asadi, Determination of Optimum Concentration of Benzimidazole Improving the Cathodic Disbonding Resistance of Epoxy Coating, coatings 8 (2018) 471. https://doi.org/10.3390/coatings8120471

[15] P. A. Sørensen, S. Kiil, K. Dam-Johansen, C. E. Weinell, Anticorrosive coatings: A review, J. Coatings Technol. Res. 6 (2009) 135-176. https://doi.org/10.1007/s11998-008-9144-2

[16] P. de Lima-Neto, A. P. de Araújo, W. S. Araújo, A. N. Correia, Study of the anticorrosive behaviour of epoxy binders containing non-toxic inorganic corrosion inhibitor pigments, Prog. Org. Coatings 62 (2008) 344-350. https://doi.org/10.1016/j.porgcoat.2008.01.012

[17] M. Zubielewicz, W. Gnot, Mechanisms of non-toxic anticorrosive pigments in organic waterborne coatings, Prog. Org. Coatings 49 (2004) 358-371. https://doi.org/10.1016/j.porgcoat.2003.11.001

[18] F. De, L. Fedrizzi, S. Rossi, P. L. Bonora, Organic coating capacitance measurement by EIS : ideal and actual trends, Electrochim. Acta 44 (1999) 4243-4249. https://doi.org/10.1016/S0013-4686(99)00139-5

[19] D. Greenfield, D. Scantlebury, The Protective Action of Organic Coatings on Steel: A review, J. Corros. Sci. Eng. 3 (2000) 1-27.

[20] G. Grundmeier, W. Schmidt, M. Stratmann, Corrosion protection by organic coatings: electrochemical mechanism and novel methods of investigation, Electrochim. Acta 45 (2000) 2515-2533. https://doi.org/10.1016/S0013-4686(00)00348-0

[21] P. H. Suegama, H. G. De Melo, A. A. C. Recco, A. P. Tschiptschin, I. V Aoki, Corrosion behavior of carbon steel protected with single and bi-layer of silane films filled with silica nanoparticles, Surf. Coat. Technol. 202 (2008) 2850-2858. https://doi.org/10.1016/j.surfcoat.2007.10.028

[22] F. Deflorian, S. Rossi, An EIS study of ion diffusion through organic coatings, Electrochim. Acta 51 (2006) 1736-1744. https://doi.org/10.1016/j.electacta.2005.02.145

[23] R. Naderi, M. M. Attar, The role of zinc aluminum phosphate anticorrosive pigment in Protective Performance and cathodic disbondment of epoxy coating, Corros. Sci. 52 (2010) 1291-1296. https://doi.org/10.1016/j.corsci.2009.12.019

[24] R. Naderi, M. M. Attar, Cathodic disbondment of epoxy coating with zinc aluminum polyphosphate as a modified zinc phosphate anticorrosion pigment, Prog. Org. Coatings 69 (2010) 392-395. https://doi.org/10.1016/j.porgcoat.2010.08.001

[25] R. Naderi, M. M. Attar, Electrochemical study of protective behavior of organic coating pigmented with zinc aluminum polyphosphate as a modified zinc phosphate at different pigment volume concentrations, Prog. Org. Coatings 66 (2009) 314-320. https://doi.org/10.1016/j.porgcoat.2009.08.009

[26] R. Naderi, M. M. Attar, Electrochimica Acta Electrochemical assessing corrosion inhibiting effects of zinc aluminum polyphosphate (ZAPP) as a modified zinc phosphate pigment, Electrochim. Acta 53 (2008) 5692-5696. https://doi.org/10.1016/j.electacta.2008.03.029

[27] R. Naderi, M. M. Attar, Application of the electrochemical noise method to evaluate the effectiveness of modification of zinc phosphate anticorrosion pigment, Corros. Sci. 51 (2009) 1671-1674. https://doi.org/10.1016/j.corsci.2009.04.015

[28] R. Naderi, M. Mahdavian, M. M. Attar, Electrochemical behavior of organic and inorganic complexes of Zn (II) as corrosion inhibitors for mild steel : Solution phase study, Electrochim. Acta 54 (2009) 6892-6895. https://doi.org/10.1016/j.electacta.2009.06.073

[29] R. Naderi, M. M. Attar, EIS and ENM as tools to evaluate inhibitive performance of second generation of phosphate-based anticorrosion pigments, J. Appl. Electrochem. 39 (2009) 2353-2358. https://doi.org/10.1007/s10800-009-9921-3

[30] R. Naderi, M. Mahdavian, A. Darvish, Electrochemical examining behavior of epoxy coating incorporating zinc-free phosphate-based anticorrosion pigment, Prog. Org. Coatings 76 (2013) 302-306. https://doi.org/10.1016/j.porgcoat.2012.09.026

[31] R. Naderi, M. M. Attar, The inhibitive performance of polyphosphate-based anticorrosion pigments using electrochemical techniques, Dye. Pigment. 80 (2009) 349-354. https://doi.org/10.1016/j.dyepig.2008.08.002

[32] C. Deyá, G. Blustein, B. Amo, R. Romagnoli, Evaluation of eco-friendly anticorrosive pigments for paints in service conditions, Prog. Org. Coatings 69 (2010) 1-6. https://doi.org/10.1016/j.porgcoat.2010.03.011

[33] M. Mahdavian, R. Naderi, Corrosion inhibition of mild steel in sodium chloride solution by some zinc complexes, Corros. Sci. 53 (2011) 1194-1200. https://doi.org/10.1016/j.corsci.2010.12.013

[34] E. Salehi, R. Naderi, B. Ramezanzadeh, Improvement in the protective performance of epoxy ester coating through inclusion of an effective hybrid green corrosion inhibitive pigment, J. Taiwan Inst. Chem. Eng. 81 (2017) 391-405. https://doi.org/10.1016/j.jtice.2017.09.049

[35] N. Asadi, M. Ramezanzadeh, G. Bahlakeh, B. Ramezanzadeh, Theoretical MD / DFT computer explorations and surface-electrochemical investigations of the zinc / iron metal cations interactions with highly active molecules from Lemon balm extract toward the steel corrosion retardation in saline solution, J. Mol. Liq. 310 (2020) 113220. https://doi.org/10.1016/j.molliq.2020.113220

[36] M. Motamedi, B. Ramezanzadeh, M. Mahdavian, Corrosion inhibition properties of a green hybrid pigment based on Pr-Urtica Dioica plant extract, J. Ind. Eng. Chem., 66 (2018) 116-125. https://doi.org/10.1016/j.jiec.2018.05.021

[37] M. Ramezanzadeh, Z. Sanaei, G. Bahlakeh, B. Ramezanzadeh, Highly effective inhibition of mild steel corrosion in 3.5 % NaCl solution by green Nettle leaves extract and synergistic effect of eco-friendly cerium nitrate additive : Experimental , MD simulation and QM investigations, J. Mol. Liq. 256 (2018) 67-83. https://doi.org/10.1016/j.molliq.2018.02.021

[38] G. Bahlakeh, M. Ramezanzadeh, B. Ramezanzadeh, Experimental and theoretical studies of the synergistic inhibition effects between the plant leaves extract (PLE) and zinc salt (ZS) in corrosion control of carbon steel in chloride solution, J. Mol. Liq. 248 (2017) 854-870. https://doi.org/10.1016/j.molliq.2017.10.120

[39] E. Alibakhshi, M. Ramezanzadeh, S. A. Haddadi, G. Bahlakeh, B. Ramezanzadeh, Persian Liquorice extract as a highly efficient sustainable corrosion inhibitor for mild steel in sodium chloride solution, J. Clean. Prod. 210 (2019) 660-672. https://doi.org/10.1016/j.jclepro.2018.11.053

[40] M. Razizadeh, M. Mahdavian, B. Ramezanzadeh, E. Alibakhshi, S. Jamali, Synthesis of hybrid organic - inorganic inhibitive pigment based on basil extract and

zinc cation for application in protective construction coatings, Constr. Build. Mater. 287 (2021) 123034. https://doi.org/10.1016/j.conbuildmat.2021.123034

[41] E. Salehi, R. Naderi, B. Ramezanzadeh, Synthesis and characterization of an effective organic/inorganic hybrid green corrosion inhibitive complex based on zinc acetate / Urtica Dioica, Appl. Surf. Sci. 396 (2017) 1499-1514. https://doi.org/10.1016/j.apsusc.2016.11.198

[42] S. Abrishami, R. Naderi, B. Ramezanzadeh, Fabrication and characterization of zinc acetylacetonate/Urtica Dioica leaves extract complex as an effective organic/inorganic hybrid corrosion inhibitive pigment for mild steel protection in chloride solution, Appl. Surf. Sci. 457 (2018) 487-496. https://doi.org/10.1016/j.apsusc.2018.06.190

[43] Z. Sanaei, G. Bahlakeh, B. Ramezanzadeh, Active corrosion protection of mild steel by an epoxy ester coating reinforced with hybrid organic/inorganic green inhibitive pigment, J. Alloys Compd. 728 (2017) 1289-1304. https://doi.org/10.1016/j.jallcom.2017.09.095

[44] D. Snihirova, S. V Lamaka, M. F. Montemor, 'SMART' protective ability of water based epoxy coatings loaded with CaCO3 microbeads impregnated with corrosion inhibitors applied on AA2024 substrates, Electrochim. Acta, 83 (2012) 439-447. https://doi.org/10.1016/j.electacta.2012.07.102

[45] M. Izadi, T. Shahrabi, B. Ramezanzadeh, Synthesis and characterization of an advanced layer-by-layer assembled Fe3O4/polyaniline nanoreservoir filled with Nettle extract as a green corrosion protective system, J. Ind. Eng. Chem. 57 (2018) 263-274. https://doi.org/10.1016/j.jiec.2017.08.032

[46] J. Sinko, Challenges of chromate inhibitor pigments replacement in organic coatings, Prog. Org. Coatings 42 (2001) 267-282. https://doi.org/10.1016/S0300-9440(01)00202-8

[47] M. L. Zheludkevich, R. Serra, M. F. Montemor, M. G. S. Ferreira, Oxide nanoparticle reservoirs for storage and prolonged release of the corrosion inhibitors, Electrochem. commun. 7 (2005) 836-840. https://doi.org/10.1016/j.elecom.2005.04.039

[48] S. K. Ghosh, Functional Coatings and Microencapsulation: A General Perspective, in: S. K. Ghosh, Functional Coatings: By Polymer Microencapsulation, Wiley-VCH, Belgium, 2006, pp. 1-28. https://doi.org/10.1002/3527608478.ch1

[49] A. J. Jadhav, S. E. Karekar, D. V. Pinjari, Y. G. Datar, B. A. Bhanvase, S. h.Snoawane, A. B. Pandit, Development of Smart Nanocontainers With A Zinc Phosphate Core and A pH-Responsive Shell for Controlled Release of Immidazole, Hybrid Mater. 2 (2015) 1-9. https://doi.org/10.1515/hyma-2015-0001

[50] B. D. G. Shchukin, M. Zheludkevich, K. Yasakau, S. Lamaka, M. G. S. Ferreira, H. Möhwald, Layer-by-Layer Assembled Nanocontainers for Self-Healing Corrosion

Protection, Adv. Mater. 18 (2006) 1672-1678.
https://doi.org/10.1002/adma.200502053

[51] B. A. Bhanvase, M. A. Patel, S. H. Sonawane, Kinetic properties of layer-by-layer assembled cerium zinc molybdate nanocontainers during corrosion inhibition, Corros. Sci. 88 (2014) 170-177. https://doi.org/10.1016/j.corsci.2014.07.022

[52] S. H. Sonawane, B. A. Bhanvase, A. A. jamli, S. K. Dubey, S. S. Kale, D. V. Pinjari, R. D. Kulkarni, P. R. gogate, A. B. Pandit, Improved active anticorrosion coatings using layer-by-layer assembled ZnO nanocontainers with benzotriazole, Chem. Eng. J. 189-190 (2012) 464-472. https://doi.org/10.1016/j.cej.2012.02.076

[53] M. L. Zheludkevich, S. k. Poznyak, l. m. Rodrigues, D. Raps, T. hack, L. F. Dick,T. Nunes, M. G. S. Ferreira, Active protection coatings with layered double hydroxide nanocontainers of corrosion inhibitor, Corros. Sci. 52 (2010) 602-611. https://doi.org/10.1016/j.corsci.2009.10.020

[54] J. Tedim, M. L. Zheludkevich, A. N. Salak, A. Lisenkov, M. G. S. Ferreira, Nanostructured LDH-container layer with active protection functionality, J. Mater. Chem. 21 (2011) 15464-15470. https://doi.org/10.1039/c1jm12463c

[55] J. Tedim, M. L. Zheludkevich, A. C. Bastos, A. N. Salak, A. D. Lisenkov, M. G. S. Ferreira, Influence of preparation conditions of Layered Double Hydroxide conversion films on corrosion protection, Electrochim. Acta 117 (2014) 164-171. https://doi.org/10.1016/j.electacta.2013.11.111

[56] E. Alibakhshi, E. Ghasemi, M. Mahdavian, B. Ramezanzadeh, S. Farashi, Active corrosion protection of Mg-Al-PO43- LDH nanoparticle in silane primer coated with epoxy on mild steel, J. Taiwan Inst. Chem. Eng. 75 (2017) 248-262. https://doi.org/10.1016/j.jtice.2017.03.010

[57] M. A. Iqbal, F. Michele, Effect of Synthesis Conditions on the Controlled Growth of MgAl - LDH Corrosion Resistance Film: Structure and Corrosion Resistance Properties, coatings 9 (2019) 30. https://doi.org/10.3390/coatings9010030

[58] S. K. Poznyak, J. Tedim, L. M. Rodrigues, A. n. Salak, M. l. Zheludkevich, L. F. P. Dick, M. G. S. Ferreira, Novel Inorganic Host Layered Double Hydroxides Intercalated with Guest Organic Inhibitors for Anticorrosion Applications, ACS Appl. Mater. Interfaces 1 (2009) 2353-2362. https://doi.org/10.1021/am900495r

[59] V. Shkirskiy, P. Keil, F. Leroux, P. Vialat, G. Lefe, K. Ogle, Factors Affecting MoO42- Inhibitor Release from Zn2Al Based Layered Double Hydroxide and Their Implication in Protecting Hot Dip Galvanized Steel by Means of Organic Coatings, ACS Appl. Mater. Interfaces 7 (2015) 25180-25192. https://doi.org/10.1021/acsami.5b06702

[60] E. Alibakhshi, E. Ghasemi, M. Mahdavian, B. Ramezanzadeh, A comparative study on corrosion inhibitive effect of nitrate and phosphate intercalated Zn-Al- layered double hydroxides (LDHs) nanocontainers incorporated into a hybrid silane layer and

their effect on cathodic delamination of epoxy topcoat, Eval. Program Plann. 115 (2016) 159-174. https://doi.org/10.1016/j.corsci.2016.12.001

[61] D. Li, F. Wang, X. Yu, J. Wang, Q. Liu, P. Yang, Y. He, Y. Wang, M. Zhang, Anticorrosion organic coating with layered double hydroxide loaded with corrosion inhibitor of tungstate, Prog. Org. Coatings 71(2011) 302-309. https://doi.org/10.1016/j.porgcoat.2011.03.023

[62] J. K. Lin, J. Y. Uan, C. P. Wu, H. H. Huang, Direct growth of oriented Mg - Fe layered double hydroxide (LDH) on pure Mg substrates and in vitro corrosion and cell adhesion testing of LDH-coated Mg samples, J. Mater. Chem. 21 (2011) 5011-5020. https://doi.org/10.1039/c0jm03764h

[63] J. Zhang, Y. Zhang, Y. Chen, L. Du, B. Zhang, H. Zhang, J. Liu, K. Wang, Preparation and Characterization of Novel Polyethersulfone Hybrid Ultrafiltration Membranes Bending with Modified Halloysite Nanotubes Loaded with Silver Nanoparticles, Ind. Eng. Chem. Res. 51 (2012) 3081-3090. https://doi.org/10.1021/ie202473u

[64] R. Subasri, K. R. C. Soma Raju, D. S. Reddy, A. Jyothirmayi, Vijaykumar S. Ijeri, Om Prakash, Stephen P. Gaydos, Environmentally friendly Zn - Al layered double hydroxide (LDH) - based sol - gel corrosion protection coatings on AA 2024-T3, J. Coatings Technol. Res.16 (2019) 1447-4163. https://doi.org/10.1007/s11998-019-00229-y

[65] N. Olya, E. Ghasemi, B. Ramezanzadeh, M. Mahdavian, Synthesis, characterization and protective functioning of surface decorated Zn-Al layered double hydroxide with SiO2 nano-particles, Surf. Coat. Technol. 387 (2020) 125512. https://doi.org/10.1016/j.surfcoat.2020.125512

[66] M. J. Anjum, J. Zhao, V. Z. Asl, M. U. Malik, G. Yasin, W. Q. Khan, Green corrosion inhibitors intercalated Mg:Al layered double hydroxide coatings to protect Mg alloy, Rare Met. 40(2021) 2254-2265. https://doi.org/10.1007/s12598-020-01538-7

[67] F. Zhang, C. Zhang, L. Song, R. Zeng, Z. Liu, H. Cui, Corrosion of in-situ grown MgAl-LDH coating on aluminum alloy, Trans. Nonferrous Met. Soc. China 25 (2015) 3498-3504. https://doi.org/10.1016/S1003-6326(15)63987-5

[68] J. Tedim, A. C. Bastos, S. Kallip, M. L. Zheludkevich, M. G. S. Ferreira, Corrosion protection of AA2024-T3 by LDH conversion films. Analysis of SVET results, Electrochim. Acta 210 (2016) 215-224. https://doi.org/10.1016/j.electacta.2016.05.134

[69] Y. Zhang, J. Liu, Y. Li, M. Yu, Fabrication of inhibitor anion-intercalated layered double hydroxide host films on aluminum alloy 2024 and their anticorrosion properties, J. Coatings Technol. Res. 12 (2015) 293-302. https://doi.org/10.1007/s11998-014-9644-1

[70] T. Wen, R. Yan, N. Wang, Y. Li, T. Chen, H. Ma, PPA-containing layered double hydroxide (LDH) fi lms for corrosion protection of a magnesium alloy, Surf. Coat. Technol. 383 (2020) 125255. https://doi.org/10.1016/j.surfcoat.2019.125255

[71] R. G. Buchheit, S. B. Mamidipally, P. Schmutz, H. Guan, Active Corrosion Protection in Ce-Modified Hydrotalcite Conversion Coatings, Corrosion 58 (2002) 3-14. https://doi.org/10.5006/1.3277303

[72] J. Uan, B. Yu, X. Pan, Morphological and Microstructural Characterization of the Aragonitic CaCO3 / Mg , Al-Hydrotalcite Coating on Mg-9 Wt Pct Al-1 Wt Pct Zn Alloy to Protect against Corrosion, Metall. Mater. Trans. A 39 (2008) 3233-3245. https://doi.org/10.1007/s11661-008-9669-0

[73] M. Saremi, M. Yeganeh, Application of mesoporous silica nanocontainers as smart host of corrosion inhibitor in polypyrrole coatings, Corros. Sci. 86 (2014) 159-170. https://doi.org/10.1016/j.corsci.2014.05.007

[74] M. Yeganeh, M. Saremi, H. Rezaeyan, Corrosion inhibition of steel using mesoporous silica nanocontainers incorporated in the polypyrrole, Prog. Org. Coatings, 77 (2014) 1428-1435. https://doi.org/10.1016/j.porgcoat.2014.05.007

[75] M. Yeganeh, A. Keyvani, The effect of mesoporous silica nanocontainers incorporation on the corrosion behavior of scratched polymer coatings, Prog. Org. Coatings, 90 (2016) 296-303. https://doi.org/10.1016/j.porgcoat.2015.11.006

[76] M. Yeganeh, N. Asadi, M. Omidi, M. Mahdavian, An investigation on the corrosion behavior of the epoxy coating embedded with mesoporous silica nanocontainer loaded by sulfamethazine inhibitor, Prog. Org. Coatings, 128 (2019) 75-81. https://doi.org/10.1016/j.porgcoat.2018.12.022

[77] M. Amini, R. Naderi, M. Mahdavian, A. Badiei, Effect of Piperazine Functionalization of Mesoporous Silica Type SBA-15 on the Loading E ffi ciency of 2 - Mercaptobenzothiazole Corrosion Inhibitor, Ind. Eng. Chem. Res. 59 (2020) 3394-3404. https://doi.org/10.1021/acs.iecr.9b05261

[78] M. Yeganeh, M. Omidi, S. H. H. Mortazavi, A. Etemad, M. H. Nazari, S. M. Marashi, Application of mesoporous silica as the nanocontainer of corrosion inhibitor, in: S. Rajendran, T, Anh Nguyen, S. Kakooei, M. Yeganeh, Y. Li, corrosion protection at the nanoscale, Elsevier, United Kingdom, 2020, pp. 275-294. https://doi.org/10.1016/B978-0-12-819359-4.00015-5

[79] M. Yeganeh, M. Saremi, Corrosion inhibition of magnesium using biocompatible Alkyd coatings incorporated by mesoporous silica nanocontainers, Prog. Org. Coatings 79 (2015) 25-30. https://doi.org/10.1016/j.porgcoat.2014.10.015

[80] A. Keyvani, M. Yeganeh, H. Rezaeyan, Application of mesoporous silica nanocontainers as an intelligent host of molybdate corrosion inhibitor embedded in the epoxy coated steel, Prog. Nat. Sci. Mater. Int. 27 (2017) 261-267. https://doi.org/10.1016/j.pnsc.2017.02.005

[81] J. Wen, J. Lei, J. Chen, L. Liu, X. Zhang, L. Li, Polyethylenimine wrapped mesoporous silica loaded benzotriazole with high pH-sensitivity for assembling self-healing anti-corrosive coatings, Mater. Chem. Phys. 253 (2020) 123425. https://doi.org/10.1016/j.matchemphys.2020.123425

[82] M. Amini, R. Naderi, M. Mahdavian, A. Badiei, Release of lanthanum cations Corrosion Protection at the Nanoscale loaded into piperazine-modified SBA-15 to inhibit the mild steel corrosion, Microporous Mesoporous Mater. 315 (2021) 110908. https://doi.org/10.1016/j.micromeso.2021.110908

[83] J. M. Falcón, L. M. Otubo, I. V Aoki, Highly ordered mesoporous silica loaded with dodecylamine for smart anticorrosion coatings, Surf. coatings Technol. 303B (2016) 319-329. https://doi.org/10.1016/j.surfcoat.2015.11.029

[84] Y. Zhao, J. Xu, J. Zhan, Y. Chen, J. Hu, Electrodeposited superhydrophobic mesoporous silica films co-embedded with template and corrosion inhibitor for active corrosion protection, Appl. Surf. Sci. 508 (2020) 145242. https://doi.org/10.1016/j.apsusc.2019.145242

[85] I. Recloux, M. Mouanga, M. Druart, Y. Paint, M. Olivier, Silica mesoporous thin films as containers for benzotriazole for corrosion protection of 2024 aluminium alloys, Appl. Surf. Sci. 346 (2015) 124-133. https://doi.org/10.1016/j.apsusc.2015.03.191

[86] C. Motte, M. Poelman, A. Roobroeck, M. Fedel, F. Deflorian, M. Olivier, Improvement of corrosion protection offered to galvanized steel by incorporation of lanthanide modified nanoclays in silane layer, Prog. Org. Coatings 74 (2012) 326-333. https://doi.org/10.1016/j.porgcoat.2011.12.001

[87] S. Bohm, H. N. McMurray, S. M. Powell, D. A. Worsley, Novel environment friendly corrosion inhibitor pigments based on naturally occurring clay minerals, Mater. Corros. 52 (2001) 896-903. https://doi.org/10.1002/1521-4176(200112)52:12<896::AID-MACO896>3.0.CO;2-8

[88] C. Deya, R. Romagnoli, B. Amo, A new pigment for smart anticorrosive coatings, J. Coatings Technol. Res. 4 (2007) 167-175. https://doi.org/10.1007/s11998-007-9021-4

[89] C. Zhou, D. Tong, W. Yu, Smectite nanomaterials: Preparation, properties, and functional applications, in: A. Wang, W. Wang, Nanomaterials from Clay Minerals: A New Approach to Green Functional Materials, Elsevier, United Kingdom, 2019, pp. 335-364. https://doi.org/10.1016/B978-0-12-814533-3.00007-7

[90] M. Massaro, G. Cavallaro, G. Lazzara, S. Riela, Covalently modified nanoclays: synthesis, properties and applications, in: G. Cavallaro, Clay Nanoparticles, INC, Italy, 2020, pp. 305-333. https://doi.org/10.1016/B978-0-12-816783-0.00013-X

[91] F. Uddin, Clays, nanoclays, and montmorillonite minerals, Metall. Mater. Trans. A Phys. Metall. Mater. Sci. 39 (2008) 2804-2814. https://doi.org/10.1007/s11661-008-9603-5

[92] N. Asadi, R. Naderi, Nanoparticles incorporated in silane solegel coatings, in: S. Rajendran, T, Anh Nguyen, S. Kakooei, M. Yeganeh, Y. Li, corrosion protection at the nanoscale, Elsevier, United Kingdom, 2020, pp. 450-469. https://doi.org/10.1016/B978-0-12-819359-4.00023-4

[93] N. Asadi, R. Naderi, M. Saremi, S. Y. Arman, M. Fedel, F. Deflorian, Study of corrosion protection of mild steel by eco-friendly silane sol-gel coating, J. Sol-Gel Sci. Technol. 70 (2014) 329-338. https://doi.org/10.1007/s10971-014-3286-8

[94] M. Fedel, M. Olivier, M. Poelman, F. Deflorian, S. Rossi, M. Druart, Corrosion protection properties of silane pre-treated powder coated galvanized steel, Prog. Org. Coatings 66 (2009) 118-128. https://doi.org/10.1016/j.porgcoat.2009.06.011

[95] M. Golomeova, A. Zendelska, Application of Some Natural Porous Raw Materials for Removal of Lead and Zinc from Aqueous Solutions, in: R. S. Dariani, Microporous and Mesoporous Materials, inTech, Croatia, 2016, pp. 21-49. https://doi.org/10.5772/62347

[96] L. Calabrese, L. Bonaccorsi, E. Proverbio, Corrosion protection of aluminum 6061 in NaCl solution by silane-zeolite composite coatings, J. Coatings Technol. Res. 9 (2012) 597-607. https://doi.org/10.1007/s11998-011-9391-5

[97] N. Asadi, R. Naderi, M. Saremi, Applied Clay Science Determination of optimum concentration of cloisite in an eco-friendly silane sol-gel film to improve corrosion resistance of mild steel, Appl. Clay Sci. 95 (2014) 243-251. https://doi.org/10.1016/j.clay.2014.04.018

[98] S. A. S. Dias, S. V Lamaka, C. A. Nogueira, T. C. Diamantino, M. G. S. Ferreira, Sol - gel coatings modified with zeolite fillers for active corrosion protection of AA2024, Corros. Sci. 62 (2012) 153-162. https://doi.org/10.1016/j.corsci.2012.05.009

[99] W. Zhang, G. S. Frankel, Transitions between pitting and intergranular corrosion in AA2024, Electrochim. Acta 48 (2003) 1193-1210. https://doi.org/10.1016/S0013-4686(02)00828-9

[100] M. L. Zheludkevich, K. A. Yasakau, S. K. Poznyak, M. G. S. Ferreira, Triazole and thiazole derivatives as corrosion inhibitors for AA2024 aluminium alloy, Corros. Sci. 47 (2005) 3368-3383. https://doi.org/10.1016/j.corsci.2005.05.040

[101] A. Boag, A. E. Hughes, A. M. Glenn, T. H. Muster, D. Mcculloch, Corrosion of AA2024-T3 Part I: Localised corrosion of isolated IM particles, Corros. Sci. 53 (2011) 17-26. https://doi.org/10.1016/j.corsci.2010.09.009

[102] A. E. Hughes, A. Boag, A. M. Glenn, D. McCulloch, T. H. Muster, C. Ryan, C. Luo, X. Zhou, G. E. Thompson, Corrosion of AA2024-T3 Part II : Co-operative corrosion, Corros. Sci. 53 (2011) 27-39. https://doi.org/10.1016/j.corsci.2010.09.030

[103] L. Rassouli, R. Naderi, M. Mahdavain, The role of micro / nano zeolites doped with zinc cations in the active protection of epoxy ester coating, Appl. Surf. Sci. 423 (2017) 571-583. https://doi.org/10.1016/j.apsusc.2017.06.245

[104] L. Rassouli, R. Naderi, M. Mahdavian, A. M. Arabi, Synthesis and Characterization of Zeolites for Anti-corrosion Application: The Effect of Precursor and Hydrothermal Treatment, J. Mater. Eng. Perform. 27 (2018) 4625-4634. https://doi.org/10.1007/s11665-018-3602-5

[105] L. Rassouli, R. Naderi, M. Mahdavian, Study of the active corrosion protection properties of epoxy ester coating with zeolite nanoparticles doped with organic and inorganic inhibitors, J. Taiwan Inst. Chem. Eng. 85 (2018) 207-220. https://doi.org/10.1016/j.jtice.2017.12.023

[106] L. Rassouli, R. Naderi, M. Mahdavain, Study of the impact of sequence of corrosion inhibitor doping in zeolite on the self-healing properties of silane sol-gel film, J. Ind. Eng. Chem. 66 (2018) 221-230. https://doi.org/10.1016/j.jiec.2018.05.033

[107] B. Singh, Why Does Halloysite Roll?-A New Model, Clays Clay Miner. 44 (1996) 191-196. https://doi.org/10.1346/CCMN.1996.0440204

[108] B. Singh, D. R. M. Ian, Experimental transformation of kaolinite to halloysite, Clays Clay Miner. 44 (1996) 825-834. https://doi.org/10.1346/CCMN.1996.0440614

[109] E. Joussein, S. Petit, J. Churchman, B. Theng, D. Righi, B. Delvaux, Halloysite clay minerals-a review, Clay Miner. 40 (2005) 383-426. https://doi.org/10.1180/0009855054040180

[110] G. Cavallaro, L. Chiappisi, P. Pasbakhsh, M. Gradzielski, G. Lazzara, A structural comparison of halloysite nanotubes of different origin by Small-Angle Neutron Scattering (SANS) and Electric Birefringence, Appl. Clay Sci. 160 (2018) 71-80. https://doi.org/10.1016/j.clay.2017.12.044

[111] P. Yuan, D. Tan, F. Annabi-bergaya, Properties and applications of halloysite nanotubes : recent research advances and future prospects, Appl. Clay Sci. 112-113 (2015) 75-93. https://doi.org/10.1016/j.clay.2015.05.001

[112] D. Fix, D. V. Andreeva, Y. M. Lvov, D. G. Shchukin, H. Möhwald, Application of inhibitor-loaded halloysite nanotubes in active anti-corrosive coatings, Adv. Funct. Mater. 19 (2009) 1720-1727. https://doi.org/10.1002/adfm.200800946

[113] Y. M. Lvov, D. G. Shchukin, H. Möhwald, R. R. Price, Halloysite clay nanotubes for controlled release of protective agents, ACS Nano 2 (2008) 814-820. https://doi.org/10.1021/nn800259q

[114] Y. Lvov, W. Wang, L. Zhang, R. Fakhrullin, Halloysite Clay Nanotubes for Loading and Sustained Release of Functional Compounds, Adv. Mater. 28 (2016) 1227-1250. https://doi.org/10.1002/adma.201502341

[115] G. Tari, I. Bobos, C. S. F. Gomes, J. M. . Ferreira, Modification of Surface Charge Properties during Kaolinite to Halloysite-7Å Transformation, J. Colloid Interface Sci. 210 (1999) 360-366. https://doi.org/10.1006/jcis.1998.5917

[116] N. G. Veerabadran, R. R. Price, Y. M. Lvov, Clay Nanotubes for Encapsulation and Sustained Release of Drugs, Nano 2 (2007) 115-120. https://doi.org/10.1142/S1793292007000441

[117] P. Yuan, P. D. Southon, Z. Liu, M. E. R. Green, J. M. Hook, S. J. Antill, C. J. Kepert, Functionalization of Halloysite Clay Nanotubes by Grafting with γ-Aminopropyltriethoxysilane, J. Phys. Chem. C 112 (2008) 15742-15751. https://doi.org/10.1021/jp805657t

[118] N. Asadi, R. Naderi, M. Mahdavian, Halloysite nanotubes loaded with imidazole dicarboxylic acid to enhance protection properties of a polymer coating, Prog. Org. Coatings 127 (2019) 375-384. https://doi.org/10.1016/j.porgcoat.2018.11.035

[119] N. Asadi, R. Naderi, M. Mahdavian, Doping of zinc cations in chemically modified halloysite nanotubes to improve protection function of an epoxy ester coating, Corros. Sci. 151 (2019) 69-80. https://doi.org/10.1016/j.corsci.2019.02.022

[120] A. R. Erdogan, I. Kaygusuz, C. Kaynak, Influences of Aminosilanization of Halloysite Nanotubes on the Mechanical Properties of Polyamide-6 Nanocomposites, Polym. Compos. 16 (2014) 1350-1361. https://doi.org/10.1002/pc.22787

[121] N. Asadi, R. Naderi, M. Mahdavian, Synergistic e ff ect of imidazole dicarboxylic acid and Zn2+ simultaneously doped in halloysite nanotubes to improve protection of epoxy ester coating, Prog. Org. Coatings 132 (2019) 29-40. https://doi.org/10.1016/j.porgcoat.2019.03.021

[122] A. Bahrani, R. Naderi, M. Mahdavian, Chemical modification of talc with corrosion inhibitors to enhance the corrosion protective properties of epoxy-ester coating, Prog. Org. Coatings 120 (2018) 110-122. https://doi.org/10.1016/j.porgcoat.2018.03.017

[123] H. Elfaki, A. Hawari, C. Mulligan, Enhancement of multi-media filter performance using talc as a new filter aid material: Mechanistic study, J. Ind. Eng. Chem. 24 (2015) 71-78. https://doi.org/10.1016/j.jiec.2014.09.010

[124] C. J. Ngally Sabouang, J. A. Mbey, Liboum, F. Thomas, D. Njopwouo, Talc as raw material for cementitious products formulation, J. Asian Ceram. Soc. 2 (2014) 263-267. https://doi.org/10.1016/j.jascer.2014.05.007

[125] M. Sprynskyy, T. Kowalkowski, H. Tutu, E. M. Cukrowska, B. Buszewski, Adsorption performance of talc for uranium removal from aqueous solution, Chem. Eng. J. 171 (2011) 1185-1193. https://doi.org/10.1016/j.cej.2011.05.022

[126] S. Şener, A. Özyilmaz, Adsorption of naphthalene onto sonicated talc from aqueous solutions, Ultrason. Sonochem. 17 (2010) 932-938. https://doi.org/10.1016/j.ultsonch.2009.12.014

[127] I. M. Ali, Y. H. Kotp, I. M. El-Naggar, Thermal stability, structural modifications and ion exchange properties of magnesium silicate, Desalination 259 (2010) 228-234. https://doi.org/10.1016/j.desal.2010.03.054

[128] R. Choudhary, S. Koppala, S. Swamiappan, Bioactivity studies of calcium magnesium silicate prepared from eggshell waste by sol-gel combustion synthesis, J. Asian Ceram. Soc. 3 (2015) 173-177. https://doi.org/10.1016/j.jascer.2015.01.002

[129] R. Mahmoudi, P. Kardar, A. M. Arabi, R. Amini, P. Pasbakhsh, The active corrosion performance of silane coating treated by praseodymium encapsulated with halloysite nanotubes, Prog. Org. Coatings 138 (2020) 105404. https://doi.org/10.1016/j.porgcoat.2019.105404

[130] E. N. Brown, M. R. Kessler, N. R. Sottos, S. R. White, In situ poly(urea-formaldehyde) microencapsulation of dicyclopentadiene, J. Microencapsul. 20 (2003) 719-730. https://doi.org/10.1080/0265204031000154160

[131] S. H. Adsul, U. D. Bagale, S. H. Sonawane, R. Subasri, Release rate kinetics of corrosion inhibitor loaded halloysite nanotube-based anticorrosion coatings on magnesium alloy AZ91D, 9 (2020) 202-215. https://doi.org/10.1016/j.jma.2020.06.010

[132] X. Xing, D. Zhou, E. Tang, S. Liu, X. Chu, X. Xu, A novel method to control the release rate of halloysite encapsulated Na2MoO4 with Ca2+ and corrosion resistance for Q235 steel, Appl. Clay Sci. 188 (2020) 105492. https://doi.org/10.1016/j.clay.2020.105492

[133] S. S. Kumar, S. Kakooei, M. C. Ismail, M. Haris, Synthesis and characterization of metal ion end capped nanocontainer loaded with duo green corrosion inhibitors, J. Mater. Res. Technol. 9 (2020) 8350-8354. https://doi.org/10.1016/j.jmrt.2020.05.110

[134] E. Abdullayev, Y. Lvov, Clay nanotubes for corrosion inhibitor encapsulation : Release control with end stoppers, J. Mater. Chem. 20 (2010) 6681-6687. https://doi.org/10.1039/c0jm00810a

[135] X. Xing, X. Xu, J. Wang, W. Hu, Preparation, release and anticorrosion behavior of a multi-corrosion inhibitors-halloysite nanocomposite, Chem. Phys. Lett. 718 (2019) 69-73. https://doi.org/10.1016/j.cplett.2019.01.033

[136] M. Izadi, T. Shahrabi, I. Mohammadi, B. Ramezanzadeh, A. Fateh, The electrochemical behavior of nanocomposite organic coating based on clay nanotubes filled with green corrosion inhibitor through a vacuum-assisted procedure, Compos. part B 171 (2019) 96-110. https://doi.org/10.1016/j.compositesb.2019.04.019

New Materials for a Circular Economy   Materials Research Forum LLC
Materials Research Foundations 149 (2023) 202-232   https://doi.org/10.21741/9781644902639-7

Chapter 7

Scrap for New Steel

Carola Celada-Casero[1], Félix A. López[1], Carlos Capdevila[1], Roberto Castelo[2], Santiago Oliver[2]

[1]Centro Nacional de Investigaciones Metalúrgicas (CENIM-CSIC), Av. Gregorio del Amo, 8. E-28040 Madrid, Spain

[2]Unión de Empresas Siderúrgicas (UNESID), C/Castelló 128, 3ª Planta; E-28006 Madrid, Spain

c.celada@cenim.csic.es, f.lopez@csic.es, ccm@cenim.csic.es, rcastelo@unesid.org, soliver@unesid.org

Abstract

Steel is the main structural material to our modern society. To guarantee the supply of sustainable steel, the steelmaking industry must achieve net-zero CO_2 emissions by 2050. Besides the replacement of coal-based energy carriers to lower the CO_2 emissions, the substitution of primary raw materials with scrap is essential to promote energy and resource efficiency in the circular economy strategy. Steel production via the scrap-based electric arc furnace (EAF) route reduces the CO_2 emissions by 80% with respect to the blast furnace-basic oxygen furnace (BF-BOF) route. Nowadays, most of the overall scrap steel is fabrication and post-consumer scrap. Scrap utilization at higher rates poses a challenge for the steel industry due to the increasing content of tramp elements (Cu, Sn, Cr, Ni and Mo) in steel since the 1990. Some steel qualities demand very strict composition requirements, as either the absences or the controlled content of tramp elements. Improving scrap processing and classification methods is essential to ensure better scrap quality for new steel.

Keywords

Decarbonization, Steelmaking Industry, Green Steel, Scrap Recycling, Metal Recovery, Electric Arc Furnace (EAF)

Contents

1.   Introduction

Steel is one of the most essential structural materials to build communities and to develop technologies in the modern world. In 2021, the global production of crude steel amounted to 1,951 million tonnes (Mt), a 21% increase compared to 2011 [1]. Steel is used in every aspect of our daily lives, i.e. construction, home appliances, food cans, cars, etc. (Figure 1). In addition, the steel industry provides high-quality products to promote the energy transition of key sectors towards a carbon-neutral society by 2050; i.e. energy efficiency and carbon dioxide (CO_2) emissions decrease in power generation, transportation of energy carriers or light weighting of modern vehicles without compromising safety [2].

Steel products naturally contribute to the conservation of resources through their durability, strength-lightweight potential, versatility and 100% recyclability, allowing for a virtual infinite reutilisation of the invested resources at the end of a product's life. The reason why steel enables the highest recovery rates from mixed waste streams is, on one side, its magnetic properties and, on the other side, the high-embodied energy that makes ferrous scrap a valuable material. Around 630 million tonnes of steel are recycled every year, which is more than the recycled share of glass, paper and plastics combined [5]. Steel is, therefore, the permanent material in a circular economy.

It is reported in the literature that the worldwide steel recycling rate is around 85%. The recycling rate is a measure of the quantity of scrap that is reprocessed with respect to the quantity of scrap available [6]. Figures by sector reported by ArcelorMittal in 2018 showed rather high recycling rates: 95% for vehicles, 80 - 95% for infrastructure and structural steel, 60% for packaging, 97% for industrial equipment and 95% for appliances [7]. These figures might be even higher, though, if the unaccounted portion belongs to unsorted or unclassified scrap by origin. In The Netherlands and in Spain, steel has become the most recycled primary packaging material in Europe, with respective recycling rates of 95%, in 2019 [8], and of 98.5%, in 2016 [9]. Technically, recycling implies turning the material that has reached the end of its life back into the same material again. So, recycling is often confused with reuse, or downcycling, i.e. making the material available for use but not as the same material as it started out as. This might be the case of ferrous scrap obtained from shredding food and drink packaging, which can contain a critical quantity of tin. Besides, recycling rates often depend on the definition of scrap used, i.e. if it considers unrecoverable discards or not. The unrecoverable portion of ferrous scrap mainly consist of material from construction foundations or civil engineering applications, material that has typically been exposed to corrosion, tear and wear during the product's useful life, and material that is oxidized during the recycling procedure and forms slag. Nevertheless, in Europe, the unrecoverable portion of scrap steel is relatively low (2%), compared to most other waste streams, since ferrous scrap recovery based on magnetic methods or Eddy currents turn out to be a rather profitable business [10]. One of the leading and efficient technologies for municipal solid waste (MSW) treatment in Europe is incineration, through waste-to-energy. Around 70% of the elemental iron is recovered from the resulting incineration bottom ash, while the remaining 30% is lost due to thermal oxidation.

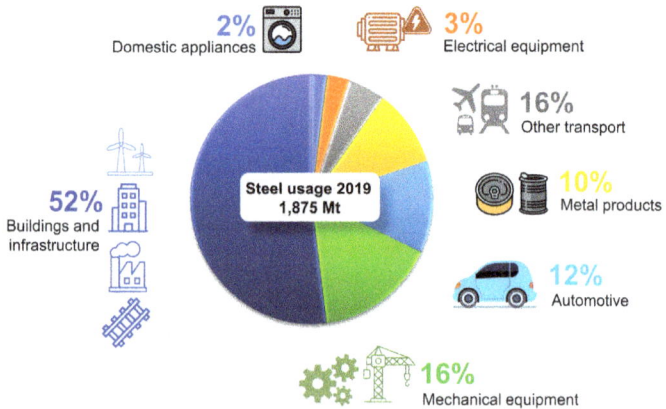

Figure 1. Worldwide distribution of steel end-usage, by sector, in 2019. Adapted from [3], using statistical data from 2019 [4].

The recoverable part of the scrap is recycled in steel mills as raw material for the production of new steel. Today, around a 56% of steel production is made from scrap in the European

Union (EU), according to data from 2019 of the European Steel Association (EUROFER) [11]. This fact has led to the development of a complete value chain for the valorisation of scrap and new markets, broadening the sources, quality and diversity of the ferrous scrap that is currently recycled. The two main steel production, and recycling, routes are [12]:

- The integrated, or primary, steelmaking route. Based on the blast furnace (BF) and basic oxygen furnace (BOF), the primary route currently represents around 70% of the global production [11]. Firstly, the iron ore is processed in a BF to obtain pig iron or hot metal. The pig iron is then input into the BOF alongside other raw materials like metallurgical coke, limestone and around 15% of recycled steel. The process emits, on average, 1.85 tonnes of CO_2 per tonne of crude steel produced [13].

- The electric arc furnace (EAF), or secondary, route. This process can be completely powered by renewable energy and it accepts different types of feedstock, such as direct reduced iron (DRI), hot metal or up to 100% of ferrous scrap. The proportion of feedstock vary significantly from one steelmaker to another and depending on the target steel quality. On average, every tonne of recycled scrap saves 1.5 tonnes of CO_2, 1.4 tonnes of iron ore and 740 kg of coal [13].

Recycling ferrous scrap through steelmaking does not only contribute to the circularity of steel products, it also decreases the energy demand, the CO_2 emissions and the need for the extraction of virgin iron ore to produce new steel. In July 2021, the European Commission adopted a packet of proposals to make the EU's climate, energy, land use, transport and taxation policies fit for reducing net greenhouse gas (GHG) emissions by at least 55% by 2030 [14]. In this sense, the electrification of the steel industry will play a key role in the long-term decarbonisation of the steel industry and the steel's circular economy as novel clean technologies are implemented and the production of clean electricity gets more competitive [15, 16]. However, the high durability of steel products, that is an average of 40 years [7], in combination with the significant amounts of ferrous scrap exported to third countries, results in a shortage of scrap [11]. A low scrap availability compromises the EU's chance to capitalize on valuable secondary raw materials [17]. Limiting the scrap exports and investing in innovation to improve the onsite production capacity for secondary raw materials of the industry are promising policy measures to increase the scrap-based steel production in the EU [18]. In addition, the potential for ferrous scrap recycling is strongly restricted by the scrap quality and industrial treatment capacity, the established steel production routes, the role of emerging technologies in steel recycling and the associated environmental, economic and societal impact. While the production of high-quality steel requires the processing of virgin iron ore in the BF-BOF route and limits the utilisation of scrap, the EAF steelmaking route allows to be powered by renewable energy and enables the utilisation of 100% scrap as feedstock. This, consequently, reduces significantly the carbon-footprint of the steelmaking process as well as the energy demand and the consumption of virgin resources. The EAF participation in the global steel production is expected to grow at the expense of the BF-BOF share; yet, the production of high-quality steel using a high portion of recycled steel in the EAF remains a major

New Materials for a Circular Economy   Materials Research Forum LLC
Materials Research Foundations 149 (2023) 202-232   https://doi.org/10.21741/9781644902639-7

challenge. Conventional sorting, classification and treatment methods of end-of-life scrap typically result in low scrap quality. A low scrap quality hampers the production of high-quality new steel and, thereby, limits the final application.

This chapter gives an overview of the factors influencing the potential of steel recycling to produce the sufficient amount of high-quality new steel to satisfy the demand and the economic growth of the society at a low environmental cost. *Section 2* defines the quality of ferrous scrap and its classification according to its origin. *Section 3* describes the main steel recycling routes and highlights key aspects affecting the energy consumption, GHGs emissions and new steel quality due to the implementation of new, more sustainable, technologies. In *Section 4*, the present scheme for steel recycling is discussed based on the scrap quality and industrial treatment capacity, the scrap export and the evolution of the scrap composition over time. Finally, S*ection 5* summarizes the identified challenges of steel recycling for the production of new steel in the future. A clear innovation demand for improved scrap processing has been identified to secure the quality of new steel in the future. The development of enhanced, quality-oriented separation methods of scrap is presented as a potential approach for the production of high-quality specific steel grades and applications. Innovation efforts should be placed on designing new processes for the recovery of alloying and tramp elements and on redesigning steel with a higher tolerance to tramp elements.

2.   Scrap steel classification and quality assessment

Steel is under continuous development, driven by the unceasing enhancement of mechanical and functional properties under increasingly tighter environmental standards. There are currently more than 3,500 different types of commercial steel grades, of which approximately 75% have been developed in the past 20 years [19]. The wide variety of chemical compositions and applications of steel products, in combination with the numerous collection and recycling pre-treatment routes at the product's end-of-life, generates different scrap grades or qualities.

In theory, steel, as well as other metallic alloys, are infinitely recyclable. In practice, recycling might be somewhat limited by factors influencing the recycling efficiency, such as the product design, the available recycling technologies or simply by the thermodynamics. In steel, alloying elements, like Mn, Si, Ni, Mo, V, P, S, etc., are carefully added to engineer mechanical or physico-chemical properties. In addition, galvanizing and tinplatting are zinc- and tin-coating methods to protect the steel surface from corrosion, which are well extended in automotive components and household appliances and in food and drink packaging, respectively. However, alloying elements and metallic leftovers from coating layers become impurities (or tramp elements) when used in steel products that do not require them [20]. To the authors knowledge, there is no clear definition of tramp element, nor it is clear whether they might differ depending on the steel grade. However, it is apparent that tramp elements are introduced with the scrap and that might complicate the process of recycling into new steel. Elements like copper, tin, nickel, molybdenum or cobalt cannot be completely removed from steel during recycling by using existing methods. Certain residual contents of these elements reduce the quality of the scrap and, thereby, of the new steel to be produced [21, 22]. For instance, copper contamination of scrap prevents its utilisation to produce new steel for structural applications as it reduces the tensile strength and toughness [23]; and the tin contents present in cans for drinks and food is responsible for adverse phenomena like surface hot shortness and temper embrittlement [24]. Then, the utilisation of such contaminated (or low-quality) scrap requires the addition of pure iron to the molten steel so that the residual contents of Cu, Sn, etc. are further diluted and the associated harmful effects are minimized in the new steel. On the other hand, whereas the dilution of alloying elements might be a temporary solution to recycle low-quality scrap, it is expected to result in an accumulation of tramp elements in a long-term perspective, i.e. after several recycling cycles. This could hinder to some extent the future applicability of recycled steel. All these factors, in addition to the fact that the scrap quality partially regulates its purchase price and the energy consumption during recycling, make the ferrous scrap quality of great importance to the steelmaker [25].

There are different detailed classifications of non-alloyed carbon scrap steel based on the source, product dimensions, physical characteristics and maximal content of undesirable elements and non-ferrous materials, i.e. EUROFER [26], UNE 36199:2013 [27], AISI. For exemplification, Table 1 describes the six main categories of scrap steel as a function of the maximum aimed total content of tramp elements Cu, Sn, Cr, Ni and S according to the EU-27 scrap steel specification. The sum of tramp elements Cr, Ni and Mo is usually preferred for assessing steel qualities over considering single elements due to their synergistic effect on the steel mechanical properties [20]. The tolerable content of Cu and Sn is fixed by standards for different steel qualities. Whereas bar steel might contain up to 0.06 wt. % Sn, cold rolled sheet steel only accepts 0.02 wt. % Sn. The material size is another parameter influencing the scrap classification since it plays an important role in the optimization of the metal extraction process. For example, the tin extraction efficiency from tinplated shredded cans through an electrochemical process is greatly influenced by the surface-to-volume ratio of the shreds [28].

Table 1. Main steel qualities according to the maximum aimed content of tramp elements, in the EU-27 Scrap steel Specification [26].

Category	Aimed content				General description
	Cu	*Sn*	*Cr, Ni, Mo*	*S*	
Old scrap	0.40	0.02	0.30		Old scrap in sizes not exceeding 1.5 x 0.5 x 0.5 m. Excludes vehicle body scrap.
New scrap (low residuals*, uncoated)	0.30				New production thin ferrous scrap (less than 3 mm thick). Must be uncoated and be free of unbound ribbons
Shredded	0.25	0.02			Shredded old ferrous scrap fragmentized into pieces not exceeding 200 mm. Free of excessive moisture, loose cast iron and incinerator material (especially tin cans)
Steel turnings	0.40	0.03	1	0.1	Carbon steel turnings of known origin, free from excessive bushy and free from contaminants such as non-ferrous metals, scale, grinding dust and heavily oxidised turnings or other materials from chemical industries.
High residual scrap	0.45	0.03	1		Old and new scrap. Free of excessive concrete or construction material, mechanical pieces and steriles**.
Fragmentized scrap from incineration	0.50	0.07			Household waste, fragmentized and incinerated, consisting partly of tin-coated steel cans. Free of excessive rust.

*Residuals: Copper, tin, lead, chromium, nickel, molybdenum.

**Steriles: non-ferrous metals and non-metallic materials, earth, insulation, excessive iron oxide, combustible non-metallic materials (rubber, plastic, wood, oil, lubricants, chemicals).

Although there is no single international standard for scrap steel, it is generally divided into three main categories: home, new and old scrap.

- *Home scrap* is generated during the steel production process in the steel mills and foundries. This form of scrap includes trimmings and rejects and rarely leaves the steelmaking production area. It is rapidly returned to the furnace on-site and melted again. The generation of home scrap accounts for approximately 20 – 30% of the total scrap. The introduction of the continuous casting steel production led to a significant decrease of the generated home scrap.
- *New scrap (also called prompt, industrial or pre-consumer scrap)* is generated within steel product manufacturing plants. It includes turnings, clippings and stampings leftovers resulting from manufacturing different semi-finished products. Figure 2 shows the iron consumption per type of semi-finished product manufactured and used in the EU, according to [29]. Hot rolling of flat and long products are the processes contributing the most to the generation of new scrap. The

chemical composition of this type of scrap is well known and, thus, it is typically returned to the mill that produced the steel. New scrap accounts for approximately 20% of the total ferrous scrap.

- *Old scrap (also called obsolete or post-consumer scrap)* results when industrial and consumer steel products reach the end of their useful life, typically after 10 – 75 years, except for food and drink cans, with a useful life of 1 or 2 years [7, 30]. Figure 1 shows the distribution of finished steel products used in the EU by sector. The content of elemental iron varies significantly from one commodity to another, i.e. 20% in batteries and accumulators, 40% in automotive electrical equipment or 60% in washing machines (see Table S18 in [29]). According to their use, the steel discards are categorized as municipal solid waste, construction and demolition debris, end-of-life vehicles, waste from electrical and electronic equipment and others (obsolete machines, ships, trains and planes) [31]. Old scrap accounts for approximately 50 – 60% of the total scrap. The chemical composition of end-of-life steel products is often unknown. Besides, the physical properties of steel products with a long lifespan might change and be degraded by the end of their life. Thus, old scrap requires cleaning, sorting and preparation before reusing and recycling, which increases significantly the difficulty and cost of the process with respect to home and new scrap forms. On the other hand, a portion of old scrap will never become available. The main reasons are either because it has been fully degraded (i.e. around 1% of construction steel is lost to corrosion and wear), or because it is dispersed into the environment in a manner or place that does not allow for retrieval or makes it too costly, particularly in third-countries to the EU [6]. This portion is usually referred to as *unrecoverable scrap*.

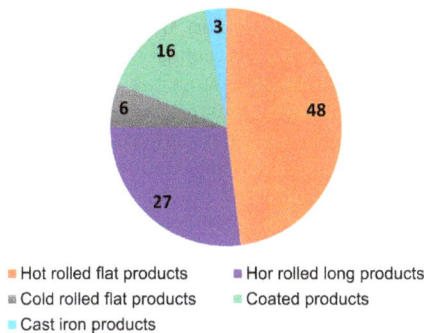

Figure 2. Iron consumption, in percentage, per type of semi-finished product manufacture in the EU. Adapted from [29].

Steel industry recycling ratios of old scrap and statistics of iron cycle for recycling often consider home scrap plus the so-called purchased scrap, defined as the addition of the recovered old scrap plus new scrap. These quantities can be well ascertained to close the iron balance. However, the unrecoverable portion of scrap is not always accounted for in the calculations of end-of-life recycling performances. Thereby, different formulae to calculate recycling rates might yield to quite dissimilar values, i.e. in the range of 31 – 75% for old scrap [29]. Using a different approach, Wang *et al.* showed that purchased scrap only contributed a quarter of the global iron and steel production in year 2000 [31].

3. Steel recycling through steelmaking

Ferrous scrap is recycled through steelmaking. In fact, recycled steel is one of the most important raw materials for the steel industry and it is gaining weight as a strategic feedstock to reduce reliance on iron ore and to cut global carbon dioxide emissions. According to data from 2017, it is estimated that the global steel industry uses, on average, about 2 billion tonnes of iron ore, 1 billion tonnes of metallurgical coal and 575 million tonnes of scrap to produce about 1.7 billion tonnes of crude steel [12]. This means that around 34% of the global new steel production is made out of ferrous scrap, although the production share varies largely among the different countries [31].

3.1 Iron value chain and main iron and steel production routes

The recycling of scrap into new steel occurs through the two main steelmaking routes: the blast furnace (BF) – basic oxygen furnace (BOF), or primary route, and the electric arc furnace (EAF), also called secondary route. The primary route is based on the utilization of fossil fuels, as both energy source and reducing agent of the iron ores, and it currently represents around the 70% of the global production [8]. The primary BOF steelmaking route is a continuous and fast process (it occurs in about 60 minutes) and the final steel usually needs to be refined through a secondary treatment to meet certain compositional requirements. Through one of the several secondary treatments in a ladle furnace or an EAF, impurities such as sulphur, oxygen, nitrogen, hydrogen or non-metallic inclusions are easily removed as the temperature and composition of the steel bath are homogenised. Nowadays, EAFs have become an essential part of integrated steel plants to supply sophisticated steel grades for continuous casting. Besides, the EAF route accepts hot metal, pre-reduced iron and up to 100% of scrap as feedstocks, providing a strategic path to improve the circularity of steel. Minor steelmaking processes include the open-hearth furnace (OHF), the direct reduction and the smelting reduction of iron ores and fines [32]. Given the scarce contribution of the OHF route to the global steel production (0.3%), only the role of the BF-BOF and EAF routes in steel recycling will be discussed in this section.

Figure 3 shows a schematic of the iron value chain, from the extraction of the iron ore to the use in society in the form of steel products and waste management, including recycling (indicated by green arrows), landfill disposal and net exports. The dashed line outlines the area of circular iron flow. Firstly, the iron ores, mainly hematite, magnetite and limonite, are extracted and transported to the industry. After an adequate preparation, typically sintering and pelletizing, the virgin iron ores are mixed with coking coal and pyrometallurgically processed to elemental iron. Producing elemental iron is called ironmaking, and it can be carried out either in a blast furnace (BF) to produce pig iron, or in a shaft furnace or fluidized furnace to produce DRI. Both are highly energy-intensive processes due to the reduction process of the oxides present in the ores to the metallic form.

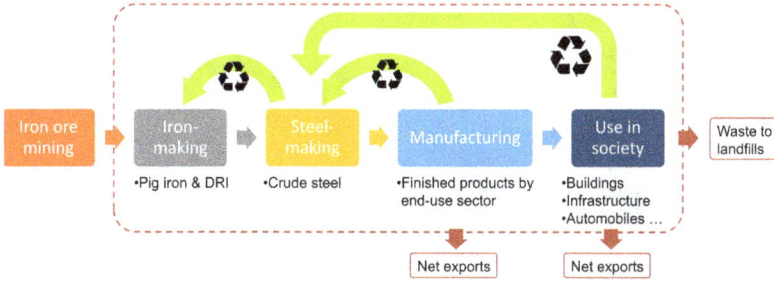

Figure 3. Schematic of the iron value chain.

DRI and pig iron are essential products to steelmaking through the primary or secondary productions routes. During the steelmaking process, pig iron or DRI are mixed with scrap steel in different proportions and other additives to produce crude steel of the desire composition. This process is done in a BOF, in the case of the primary steelmaking route, or in an EAF, in the case of the secondary route. Table 2 provides an overview and a description of the metallurgical processes involved in each step of both primary and secondary iron and steel making routes. The combined interpretation of Table 2 and Figure 3 is advised for a correct understanding of the iron value chain. The scrap generated during steelmaking (*home scrap*), which has a well-known composition, rapidly returns to the on-site furnace to be re-melted. Subsequently, the crude steel is commonly continuously cast into different semi-finished products (slabs, billets, blooms), which are further manufactured through thermomechanical treatments and forming operations such hot-rolling, cold rolling, pressing, cutting and bending. Some finished products might require successive rolling passes, coating, or additional thermal treatments. Key finished products include coils, sheets, strips, wire, rods, tubes, pipes, and their coated versions. The scrap resulted from steel product manufacturing (*new scrap*) has also a well-characterised composition. Thus, it can be returned to the steelmaking step, in an EAF, to produce fresh molten steel. Finally, when a steel product reaches its end-of-life (*old scrap*), it is recycled into new steel, reducing the need of pig iron or DRI.

Table 2. Overview of main iron and steel making routes.

1) RAW MATERIAL PREPARATION		
Coking coal, sintering and pelletizing of iron ores		
Recycled steel preparation		

2) IRONMAKING	
Blast Furnace (BF)	Direct reduction (DR) and Smelting

Blast Furnace (BF):

The iron ore, coke and limestone are charged into the BF from the top. Hot air is blown from the BF's bottom to the top to ignite a porous bed of coke as it passes through. The ignition produces intense heat, which melts the charged materials, and CO gas, which reduces the iron oxides. This results in hot metal, which flows to the BF's bottom and is regularly tapped.

Direct reduction (DR) and Smelting:

The iron oxides present in the ores are reduced into elemental iron. Natural gas and coal (C, H and CO) are common reducing agents of DR and smelting reduction routes, respectively. The solid-state reduction sequence in DR is: Fe_2O_3 (hematite) $\rightarrow$ Fe_3O_4 (magnetite) $\rightarrow$ Fe_xO (wüstite) $\rightarrow$ Fe (iron). The result is direct reduced iron (DRI), whereas hot metal is obtained in the smelter.

3) STEELMAKING	
Basic Oxygen Furnace (BOF)	Electric Arc Furnace (EAF)

Basic Oxygen Furnace (BOF):

Hot metal is converted into steel in the BOF at 1200-1700 °C by blowing oxygen into the melt. This oxidizes Mn, Si, P and Fe, forming slag. The addition of slag formers like lime, dolomite or silicon carbide aids the process and the desulfurization of the melt. At a later stage, C is also oxidized as CO and CO_2. The metal-slag interaction leads to the exchange of iron through reduction-oxidation reaction. Improving the steel quality requires extensive metallurgical reactions. The melt is tapped once it reaches the target composition.

Electric Arc Furnace (EAF):

The charging mix (DRI or hot metal, scrap and slag formers) is molten at around 1800 °C by the electric arc generated between the graphite electrode and the metal. Oxygen is injected to accelerate the melting process and to remove the impurities (P, C, Al, Si, Mn) that deteriorate the steel quality. Such oxidation leads to the formation of a slag layer above the melt. The melt is tapped once it reaches the target composition. The average energy and electrode consumption are of 550 kWh and of 4 kg per ton of crude steel [33].

3.2 Key aspects of steelmaking influencing steel recycling

Steel is a permanent circular material in the society; however, steel recycling implies steel making, and the iron and steel making industries are today responsible for the 8% of the CO_2 emissions on the planet [11]. To drastically reduce the carbon footprint of the steelmaking industry, in July 2021, the European Commission adopted a packet of proposals to make the European Union's (EU) climate, energy, land use, transport and taxation policies fit for reducing net greenhouse gas emissions by at least 55% by 2030 [14]. In steel production, essential measures are the replacement of coal-based with green energy carriers, like biomass or green hydrogen, the development of energy- and resource efficient clean technologies and the replacement of primary raw materials, like DRI, with an increased and optimised utilisation of recycled steel. In this sense, the development of sustainable processes will play a central role. Key aspects are analysed and contrasted per main steelmaking route in the following paragraphs.

3.2.1 Type of feedstock and steel quality

High-quality flat products are typically produced through the BF-BOF route. Based on realistic date from 21 existing BF and BOF steelmaking plants in different EU member countries, the BF uses 1,600 kg of iron ore, sinter and pellets, 360 kg of coke and 26 kg of limestone to produce 1 ton of hot metal (pp. 304, in [33]). Additionally, the BOF uses an average of 800 kg of hot metal, 220 kg of scrap, 10 kg of iron ore, 30 kg of additional Fe material, 0.2 kg of coke and 65 kg of lime and dolomite to produce 1 ton of crude steel (pp. 369, in [33]). Only strictly measured volumes of selected home and new scrap compositions are mixed with the hot metal due to the difficulty to remove contaminants in the BOF. Aside from steel quality reasons, a 25–35 % of scrap is the theoretical maximum that might be dosed up in the BOF not to risk the stability of the operation due to temperature decrease.

The EAF exhibits more versatility for feedstock, as it can be fed with DRI, hot metal or up to 100 % of scrap. On average, the recycled steel-EAF route uses 1100 kg of scrap, up to around 150 kg of hot metal, up to 215 kg of DRI, 16 kg of coal (including anthracite and coke) and 80 kg of limestone to produce 1 ton of crude steel (pp. 429, in [33]). The selected proportion of feedstock will depend on the target steel quality and on the quality and the amount of available scrap. Current established scrap sorting and pre-treatment methods usually result in low scrap qualities. Thus, for the production of highly demanding steels, up to 50% - 40% DRI is typically input to ensure the dilution of the tramp elements in a large mass of material. Larger amounts of scrap in the EAF leads to steel qualities that are mainly used to produce large construction steel shapes. Since the content of residual elements introduced with the scrap can be difficult to control, high quality steels with stringent requirements on impurities can be more difficult to manufacture using the EAF process than the BOF process.

New Materials for a Circular Economy   Materials Research Forum LLC
Materials Research Foundations 149 (2023) 204-235   https://doi.org/10.21741/9781644902622-7

3.2.2 Energy consumption

With respect to the preparation of raw materials, it is estimated that the energy consumption for sintering the iron ores is in the range of 0.82 – 1.54 GJ/ton. As for the production of coking coal, coking facilities at integrated steel plants spend around 3.5 GJ/ton (pp. 224, in [33]).

According to data collected from different EU member countries, about 16 % of the input energy in the BF-BOF route comes from natural gas, about 12 % from electricity and the remaining 52 % from coke oven gas (COG), BF and BOF gas recovery [33]. Coal and natural gas are used as energy sources for heating and as reductant agents of the iron ore into hot metal [34]. The variation of energy inputs depends on the energy management at different sites. It is important to mention that energy recovery through converters gas recovery lowers the CO_2 generation from the use of fossil fuel and electricity by around 0.05 t_{CO2}/t of steel. The BF and BOF iron and steel making energy intensities are estimated in 1.54 GJ/ton of hot metal and 0.67 GJ/ton of crude steel, respectively. The large energy consumed in the BF is mainly due to the high chemical energy needed to reduce the iron ore to pig iron using carbon-based reducing agents [28]. Though, the BOF energy consumption might vary significantly depending on the refining operation [35]. The BOF process comprises complex metallurgical reactions acting simultaneously through intense heat production. Since all the reactions are exothermic, no external heating sources are needed. In fact, scrap and coolants are usually added to avoid overheating and to keep the temperature in the range of 1600 – 1700 °C.

In the case of the EAF route, the energy input from coal is 48 % from electricity and 52 % from natural gas and liquid fuels (pp. 429, in [33]), although it offers the possibility to be completely powered by renewable electricity. On average, the recycled steel-EAF route uses 2.16 GJ of electricity to produce 1 ton kg of crude steel [35]. That is around one-eighth of the energy used through the BF-BOF route from iron ore [36]. Nevertheless, the EAF energy demand might increase with the decrease of the scrap quality, i.e. with increasing contents of P, S, Sb and Cu [25] or with the increase of the DRI portion in the feedstock mix since melting DRI requires higher temperatures than melting scrap. It should be born in mind that the production of DRI is also a highly energy demanding process, which requires on average around 2.7 GJ/ton.

3.2.3 CO₂ emissions

The BF-BOF steelmaking process emits an average of 1.85 tonnes of CO_2 per tonne of crude steel produced [13], although the CO_2 emissions vary significantly per territory and country. Figure 4 shows the CO_2 intensity of the BF-BOF route (top) and EAF route (bottom) for different regions. For instance, whereas the BF-BOF process emits of average 3.0 t_{CO2}/t_{Steel} in India, the BF-BOF process powered with biomass (charcoal) in Brazil emits around 1.5 t_{CO2}/t_{Steel}. However, this figure increases above 2 t_{CO2}/t_{Steel} if the charcoal is not considered as a carbon neutral fuel. Indirectly, the iron ore extraction and transportation also add up a carbon-footprint and extra energy expense to the whole process [37].

Figure 4 highlights that the EAF steel production route offers opportunities for CO_2 mitigation within the EU's circular economy strategy, although it might largely vary depending on the feedstock. Current figures of the combined direct and indirect emissions of the EAF route show ranges of 0.6 – 1.5 t_{CO2}/t_{Steel} for DRI-based raw steel production, and 0.4 – 0.5 t_{CO2}/t_{Steel} for the scrap-based raw steel production [34, 37]. By comparing the CO_2 emissions of the key established processes (BF-BOF, DRI-EAF and scrap-EAF), there is potential for a 30 – 80% CO_2 abatement with scrap-EAF. However, interchanging the processes might be not possible or economically feasibility as the DRI-EAF and scrap-EAF processes are linked to the availability of scrap and carbon lean fuels, like natural gas.

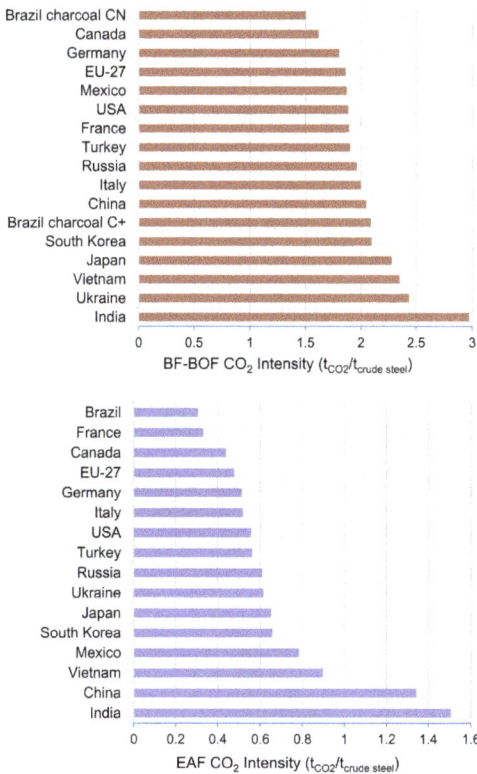

Figure 4. Intensity in direct plus indirect emissions of CO_2 of the BF-BOF (left) and EAF (rigth) steel production in the several countries in 2019 according to data from [38]. Brazil charcoal CN and C+ represent the data considering charcoal carbon neutral or not, respectively.

3.2.4 Penetration of new sustainable technologies

The BF generates the largest CO_2 flow of a conventional integrated steel mill due to the substantial dependency of coking coal as energy source and reductant. Although the process itself makes it difficult to use other reductants, the utilisation of pulverized coke, oil or natural gas, is an established practice to reduce the coal consumption in the blast furnace. The additional injection of hydrogen as auxiliary reductant, through the so-called Hot Reducing Gas (HRG) injection; i.e. a mixture of H_2 and CO, has been also evaluated [39, 40]. Yet, it is claimed that a theoretical minimum of 200 kg of coke per ton of produced pig iron is necessary to ensure an adequate furnace operation [39]. In this light, the EAF route rises as the main alternative to the BF-BOF route since it has the potential to be powered by 100% renewable electricity and to produce new steel from 100% recycled steel, implying an abatement of CO_2 emissions in up to five times and a reduction of energy consumption with respect to the BF-BOF route.

Nevertheless, improved energy efficient and heat-recovery technologies, such as coke dry quenching (CDQ) for the coking process, top-pressure recovery turbines (TRTs) for BFs and continuous casting has helped process optimization and reducing the energy intensity of the BF-BOF steel production [41]. In 2004, a group of European steel companies with the support of the EU, started the Ultra-Low CO_2 Steelmaking (ULCOS) project [42], aiming at identifying technologies to decrease by half the CO_2 emissions per tonne of steel by 2050. Meeting this goal requires the deployment of anthropogenic carbon dioxide removal (CDR) technologies [43]. Recent developments include innovative BF, smelting, direct reduction and carbon capture storage and utilisation (CCS/U) technologies, as well as of new carbon-lean forms of energy. Besides their great potential for GHG emissions abatement, these technologies ensure energy efficiency and a smart utilisation of resources without compromising the quality of the final steel product [41]. The most important advances are described in the next paragraphs:

- **CCS/U:** CO_2 gas separation-purification technologies are strategies commonly applied to exhausted gases of different processes of existing steel plants. These technologies can be based on chemical or physical absorption, membrane separation, carbonation or cryogenic separation. Once captured, the CO_2 might be compressed, transported and stored, typically in geological formations [44]; or it might be utilised, generally for chemical reactions leading to the formation of carbonates, methane, methanol, urea for fertilizers or to photosynthesis processes in greenhouses [45]. Around a 30% of the BF's CO_2 emissions could be captured applying current end pipe technologies to existing BF and other gas flows. Replacing or retrofitting conventional BF with top gas recycling blast furnaces (TGR-BF) offers another promising opportunity for CO_2 gas sequestration and separation from the CO-rich stream, which is returned to the BF [46].

- **Smelting reduction processes:** The smelting process might include a pre-reduction of the ores, such as the FINEX [47] and COREX [48] processes in a shaft furnace, or not, such as the HIsmelt [49] process in fluidized bed or HIsarna [50]. These technologies are already in a commercial scaling stage. The CO_2 intensity of FINEX and COREX is only slightly lower than that of the BF. However, the resulting hot metal quality compares well

New Materials for a Circular Economy   Materials Research Forum LLC
Materials Research Foundations 149 (2023) 204-235   https://doi.org/10.21741/9781644902622-7

with that obtained in the BF, even if lower quality raw materials are used (higher scrap portions), thus reducing the process costs. A relevant example is POSCO's Finex pilot plant in Pohang (South Korea) [47]. In contrast, the HIsmelt and HIsarna processes directly use iron ore fines and coal without pre-treatment through sintering and coking. The HIsarna Tata Steel's pilot plant in IJmuiden (The Netherlands) uses a smelting cyclone to simultaneously pre-reduce and melt the injected particles of unagglomerated iron ore, which instantly melt and drop to the bottom. Temperatures above the melting point of iron are reached, and oxygen is injected to increase further the temperature of the process gasses and to react with the CO present. Coal is injected into the bottom of the vessel for the final reduction of the iron oxides. This leads to the formation of slag and pure liquid iron, which can be then tapped [51]. HIsarna offers up to a 20% CO_2 reduction, which can be increased up to 80% in combination with CCS/U technologies, as well as an important reduction of other GHGs like fine dusts, CO, NO_x and SO_x. Additionally, the process eliminates the need for coking and sintering and allows for less quality requirements of the raw materials, i.e. accepts larger portions of scrap (up to 35%) [52].

• **Production of DRI:** Several coal and natural gas-based processes have been developed as alternatives to the blast furnace during the last two decades. Representative examples are the MIDREX [53] and TENOVA-HYL [54] processes. Nevertheless, the location of these plants is restricted to natural gas producing areas. The direct reduction (DR) process of the iron ore pellets using natural gas has attracted great attention, as it makes the production of DRI a 70% less CO_2-intensive than the production of molten steel by the BF-BOF route [55]. The reduction of the iron ores occurs in solid state at temperatures well below the metal's melting point. Thus, there is not melting and no slag formation involved. Instead, a porous sponge iron (DRI) is obtained. Intermediate iron oxide variants form during the reduction reaction, following the sequence: Fe_2O_3 (hematite) $\rightarrow$ Fe_3O_4 (magnetite) $\rightarrow$ Fe_xO (wüstite) $\rightarrow$ Fe (iron). The first reduction steps are quite efficient, being the conversion of Fe_xO into Fe the rate limiting step. The obtained DRI can be directly sent to the EAF to be converted into steel.

• Recently, the spotlight has been placed on the switch to a hydrogen-based DRI production by adapting existing plants, powered by natural gas. The hydrogen-based ironmaking technology is currently being demonstrated by initiatives like HYBRIT[1] [56]. The basic concept behind is to use green hydrogen, i.e. generated through water electrolysis using renewable energy, as energy source and as a reducing agent to produce DRI [57-60]. In addition to the iron ore pellet size and porosity, the reduction rate by molecular hydrogen is strongly influenced by the reduction kinetics of Fe_xO, which is endothermic [61]. This is not the case when plasma hydrogen (ionized species, electrons, molecules) participates in the reduction process. The reduction of molten wüstite by plasma hydrogen becomes exothermic since the high energy content of the plasma is transferred to the reaction interface [62, 63]. In fact, the plasma hydrogen reduction (HPR) process is much more efficient than the commercial direct reduction conducted in shaft furnaces using reformed natural gas or other hydrocarbon gases (Midrex process) [64]. Firstly, the reduction kinetics

---

[1] HYBRIT project: Hydrogen Break-through Ironmaking Technology

New Materials for a Circular Economy   Materials Research Forum LLC
Materials Research Foundations 149 (2023) 204-235   https://doi.org/10.21741/9781644902622-7

is faster and it allows for the simultaneous melting of the feedstock (both, iron ore and scrap). In addition, the production of hydrogen-based DRI emits water vapour instead of CO_2, enabling up to a 100% CO_2 abatement with respect to the conventional process. Although the hydrogen plasma reaction is in exploratory stage yet, many technical aspects are already established in existing EAFs, i.e. operational conditions of electric and plasma arcs (see pp. 303–375 in [41]). On the other hand, the infrastructure and technology needed for the competitive and mass production of green hydrogen is at a low technology readiness level, and it will remain a bottleneck for the steel industry during the next two decades [36]. For exemplification, the production of a ton of crude steel requires around 60 kg of hydrogen [16]. Thus, satisfying the current annual production of 1,900 Mt of crude steel would require around 114 Mt of green hydrogen. One proposed solution for the hydrogen transition is the hybrid DR and HPR strategy, which would turn out to be a more efficient strategy to produce hydrogen based DRI than the conventional DR and HPR processes alone. A conventional DR step is, firstly, applied to the iron ore to obtain a semi-reduced product (mainly FeO), which is then transferred to a hydrogen plasma furnace to complete the sluggish reduction of FeO into Fe [65].

- **The electric arc furnace (EAF):** The EAF process is becoming increasingly cost and quality competitive, and is gaining market share in the steelmaking industry due to its higher energy efficiency and versatility to produce new steel from a variety of feedstock mixtures (hydrogen-based DRI, hot metal and scrap) with respect to the BF-BOF route. Besides, the EAF process itself allows for a reduction of the CO_2 emissions and raw material extraction and it can be used to recover iron and other valuable metals from steelmaking by-products (slag, dust, sludge or rolling mill), metallurgical and social wastes (plastics and refractory materials) [66-69]. For instance, the EAF dust collected by filtering the off-gas contains heavy metals and oxides that vaporize during the melting process, i.e. up to 40 wt. % Zn, Cd, Pb, Cr and Ni. The leaching potential of these metals leads to their accumulation in groundwater and soil. Emerging pyrometallurgical and hydrometallurgical processes partially allow for the recovery of those metals, providing new avenues for raw material procurement and solutions to landfilling of by-products and wastes and to environmental issues caused by heavy metals [70-73].

Although the continuous development and improvement of production process have helped boosting the efficiency and productivity of the EAF furnace, while reducing the electrode consumption (see [41], pp. 309-310), reducing the energy consumption is one of the main current challenges. The EAF energy intensity is largely impacted by the quality of the scrap input as raw material and by the DRI portion [21, 25]. In this sense, the development of new scrap sorting and pretreatment methods based on strict compositional analyses will play a key role, as it will be shown in the next section.

4. Present scheme for steel recycling

The global production of steel is forecast to keep continuously growing since steel is the essential structural material to our modern society [1]. The previous sections of this chapter highlight that recycled steel is, today, the strategic resource of the steel industry. In this sense, the secondary EAF steelmaking route has the potential to play a decisive role in the future of steel in the European circular economy context. The EAF steelmaking route reduces the energy and CO_2 intensity with respect to the primary route, as it can be completely powered by renewable energies and it allows for the utilisation of a variety of feedstock, including up to 100 % scrap steel. For these reasons, and driven by active legislation linked to climate change mitigation actions to reduce the reliance on virgin raw materials and on fossil fuels combustion, it is expected that the scrap share in steel production keeps increasing [2, 14, 74]. Nevertheless, due to current scrap exports and the mismatch between scrap delivery and current crude steel production, there will not be sufficient scrap available to satisfy the steady growing global steel demand in the future. Besides, the delay between old scrap delivery and the composition requirements of the new steel grades along with the limited scrap treatment capacity of the industry, results in scrap of low quality, which can only be used after dilution with pure iron to produce new steel. These aspects are reviewed in this section with the aim to pinpoint future challenges of the recyclability of steel.

4.1   Scrap steel quality and treatment capacity

The numerous sources and forms of old scrap require the use of a number of sorting and preparation techniques to remove contaminants and non-metallic parts, and to recover valuable metals prior to entering the steelmaking process. Old scrap, as well as new scrap, frequently contains alloying elements and corrosion protective metallic coatings to grant specific mechanical and in-use properties. However, main collection and classification methods are only based on the scrap shape and origin. Established methods include manual sorting, shredding, scrap size reduction, magnetic and Eddy current separation, decoating, dezincing of galvanized scrap, detinning and decopperization and radioactivity detection [75]. Although characterisation techniques, like inductively coupled plasma optical emission spectrometry (ICP-OES), X-ray diffraction or energy dispersive X-ray spectrometry (ED-XRF) [28, 76], allow for the compositional analysis and sorting of the scrap, their utilisation is not extended or commercially available yet. From existing analyses of the overall steel and scrap flows in the literature, it can be drawn that the established scrap classification and recycling practices result in an uncontrolled mixture of different metals [29, 77, 78]. This results in the dissipation of such metals and in the decrease of the scrap quality.

Table 3 summarizes the content of alloying elements in scrap steel from different sources and qualities after conventional treatment. It is striking the relatively high Cr content present in end-of-life vehicles (ELV) derived scrap and combustion engines, which in addition to the Ni and Mo contents is above the maximum aimed composition specified by the EU-27 Scrap steel Specification in Table 1. These elements cannot be removed by simple metallurgical processes and, thus, such scrap must be used by dilution. Around 40 % of the ELV combustion engine scrap is typically fed in the EAF for steel recycling by dilution with pure iron. This fact has been evidenced and quantified by studies carried out in Japan. Ohno *et al.* [21, 22] showed that 93% of steel old scrap was recycled to produced carbon steel in Japan in 2005, hence diluting the content in alloying elements into carbon steel and steelmaking by-products. Whereas admissible levels of alloying elements within carbon steel causes no harm, it makes it virtually impossible to recover them or capitalize on them. Tracking methodologies of alloying elements over multiple steel recycling cycles demonstrate that the functionality of elements, like Cr and Ni, could be retained over long periods of time only under a high level of scrap sorting [79]. For instance, a parts-based scrap sorting approach according to industrial standards of steel grades exhibited maximum potential to quality-oriented recycling of end-of-life vehicles (ELV) and to high-yield recovery of alloying elements like Mn, Cr, Ni and Mo [80].

Similar to Cr, Ni and Mo, the tin content present in food cans after detinning (0.20 ± 0.12, in wt. %, in Table 3) is also well above the maximum aimed Sn content specified in Table 1. Although detinning and dezincing processes have been in commercial use for a long time, the recovery of low zinc and tin contents through these methods is not cost-effective. Dezincing might be performed by leaching in either acid or basic solutions, which after few hours results into recovery rates higher than 90%. Alternatively, vacuum treatment at 700 °C yields around 97% zinc recovery. In a similar manner, commercial electrolytic detinning could decrease the content down to 0.02 wt. % Sn; however, it is only suitable for scrap where the tin is present as a surface coating and it requires quite large volumes of scrap to be economical [75]. The highest contents of tin are found in tinplated shredded cans. Current industrial processes treat shredded cans by static tanks or rotatory drums, which reduce the effective tinned surface expose to the electrolytic process and, thus the Sn yield recovery efficiency to below 70% [81].

Tin, as well as copper, are nobler metals than iron and, thus, cannot be removed from the melt by oxidation in the EAF since the iron oxidizes preferentially and the copper ends up at the oxide/steel interface [25]. During hot rolling processes, a Cu-rich liquid layer forms once the Cu enrichment in the austenite phase exceeds its solubility. The liquid Cu might then penetrate into the steel through the austenite grain boundaries, causing intergranular decohesion. This fracture failure is augmented by the presence of Sn, Sb and As since these elements decrease the melting point of Cu and its solubility in the austenite phase [24]. Similar difficulties are found to remove metals such as Zn and Pb [82].

Table 3. Estimated contents of alloying elements present in different scrap steel types and qualities after treatment, from studies performed in Japan and in the European Union. ELV stands for end-of-life vehicles.

Scrap source	S	P	Cu	Sn	Cr	Ni	Mo	Mn	Ref.
ELV derived scrap					0.67	0.16	0.02	0.38	[22]
ELV combustion engine					0.71	0.20	0.03		[21]
Cars, trucks, transport				< 0.18					[83]
Tinplated food cans				0.20 ± 0.12					[28]
Packaging				< 0.18					[83]
Municipal solid waste	0.19 ± 0.02	0.03 ± 0.01	0.64 ± 0.08	0.04 ± 0.01					[25]
Shredded scrap	0.06	0.02	0.40 ± 0.01	0.02					
Turnings	0.07	0.04	0.30 ± 0.05	0.02 ± 0.01					
Sheet metal	0.02		0.09 ± 0.01	0.01					
Electrical strip				< 0.18					[83]
Domestic appliances				< 0.18					
Buildings & infrastructure (hot rolled strips and tubes)				0.18 - 0.25					
Buildings & infrastructure (wire rods)				0.25 - 0.35					
Buildings & infrastructure (Reinforcing and hot rolled bars; heavy, light and rail sections)				> 0.35					
Steel bars (new and old scrap)	0.028 ± 0.04	0.032 ± 0.01	0.32 ± 0.10	0.018 ± 0.07	0.22 ± 0.15	0.10 ± 0.03	0.014 ± 0.01	0.80 ± 0.10	[20]

Nowadays, scrap steel is yet regarded as an iron source, even in regions like the EU and Japan, where automobile recycling laws under 3E policies and ELV directive[2] apply [84]. Nevertheless, scrap steel also shows great potential as secondary source of alloying elements, lately termed as *urban mining* in contrast to virgin raw materials mining. Urban mining involves additional benefits, such as a reduction of the carbon footprint and the costs associated to mining activities of virgin materials. In a similar manner, steelmaking wastes and by-products, like slag, dust or mill scale, offer another alternative raw material source, given their quite high content in iron and other valuable metals [66, 68, 70, 85-87].

---

[2] European Commission. Directive 2000/53/ec of the European parliament and of the council of 18 September 2000 on end-of-life vehicles. 2000.

According to the action plan for a circular economy released by the European Commission in 2015, a 75% of the packaging waste generated must be recycled or prepared for reuse by 2030, while landfilling of all wastes should be reduced to 10% [88]. Thus, the development of new valorisation routes and sustainable processes for metal recovery from scrap steel, steelmaking by-products and wastes would enhance the circularity and value chain of iron while reducing unnecessary landfilling to the environment.

In addition to the steel-quality oriented benefits, scrap quality-oriented recycling is also of great importance to the recycling process itself since the scrap quality strongly correlates with the electricity demand of the melting process [25]. Low-quality scrap recovered from bottom ashes of municipal solid waste incineration (MSWI), which usually contains sulphur, tin and copper concentrations (Table 3), demanded up to a 45% higher electricity in the electric arc furnace (EAF) than high-quality scrap[3]. In the case of sulphur and phosphorous, higher electricity is demanded as longer operation times are needed to reduce their concentration in the melt [25]. Therefore, old scrap exceeding the maximum contents of tramp elements (Table 1) can only be utilized in the European steel industry by dilution with primary iron sources. Instead, scrap steel from sheets has a high quality and exhibits the best recycling performance. In general, EAF allows for a more intensive removal of contaminants than BOF, although in both cases, BOF and EAF, around an 8 wt. % of the removed contaminants are lost, either through stack emissions or through incorporation to the slag [79]. Common air hazardous pollutants are Cr, Mn, Pb, Ni.

In summary, the development of recycling approaches combining quality-oriented recycling and new urban mining methodologies would prevent the loss of valuable alloying elements by dilution in the scrap and their accumulation in recycled steel over time. This is extremely important to increase the quality of new steel and to assure the circular economy of steel and valuable metals in the future, while decreasing the energy consumption of the recycling process and its carbon footprint.

4.2   Scrap steel availability and exports

The availability of scrap at a certain point in time is defined by the past production and the ongoing recycling rate. Steel often remains in products for an average of 40 years [7]. Thus, the amount of steel that is up for recycling today is proportional to the amount of steel produced few decades ago. In addition to the past production and durability of steel products, the availability of scrap is linked to the exports' regulations by territory. Figure 5 shows the evolution of total scrap consumption in the EU and the net exports from the EU during the last decade. The destiny of the exports of year 2019 is indicated in the dash-lined graph. Significant amounts of ferrous metal scrap and end-of-life steel products are not utilized by the European industry, but exported to third countries, usually, of lower environmental, climate, labour and social standards [11].

---

[3] Quality ranking according to limit values for copper, tin and chromium-nickel-molybdenum and bulk weight in Swiss scrap steel nomenclature (Stahl Gerlafingen AG and Swiss Steel AG 2010).

Finished products manufactured and used in the EU

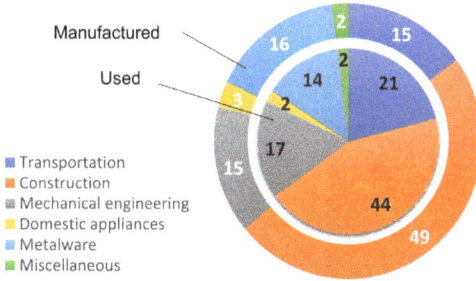

Total scrap consumption and net export in the EU

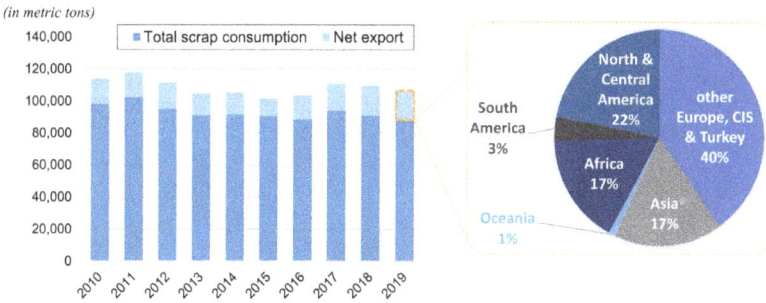

Figure 5. Finished steel products manufactured and used in the EU (top). Evolution of the total scrap consumption and net exports, in metric tons, in the EU from 2010 to 2019, according to data from [11] (bottom).

The scrap prices may vary substantially on a monthly basis [89], being sufficiently high to cover transport costs and scrap processing in third countries. According to Eurostat, EU total exports of ferrous scrap in 2021 (26% in volume) represented 48% of all exported recyclable materials, being Turkey the first export destination.

Scrap is often exported for quality reasons. Current export practices cause a rather limited scrap stock in the EU, particularly of high-quality scrap. Quality constrains and limited onsite production capacity of secondary raw materials of the industry are, therefore, main reasons for the current export figures. Low quality scrap has a low recycling yield due to its low bulk density and high amounts of impurities and tramp elements, like copper and tin, which reduce the steel quality. However, as warned by EUROFER in May 2022, current provisions on waste exports risk to undermine the EU's Circular Economy and Green Deal objectives by compromising our chance to capitalize on valuable secondary raw materials [17]. Although, up-to-date data and analyses of global crude steel and scrap steel flows are scarce in literature, an analysis from 2018 estimated that terminating the scrap steel exports could potentially increase the secondary steel production share in the EU by almost 10% [18]. Yet, according to EUROFER data, the amount of ferrous scrap exported from the EU has kept rising and the total scrap consumption has kept decreasing during the past decade (Figure 5).

In summary, the future expansion of scrap-based steel production will depend on the availability of high-grade scrap. However, it is expected that the limited overall availability of scrap causes a decrease in the scrap quality and an increase of the scrap costs. Limiting the export of scrap to non-EU countries, promoting the use of best available technologies (BATs) and fostering innovation of scrap refining and recycling solutions are promising policy options to ensure the availability of a sufficient amount of high-quality scrap in Europe [90].

4.3 The evolution of scrap steel composition over time

The development of modern steels during the last decades have demanded the addition of numerous alloying elements (Si, Mn, Al, Cr, Ni, V, Cu, Co, Mo) to control the microstructure development during the thermo-mechanical processing and to grant certain mechanical and functional properties to the final steel products. Steel grades for highly demanding applications are good examples, such as the last generations of advanced high-strength steels for the automotive sector, which typically contain Mn, Si, Mo, Al [91], or the creep-strength-enhanced ferritic steels for power generation components, containing microadditions of V, Nb, Ti and Cr to provide grain size stability at temperatures around 600 °C [92]. It is also important to highlight the role of coatings as one of the most common methods of improving the corrosion resistance of steels. Typical coatings are made of a Zn or Zn-Sn alloys or combine Al, Mg and Si; but the application of organic compounds and paints is also common.

Considering the presence and tolerance levels of major tramp elements in steel since 1946 (Cu, Sn, Cr, Ni and Mo), a material pinch analysis in the EU revealed a surplus of scrap steel with higher levels of tramp elements since the 1990ies [83]. Whereas until 1980 most of the scrap generated was new scrap, nowadays the amounts of old scrap are dominant. The generation of higher shares of old scrap relates to the increment in the content of tramp elements. Since the steel production in the EU is mainly dedicated to high-quality flat steel products, with a low tolerance to tramp elements, the surplus of low-quality scrap is largely exported to third countries.

The unceasing increment in tramp element concentrations in the scrap steel observed in the EU since the 90ies is also found in the scrap steel flows in the United States (US) [83]. However, the US steel industry is capable of producing larger amounts of flat products from scrap/hot metal/DRI mixtures through the EAF route than the EU. A similar scrap surplus and increment in tramp element concentrations in the scrap to that observed in the EU and the US can be expected in the next decades at a global level, as more steel products reach the end of life in the countries under development. This will create instabilities in the established export practices, making current international trade of scrap unsustainable.

5.   Summary and present challenges of steel recycling

The production of new steel from scrap steel goes hand in hand with the decarbonization of the steel industry. Steel recycling, particularly through the electric arc furnace (EAF) steelmaking route, will play a decisive role in the decarbonization and future of the steel production in the European circular economy context. However, the limited availability of scrap, especially of high-quality scrap, However, the limited availability of scrap, especially of high-quality scrap, along with the global increasing demand of steel, makes it not possible to satisfy the continuously growing global demand for steel based only on recycled steel. In addition, it is forecast that several steel recycling cycles will lead to an accumulation of tramp elements (like Cu, Sn, Zn, Cr, Ni and Mo) in the steel over time, which would be further magnified by the steady increment of alloying elements detected in steel since the 1990ies. These factors, along with the limited scrap treatment capacity of the industry, result in low scrap quality, which can only be used after dilution with pure iron to produced new steel. Particularly, the utilization of old scrap at higher rates will challenge the local and global steel industry.

In this light, *the viability of steel production through different routes and from different feedstocks must be preserved and improved to ensure the EU steel sector's circularity, competitiveness and capacity for delivering high-quality steel at low environmental cost.* The modernization of the steel industry is right ahead and demands a systemic change involving the full value chain of steel, from production to recycling. To achieve this main goal the following scientific and technology challenges of the steel industry must be tackled:

To enhance the onsite industrial treatment capacity for secondary raw materials. The improvement of the circular economy of steel is only possible through the development of enhanced, quality-oriented separation methods of old scrap to increase the quality for the production of specific steel grades and applications. One of the most important steel recycling challenges is to ensure the adequate quality of the new steel products by minimizing contamination with alloying elements nobler than iron, since they cannot be removed during the melting process. Preparation techniques oriented to a quality and composition-based scrap sorting methodology, as well as metal recovery treatments, would greatly benefit steel recycling by:

1) Increasing the quality of the new steel, as only selected scrap grades of similar composition would be used to produce new steel, and preventing the further dilution of alloying elements over time, making impossible their recovery in the future. This is particularly important for scarce alloying elements, like Ni.

2) Improving the efficiency and reducing the GHG emissions of steel recycling since low-quality scrap, whose conversion into steel in the EAF is highly energy demanding, would be separated from high-quality scrap beforehand.

3) Turning scrap, as well as steelmaking by-products and wastes, into indispensable sources for valuable metals (*urban mining*), as scrap grades of similar compositions could be processed together in the EAF, yielding higher metal recovery rates.

Recycling methods at the end of the steel product's life will require a continuous optimisation over time in order to keep up with the unceasing development of new steels and the penetration of novel sustainable technologies.

To drastically decrease the energy, resources and CO_2 intensity of the iron and steel making industries. The future of steel recycling requires the optimisation and adaptation of the primary and secondary steelmaking routes through the implementation of innovative clean technologies, allowing for the utilisation of different energy vectors and reductants with increasing carbon-to-hydrogen content ratios (H_2, CH_4, biomass) and enabling different types of iron feedstocks (scrap steel, iron ores, wastes or steelmaking by-products). Besides recycling scrap, electric arc furnaces might also be used for recovering internal wastes, like electric arc furnace dust, slags, and refractory materials. The development of new valorisation routes and sustainable processes for metal recovery from scrap, by-products and wastes would enhance the circularity and value chain of iron while reducing unnecessary landfilling to the environment.

References

[1] Worldsteel. World Steel in Figures 2022: https://worldsteel.org/steel-topics/statistics/world-steel-in-figures-2022/#steel-production-and-use-geographical-distribution-2021. 2022.

[2] EuropeanCommission. Paris Agreement 2015: https://ec.europa.eu/clima/eu-action/international-action-climate-change/climate-negotiations/paris-agreement_es. 2015.

[3] Worldsteel. Steel in the circular economy: A life cycle perspective: https://worldsteel.org/publications/bookshop/circular-economy-life-cycle-steel/ 2015.

[4] Statistica: Distribution of steel end-usage worldwide in 2019, by sector. https://www.statista.com/statistics/1107721/steel-usage-global-segment/#:~:text=In%202019%2C%20the%20building%20and,percent%20of%20total%20steel%20demand.

[5] Broadbent C. Steel - the surprising recycling champion: https://worldsteel.org/media-centre/blog/2018/steel-surprising-recycling-champion/. In: Association W, editor, 2018.

[6] Bowyer J, Bratkovich S, Fernholz K, Groot H, Howe J, Pepke E. Dovetail Partners Outlook 2015.

[7] AcerlorMittal. By-products, scrap and the circular economy: https://corporate.arcelormittal.com/sustainability/by-products-scrap-and-the-circular-economy. vol. 2021.

[8] NL TS. Sustainability in packaging.
https://www.tatasteeleurope.com/packaging/sustainability. 2019.

[9] Ecoacero. Ecoacero: Ecología para el reciclado de la hojalata.
https://ecoacero.com/informes/reciclado-de-acero-datos-2016/. 2016.

[10] Company M. The future of the European steel indutry: a roadmap toward economic and environmental sustainability. 2021.

[11] EUROFER. The European Steel Association. European Steel in Figures (2011-2020): https://www.eurofer.eu/assets/Uploads/European-Steel-in-Figures-2020.pdf. 2020.

[12] Worldsteel. Raw materials: Maximising scrap use helps reduce CO2 emissions. https://worldsteel.org/steel-topics/raw-materials/. 2017.

[13] Worldsteel. https://www.worldsteel.org/publications/position-papers/climate-change-policy-paper.html. 2020.

[14] EuropeanCommission. Delivering the European Green Deal: https://ec.europa.eu/info/strategy/priorities-2019-2024/european-green-deal/delivering-european-green-deal_en. 2021.

[15] Fischedick M, Marzinkowski J, Winzer P, Weigel M. Journal of Cleaner Production 2014;84:563. https://doi.org/10.1016/j.jclepro.2014.05.063

[16] Bhaskar A, Assadi M, Somehsaraei HN. Energy Conversion and Management: X 2021;10:100079. https://doi.org/10.1016/j.ecmx.2021.100079

[17] EUROFER. Stop waste and scrap export to countries not meeting EU environmental and social standards, asks EUROFER. 2022.

[18] Fellner J, Laner, D., Warrings, R., Schustereder, K., Lederer, J. Detritus 2018;2:16. https://doi.org/10.31025/2611-4135/2018.13666

[19] Worldsteel. About steel: https://worldsteel.org/about-steel/about-steel/.

[20] Daigo I, Tajima K, Hayashi H, Panasiuk D, Takeyama K, Ono H, Kobayashi Y, Nakajima K, Hoshino T. ISIJ International 2020;advpub.

[21] Ohno H, Matsubae K, Nakajima K, Nakamura S, Nagasaka T. Journal of Industrial Ecology 2014;18:242. https://doi.org/10.1111/jiec.12095

[22] Ohno H, Matsubae K, Nakajima K, Kondo Y, Nakamura S, Nagasaka T. Resources, Conservation and Recycling 2015;100:11. https://doi.org/10.1016/j.resconrec.2015.04.001

[23] Olatunde I. Sekunowo, Stephen I. Durowaye, Gbenebor OP. World Academy of Science, Engineering and Technology (WASET)

International Journal of Structural and Construction Engineering 2014;8.

[24] Yin L, Sridhar S. Metallurgical and Materials Transactions B 2011;42:1031. https://doi.org/10.1007/s11663-011-9528-z

[25] Haupt M, Vadenbo C, Zeltner C, Hellweg S. Journal of Industrial Ecology 2017;21:391. https://doi.org/10.1111/jiec.12439

[26] EuRIC. EU-27 Steel Scrap Specification. 2007.

[27] UNE 36199:2013: Clasificación de chatarras férricas no aleadas para uso general. https://www.en-standard.eu/une-36199-2013-clasificacion-de-chatarras-ferricas-no-aleadas-para-uso-general/. 2013.

[28] Mombelli D, Buonincontri M, Mapelli C, Gruttadauria A, Barella S, Fusari F, Rinaldini D. Journal of Materials Research and Technology 2022;19:1217. https://doi.org/10.1016/j.jmrt.2022.05.114

[29] Passarini F, Ciacci L, Nuss P, Manfredi S. Material Flow Analysis of Aluminium, Copper, and Iron in the EU-28. EUR 29220 EN. Joint Research Centre, Luxembourg, 2018.

[30] Gauffin A, Pistorius PC. Metals 2018;8:338. https://doi.org/10.3390/met8050338

[31] Wang T, Müller DB, Graedel TE. Environmental Science & Technology 2007;41:5120. https://doi.org/10.1021/es062761t

[32] Shamsuddin M. Secondary Steelmaking. In: Shamsuddin M, editor. Physical Chemistry of Metallurgical Processes, Second Edition. Cham: Springer International Publishing, 2021. p.293. https://doi.org/10.1007/978-3-030-58069-8_8

[33] Joint Research Centre IfPTS, Remus, R., Roudier, S., Delgado Sancho, L., et al.

[34] Kirschen M, Badr K, Pfeifer H. Energy 2011;36:6146. https://doi.org/10.1016/j.energy.2011.07.050

[35] Cavaliere P. Basic Oxygen Furnace: Most Efficient Technologies for Greenhouse Emissions Abatement. In: Cavaliere P, editor. Clean Ironmaking and Steelmaking Processes: Efficient Technologies for Greenhouse Emissions Abatement. Cham: Springer International Publishing, 2019. p.275. https://doi.org/10.1007/978-3-030-21209-4

[36] (IEA) IEA. Iron and Steel Technology Roadmap. 2020.

[37] Worldsteel. Energy use in the steel industry: https://www.steel.org.au/resources/elibrary/resource-items/worldsteel-fact-sheet-steel-and-energy/download-pdf.pdf/. 2019.

[38] Hasanbeigi A. Steel Climate Impact. An International Benchmarking of Energy and CO2 Intensities. Global Efficiency Inteligence, 2022.

[39] Babich AI, Gudenau HW, Mavrommatis KT, Froehling C, Formoso A, Cores A, García L. Rev. Metal. Madrid 2002;38:285. https://doi.org/10.3989/revmetalm.2002.v38.i4.411

[40] Yilmaz C, Wendelstorf J, Turek T. Journal of Cleaner Production 2017;154:488. https://doi.org/10.1016/j.jclepro.2017.03.162

[41] Cavaliere P. Clean Ironmaking and Steelmaking Processes: Efficient Technologies for Greenhouse Emissions Abatement. Springer Nature Switzerland AG 2019: Springer Cham, 2019. https://doi.org/10.1007/978-3-030-21209-4

[42] ULCOS: Ultra-Low CO2 steelmaking. https://cordis.europa.eu/project/id/515960/es

[43] (EESC) EEaSC. Role of carbon removal technologies in decarbonising the European industry. https://www.eesc.europa.eu/en/our-work/opinions-information-reports/opinions/role-carbon-removal-technologies-decarbonising-european-industry. 2022.

[44] Ajayi T, Gomes JS, Bera A. Petroleum Science 2019;16:1028. https://doi.org/10.1007/s12182-019-0340-8

[45] Cuéllar-Franca RM, Azapagic A. Journal of CO2 Utilization 2015;9:82. https://doi.org/10.1016/j.jcou.2014.12.001

[46] Europea C, Innovación DGdIe, Feiterna A, Zagaria A, Feilmayr C, Ansseau O, Hirsch A, Sert D, Boden A, Zeilstra C, Simoes J, Pettersson M, Babich A, Grant M, Stel J, Lin A, Sundqvist L, Lövgren J, Born S, Sköld B, Schott R, Küttner W, Bürgler T, Edberg N, Louwerse G, Delebecque A, Adam J, Diez-Brea P, Kerkkonen O, Janhsen U, Hattink M, Sihvonen M, Eklund N. ULCOS top gas recycling blast furnace process (ULCOS TGRBF) : final report: Publications Office, 2014.

[47] FINEX®: https://newsroom.posco.com/en/discover-the-tech-making-steel-more-sustainable-finex/

[48] SIEMENS VAI. "SIMETAL Corex technology": https://silo.tips/download/simetal-corex-technology

[49] Productivity TIfI. HIsmelt: http://www.iipinetwork.org/wp-content/Ietd/content/hismelt.html.

[50] HISARNA: Building a sustainable steel industry (2020). https://www.tatasteeleurope.com/sites/default/files/TS%20Factsheet%20Hisarna%20ENG%20jan2020%20Vfinal03%204%20pag%20digital.pdf

[51] Steel T. Sustainable in every sense: Tata Steel; 2018. Available from: https://www.tatasteeleurope.com/en/sustainability/hisarna. 2018.

[52] Abdul Quader M, Ahmed S, Dawal SZ, Nukman Y. Renewable and Sustainable Energy Reviews 2016;55:537. https://doi.org/10.1016/j.rser.2015.10.101

[53] MIDREX®: https://www.midrex.com/.

[54] TENOVA®: https://www.tenova.com/product/iron-reduction-technologies/

[55] Ariyama T, Takahashi K, Kawashiri Y, Nouchi T. Journal of Sustainable Metallurgy 2019;5:276. https://doi.org/10.1007/s40831-019-00219-9

[56] SSAB, LKAB, Vattenfall. HYBRIT Fossil-free steel, Sumary of findings from HYBRIT pre-feasibility study 2016-2017. 2017.

[57] Otto A, Robinius M, Grube T, Schiebahn S, Praktiknjo A, Stolten D. Energies 2017;10:451. https://doi.org/10.3390/en10040451

[58] Naseri Seftejani M, Schenk J, Zarl MA. Materials 2019;12:1608. https://doi.org/10.3390/ma12101608

[59] Vogl V, Åhman M, Nilsson LJ. Journal of Cleaner Production 2018;203:736. https://doi.org/10.1016/j.jclepro.2018.08.279

[60] Souza Filho IR, Ma Y, Kulse M, Ponge D, Gault B, Springer H, Raabe D. Acta Materialia 2021;213:116971. https://doi.org/10.1016/j.actamat.2021.116971

[61] Spreitzer D, Schenk J. steel research international 2019;90:1900108. https://doi.org/10.1002/srin.201900108

[62] Naseri Seftejani M, Schenk J. Metals 2018;8:1051. https://doi.org/10.3390/met8121051

[63] Naseri Seftejani M, Schenk J. Metallurgia Italiana 2018;n. 7/8 2018:5. https://doi.org/10.3390/met8121051

[64] Patisson F, Mirgaux O. Metals 2020;10:922. https://doi.org/10.3390/met10070922

[65] Souza Filho IR, Springer H, Ma Y, Mahajan A, da Silva CC, Kulse M, Raabe D. Journal of Cleaner Production 2022;340:130805. https://doi.org/10.1016/j.jclepro.2022.130805

[66] López FA, López-Delgado A. Journal of Environmental Engineering 2002;128:1169. https://doi.org/10.1061/(ASCE)0733-9372(2002)128:12(1169)

[67] Dankwah JR, Koshy P, Saha-Chaudhury NM, O'Kane P, Skidmore C, Knights D, Sahajwalla V. ISIJ International 2011;51:498. https://doi.org/10.2355/isijinternational.51.498

[68] Martín MI, López FA, Torralba JM. Ironmaking & Steelmaking 2012;39:155. https://doi.org/10.1179/1743281211Y.0000000078

[69] Mensah M, Das A. Environmental technology 2021:1.

[70] López FA, Balcázar N, Formoso A, Pinto M, Rodríguez M. Waste Management & Research 1995;13:555. https://doi.org/10.1016/S0734-242X(05)80034-5

[71] Gómez FAL, Hernández MIM, Pérez C, López-Delgado A, Alguacil FJ. Adsorción de metales pesados sobre cascarilla de laminación. CSIC - Centro Nacional de Investigaciones Metalúrgicas (CENIM), 2003.

[72] Martín MI, Gómez FAL, Alguacil FJ. Posibilidad de usar subproductos de la industria del acero para eliminar. CSIC - Centro Nacional de Investigaciones Metalúrgicas (CENIM), 2008.

[73] Lin X, Peng Z, Yan J, Li Z, Hwang J-Y, Zhang Y, Li G, Jiang T. Journal of Cleaner Production 2017;149:1079. https://doi.org/10.1016/j.jclepro.2017.02.128

[74] EUROFER. The European Steel Association. A steel roadmap for a low-carbon europe 2050: https://www.eurofer.eu/assets/publications/archive/archive-of-older-eurofer-documents/2013-Roadmap.pdf. 2013.

[75] Björkman B, Samuelsson C. Chapter 6 - Recycling of Steel. In: Worrell E, Reuter MA, editors. Handbook of Recycling. Boston: Elsevier, 2014. p.65. https://doi.org/10.1016/B978-0-12-396459-5.00006-4

[76] Kashiwakura S, Wagatsuma K. ISIJ International 2015;55:2391. https://doi.org/10.2355/isijinternational.ISIJINT-2015-316

[77] Cullen JM, Allwood JM, Bambach MD. Environmental Science & Technology 2012;46:13048. https://doi.org/10.1021/es302433p

[78] Zhu Y, Syndergaard K, Cooper DR. Environmental Science & Technology 2019;53:11260. https://doi.org/10.1021/acs.est.9b01016

[79] Nakamura S, Kondo Y, Nakajima K, Ohno H, Pauliuk S. Environmental Science & Technology 2017;51:9469. https://doi.org/10.1021/acs.est.7b01683

[80] Ohno H, Matsubae K, Nakajima K, Kondo Y, Nakamura S, Fukushima Y, Nagasaka T. Environmental Science & Technology 2017;51:13086. https://doi.org/10.1021/acs.est.7b04477

[81] Linley BD. Resource Recovery and Conservation 1977;2:225. https://doi.org/10.1016/0304-3967(77)90013-0

[82] Mustafa S, Luo L, Zheng B-T, Wei C-X, Christophe N. Metals 2021;11:407. https://doi.org/10.3390/met11030407

[83] Dworak S, Fellner J. Resources, Conservation and Recycling 2021;173:105692. https://doi.org/10.1016/j.resconrec.2021.105692

[84] Commission E. End-of-Life Vehicles. EU rules aim to make the dismantling and recycling of end-of-life vehicles more environmentally friendly., 2022.

[85] Sohn I, Jung SM. steel research international 2011;82:1345. https://doi.org/10.1002/srin.201100144

[86] Huaiwei Z, Xin H. Resources, Conservation and Recycling 2011;55:745. https://doi.org/10.1016/j.resconrec.2011.03.005

[87] Shen H, Forssberg E. Waste Management 2003;23:933. https://doi.org/10.1016/S0956-053X(02)00164-2

[88] (EEA) EEA. Waste recycling in Europe. https://www.eea.europa.eu/ims/waste-recycling-in-europe. 2021.

[89] Kallanish Commodities: https://www.kallanish.com/en/prices/list/ferrous/product-type/raw-materials/.

[90] Commission E, Agency ERE, Vu H, Cecchin F, Iacob N. Climate-neutral steelmaking in Europe: decarbonisation pathways, investment needs, policy conditions, recommendations, 2022.

[91] Celada-Casero C, Vercruysse F, Linke B, Smith A, Kok P, Sietsma J, Santofimia MJ. Mater. Sci. Eng. A 2022;846:143301. https://doi.org/10.1016/j.msea.2022.143301

[92] Vivas J, Celada-Casero C, San Martín D, Serrano M, Urones-Garrote E, Adeva P, Aranda MM, Capdevila C. Metall. Mater. Trans. A 2016;47:5344. https://doi.org/10.1007/s11661-016-3596-2

New Materials for a Circular Economy
Materials Research Foundations 149 (2023) 233-310

Materials Research Forum LLC
https://doi.org/10.21741/9781644902639-8

Chapter 8

Nanomaterials and their Synthesis for a Sustainable Future

Helena Gavilán*, María B. Serrano*, Juan Carlos Cabanelas

Departamento de Ciencia e Ingeniería de Materiales e Ingeniería Química (IAAB), Universidad Carlos III de Madrid, 28911 Leganés, Spain

hgavilan@ing.uc3m.es, berna@ing.uc3m.es

Abstract

Nanomaterials are structured materials whose dimensions lie in the nanoscale, at least in one dimension. Their small size and high surface area lead to properties not observed in their bulky state, some of which have revolutionized different fields in the last decades. While it is acknowledged that nanomaterials have been obtained or created since ancient times, with little or no knowledge about nanotechnology itself, it was not until this century that the development of nanomaterials was done on purpose, achieving a high level of sophistication in terms of fine-tuning the nanomaterial's properties, including size, shape, chemical composition, and structure. As such, nanomaterials are used in many industries as advanced materials with high strength while being light, superhydrophobicity, and antimicrobial properties, to name a few. Some of the nanomaterials with high value, given their outstanding properties, are quantum dots (superior luminescence properties), gold nanoparticles (localized surface plasmons), layered perovskites (optimal band gaps for materials like solar cells), and carbon nanotubes (very high tensile strength, electrical conductivity). Consequently, there has been a tremendous boom of nanomaterials in the industry, so they have been introduced into our daily lives. Despite the little knowledge available about their impact on the environment and our health, such intensified use has raised some concerns about the safe use of nanomaterials. Furthermore, due to the extended use of resources and current pollution levels, given that access to energy, food, clean water, and health is not guaranteed to future generations, the concept of "sustainability" and the transition from a linear to a circular economy is becoming more important in the manufacturing of products. As a result, society is making efforts to implement the 3Rs 'reduce', 'reduce', and 'recycle' in our community. In addition, other Rs are of utmost importance: 'Recover', 'Redesign', 'Remanufacture', etc., so that products, materials, and resources are maintained in the economy for as long as possible, and the generation of waste is minimized. This book chapter tackles all these aspects for nanomaterials and "nano-products" (nanomaterials already introduced in specific markets or industries). In particular, it analyzes and collects data available in the literature, where it was possible to implement the sustainability concept in different steps of the life-cycle of nanomaterials:

New Materials for a Circular Economy   Materials Research Forum LLC
Materials Research Foundations 149 (2023) 233-310   https://doi.org/10.21741/9781644902639-8

from their synthesis to subsequent remanufacturing processes. In this line, this chapter discusses the 'green' synthesis of nanomaterials, which are environmentally friendly processes that take place in natural environments (i.e., processes where nanoparticles are produced by microorganisms), or techniques that eliminate toxic reagents, minimize waste, reduce energy consumption and use ecological solvents. In addition, a section of the chapter covers reported strategies where the recovery, reuse, and recycling of nanomaterials were successful. The chapter has been structured into five parts. First, a general introduction to nanomaterials is provided. Then, different green synthesis methods are described, focusing on the biosynthesis of metal/metal-based oxide nanoparticles. After, the definition and classification of nanowastes are given, as well as a general overview of nano-toxicity and the different management procedures applied to nanomaterials after their end-of-life. Then, the book chapter covers the reuse and recycling of nanomaterials. In the fourth section of the book chapter, we provide data on 'safe- and sustainable-by-design' (SSbD) synthesis methods of nanomaterials. SSbD is a key concept for implementing a circular economy on nanomaterials. Finally, we provide some conclusions and final remarks about nanomaterials and their synthesis for a sustainable future.

Keywords

Nanomaterials, Green Synthesis, Ecotoxicology, Nano-Wastes, Nanomaterial Recycling and Reusing, Sustainable Nanomaterials

Contents

1. Introduction to nanomaterials

1.1 Definition and origins of nanomaterials

Some basic and essential concepts that need to be introduced in this chapter are the definition and classification of nanomaterials. Dimensions in the nanoscale lie between approximately 1 and 100 nanometers. How small is that? A valuable and easy comparison to visualize how "tiny" an object in the nanoscale is would be that a single human hair is ca. 80,000 to 100,000 nm wide. But technically speaking, an ISO norm (ISO/TS 80004-4:201) defines *nanostructured materials as one of the leading products of nanotechnologies and as a material with one, two, or three external dimensions in the nanoscale.*[1] Although much discussion still exists around the various definitions of "nanomaterials",[2] some relevant and accepted characteristics are their small size, high surface area, as depicted in **Fig. 1**, and enhanced reactivity over bulk materials (i.e., materials that have their size above 100 nm in all dimensions).

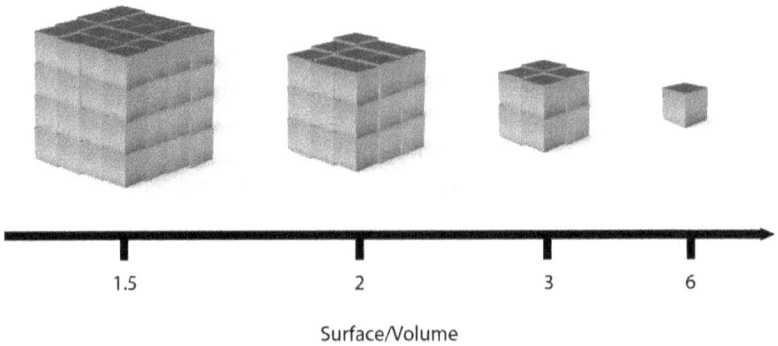

Figure 1. Schematic representation of the surface/volume ratio increase that takes place in materials with reduced dimensions. This ratio is indeed maximized for the case of nanoparticles whose external dimensions are in the nanoscale.

Moreover, "Nanotechnology" involves the understanding and control of matter at the nanometer scale. Indeed, we generally assume an understanding and control by scientists, but natural nanomaterials have been abundant on earth over the past 4.54 billion years in different (bio)geochemical or mechanical processes,[3] and scientists drew much inspiration from these existing nanomaterials. Relevant examples of nanomaterials in nature are found in soil microbial communities. Among them magnetotactic bacteria, which produce iron oxides and other iron-based nanoparticles to orient themselves in soils directionally, and these natural nanomaterials may even affect soil's magnetic properties.[4] Other well-known examples of nanomaterials in nature are cellulose in wood, chitin in crustaceans and insects, and fibroin in spider and silkworm silk, which exhibit hierarchical nanostructures.[5] These latter nanomaterials are attractive due to their extraordinary mechanical properties.

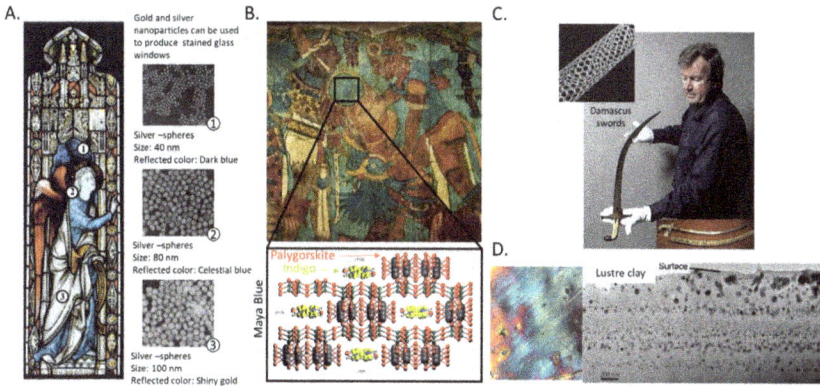

Figure 2. Examples of "ancient nanotech" artifacts: (A) Medieval artisans stained glass windows and examples of how gold and silver nanoparticles of different sizes were applied to stained glass windows;[6] Copyright © 2015, Elsevier Books (B) Archaeological artifacts containing Maya blue and its nanostructure made of indigo (a natural dye) and palygorskite (a microfibrous clay); (C) Damascus steel swords whose structure contains iron carbide nanowires and carbon nanotubes, and (D) an image of luster and its shine and a TEM image showing the metal nanoparticles used in a piece of luster.[7]

It is hard to determine when the history of nanotechnology began since **humans regularly implemented nanomaterials in the ancient past**. The Lycurgus cup, currently in a British Museum collection, is an amazing feat for Romans and one of the most well-known "ancient nanotech" artifacts. It is a 4th-century Roman glass chalice made of dichroic glass. Dichroic glass displays different colors based on the amount of light passing through it. This phenomenon is achieved by making the glass with tiny proportions of nanoparticles of gold and silver. Medieval artisans likewise exploited the integration of metal nanoparticles in glass to produce different colors in stained glass windows **(Fig. 2A)**.[6] Other awe-inspiring "ancient nanotech" works are depicted in **Fig. 2B-D**: *Maya Blue*,[8] a corrosion-resistant pigment made of indigo (a natural dye) and palygorskite (a microfibrous clay), first produced in AD800 in the pre-Columbian Mayan city of Chichen Itza; *Damascus steel swords*,[9] produced in AD300 and AD1700, whose impressive strength is due to its nanostructure, based on oriented iron carbide nanowires and carbon nanotubes, and *Islamic lust*er, produced in the 13th century AD, it is clay with a unique metallic glaze that gives the effect of iridescence, and it is made of metal copper or silver nanoparticles embedded in a silica-based glassy matrix.[10] However, most probably, the artisans who obtained or crafted these materials had little or no knowledge about nanotechnology, and it was not their intention to control matter at the nanometer scale.

The modern history of nanotechnology began in the 20th century, many years after discovering these artifacts. It was in 1925 when *Richard Zsigmondy, Nobel Prize Laureate in chemistry, coined the term "nanometer",* when the size of gold nanoparticles coming from a colloid was determined using a microscope. Nevertheless, the American physicist **Richard Feynman (1965 Nobel Prize Laureate in physics) is considered the real father of nanotechnology. He introduced the concept of manipulating matter at the atomic level during his most known lecture, titled "There is Plenty of Room at the Bottom"** which was given in 1959 at an American Physical Society meeting at Caltech.[11] Another significant advance in science that has boosted the knowledge and development of nanomaterials is the creation of the scanning tunneling microscope. It was developed by Gerd Binnig and Heinrich Rohrer (Nobel Prize in Physics in 1986). This microscope can create an image of a surface by using a very sharp conducting tip. It has a lateral resolution of 0.1 nm and a depth resolution of 0.01 nm, allowing the possibility to image atoms. In addition, this technique can manipulate them.[12] In fact, the IBM company used this microscope to arrange 35 individual xenon atoms on a substrate to spell out the company's initialism. Later, they released even a stop-motion animated short film called "A Boy and His Atom: The World's Smallest Movie".[13] By the end of the 20th century, many companies and governments were investing in nanotechnology and in less than 50 years, nanotechnology and nanomaterials have become the foundation for remarkable industrial applications and exponential growth.[14] Actually, **nanomaterials are currently used in a never-ending list of applications, which can be grouped into different sectors**: they are used as materials with improved properties, as coating agents, i.e., in textiles,[15] in the life science sector for therapy techniques, diagnostics and complex drug-delivery systems,[16] in food science,[17] agriculture,[18] electronics,[19] energy,[20] water, soil, and air treatment.[21]

1.2   Classification of nanomaterials

The classification of nanomaterials can be done with respect to several categories, including the dimensions in each of their axes (0D, 1D, 2D, and 3D), their chemical composition (metal, metal oxide, carbon-based, organic, and composites), their structure (homo and heterostructures) and depending on the properties they display in the nanoscale (optical and electronic, magnetic, mechanical, catalytic, thermal and antimicrobial).[22-24] Table 1 summarizes the different categories of nanomaterials and provides different examples in each case. In particular, this book chapter focuses on the category of dimensions, as depicted in **Fig. 3**, and pays attention to different nanomaterials that are currently exploited in the industry:

Table 1.   *Classification of nanomaterials attending to their dimensions, chemical composition, structure, and properties.*

Categories	Classification	Examples
Dimensions	0D	Quantum dots (QDs), fullerene, gold NPs
	1D	Metal nanorods, carbon/metallic nanotubes, gold nanowires, polymeric nanofibers
	2D	Layered perovskites and graphene/MXene sheets
	3D	Liposomes, polycrystals, and dendrimers
Composition	Metal	Au, Ag, Cu NPs
	Metal oxide	Fe_2O_3, Al_2O_3, NiO, ZrO_2, SiO_2
	Carbon-based	Fullerene, carbon nanodot (CND), nano-onion (CNO), nanodiamond (ND), nanotubes (CNTs), graphene
	Organic	Molecules, linear/block /branched polymers, dendrimers, biomolecules, hydrogels
	Composites	**Organic:** Hydrogels, Layer-by-layer (LbL) materials, polymer brushes **Inorganic:** NPs, CNTs, quantum dots, hydroxyapatite
Structure-domain	Homo-structures	Nanospheres, nanocubes
	Hetero-structures	Core-shell NPs, dumbbells, trimers, Janus NPs
Properties	Optical and electronic	Au and Ag NPs, CdSe, ZnS, CdTe, ZnO, and HgSe QDs
	Magnetic	Fe_3O_4, Fe_2O_3, Co_3O_4, FeC, Fe, Ni, Co
	Mechanical	CNT-reinforced polymer composite fibers
	Catalytic	Au, Pd, AuPt, zeolites, metal-organic frameworks (MOFs)
	Thermal conductivity,	CNTs
	Heat dissipation	Fe_3O_4
	Antimicrobial	Ag, MgO, and TiO_2 NPs

Materials Research Forum LLC
https://doi.org/10.21741/9781644902639-8

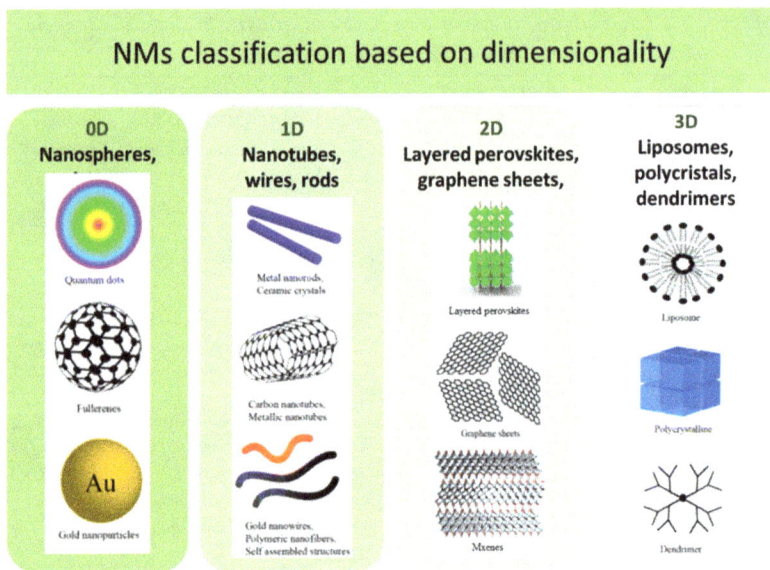

Figure 3. Nanomaterials (NMs) classification based on dimensionality: 0D, 1D, 2D, and 3D materials and relevant examples in each case. Adapted from literature.[24] Copyright © 2022, Elsevier Books.

1.3 Current uses of nanomaterials in industry and hazard risks associated to released/used nanomaterials

Once showed the classification of nanomaterials, it is worth mentioning that despite being 'relatively new' materials, they are widely applied in industry. Some of the most common uses of NPs in the industry are the following:

- Catalysis
- Energy
- Electronics
- Biomedicine
- Environmental Industry
- Materials with improved properties: hardness, antimicrobial
- Coatings
- Other: textile, pharma & cosmetic industry

Depending on the properties of the NPs, it is probable that they display an abrupt change in their properties when passing from the bulky form to the corresponding nano-sized material. For instance, depending on their chemical composition and structural properties, they can be increase in the surface area, have antibacterial properties, hardness, electron band gap, and electric/magnetic properties. Consequently, there are helpful for diverse applications, including catalysis, antibacterial properties, renewable energies, smart coatings, and magnetic separation, as summarized in **Fig. 4**.

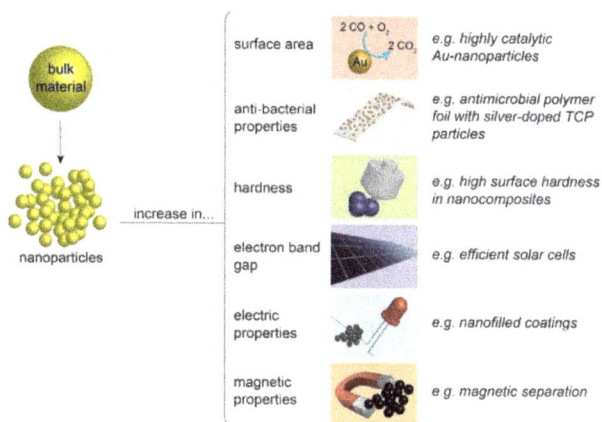

Figure 4. Effect of passing from bulk material to the corresponding nanoparticles, which, depending on their chemical composition and structural properties, can increase the surface area, have antibacterial properties, hardness, electron band gap, and electric/magnetic properties. Derived from these properties, there are diverse applications: catalysis antibacterial properties, high surface hardness, renewable energies, smart coatings, and magnetic separation. Reproduced from Ref. 25 with permission from the Royal Society of Chemistry.[25] Content available at: https://doi.org/10.1039/C4CS00362D

Among these, some of the well-established industrial applications of nanoparticles are:

- Particles as chemically inert additives (carbon, titania, and silica polymer fillers; 2D particles as fillers; effect pigments, dyes, and UV protection; processing aids; functional surfaces)
- Chemically active particles: catalysts, biomaterials, and antimicrobial additives (heterogeneous catalysis, biomaterials, antimicrobial additives)
- Replacement of traditional pigments
- UV protection for cosmetic products (inorganic/organic sub-micrometer UV filters)
- Novel organic color filters for LCD-technology

Nanotechnology has improved the design of numerous products, such as light bulbs, paints, coatings, computer screens, fuels, and construction materials. Furthermore, nanotechnology has also had a significant impact on **renewable and conventional**

energies. For example, solar cells and light-emitting devices were improved by nanotechnology thanks to the development of nanomaterials like quantum dots and perovskites.[26] In contrast to big and bulky traditional photovoltaic cells, usually made with crystalline semiconductor material assembled as a series of large flat panels, nanotechnology-based solar cells have improved the efficiencies and costs associated to solar energy. Further, experimental solar panels have been made in flexible rolls rather than rigid panels.[27] Another example are **carbon nanotubes**, which have become crucial to diverse technologies, including **wind power**. Turbines consist of a composite material containing carbon nanotubes, which improve the turbine's strength and weight.[28] Other **impact industries of carbon nanotubes are electronics and computers**[29] and engineered materials (they are used, for example, to make lighter, thinner, and more resilient bicycle frames, tennis rackets, or hockey sticks). Another relevant **2D nanomaterial is MXene**. MXenes are ceramics and are made from a bulk crystal called MAX. 2D layered materials derived from MAX phases. Unlike most 2D ceramics, MXenes have inherently good conductivity and excellent volumetric capacitance because they are molecular sheets made from the carbides and nitrides of transition metals like titanium. Since its discovery in 2011, the continued exploration has revealed their **exceptional ability to store energy,**[30], and their high interest in other applications.

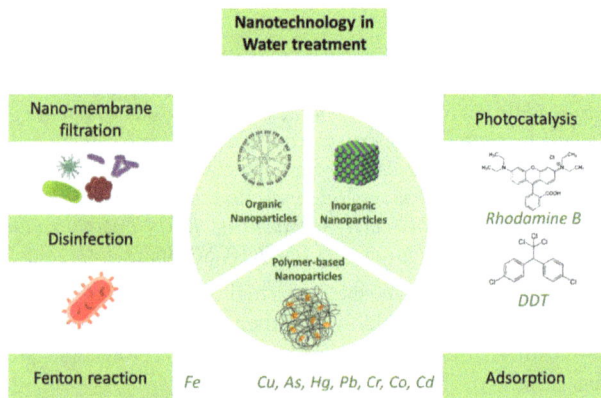

Figure 5. Water remediation by nanotechnology. The systems exploited for water remediation are organic nanoparticles, such as liposomes and micelles; inorganic nanoparticles, like SiO$_2$; and polymer-based nanoparticles, like polymersomes and fibers. These systems were used for photocatalysis, adsorption, and disinfection processes; they have been involved in the catalysis of Fenton reaction, and they have been integrated into membranes to filtrate water and reduce the level of bacteria, heavy metals, and other elements. Adapted from Ref 31.[31] Copyright © 2021, Elsevier.

New Materials for a Circular Economy   Materials Research Forum LLC
Materials Research Foundations 149 (2023) 233-310   https://doi.org/10.21741/9781644902639-8

Nanotechnology has also made a significant impact in the field of **water remediation**. Indeed, access to clean water has become a problem in many parts of the world due to a non-uniform distribution of water and increased pollution.

In this context, nanotechnology has resulted in advanced water remediation processes. Thus, nanotechnology has facilitated water reuse, recycling, and remediation. **Nanomaterials have been used as adsorbents, photocatalysts, Fenton reaction agents, and disinfection agents; and also to develop membranes for water remediation**, as summarized in **Fig. 5**.[31] For instance, nanoparticles with high surface area can help to eliminate micropollutants like toxic metals or organic compounds like dyes. Nanoparticles have accounted for the adsorption of heavy metals and other elements like Cr, Pb, Cd, Cu, As, Ni, Hg, and Co. In addition, nanoparticles have catalyzed the degradation of Rhodamine B (RB), Bisphenol A (BPA), acridine orange (AO), and insecticides such as dichlorodiphenyltrichloroethane (DDT) and dichlorodiphenyltrichloroethylene (DDE). Also, nanoparticles with specific chemical compositions, like Ag, help lower the concentration of bacteria/virus in water. Additionally, there have been some efforts to turn saltwater into freshwater in a process called **desalination** through nanocomposite membranes.[32] Along this line is the isolation and removal of high-value compounds such as oil. Oil and gas resources are still the world's major contributors to energy supply. **Enhanced oil recovery (EOR) methods** based on nanotechnology were developed to enhance the overall oil displacement efficiency. The use of magnetic nanoparticles has allowed oil recovery from seawater.[33] Then, the nanoparticles can be regenerated, and oil can be further used.

Another important industry where nanoparticles are exploited is **the textile industry**. Coating fabrics with a thin layer of zinc oxide (ZnO) nanoparticles gives protection from UV radiation and microbes.[34] On the other hand, coating fabric with a thin layer of ZnO and TiO_2 has also accounted for improved wetting behavior achieving superhydrophobic fabric.[35] The US company Nanotex exploited the integration of cellulose nanowhiskers in fabric (Nanotex®), achieving wrinkle resistance, stain resistance, and water repellency.[36]

The following industry in which nanotechnology has had a significant impact is the **cosmetic industry**, achieving products for human consumption that have high clarity, coverage, cleansing, or absorption. Notably, an increasing number of nanocarriers are used to encapsulate active ingredients (i.e., hyaluronic acid, vitamin C, retinol, etc.) for their efficient delivery through the skin barriers. Among these are liposomes, nanoemulsions, nanoparticles, graphene, fullerene, carbon nanotubes, dendrimers, and nanospheres.[37]

In **the food industry**, the use of nanotechnology has had a significant impact on the packaging and agricultural sectors. Layered silicate nanoclays (such as montmorillonite and kaolinite) appear to be the most promising "nanoscale fillers". They can form individual platelets when dispersed in a polymer matrix, which do not allow diffusion of small gases, producing a tortuous path that works as a gas barrier structure. This nanotechnology is used to extend the shelf life of products.[38]

Therefore, it is clear that nanomaterials have significantly impacted different industries. In addition, nanotechnology is expected to revolutionize other sectors like medicine.[39] "Smart nanomaterials" are ideal candidates since they can respond to environmental changes at the most optimum condition and manifest their own functions according to these changes. In particular, smart nanomaterials can be engineered to respond to physical (i.e., light, mechanical forces, temperature, or a magnetic field), chemical (pH, electrochemical, solvent), or biological (enzymes) stimuli, as depicted in Fig. 6. This makes them ideal candidates for the area of biomedical applications such as diagnosis, therapy, and theranostics. Furthermore, smart nanomaterials have been recently used as biodegradable implanted devices, with a unique capability of multimodal functions and enabling large surface electrophysiological mapping. They can be built to be powerful probes for the diagnosis and treatment of neurological disorders, cardiac diseases, and many other health issues. Among "nanotherapies", artificial targeted drug release[40], and selective killing of cancer/bacteria cells through hyperthermia (activated through various stimuli),[41, 42] are highlighted. Nevertheless, other biomedical applications are muscle actuation, energy transduction, and many others. For example, in contrast to conventional cancer therapies like chemotherapy and radiation, which can damage both healthy and diseased tissue, magnetically-triggered hyperthermia, for instance, can be applied using Fe_3O_4 nanoparticles, which are injected into the tumor. They produce heat under an alternating magnetic field in a remote way, destroying cancer cells selectively. This modality was thought to ablate metastatic tumors preferentially.[43] Some advantages of Fe_3O_4 nanoparticles are that they can safely degrade in the body and that the alternating magnetic field used in this therapy is safe for healthy tissues. It also has the advantage of magnetic fields penetrating the body, unlike other physical stimuli like light employed in phototherapy. Other types of nanoparticles used for nanotherapies are **liposomes,[44] gold nanoparticles,[45] polymeric, and viral nanoparticles.**[46] Finally, **fullerene (also named C_{60})** has been exploited in nanomedicine. Although it is insoluble in water, impeding its biological applications, C_{60}-derivatives, such as fullerol or fullerenol (hydroxylated fullerene), have been shown as promising therapeutic candidates for conditions related to oxidative stress and inflammation, such as cancer, diabetes mellitus, cardiovascular anomalies, neurodegeneration, skeletal muscle degeneration, osteoarthritis, and intervertebral disc degeneration.[47]

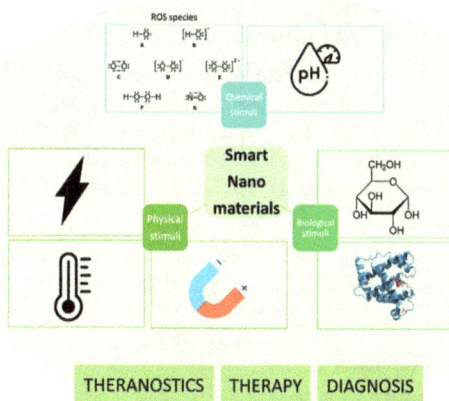

Figure 6. Smart material's concept: they can be engineered in terms of chemical composition and surface functionalization to respond to physical (i.e., light, mechanical forces, temperature, or a magnetic field), chemical (pH, electrochemical, solvent) or biological (enzymes) stimuli, which makes them ideal candidates for the area of biomedical applications such as diagnosis, therapy and theranostics (a term derived from the combination of therapeutics and diagnostics, which is an area emerging in the field of medicine). Adapted from Ref 42.[42]

Another application that "smart nanomaterials" have allowed is drug delivery, which involves transporting a pharmaceutical compound to its target site to achieve a desired therapeutic effect. In this sense, **dendrimers**, i.e., macromolecules that contain symmetric branching units built around a small molecule or a linear polymer core, are ideal for drug delivery since it endows a high degree of entrapment of drugs by non-covalent interactions.[48]

The reality is that we are actively exposed to nanoparticles through the use of different products, such as clothes, cosmetics, drugs, or food, and passively, since the presence of nanoparticles has been detected in water and air. Indeed, one of the significant concerns relating to nanomaterials is their toxicity, which is to date poorly understood, **and this is a severe concern relating to their environmental, domestic, and industrial use. Human exposure to nanomaterials can occur through different pathways, such as skin, gastrointestinal tract, and respiratory tract.** However, to what extent and how nanomaterials can cause cellular toxicity is still unclear. Depending on the nanomaterials characteristics, they will have a bio-distribution profile, being translocated and distributed to different sites, including bone marrow, kidney, spleen, and heart.[49] Finally, the excretory pathways of nanoparticles are, among others, urine and feces.[50] However, other routes, such as sweat and breast milk, may exist.[51] There is a demand but also a need to deepen their possible *in vivo* toxicity and acquire knowledge about their impact on the environment before introducing nanomaterials definitively in our daily lives.

Within the last decade, the emerging field of nano(eco)toxicology has evolved into an established field of research. Here, non-classical parameters play a key role in chemical toxicology: nanoparticle dispersion stability, dissolution, surface activity, and complex bio-distribution pattern complicate the already demanding toxicology investigations of (traditionally) molecular compounds as well-defined solutions.

Thus, many scientists and engineers continue to voice their concerns about nanotech's future, which will depend on their sustainable synthesis, use, safety, and regulation. In sections 3.3-3.4, diverse data about the toxicity and the impact of nanomaterials in the environment is provided.

1.4   Synthesis of nanomaterials

Several methods of synthesizing nanomaterials exist, which are subgrouped in two approaches: "Top-down" or "Bottom-up". The first uses bulk material to get nanosized particles, whereas the latter employs molecular/atomic "building blocks". Within the top-down approaches, we can find mechanical milling, etching, laser ablation, or sputtering.[52] Within the bottom-up approach, there exist different routes such as supercritical fluid synthesis, spinning, sol-gel process, laser pyrolysis, chemical vapor deposition, molecular condensation, chemical reduction, and green synthesis or "biosynthesis",[52] as depicted in **Fig. 7**. Among the top-down method, **mechanical milling (MM)** is the most extensively used to produce nanoparticles such as metal, oxide, and polymer-based. Different elements (precursors) are milled in an inert atmosphere. Among other factors influencing the nanoparticle's structural properties, we can find plastic deformation, fracture, and cold welding. MM has the advantages of: i) working at low temperatures, ii) without solvent, and iii) with a wide variety of matrixes, i.e., polymeric. Other advantages directly related to the topic of this chapter are that MM strongly reduces environmental disposal and the possibility of treating waste disposal and recycling materials.[53]

Regarding the Bottom-up approaches, at large scale, **vapor and liquid phase techniques** are the leaders for the large-scale production of inorganic nanoparticles.[54] **Laser pyrolysis** is the most commonly used process in industries for large-scale production of nanoparticles. It is based on using a laser to easily evaporate the precursor. Then, the precursor is fed into a furnace at high pressure. The generated nanoparticles are then recovered. Some of the furnaces use a flame or plasma instead of flame to produce high temperatures. Indeed, it is also possible to use the precursor in the vapor state directly. The most common nanoparticles produced through laser pyrolysis are carbon[53], metal oxide-based, and ceramic.[55] Finally, the advantages of this method are its simplicity, high efficiency, low cost, and the possibility of continuous processes with high yield. Despite the advantages of these methods, specific applications require nanomaterials with well-define properties in terms of chemical purity, crystallinity, size distribution, and well-defined shape. The variation of one of these parameters may lead to a change in the expected properties. In terms of fine-tuning all these features of nanoparticles, **Bottom-up approaches, particularly those in liquid phase such as sol-gel, are the preferred ones**. Most of the nanoparticles' composition can be synthesized using this method. It is a wet-chemical process (therefore, it occurs in solution, just like chemical co-precipitation or hydrothermal syntheses). The precursor(s) is/are dispersed/dissolved in a liquid, i.e., the solvent, and react and evolve, generating the nanoparticles. Thus, the system contains a liquid and a solid phase (the nanoparticles). The nanoparticles can then be recovered through sedimentation, filtration, or centrifugation. The obtained nanoparticles can be either dried when intended to be handled in powder state or redispersed in a new solvent. The most common nanoparticles produced through sol-gel are metal oxides.[56] Worth note is that there exist aqueous and non-aqueous sol-gel processes depending on the solvent. In the case of aqueous sol-gel processes, the synthesis process frequently involves the following steps: hydrolysis, condensation, aging & drying, and calcination. After, the final product is obtained. After the hydrolysis, a Sol (containing colloidal structures) is formed. After their subsequent condensation, a gel (connected porous structure) is formed. Aging and drying processes include: supercritical/thermal or freeze drying, after which an aerogel, xerogel, or cryogel are formed, respectively, which are the final products[56].

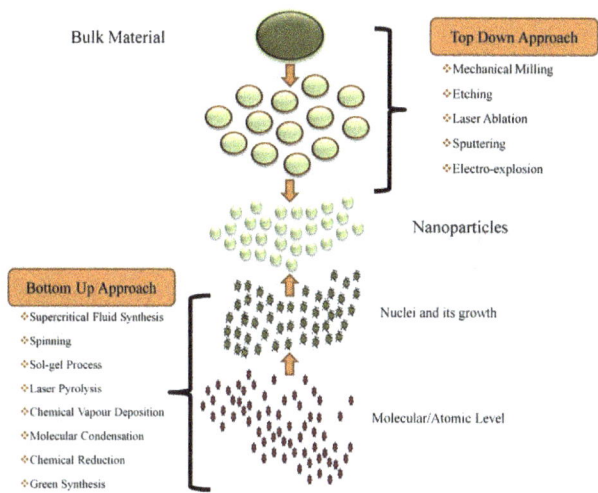

Figure 7. Synthesis approaches of nanomaterials. Reproduced from Ref 57.[57]

Despite the high-quality nanomaterials obtained through all these routes, their implementation and widespread application in the industry often endanger the environment (due to toxic reagents or harsh synthesis conditions). Moreover, most routes are not sustainable, having a negative economic and ecological impact. **Green nanoparticle synthesis aims to minimize generated waste and implement sustainable processes.** It pursues the controlled synthesis of nanoparticles using bio-organisms as reactors or conventional reactors but with mild reaction conditions and non-toxic precursors, promoting environmental sustainability. In the next section, we focus on this route, describing its different approaches, trends, and future perspectives.

2.   Green chemistry-based synthesis of nanoparticles

A green chemistry-based synthesis requires new tools to create sustainable, healthy, and economically viable processes and products. According to 12 Principles of Green Chemistry or sustainable chemistry,[58], the processes involved must be environmentally benign and sustainable, reducing or eliminating the use of hazardous chemicals with zero contaminants and minimizing or reusing by-products. Indeed, "green chemistry" is a different way of thinking about how chemistry and engineering can be done, moving far beyond the mere understanding of chemical toxicity. In addition to that, **scientists and engineers must find creative and innovative ways to reduce waste, conserve energy, and discover renewable raw materials** to create a sustainable future for materials production.

Indeed, this book chapter focuses on nanomaterials and their green synthesis. Nanotechnology is today one of the faster-developing and more innovative fields in chemistry, and despite its considerable progress, the toxicity and danger associated with nanomaterials have a negative impact. The green synthesis of nanoparticles or green nanotechnology has the ability to change the non-ecofriendly traditional nanomaterials production. This green approach pretends to use innovation in materials science and engineering to support the fabrication and generate energy-efficient products and processes that are environmentally and economically sustainable. As said by Dr. James Hutchison, pioneer in green nanotechnology, *"green chemistry is a terrific way to do nanotechnology responsibly"*. Thus, in terms of how nanotechnology can benefit from green chemistry in the process, **scientists and engineers try to find alternative approaches for the fabrication of nanoparticles** taking advantage of natural sources, waste, and clear technologies.[59]

To date, numerous nanomaterials, such as metallic nanomaterials, metal oxide, and quantum dots, have been successfully produced using biological synthesis. A search in the web of science database, with "biosynthesis nanoparticle" as the topic, **detected more than 5,000 results,** 1,105 belonging to 2021 and 2022. **Among them, 428 are** *reviews*. Research areas are headed by Science Technology, Materials Science, Pharmacology, Pharmacy, Biochemistry, Molecular Biology, Chemistry, Environmental Sciences, Ecology, and so on, up to more than 50 areas. **Changing "biosynthesis" to "green chemistry" resulted in 9,500** references. Both searches only consider the publications referred to as "articles" and "reviews", discarding patents. **Fig. 8** shows the number of publications for the first 15 areas. These data give an idea of the importance of green synthesis in research, covering a wide range of scientific and technological areas.

Green synthesis represents a sustainable alternative to chemical and physical methods without compromising the quality and properties of the nanomaterial produced.[60] Green synthesis aims, in particular, to decrease the use of toxic chemicals by using biological materials such as plants and microbes and by waste as well. This section includes the most relevant aspects of "green chemistry" synthesis of nanosized materials divided into two sections: *biosynthesis using plant extracts and microbial organisms as natural-biological sources, and sustainable and green synthesis of carbon nanoparticles from biomass.*

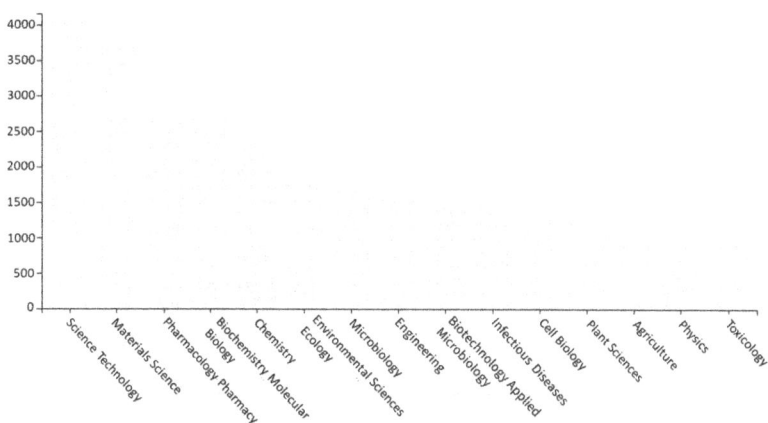

Figure 8. Analysis of results found from all databases in WoS (October 2022) with "biosynthesis nanoparticle" as a topic, showing the first 15 areas.

2.1 Biosynthesis using plant extracts and microbial organisms

Keeping in view the aforementioned environmental drawbacks, the scientific community has directed much of its research toward "green protocols" using biological constituents which are economical, biocompatible, non-toxic, and eco-friendly. With the increasing need for efficient green chemistry, **biosynthetic routes using natural sources** for synthesizing nanoparticles have aroused much interest, primarily in obtaining metal and metal oxide nanoparticles.[61-63] The biological synthesis (or biosynthetic) process is similar to the chemical reduction process but with the expensive and non-toxic reagents substituted by bioagents extracted from viable natural sources such as organisms and plants. **These natural sources act as templates or scaffolds, playing the role of reducing, as well as capping and stabilizing agents of nanoparticles.** Additionally, external experimental conditions like high energy and pressure are not required.[64]

Indeed, compared to conventional synthesis, **green synthesis of nanoparticles has lower production costs as raw materials are readily available (mainly plant residues), and additional purification processes are not commonly required.** In addition, these methods possess unique and enhanced properties with many applications in bioengineering because of their use of biological species.[59] Single-element and multi-element (mainly metal oxide, but also sulfides) covering up 55 elements in the periodic table have been prepared in this way,[65] including alkali metals, alkaline earth metals, transition metals, post-transition metals, metalloids non-metals, even several lanthanides, and one actinide. Recently, it has been published a complete review that discussed several biosynthesis approaches developed over the last decade.[66] The review also provides insights into the molecular aspects of the synthesis mechanisms and several biomedical applications. As indicated in the previous section, biosynthesis can be categorized as a "bottom-up" approach where the metal atoms assemble to form nanoparticles. Several synthesis approaches are also considered "top-down" strategies because some processes start from bulk materials (metal precursors) to generate nanosized particles by reducing their size.

One of the main aspects of biosynthesized nanoparticles is their remarkable potential application in numerous fields, including catalysis,[67] energy harvesting and storage,[68] electronics,[69]   antimicrobials,[59] [especially in environmental and biomedical applications.[70, 71] In particular, **biocompatibility is an essential factor to consider when using inorganic nanoparticles in biomedical applications** and biosynthesized nanoparticles possess biomolecules as surface functional groups, which reduce their toxic nature. In comparison, nanoparticles synthesized via conventional approaches must be further engineered to reduce the potentially toxic groups on their surface, making them unviable in many biomedical applications.

In the biological synthesis of nanomaterials, microorganisms such as bacteria, yeasts, fungi, algae, and plants are the principal natural sources. In general, the biosynthesis of nanoparticles is classified into two main types, **microbe and plant-mediated (phytosynthesis) methods.**[72] The biomolecules and phytochemical constituents extracted from microbes and plants can act as reducing and/or stabilizing agents for forming metallic and metal oxide nanomaterial. This synthesis can be carried out with fungi, algae, bacteria, and plants. Some parts of plants, such as leaves, fruits, roots, stems, and seeds, are the most commonly used to synthesize nanoparticles.[73] In biological synthesis, no capping or stabilizing agents are added as the biomolecules can perform this function themselves. **Instead, the constituents contained in them acts like stabilization and reducing agent.** For example, alkaloids, flavonoids, saponins, steroids, terpenoids, tannins, and other bioagents contained in plants (also named phytochemicals), as well as biomolecules such as DNAs, proteins, enzymes, peptides, or amino acids, act as bioagents in biosynthesis.[62] **The properties of these nanoparticles (such as shape and size) are governed by these biomolecules. Fig. 9** represents a scheme with a graphical representation of the synthesis of metal nanoparticles from plants and microbes, indicating the most common bioagents in each type of synthesis. Also, agricultural waste plants have been considered potential biomolecule sources for nanoparticle biosynthesis.[74]

New Materials for a Circular Economy   Materials Research Forum LLC
Materials Research Foundations 149 (2023) 233-310   https://doi.org/10.21741/9781644902639-8

However, after the organism dies, it will no longer be stable and begin to degrade immediately, leading to improper nanoparticle formation. These sources will be covered in a separate section at the end.

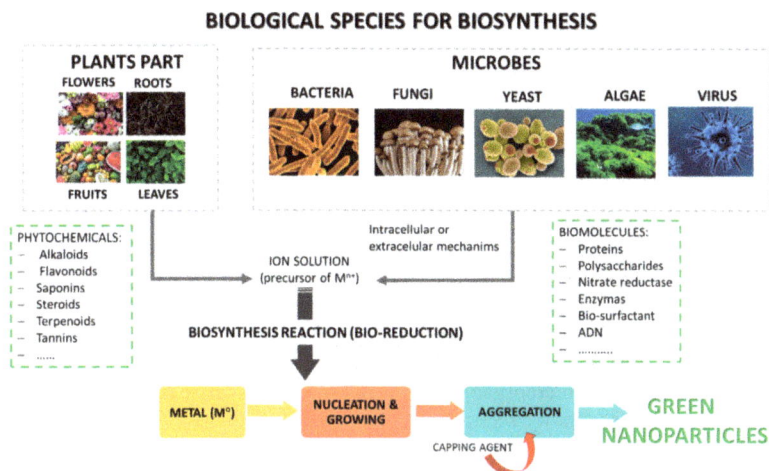

Figure 9. A general mechanistic scheme with a graphical representation of the synthesis of metal nanoparticles from plants part and microbes.

The use of plants and microbes for biosynthesis has been discussed in detail previously by numerous groups; including scientists worldwide (see discussion at the beginning of the section). We encourage the reader to consult the reviews compiled of different biosynthesized metal and metal oxide nanoparticles using multiple natural sources. Ref. 66[66] lists several microbes (bacteria, fungi, and algae) and their biomolecules involved in the synthesis of metallic nanoparticles, as well as a summary of various metal and metal oxide nanoparticles synthesized from plant extracts (Tables 1-5). In Ref 73,[73] the reader can find an exhaustive list of synthesized biogenic nanoparticles using plant extracts. Chapter 15 (Microbially synthesized nanoparticles as next generation antimicrobials: scope and applications) from the book Nanoparticles in pharmacotherapy.[75] presents a list of metal-based nanoparticles synthesized from a wide range of bacteria, fungal and algae species. Ref. 76[76] reports a list of microbial enzymes used in extracellular and intracellular bio reducing organisms and their resulting nanoparticles.

Some of the main aspects of both microbe and plant-mediated (photosynthesis) methods are briefly commented on in the following sections.

New Materials for a Circular Economy   Materials Research Forum LLC
Materials Research Foundations 149 (2023) 233-310   https://doi.org/10.21741/9781644902639-8

2.2   Biosynthesis of Nanoparticles from microorganisms

Metal nanoparticles such as gold, silver, copper, and platinum are the most common metal nanoparticles fabricated via microbial approaches. Iron, zinc, silicon, and rare earth metals have also been synthesized. **Microorganisms such as bacteria, fungi, yeasts, seaweed, and microalgae have been the most considered microbial precursors** to fabricate metal and metal oxide nanoparticles via green chemistry.[76]

The most significant aspect of microbe-mediated nanoparticle synthesis is the reduction in their toxicity and increment in their biological activity, such as bioavailability, biocompatibility, and bioreactivity.[77] Microbial synthesis can be classified as intracellular and extracellular approaches depending on metal ions trapped inside the microbial cells or adsorbed on their surface.[78] **Fig. 10** shows a scheme of both routes. **The intracellular method involves transporting ions into the microbial cell to form nanoparticles in the presence of enzymes** (provide electrons to metal ions during microbial synthesis). **The extracellular synthesis of nanoparticles involves trapping the metal ions on the surface of the cells** and reducing ions in the presence of enzymes (with the same role as donor electrons). Many biotechnological applications, such as remediation of toxic metals, employ microorganisms like bacteria and yeast by a mechanism where detoxification often occurs by reducing the metal ions in the cell. However, it is only recently that the use of microorganisms for synthesizing nanomaterials has been viewed with interest.[79] Biotransformation-based nanoparticle synthesis strategies **have had greater commercial viability since the nanoparticles are synthesized extracellularly directly in the aqueous medium**. The main advantage of extracellular mediated nanoscale material synthesis is that it is devoid of impurities such as intracellular proteins. Treatment with detergents or ultrasound is not required, is cheaper, and favors large-scale production. Thus, it is preferable in comparison to the intracellular method.[80]

The microbial approach begins with microbes trapping metal ions from the environment by binding to nucleotides and phosphates, which are molecular components of DNA, proteins, and enzymes. Such synthesis processes may be a direct or indirect consequence of metabolism and/or the reactivity of structural components. In fact, **the microbe act as biofactories where the metabolites secreted participate as reducing and stabilizing agents** forming primary nanocrystals and, finally, the nanoparticles by aggregation.[81] One example is the exopolysaccharide secreted by several bacteria that act as reducing and stabilizing agents in nanoparticle biosynthesis.

Materials Research Forum LLC
https://doi.org/10.21741/9781644902639-8

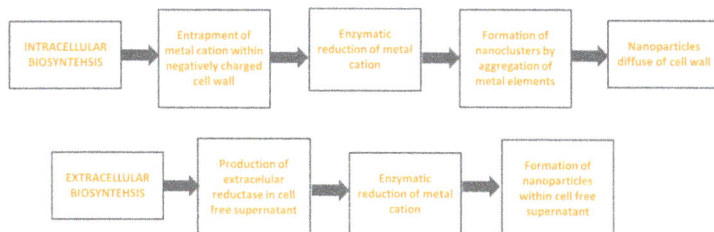

Figure 10. Scheme of intracellular and extracellular routes (adapted from Ref. 81).[81] Copyright © 2017, Elsevier.

Due to macromolecular structure, the proteins are considered as both template and stabilizing agents, whereas enzymes act as stabilizing agents and as reducing agents (reductase) through the production of H_2O_2 and superoxide anion (O_2^-) generated in enzymatic reactions.[82] Namely, nanoparticle-mediated microorganism by extracellular mechanisms is mainly an enzymatic reduction. Studies show that cofactors such as nicotinamide adenine dinucleotide (NADH) and reduced form of nicotinamide adenine dinucleotide phosphate (NADPH) act as reducing agents via electron transfer from NADH by NADH-reliant enzymes, which act as electron carriers and reducing metal ions such as Ag^+.[83] Enzymes are the major biomolecules responsible for reducing metal and non-metal ions. Thus, reductases such as NAD(P)H-dependent reductase, nitrate and nitrite reductases, and sulfate (SO_4^{2-}) and sulfite (SO_3^{2-}) reductases play major roles in inorganic nanoparticle biosynthesis.[65] Besides enzymes, several compounds, including naphthoquinones, anthraquinones, and hydroquinones, are also involved in the synthesis of nanoparticles.

In the intracellular mechanism, bacterial and fungal cells, along with polysaccharide molecules, are responsible for the fabrication of nanoparticles. The interactions of intracellular enzymes and positively charged metal ions are utilized for the entrapping and subsequent reduction inside the cell.[81] When the cell is observed under a microscope, the nanoparticles are found in the periplasmic space, the cytoplasmic membrane, and the cell wall.

Bacteria are the most extensively microorganisms used for the biosynthesis of nanoparticles and are generally known to synthesize metal nanoparticles either by extracellular or intracellular mechanisms. The major advantages of bacteria-based biosynthesis are their abundance in nature, exceptional adaptability, large-scale sustainable production with minimal use of toxic chemicals, and the facility with which the bacteria can be manipulated.[84] As opposed to other microbes, bacteria can be easily manipulated genetically for the biomineralization of metal ions. However, there are certain limitations, like laborious bacterial culturing processes and less control over their size, shape, and distribution size compared with fungi. Bacteria are continuously exposed to harsh and toxic environmental conditions. Thus, bacteria have created natural defense mechanisms such as intracellular sequestration, efflux pumps, change in metal ion concentration, and extracellular precipitation to minimize these stress conditions, that efficiently are utilized for the synthesis of nanoparticles. *Prokaryotic bacteria* and a*ctinomycete*s are the most commonly used among the various bacterial species for the synthesis of metal and metal oxide nanoparticles.[62, 85]

Fungal biosynthesis of nanoparticles is another simple approach that has been explored extensively for the fabrication of nanoparticles. **Compared with bacteria, fungi possess higher bioaccumulation ability toward metal ions** resulting in efficient and cost-effective production of nanoparticles.[86] Compared to bacteria, fungi possess enzymes and produce metabolites capable of fabricating mono-dispersed nanoparticles with desired morphologies. It has been reported that extracellular proteins and enzymes such as glucosidase, acetyl xylem esterase, hemicellulase, protein cellulose, 3-glucanase, β-glucosidase, and glucose, among others, act as reducing agents and capping agent synthesis of nanoparticles in many fungal cells. In the intracellular route, the fungal cell walls and their water-soluble polysaccharides, such as chitin, containing negatively charged groups, can interact with metal ions to biosynthesize numerous nanoparticles.[87] The large surface area of the fungal mycelium aids in the production of nanoparticles in larger quantities.[75] The advantages of using fungi in biosynthesis also lie in the ease of scale-up and downstream processing. Bhargava et al. studied the effect of pH, salt concentration, and reaction time on the particle size and yield of fungi *Cladosporium oxysporum* to the synthesis of nanoparticles of gold (AuNP)[88]. They found the maximum yield with a biomass-to-water ratio of 1:5 at 1 mM salt concentration and neutral pH.

Different research groups have documented the successful synthesis of nanoparticles via yeast.[65] **The increased pH in the internal environment of yeast causes the activation of reductases which reduce the metal ions and lead to nanoparticle formation**.[89] The process can be explained by the generation of the stress response on the presence of metals in the nutrient medium, initializing a metabolic cascade of reactions that lead to the production of some biomolecules responsible for internal stress elimination. This biomolecule presents redox and nucleophilic properties capable reducing metal ions due to their ability to bind ions of silver, gold, zinc, cadmium, copper, nickel, and other metals.[90]

Microalgae are autotrophic and aquatic photosynthetic microorganisms that can transform metal and non-metal ions into inorganic nanoparticles. The polysaccharides in microalgal cell walls contain –OH and –COOH groups involved in the bioreduction processes. Mainly, microalgae have been shown to fabricate metal nanoparticles such as metals Ag[91] and Au,[92] and metal oxides, for example, Cu.[93]

Although several nanoparticles have been successfully biosynthesized using a significant number of microorganisms, **the mechanism of biosynthesis remains controversial**. Ref. 76 describes in detail the possible mechanisms of enzyme production in microbes, as well as metabolite species in plants involved in the biosynthesis of nanoparticle.[76]

The exact mechanisms associated with inorganic nanoparticle biosynthesis in microbial cells are unclear. Some microorganisms develop protective mechanisms to eliminate toxic ions either by ion expulsion using efflux pumps, changing their solubility, or by reduction of inorganic ions to their elemental form.[94] On the other hand, the reduction of metallic ions into nanoparticles is considered dependent on a variety of essential factors, such as functional biomolecules present on the cell wall that induce biomineralization; and environmental conditions such as pH, the composition of the medium, metallic salt concentration and temperature. For example, it is well-known that high concentrations of transition-metal ions affect cell viability and even cause cell death.

2.3   Biosynthesis of Nanoparticles from plant extracts

The second considered method for biosynthesis is from plant extracts. Note that the microbe-based biosynthesis approaches started to be reported in 1989 or earlier. In contrast, the plant-mediated biological synthesis of nanoparticles began two decades ago (2002).[95] Nanoparticles produced by plants are more stable, and the synthesis rate is faster than in the case of microorganisms. Compared to other biological synthesis, there are some advantages and benefits. For example, **plants possess the ability to actively take up and reduce metal ions in soils during the detoxification process.**[96] Another advantage is the elimination of elaborate process of maintaining microbial culture, since it is completed within a few minutes or hours, depending on the types of plant used and the concentration of phytochemicals.[73] On the contrary, the nanoparticles are more variable in shape and size than those produced by other organisms.

Different parts of the plants are used to prepare the plant extract for the biosynthesis of nanoparticles, as fresh or dry material such as fruit, leaf, stem, and root. The phytocompounds present in these parts, such as polyols, terpenes, polyphenols, alkaloids, saponins, and proteins, are responsible for metallic ions reduction.[97] **The phytocomponents of plant extracts act as reducing and stabilizing agents.** Although many nanoparticles such as gold, silver, zinc oxide, and iron[98] are synthesized by adopting this green approach, the specific mechanism of how phytosynthesis acts in nanoparticle synthesis have not been well understood. It has been proposed different tentative mechanisms for their synthesis.[97] The reader can find some examples of plants used for biosynthesis in the recent bibliography.[73, 99] Plants such as *Acalypha indica, Ficus benghalensis, Zingiber officinale, Plumbago zeylanica, Centella asictica, Parthenium hysterophorus, Sapindus rarak, Passiflora foetida* are typical examples.

Other examples very interesting in the biological and pharmaceutical field are medicinal plants, such as *Saccharum officinarum, Helianthus annus, Cinamomum camphora, Oryza, Aloe vera, Capsicum annuum, Medicago sativa, Zea mays*, and *Magnolia Kobus*, can be found in the revision of Jadoun *et al.*[73]

In summary, **Table 2** shows the main aspects of probable mechanisms of metallic nanoparticle formation by microbes and plant-mediated biosynthesis.

Table 2.   Mechanism of metallic nanoparticle formation via various biological sources (adapted from Ref. 100).[100]

Biological source	Mechanism
Bacteria	Metal ion reduction by exclusive proteins, including nitrate-dependent reductase or NADH dependent reductase.
Fungi	Mineralization via biomimetic approach and exclusive enzymes for metal ion reduction.
Yeast	Presence of membrane bound (cytosolic) oxidoreductases and quinones.
Algae	Bio-reduction and nanoparticle stabilization via hydroxyl as a functional group.
Plants	Secondary metabolites, which include saponins, alkaloid tannins, steroids, and flavonoids as reducing and stabilizing agents.

2.4   Control of Shape, Size, and Composition

The nanoparticle size is mainly controlled by the strong affinity between metal ion and electron donor groups of biomolecules, like proteins. It is well known that the shape of metal nanoparticles changes their optical and electronic properties.[101] For example, concerning antimicrobial activity, morphology plays a critical role. In general, the cytotoxicity of nanomaterials depends on their size, shape, coating/capping agent, and the type of pathogens against which their toxicity is investigated. For example, smaller particles possessing larger surface-to-volume ratios have greater antibacterial activity.[102] Indeed, the smaller the nanoparticles are, the greater compatibility with biomolecules, making them a possible candidate for application as biosensors, drug carriers, and in bioimaging.

In order to control the properties of nanoparticles by varying the composition of the reaction mixture, different concentrations of biomass and cell extract could be used to produce nanoparticles of desired shapes and sizes. It appears that the particle size decreased with an increase in the leaf broth concentration or a decrease in metal ions concentration. It was discovered, for example, that metal nanoparticles formed at a slower rate at the lowest concentration, and that temperature affected the metal reduction process.[103] They reported banana peel extract (*Musa paradisiaca*) as a new source to obtain silver nanoparticles (AgNP). The banana peels were washed and boiled in distilled water for 30 min at 90 °C. The reactants were consumed rapidly when the temperature was increased, eventually producing smaller nanoparticles.

Nanoparticle size, morphology, and composition can also be significantly affected by environmental parameters such as pH and temperature.   For example, Ramanathan et al. studied the effect of growth kinetics parameters on AgNPs morphology using *Morganella psychrotolerans*, varying the temperature.[104] At an optimum growth temperature of 20 °C, spherical AgNPs were produced with an average diameter of 2–5 nm. However, a mixture of triangular and hexagonal nanoplates and spherical nanoparticles were obtained at 25 °C. Moreover, they found a mixture of shapes larger in size (70–100 nm) above and below the optimum temperature.

Other examples have shown this dependence. This is the case of the antimicrobial activity of AgNPs that undergo size-dependent and shape-dependent bactericidal properties against Gram-negative bacteria like *E. coli*.[105] In comparison with spherical and rod-shaped nanoparticles, triangular silver nanoplates showed better bactericidal activity against the Gram-negative bacterium *Escherichia coli*.[106]

In addition to the environmental conditions and bioagents present, the type of mechanism also affects the morphology obtained. It has been suggested that the particles synthesized intracellularly are usually smaller and monodispersed than those synthesized extracellularly. Moreover, in the case of extracellular synthesis, the desired shape and size may be obtained by tuning the reaction conditions.[107]

Finally, it is important to highlight in this section the work of Yoojin Choi et al.,[65], which proposes a flowchart for developing a strategy for inorganic-nanomaterial synthesis using microbial cells or bacteriophages. **Fig. 11** can be extended to the biosynthesis of nanoparticles from plant extracts. The process is divided into 10 stages, starting with selecting the inorganic nanomaterial to be biosynthesized (metal, non-metal, oxide, hydroxide, phosphate, carbonate) and ending with the final application (catalyst, energy storage and harvesting, electronics, antimicrobial agents, biomedical, etc).

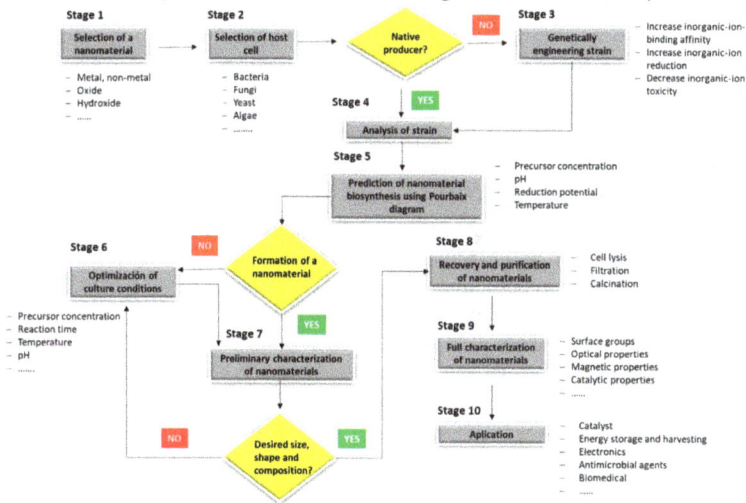

Figure 11. Flowchart to develop a strategy for inorganic-nanomaterial synthesis using microbial cells or bacteriophages adapted from Ref. 65.[65] Copyright © 2020, Springer Nature.

Obviously, one does not need to make these decisions precisely in the order as reflected in the flowchart since there are diverse biosynthetic methods and many experimental variables involved. Nevertheless, the authors suggest ten decisions that must be considered regarding the strategy to be used. This flow chart will serve as a general guide and a wise attempt to prepare inorganic biosynthetic nanoparticles.

The following sections describe some examples of gold, silver, and magnetite nanoparticle biosynthesis.

2.5   Gold and silver nanoparticles

Other greener synthetic efforts reported earlier have been dedicated to Ag and Au nanoparticles, due mainly to their importance in antimicrobial applications. **Together with gold, silver is considered a model metal to study the feasibility of nanoparticle biosynthesis using biological resources.** Biosynthesis Ag and Au nanoparticles have been synthesized using biological agents, both microbes (bacteria, fungi, yeast, algae) and plant extracts. Some examples are shown below.

Gold nanoparticles (AuNPs) are the most commonly biosynthesized metal nanoparticles due to their wide range of biological applications.[108] AuNPs have been generated using various living organisms (bacteria, fungi, and algae). Beveridge and Murray were the first to report the deposition of AuNP extracellularly on *Bacillus subtilis* cell wall using a gold chloride solution.[109] Since this pioneering work, numerous bacteria, such as *Rhodopseudomonas capsulata, Pseudomonas aeruginosa, B. subtilis, E. coli DH5, B.licheniformis*, and many more, have been widely studied.[110]

Microorganisms can be used for biosynthesis of gold nanoparticles either extracellularly or intracellularly. Sastry and coworkers reported the extracellular synthesis of gold nanoparticles by fungus *Fusarium oxysporum[95]* and actinomycete *Thermomonospora sp.*[95] and also the intracellular synthesis by fungus *Verticillium sp.[111]*

Biosynthesis intra and extracellular of gold nanoparticles (AuNPs) using several bacterial species such as Bacillus megateriumD01, Desulfovibriodesulfuricans, E. coli K12, Shewanella alga, Rhodopseudomonascapsulata, and PlectonemaboryanumUTEX 485 (Kaur et al. 2019), Paracoccus haeundaensis sp have been used, with narrow size about 20 nm. For example, 50 nm AuNPs were synthesized at room temperature using Escherichia coli cells via extracellular approach. Membrane-bound proteins were involved in the bioreduction and subsequent stabilization of AuNPs.[112] However, bacterial synthesis of AuNPs is not recommended since the binding affinity of bacterial metabolites toward gold ions is too high.

Nagajyothi et al.,[113] used *Lonicera japonica* flower extract in the bioreduction of chloroauric acid to synthesize gold nanoparticles. Amide, alkane, amino, and alcohols present in this extract were involved in the stabilization and bioreduction of gold nanoparticles. An extensive list of bioreductants involved in the biosynthesis of gold nanoparticles, along with size and shape, are listed in Table 1 of Ref. 114.[114] Due to their excellent antimicrobial properties, silver nanoparticles (AgNP) have been widely

New Materials for a Circular Economy   Materials Research Forum LLC
Materials Research Foundations 149 (2023) 233-310   https://doi.org/10.21741/9781644902639-8

synthesized using biosynthesis approaches.[115] However, besides as antimicrobial agent, Ag NPs are used as electrochemical sensors, biosensors, in medicine, health care, agriculture, and biotechnology. They have great bactericidal potential against both gram-positive and gram-negative pathogens.

It has been demonstrated that AgNPs inhibit the growth and proliferation of many bacteria by binding Ag/Ag^+ with the biomolecules present in the microbial cells (to name a few: *Bacillus cereus, Staphylococcus aureus, Citrobacter koseri, Salmonella typhii, Pseudomonas aeruginosa, Escherichia coli*). Various microbes have been employed to reduce Ag^+ to metal Ag to form silver nanoparticles, involving bacteria, fungi, and yeast.[77] Intracellular and extracellular approaches have been used using Gram-positive and Gram-negative bacterial strains. In general, the genera *Bacillus* is widely used in the pharmaceutical and enzyme industry due to its potential bioactive compounds. For example, *B. megaterium, B. cereus, B. licheniformis, B. amyloliquefaciens, B. marisflavi, B. flexus*, and *B. subtilis[77]* have been sustainably used for AgNPs biosynthesis. Recently Saravanan et al.,[116] prepared ~50 nm sized, spherical, AgNPs using the cell-free extract of Gram-positive, spore-forming, aerobic *Bacillus brevis* bacteria. *Pseudomonas stutzeri* is a bacterial strain capable of reducing Ag^+ ions when placed in a concentrated aqueous solution of silver nitrate and forming AgNPs of well-defined size using an intracellular mechanism.[117] Well control of size and morphology by biosynthesis has been reported in numerous studies. The AgNPs have been biosynthesized as spherical, disk, cuboidal, hexagonal, and triangular shapes.[118] They have been fabricated using culture supernatant, aqueous cell-free extract or cells. For example, Das et al. reported the synthesis of spherical AgNPs, of 42–92 nm in size using the supernatant of *Bacillus strain CS11* isolated from an industrial zone.[119]

The biosynthesis of AgNPs has been carried out by utilizing several plants ethanol extracts, such as *Cardiospermum halicacabum L.* leaves[120] and *Impatiens balsamina L.* leaves.[121] For example, the leaves extract of *Croton sparsiflorus* were used to synthesize spherical AgNPs with diameters of 22–52 nm, nd such NPs showed effective antibacterial activity against *S. aureus, E. coli*, and *B. subtilis*.[122] Another example using plant extracts, well-defined silver nanoparticles (16–40 nm) were rapidly synthesized using *Datura metel* (a plant in the family Solanaceae).[123] Leaf extracts of this plant contain biomolecules such as alkaloids, proteins, enzymes, amino acids, alcoholic compounds, and polysaccharides which could be used as suitable bioagents with Ag^+ and as scaffolds also, to direct the formation of AgNPs in solution. For this example, alcoholic compounds, such as quinol and chlorophyll pigment, were responsible of reducing silver ions and stabilizing the resultant nanoparticles. AgNPs have also been synthesized by Fungus mediated synthesis in the form of a film or produced in solution or accumulated on the surface of its cell using *Verticillium, Fusarium oxysporum*, or *Aspergillus flavus.[124, 125]*

The use of fruit extracts has also been widespread. For example, biosynthesis of silver and gold nanoparticles using *Emblica officinalis* fruit extract produced highly stable Ag and Au nanoparticles extracellularly.[126] Pear extract contains essential phytochemicals consisting of organic acids, peptides, proteins, and amino acids. Also, it contains saccharides that provide synergetic reduction power, for example, for reducing chloroaurate ions ($AuCl_4^-$) to Au metal. Great control over the size of silver and gold nanoparticles with spherical and triangular shapes was rapidly and efficiently obtained by fruit extract of *Tanacetum vulgare*.[127] Ag and Au nanoparticles were 16 and 11 nm in size, respectively. Furthermore, it was reported the biosynthesis of bimetallic core-shell nanoparticles of gold and silver by simultaneous reduction of aqueous Ag^+ and $AuCl_4^-$ ions with the wash water of neem leaves (*A. indica*).[128] For example, with *Bacillus brevis (NCIM 2533)*, spherical silver NPs with sizes between 41–68 nm were obtained using proteins 57 as capping and stabilizing agents.[116] More details of biosynthesis from plants and their parts containing biomolecules and several types of secondary metabolites are provided in Table 1 and Table 2 of Ref. 115.[115]

2.6   Iron oxide (Fe₃O₄) nanoparticles

Like in Ag and Au nanoparticles, the conventional methods of preparing magnetic nanoparticles (MNPs) (hydrothermal process, sonochemical method, micro-emulsion technique, electrochemical route, and co-precipitation method,[129] may still be either toxic, evolve hazardous gases, or time-consuming and expensive. The high potential in environmental and catalysis and biomedical applications,[130, 131] have caused the active investigation of the synthesis and control of magnetic nanoparticles. A variety of systematic studies have been conducted on the synthesis of MNPs and their applications in technology.

Biosynthesis of MNPs by microbes and plants has been reported with the same advantages: non-hazardous, eco-friendly, cost-effective method, compared to other nanoparticles. Indeed, biological synthesis can reproduce magnetic nanoparticles in a reasonably good quantity and can even exhibit superparamagnetism due to their specific chemical structure. Many different biological molecules have been used to synthesize MNPs, including microorganism DNA-RNA templates, plant extracts, or enzymes,[132, 133] however, relatively few reports exist. Several successful reports for plant-mediated synthesis of Fe₃O₄ nanoparticles from various plant extracts have been reported using seed extract of Grape proanthocyanidin, leaf extract of *Tridaxprocumbens, seaweed* of *Kappaphycu salvarezii, Carica papaya*, peel extract of plantain, and seaweed of *Sargassum muticum* and *Zanthoxylum armatum DC*. For example, Fe₃O₄ nanoparticles were successfully synthesized using Carica papaya leaf extract.[134] Showing saturation magnetization as high as 60 A-m^2 kg^{-1}, comparable to that obtained with nanoparticles obtained by chemical methods. The addition of ferric nitrate solution to the plant extract which contains carbohydrates, induces the reduction of Fe^{3+} to produce Fe₃O₄ nanoparticles with spinel ferrite phase after microwave heating. Moreover, the size, shape, and saturation magnetization led to quite good SAR value. In general, co-precipitation method in the presence of aqueous plant extracts as a reducing and capping agents is the most common method used to biosynthesize Fe₃O₄ nanoparticles.

Regarding microbe-mediated biosynthesis, unicellular magnetotactic bacteria are known for their natural ability to produce magnetite (Fe₃O₄).[135, 136] These microorganisms biomineralize magnetic nanoparticles composed of magnetite or greigite embedded in a biomembrane in microaerobic or anaerobic conditions.

2.7 Synthesis of nanoparticles from waste

Considering the present situation, using waste materials to synthesize nanoparticles will help researchers design safer nanomaterials, and our society will be better. Bio and industrial waste as feedstock provide a solution for environmental burdens while simultaneously reducing the negative impact of the waste on the environment. It is very attractive because they are readily available and inexpensive. Since mechanical recycling processes are often costly, both from an economic and energy point of view, it is inevitable to recycle and valorize waste materials into products with high economic value. Therefore, many "green processing routes" have been explored to minimize the recycling process's associated cost and environmental impact.

Waste materials can be classified into two categories: biomass (biological waste) and industrial waste. Some examples of waste materials appear in **Fig. 12**. Biomass contains various primary and secondary metabolites such as phenolic acids, terpenoids, alkaloids, and flavonoids involved in the redox reaction to produce green nanoscale particles, such as metallic nanoparticles (see the previous section), but also for fabricating carbon nanoparticles. Regarding industrial wastes such as batteries, plastics, and tires have been employed as primary sources for nanoparticle synthesis. For example, the fabrication of Pb nanoparticles from lead batteries.

Various treatment methods have been developed and applied for the conversion of wastes into valuable forms of nanomaterials, especially carbon nanoparticles, such as carbon nanotubes and nanosheets (graphene) via chemical and thermal activation, arc discharge, vacuum evaporation-inert gas condensation, sodium borohydride reduction and solvent thermal approaches[137] that are included in the next section. Due to the diverse types of waste, the first step in preparing the nanoparticle from waste materials is its pretreatment. After sample preparation, several techniques can be employed to obtain the final product. Various treatment methods have been developed and applied for converting wastes into useful nanomaterials, especially carbon nanoparticles (carbon nanotubes and nanosheets (graphene)), via chemical and thermal activation, arc discharge, vacuum evaporation-inert gas condensation, reduction, and solvent thermal approaches. A scheme of various methods for preparing nanomaterials from biomass and industrial wastes is shown in **Fig. 12**.

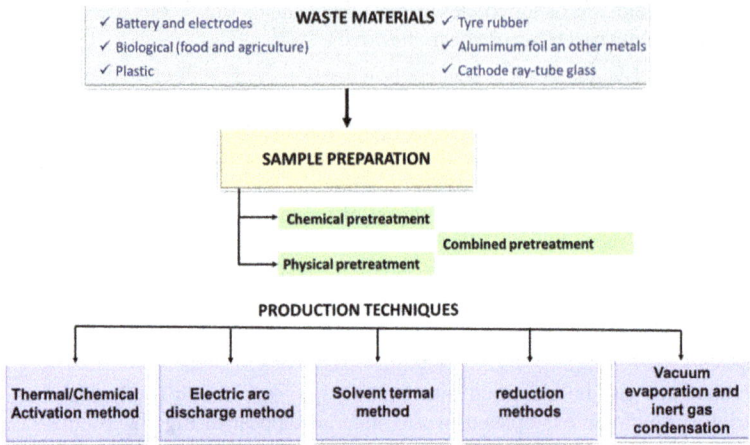

Figure 12. preparation of nanomaterials from biomass and industrial (adapted from Ref 137).[137] Copyright © 2018, Elsevier.

Table 3.   List of waste-derived nanomaterials and their application in various fields.

Waste material	Nano material	Applications
Rice husk waste	Silica nano particle	Food additive, biomedical and energy applications
Food waste	Green carbon nanodots	Biomedical
Stainless steel slags	Pt/SiO_2	Catalysis
Breynia rhamnoides	Au and Ag nanoparticles	Catalysis
Biomass	Au and Ag nanoparticles	Catalysis
Polystyrene waste	Solid SiO_2 -PS nanocomposite	Catalysis
Plastic bag wastes	Carbon nanoparticles	Sensing applications
Pomelo peel wastes	Carbon nanoparticles	Sensing applications

Cathode ray-tube funnel	Pb nanoparticles	Energy applications
Glass waste	Pb nanoparticles	Energy applications
Plant waste	Carbon nanosheets	Energy applications
Orange peel waste	TiO_2	Photocatalysis and nanocomposite filler
Egg-shell waste	Nano-hydroxyapatite	Biomedical
Potato peel waste	Cellulose nanocrystal	Biomedical
Sesame oil waste	MWCNTs and graphene	Electrical and energy applications
Pig bone	Hicrarchical nanoporous carbon	adsorption and energy applications
Leaf extract	Fe nanoparticles	Catalysis

One of the most significant advantages found in nanoparticles synthesized in this way is that several properties related to the surface area (for example, for energy storage applications, and for catalytic or adsorptive activity) are often greatly improved relative to *virgin* nanoparticles. Therefore, the selection of strategies for biomass-derived nanoparticles will increase the benefit derivable from the development of electrode capacity, catalytic capacity, and sensing efficiency.[137] **Table 3** includes a list of nanoparticles synthesized from waste material (also indicated), their synthesis method (more information can be found in Ref 137), and some applications.

2.8   Sustainable and green synthesis of carbon nanomaterials

Biomass may be converted into carbon nanomaterials. Biomass waste management is a big challenge. Therefore, serious efforts have been made to create added value. This includes mainly transformation in biogas and biofuel, but during the last 15 years, several procedures have been developed to transform into valued carbon nanomaterials for use in supercapacitors, batteries and fuel cells, catalysis, drug delivery, water treatment or as part of nanocomposites for many applications.[138]

Production methods involve multistep procedures with energic conditions, including oxidation-reduction, fermentation, and degradation, which are unable to reach high yields on nanoparticles. Recent studies focused on improving the existing methods. We may mention microwave-assisted pyrolysis, hydrothermal synthesis, molten salt processes, and combinations within and with other methods.

It is possible to obtain from biomass a wide variety of nanocarbons, including nanoporous carbons, carbon dots, graphene and graphene oxide, and carbon nanotubes, often doped with heteroatoms present in the biomass. We can mention some random examples. Nanoporous carbons with very high specific areas have been prepared from peanut meal for supercapacitors devices. The presence of N, P, and S impurities in the material improve the pseudocapacitive behavior and increase the capacitance.[139] Two-dimensional highly porous graphene structures have been obtained from pomelo peel as the precursor and through an H_2O_2/HCl treatment,[140] showing a good performance in EMI shielding. N-doped high-quality graphene was obtained by pyrolysis of chitosan derived from biomass waste, showing a good performance in the conversion of solar energy to chemical hydrogen energy when treated with R_xCrO_x co-catalyst.[141]

In summary, **numerous carbon nanomaterials derived from biomass waste have been developed, and some are already on the market**. However, some issues need to be solved for a large-scale commercial application: i) It is necessary proper management of biomass wastes in order to separate the different kinds of waste because each type of nanocarbon material will need a selected method with a specific feedstock; ii) usual procedures involve high temperatures and pressures for long processing times, leading to degradation of carbon and reducing yield; iii) synthesis procedures make use of hazardous chemicals (strong bases and acids, oxidants) and heavy metal catalysts, which are intrinsically hazardous to the environment; iv) not all the routes are competitive at commercial scale respect to synthesis from other sources, and its viability will depend on energy policies and taxes of regulatory bodies. Therefore, the practical implementation of nanocarbons derived from biomass is limited by the high cost and hazard challenges.

Carbon nanotubes (CNT) are among the countless engineered nanomaterials (ENMs) with advanced properties. CNT are among the strongest materials ever discovered, with excellent properties (outstanding thermal, mechanical, electrical, and chemical properties: 100 times stronger than steel, best field emission emitters, can maintain a current density of more than 10^{-9} A/cm^2, thermal conductivity comparable to that of diamond).[142] that, however, engineers are just starting to unlock their full potential. CNTs are synthesized using different methods leading to the formation of the well-known types of CNTs (i.e., SWCNTs, DWCNTs, FWCNTs, and MWCNTs), with distinct properties such as size and purity depending on the technique. Several examples are chemical vapor deposition (CVD), laser ablation, electrochemical synthesis, and arc discharge.[143] Among them, chemical vapor deposition (CVD) is the commonly used method for synthesizing CNTs.[144] In this technique, the size and properties of the catalyst play a significant role in the growth of SWCNTs or MWCNTs, where smaller particle size (a few nm) leads to the growth of SWCNTs, whereas MWCNTs are formed when the particle size is larger (tens of nm). The SWCNTs have higher formation energy than MWCNTs since they are easily grown even at low temperatures (~600 °C), while SWCNTs require high temperatures (~1000 °C).[145] Therefore, hydrocarbons such as benzene and carbon monoxide, which are stable at high temperatures, are the most used for the synthesis. However, the large-scale production of CNTs requires the establishment of greener precursors and sustainable methods to minimize costs and negative environmental impacts. To obtain high-quality CNT, the optimization of the preparative parameters, such as carbon source, catalyst, and support, is the key. Since they are man-made materials produced via the principles of nanoscience and nanotechnology, it is possible to control their properties to make them applicable in the water-energy-food sectors, for example, through green and sustainable synthesis. Indeed, CNTs have surfaces that can easily be modified to make them more biocompatible.

To make the production processes green, the carbon source, catalysts used, purification process and gas emissions must meet the principles of green chemistry and engineering. As such, the use of natural carbon sources, microwave-assisted methods, green catalysts, and support materials must be considered. The growth mechanism of CNTs synthesized from green precursors is not thoroughly understood. It is believed to follow a mechanism like that of conventional precursors. For example, high-purity carbon nanotubes (CNTs) are produced by CVD using environment-friendly hydrocarbons like camphor or cellulose. The active part decomposes to form carbon at a specific temperature in an inert environment, followed by deposition on the surface of the natural/ waste catalyst, which provides nucleating sites for the growth of CNTs.[146] Other natural sources as rice straw, several oils such as olive oil, turpentine oil, coconut oil, sesame oil, palm oil, and hydrocarbons such as camphor, petroleum, and plastic waste (mainly LDPE and PP), have been used to prepare CNT (single and multi-wall) from CVD, spray pyrolysis and pyrolysis techniques. Using the catalysis/support adequate (for example, Fe-Ni/Al$_2$O$_3$, Fe-Co/zeolite, Ferrocene/silicon, Ni, Fe/silica, among others), it has been possible to obtain high-quality CNTs, with lengths ranging from the nano to micron scale. More details of each of these syntheses are referenced in Ref. 147.[147]

Graphene is a remarkable material, like carbon nanotubes, incredibly strong, with ultra-high intrinsic charge mobility and thermal conductivity of 2.5×10^5 cm^2 V^{-1} s^{-1} and 5.0×10^3 W m^{-1} K^{-1}, respectively.[147-149] Indeed, a graphene sheet, one atom thick (2D), is entirely invisible to the human eye.

Due to its incomparable electrical, mechanical, thermal, and excellent physical properties, graphene has great potential in the green chemistry domain in terms of synthesis and application. For the preparation of graphene oxide (GO), modified Hummer's method is typically used, in which graphite powder, as a starting material, is oxidized in the presence of a concentrated acid solution and KMnO$_4$.[150] Them, chemical o thermal reduction is carried out to produce reduced graphene oxide (RGO) in an effective way. Several authors considered thermal reduction a green technique because no chemicals are used. Nevertheless, chemical reduction has taken priority over non-chemical reduction methods because of the high quality and productivity of RGO obtained. Among the large number of chemical reducing agents used to prepare RGO, the most efficient is hydrazine,[151] well-known reagent as toxic and harmful to the environment. Therefore, in "green terms" RGO synthesis requires the use of 12 principles, including looking for green, economical, and efficient biomass reducing agents. For example, RGO has been green synthesized using borohydride, biomass, plant extracts, and even animals as reducing agents.[152, 153] GO

New Materials for a Circular Economy   Materials Research Forum LLC
Materials Research Foundations 149 (2023) 233-310   https://doi.org/10.21741/9781644902639-8

is not a naturally occurring compound. **Fig. 13** displays a representative diagram of the synthetic route of GO from graphite and the different methods for obtaining GO. Subsequent chemical reduction is made with several common chemical reducers, and most recently, green-reducing ones such as ascorbic acid, a non-toxic, inexpensive organic acid. Other conventional methods, such as the micromechanical exfoliation technology[154] and CVD could be considered green synthesis.[143] However, yield is greatly limited in the micromechanical exfoliation method, leading to industrial production limitations. As mentioned before, the chemical reduction of graphene oxide requires oxidizing graphite with strong acid or oxidizing agents.[143] Thus, to counteract the problems associated with toxic and extreme conditions, scientists have searched for eco-friendly reducing agents similar to hydrazine in yield and efficiency.[155] As a result, many research works are based on developing different "green reduction strategies" on GO. Evaluating such strategies within the last decade has contemplated numerous "green reducing agents" (i.e., non-toxic and environmentally friendly). They include organic acids, plant extracts, microorganisms, sugars, antioxidants, amino acids, or proteins. Many are already used and contrasted in the biosynthesis of metal nanoparticles.

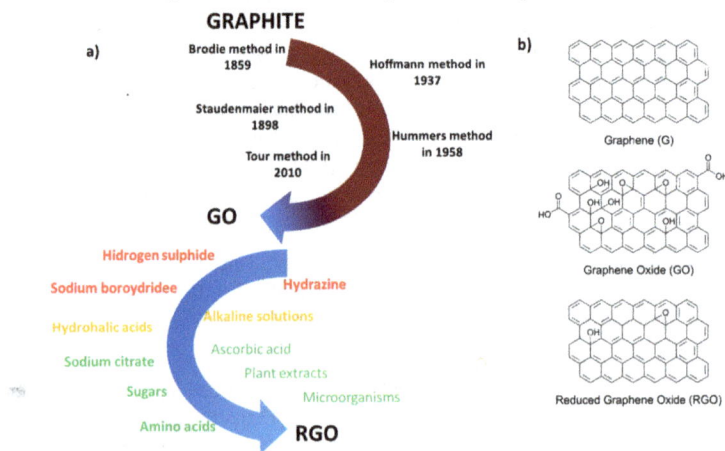

Figure 13. Schematic representation of a) the major oxidation methods of graphite to graphene oxide and the chemical reduction of graphene oxide by several reducing agents (adapted from Ref. 155.[155] Including chemical reducers (red) and more eco-friendly reducers (yellow and green) b) schematic structures of graphene, graphene oxide prepared from several indicated methods and reduced graphene oxide (RGO) resulting from reducing methods. Copyright © 2017, Elsevier.

Among the numerous agents used for GO reduction is hydrazine the most used. Hydrazine is known to be the most efficient reductant in terms of giving RGO with good qualities.

However, the toxic nature of hydrazine has led to the proposal and development of new green routes to synthesize RGO. As a green alternative, the first environmentally friendly reducing agent was ascorbic acid (AA) (also known as Vitamin C),[156], a non-toxic, inexpensive organic acid. Since that date, the use of AA has been massively used by researchers. Regarding quality, RGO obtained by the reduction of GO via AA competes with RGO obtained by hydrazine reduction, and it is a safe and effective method in the large-scale production of RGO.

With the evolution and need for "green reduction", many researchers have focused on finding novel reducing agents. As a result, many reports on plant extracts and microorganisms have been published and still counting. To illustrate a few, Na-citrate, caffeic acid, green tea, pomegranate juice, glucose, yeast, and L-cysteine, among others, have been used to produce RGO with excellent C/O ratios. In ref 155 it is discussed a few green reductants[155]. They have proven to be environmentally friendly, and the product obtained is highly dispersible, biocompatible, and useful for applications such as electrochemical, bioremediation, or catalysis. Furthermore, almost all bioagent reducers from the plant extracts are large biomolecules that can interact superficially with the RGO sheets resulting in stable dispersions of RGO, making it more biocompatible. That means the **surfaces of these carbon nanoparticles can be highly functionalized and do not need any further modification**. However, for some applications, like materials for conductive applications, the presence of these biomolecules can be disadvantageous.

3. Nano-waste

In a simple definition, **nano-waste is waste containing nanomaterials**. The high growth of the nanotechnology market also involves a significant challenge related to the monitoring, use, and disposal of nanoparticles once they finish their life cycle. Nanomaterials demonstrate remarkable physical and chemical properties, but their toxicity and environmental impact may also be increased concerning their corresponding bulk materials. Their inherent nano dimensions cause nanoparticle penetration into the biota, often showing higher bioaccumulation and long-range effects. In addition, many nanoparticles contain components, such as heavy metals, potentially hazardous to organisms and ecosystems. Furthermore, very fine nanoparticles are difficult to track and monitor in the body and the environment. Current studies have raised serious concerns about nanoparticle reactivity and toxicity, although the adverse effects of nano-waste are still under debate. Their physicochemical behavior regarding human health risks and environmental effects is not fully known. In addition, for a given particle, the hazard potential varies significantly with its solubility in water, chemical stability, size, and tendency to agglomerate. Many ecotoxicity studies have been carried out in cell culture systems under controlled conditions (laboratory scale), with controversial results. Some studies show biocompatibility, while others claim their potential hazard. There is also a lack of knowledge about the behavior under realistic conditions or long-range effects on bodies and the environment, so this field is still under study.

New Materials for a Circular Economy
Materials Research Foundations 149 (2023) 233-310

Materials Research Forum LLC
https://doi.org/10.21741/9781644902639-8

Concerning the origin of nanoparticles, there are **natural sources** of nanomaterials, not created through human actions and usually integrated into the life cycle of the ecosystems, and anthropogenic nanomaterials, present in nanowastes products of human activities. This includes **incidental nanomaterials** unintentionally released into the environment and generated from materials and devices not belonging to the nanotechnology, and synthetic nanomaterials created for specific applications, usually denominated with the term **engineered nanomaterials**. In addition, it is worth mentioning that synthetic nanoparticles are sometimes deliberately released in agriculture or bioremediation. For example, some nanoparticles promote the germination and growth of plants, while others act as nano-pesticides or pollutant degradation agents in contaminated soils or oil spills. However, even those nanoparticles can potentially affect on ecosystem balance or enter the food chain.[157] This section will focus on engineered nanoparticle release and its effects on the environment, human health and ecosystems, and the disposal and treatment of nano-waste.

3.1 Classification and sources of nano-wastes

During manufacturing and handling, the life cycle, or at the end-of-use of the engineered nanomaterials and nanodevices, the constituent nanoparticles may enter the waste streams (solid, liquid, suspensions, and aerosols) in the form of specific substances (organic or inorganic) or even in complex combinations. Another group of nano-wastes would constitute the end-of-life products contaminated by nanoparticles of any nature (i.e., solid matrices waste containing nanomaterials).

A brief classification of nano-wastes may be done by attending to their chemical nature, composition, or dimensions, in a similar way as indicated in Table 1 for nanomaterials. Nanowastes can be composed of carbon-based materials (fullerenes, carbon nanotubes, nanofibers, graphene nanosheets), metal-based materials (metals such as Ag, Fe, Au, Cu, Pt; metal oxides such as TiO_2, SiO_2, Al_2O_3, ZnO, Bi_2O_3; chalcogenides such as MoS_2, CdSe), polymer-based materials (dendrimers, coated particles) quantum dots, nanoclays and mixtures of all them. A crucial consideration is whether these nanoparticles are free-standing or supported in a matrix. For instance, free-standing 2D and 3D nanoparticles pose a higher potential health risk concerning 1D nanosheets, but they all show much higher hazards than supported nanoparticles or matrix-embedded nanocomposites. Therefore, a more effective and helpful classification should be based on the toxicity of the nano-waste. However, due to the multiplicity of nano-waste streams from the same product and the lack of risk exposure data, the **current risk classifications do not specifically refer to risk-associated nanoparticles**. Considering this, nano-waste can be broadly classified into five categories, summarized in **Table 4**.[158, 159]. It must be noted that **a given nanomaterial cannot be ascribed to a single class** because it can show a wide variety of risk profiles depending on the nano-waste carrier and its transfer coefficient, among other factors influencing its hazard. For example, TiO_2 is usually ascribed to Class I for almost any application. However, Ag nanoparticles can be ascribed to Class I (not toxic) in sports equipment wastes but Class II-III (Harmful/Toxic) in personal care disposal products.

Several characteristic properties are relevant when considering nanoparticle risks. Surface-related properties (composition and purity, surface charge, zeta potential, solubility, porosity...) and dispersion state (size distribution, agglomeration, dispersibility...) are key factors to consider in risk assessment frameworks.

Table 4.   Classification of nano-wastes based on their toxicity (adapted from Ref. 158).[158]

Class	Hazard	Exposure	Risk	Example sources
I	Nontoxic	Low to High	None or very low	Solar panels, polishing agents, nanocircuits, electrodes, polishing agents.
II	Harmful	Low to medium	Low/medium	Solar panels, nanocircuits, surface coatings, paints, polishing agents.
III	Toxic	Low to medium	Medium/high	Food packaging, food additives, cosmetics, pesticides, personal care products, polishing agents.
IV	Very toxic	Medium to high	High	Paints, surface coatings, cosmetics, pesticides, personal care products.
V	Extremely toxic	Medium to high	Very High	Sunscreen lotions, pesticides.

3.2   Sources and routes of nano-waste contamination

Nano-waste sources can be classified into two main categories: i) **point sources**, i.e., direct release of nano-waste from industry manufacturing units, research laboratories, power generation, incinerators, and household nano-containing products discarded on home sewage. For example, current personal care products incorporate increasing amounts of nanoparticles. In most cases, nano-waste is discharged into the sewage system and ends up as part of wastewater treatment facility sludges; ii) **non-point sources**, i.e., nano-waste generated from deterioration of nanoparticle-containing products. Paints and pigments are among the highest contributors to nano-waste release during its life cycle when confronting environmental conditions. In optoelectronics, a large amount of nano-waste is also generated that ends up being incorporated into the environment through landfilling and subsequent leaching. Nanotechnology applied to bioremediation or agriculture may also represent a hazard through soil accumulation and transfer to water bodies. Finally, combustion engines and aerosols are important sources of nanoparticles incorporated directly into the atmosphere. Waste sources can also be classified as stationary, for example, chemical factories, incinerators, or boilers; or dynamic, mainly combustion engines in the transportation industry.

Nanoparticle contamination may occur during manufacturing processes, during the service life of devices containing nanomaterials, and during nano-waste management at the end of life, contaminating the environment.[160] For example, quantum dots are small-size semiconductors with unique luminescent properties derived from quantum effects, with continuously increasing production. A recent report showed that the main release sources are during the synthesis and at the final disposal stage after use.[161] Other nanoparticles, such as Ag or TiO_2, extensively used in personal care and cosmetics, catalysis, or agriculture, enter the wastewater streams during the whole life cycle.

Concerning the manufacturing stage, nano-wastes are generated during the production of nanoparticles, quantum dots, nanotubes, and nanowires. There are also produced during the fabrication and manufacturing of nano-based products, especially in electronics, energy, medicine, cosmetics, pesticides, packaging, paints and coatings, and textiles. Plant operations and handling may generate occupational health risks in all these cases.[162]Nano-wastes are often present in the residues of R&D laboratories. However, this represents a very low fraction of nano-wastes and can be subject to specific treatment since the type of nanoparticles and the source of contamination are well-defined. The main source of nano-waste in this stage is undoubtedly the nanomaterials manufacturing industry, especially in "top-down" processes such as mechanical cutting and shredding, laser ablation or lithographic processes, and the release of nanoparticles depends on the technology and synthesis routes used. Nano-waste from production residues are relatively pure streams in the form of suspensions in liquids or air/gas, solids, and being part of items such as containers, wipes, and disposable parts contaminated by nanoparticles.

Nanoparticle release from nanocomposites during use has also been reported due to wear under external conditions, often in the form of aggregates of nanoparticles with binder or matrix.[163]

Indeed, nanocomposites may release fragments of variable sizes containing former nanoparticles in their whole life cycle. As stated before, nanomaterials are considered safe when they are embedded or incorporated into a solid matrix but potentially hazardous when returned to dispersed form. In principle, there is a low probability of the free release of nanoparticles when they are incorporated into a well-designed composite. The main causes of degradation are machining (handling, cutting, drilling) at the nanocomposite fabrication stage; degradation during use (UV, heat, moisture, chemical, biological); and at the disposal stage as a function of the nano-waste management. For example, TiO_2 is often incorporated into paints and pigments and may be released due to paint degradation or abrasion.

Finally, the nanomaterials may be part of household effluents from end-of-life products as part of the waste contamination. **Waste handling protocols** (landfilling, incineration, etc) will determine the possible release of nanoparticles during disposal due to degradation by interaction with light, moisture, or microorganisms. It is estimated that around 80-90% of the 10 most common nanoparticles are landfilled, in some cases as part of sewage sludges of wastewater treatment plants. The rest will end up in water bodies, and only around 1%

end up in the atmosphere. A particular issue is the airborne release of nanoparticles during waste incineration.[164]

3.3   Toxic effects of nano-wastes

The main causes of the toxicity of the nanoparticles have been established as: i) **inherent toxicity** of the materials, such as heavy ions, which can be released due to the high surface-volume ratio of the nanoparticles; ii) due to their **small size** compared with biological structures, nanoparticles can access and enter tissues and cells and cause bioaccumulation; iii) the **shape** of some nanoparticles, for example, nanotubes may pierce cell membranes. Many nanoparticles induce **oxidative stress**, altering the redox potential and inducing protein denaturation or other cell injuries.

Most studies on nanoparticle toxicity have been done in cell culture systems (in vitro) with well-defined free-standing nanoparticles. The health effects depend on their composition, size, exposure duration, and concentration. Occupational exposure in synthesis and processing workplaces is the highest potential hazard, being airborne exposition the most important. Other ways include public exposure (cosmetics, personal care, textiles, food, and beverages) and pollution from combustion engines, incineration, leaks, and wear of nanoparticle-containing materials.

Nanoparticles may enter the body through three main routes: inhalation, swallowing, and absorption through the skin.[165] Once inside the body, nanoparticles are highly mobile, can be persistent, and show organ bioaccumulation, leading to neurotoxic effects or causing long-term tumors. **Particle size is very relevant for particle toxicity**. This was well stated from the early studies on asbestos, as its fibers were cataloged as carcinogens,[166] and recent reviews focus on the carcinogenicity of other 2D nanomaterials such as carbon nanotubes.[167]

Human skin acts as a barrier toward nanoparticles. However, small particles can penetrate hair follicles, sweat glands, or injured skin. This is especially relevant for personal care and sunscreen lotions. Nanoparticles can penetrate the lungs by inhalation, causing inflammation and long-range chronic effects, leading to cell destruction. Particle-tissue interaction may promote more adverse health effects. Nanoparticles can cross the blood-air tissue barrier and reach the organs. For example, carbon nanotubes are more toxic when inhaled, as compared to oral or dermal exposure, which has mild effects.[168] Even the harmless and inert nanosilica (50 nm or less) may induce silicosis, pulmonary emphysema, and tumors under high airborne exposure.

Nanoparticles can also be consciously swallowed, as part of food additives and drugs, or unconsciously, because nano-waste may accumulate in livestock, especially in fish in which nanowaste present in water bodies accumulates. In all cases, nanoparticles get absorbed into the gastrointestinal tract. Absorption depends on size and geometry and the chemical nature of the nanoparticle. Ingested nanoparticles may damage digestive cell membranes through oxidative stress mechanisms and modify the metabolism and absorption capacity. In the worst case, nanoparticles can be hepatotoxic and cause

New Materials for a Circular Economy   Materials Research Forum LLC
Materials Research Foundations 149 (2023) 233-310   https://doi.org/10.21741/9781644902639-8

obstruction or gradual fibrosis. A more detailed view of the studies on the toxic effects of nanoparticles on human health can be found in many recent reviews.[162, 165, 169-171]

3.4 Impact of nano-wastes on the environment

Once the nanoparticles are released into the environment, they spread to broader areas through the air, the water, or deposit in soil sediments, although transferences may occur. For example, water irrigation can be a carrier for soil nano-waste entering crops and eventually into the food chain, and nanoparticles in aerosols may end up in water bodies or soils far away from the source. However, as was stated before, most of the nano-waste end disposal occurs in soils, compared to water and air.

Contrary to the investigation on human health, there are few studies about potential risks in other organisms. Despite this, enough evidence confirms that nano-waste may carry environmental impacts. Chemical stability, surface chemistry, size and shape, solubility, and dispersibility (agglomeration state) will determine the environmental behavior.[172] Nanoparticles can disturb the ecosystem balance, and several examples may be found in the bibliography.[173] Ag (10-100 nm) may alter aquatic organisms' immune response and induce cell apoptosis and oxidative stress. Cadmium and Arsenic, former of quantum dots and solar cells, are extremely toxic to biota, causing respiratory and renal function problems, oxidative stress, and are carcinogenic. Non-metallic particles are also hazardous, as shown by carbon nanotubes, which indirectly affect soil and aquatic organisms, slow algal growth, and are toxic to aquatic invertebrates, and so on.

Metal and metal oxide nanoparticles enter microbial cells and alter the catalytic activity of enzymes. ROS (reactive oxygen species) stress production induced by metal ions leads to oxidative stress, disturbing the normal redox state of the cell. Metallic particles also interact with cell antioxidants. In aquatic ecosystems, nanoparticle accumulation in major organs is a major cause of organism affectation, malformations, and mortality. Studies of bioaccumulation of metals, metal oxides, and carbon nanotubes are conclusive.[160, 173] Regarding nanoparticle phytotoxicity in plants, nanoparticles can also induce ROS leading to the oxidation of lipids, proteins, and nucleic acids. This causes apoptosis and cell death, disrupt phytohormone synthesis, and alters photosynthesis, reducing crop production and causing plant death.

In summary, the environmental hazard of nanowaste release has been demonstrated, but the impact of nanoparticles on ecosystems needs to be further studied. For example, many nanoparticles have long-life deposition times, and their impact on the biota is not well known. There is also a lack of practical techniques for nano-waste detection in soils and aquatic ecosystems.

3.5 Nano-waste treatment strategies

Because of the wide variety of nanomaterials, there is no unique method of nano-waste treatment. In addition, typical waste stream treatments may not be optimal for nano-waste.

New Materials for a Circular Economy   Materials Research Forum LLC
Materials Research Foundations 149 (2023) 233-310   https://doi.org/10.21741/9781644902639-8

Current nano-waste treatments are included in four categories: recycling, incineration, landfilling, and liquid effluent treatments.

Recycling is the desired goal for any type of waste from a sustainable approach. Besides pollution, some engineering nanomaterials contain high-added-value components such as rare earth elements that should be recovered. However, this possibility is often tricky and a big challenge in nano-wastes. First, it is necessary to identify and separate nanoparticles from the nano-waste stream. This is usually done by size-minimizing procedures (grinding, shredding, cutting), and nanoparticles can become airborne or part of wastewater streams. Earlier conventional separation techniques are being substituted by more effective and economical alternatives specific for each type of nanoparticle, such as magnetic fields, pH and thermal response of nanoparticles, or selective extraction using colloidal solvents. There are also concerns with occupational risks and environmental impact associated with the residues generated during the recycling process, and finally, how to introduce the recycled nanoparticles into new products. The effectivity of recycling will be strongly related to the type and appearance of the nanoparticles (pure, suspension in water, solid matrices, presence of other nanoparticles or pollutants).

Incineration is a thermal treatment in which flammable parts of nano-waste are fully degraded, and nanoparticles must be recovered. This treatment is often performed as an energy valorization process of wastewater treatment sewage sludges containing nano-wastes. The implications of the incineration of nano-wastes are not well understood, although the prevalence of the nanoparticles after thermal treatment is known in the combustion trash and waste ash emitted by incineration plants. Therefore, they must be subsequently captured by filters and scrubbers for the exhaust gases to avoid the release of airborne nanoparticles. Inorganic nanoparticles may remain in the ash, but carbon-based nanoparticles are considered combustible. Nevertheless, materials such as carbon nanotubes have been observed in the exhaust gas phase becoming from incinerated CNT/polymer nanocomposites.[174] There are not enough data about the influence of the presence of the nanoparticles in the combustion chamber and their interaction with combustion products. It has been reported, for example, that nanoparticles may increase the formation of polycyclic aromatic hydrocarbons and dioxins during combustion, probably due to catalytic activity and the high surface area of the nanoparticles. Therefore, further studies are necessary to make incineration an optimal end-of-life nano-waste treatment.

Landfilling is undoubtedly the most common practice for solid nano-waste management. However, there are some concerns to take into consideration. As indicated before, most nanowaste ends up in landfills, which is the most used management technique. However, nanomaterials may interact with other wastes, aggregate in leachates, and therefore transmit through bottom lines, spreading pollution to water and soils. Physical, chemical, and hydraulic conditions of the landfill (composition, porosity and thickness of soil liners, and temperature and pH) will influence the drainage and leaching of nanoparticles. For example, organic acids reduce carbon nanotube leaching, impeding their diffusion through an HDPE membrane.[175] Therefore, landfilling should be carefully analyzed because nanoparticles can move through solid waste, discharging into the atmosphere, water, and soil. In addition, nano-waste can affect anaerobic and aerobic processes in leachates and landfills.

Liquid effluent treatments. A high amount of nano-waste belongs to water streams containing nanoparticles released from devices used in contact with water, industrial process effluents, or household effluents (personal care and cosmetics, but mainly detergents). Landfilling leachates may also contain high amounts of nano-wastes dispersed in water. Most of those streams end up in wastewater treatment plants (WWTP), where processes have not been designed for specific nano-waste treatments. Even so, studies show that conventional methods can capture 90% or more of nanomaterials into solid sewage sludges thanks to their surface activity.[176] The main factors influencing sewage sludges' nanoparticle entrapment are aggregation, transport, and sorption/desorption from the biomass. Although the high transfer yield to sludge is promising, the remaining nanoparticles will stay in water effluent. Coagulation and flocculation with iron-based and aluminum-based coagulants is a valuable procedure to precipitate nano-waste from WWTP effluents. The best results are found when two or more techniques are used, for example, coagulation and filtration or coagulation and emulsification. The type of nanoparticles will determine the best practice.

Another issue to consider is the disposal of WWTP sludges containing nanoparticles. They are usually used as fertilizer or incinerated, which may introduce nano-waste pollution into soils, air, and the food chain. Also, nanoparticles may affect the quality of compost. Therefore, strategies to separate nanoparticles from sludges must be developed.

3.6   Regulatory bodies of nano-waste generation

As stated before, there are serious concerns about the safety of using nanoparticles and engineered nanomaterials, and even though their hazards are still uncertain, the emergence of new applications for nanomaterials is continuous, and the nanotechnology field growth does not show a visible limit. There are regulations for bulk materials, but there is insufficient knowledge about nano-waste treatment and its effect on the environment and human health. The lack of specific regulations about the use and disposal of nanoparticles is a great challenge for the future.

The current regulatory framework for hazardous materials is based on the mass of bulk material (*Registration, Evaluation, Authorization and Restriction of Chemicals, REACH*) and not on the nanoscale physicochemical characteristics of nanomaterials. The toxicity of nanomaterials depends on factors such as their chemical reactivity, shape, scale, and surface. **All those factors should be considered to set up helpful hazards/toxicity indices for nanomaterials in biota**. These indices will help to create policies to minimize the impact of nano-wastes and will affect the development and use of new nanomaterials.

Worldwide organizations have been working independently on this issue for years. The first ISO-specific standard for nanomaterial risk evaluation dates from 2011,[177], and for example, different committees of ISO (TC 229, TC 201) published 114 nanotechnology

New Materials for a Circular Economy   Materials Research Forum LLC
Materials Research Foundations 149 (2023) 233-310   https://doi.org/10.21741/9781644902639-8

standards until 2021, most of them about characterization, metrology, and toxicity evaluation of nanoparticles, although there is still work to do. Higher efforts may be made to regulate engineered nanomaterials and nano-containing products. In particular, there is a need for more ecotoxicological data for specific nanomaterials. The lack of adequate characterization and analysis tools are other drawbacks.

In the USA, nanomaterials legislation frameworks are regulated by Environmental Protection Agency (EPA), mainly under the Toxic Substances Control Act (TSCA). The Nanoscale Materials Stewardship Program (NMSP) was created in 2008 to gather available knowledge on nanotechnology. According to available data, EPA evaluates and authorizes nanomaterials' production and use and urges manufacturers to voluntarily report feedback data about manufacturing, processing, use, exposition, and environmental data. In the European Union, there have been specific recommendations since 2011 for nanomaterials in different regulations such as REACH, Scientific Committees on Emerging and Newly Identified Health Risks (SCENHIR), and Classification Label and Packaging (CLP). The European Chemicals Agency (ECHA) regulatory policies on nanotechnology are based on existing frameworks (the so-called adaptative management) for chemical substances but, in collaboration with the mentioned committees, are continuously revised when needed. Several UK committees, including the Advisory Committee on Hazardous Substances (ACHS), counsel the Environmental Agency and Health Prevention Agencies. In Australia, different bodies regulate nanotechnology, such as the National Industrial Chemicals Notification and Assessment Scheme (NICNAS), which asses nanomaterials used in products...

An extensive description of regulatory bodies is beyond the scope of this document. Beyond those regulatory frameworks, a more collaborative approach between agencies will be desirable to develop guidelines for the synthesis, use, and disposal of nanomaterials and nanotechnology products. Current tendencies are in the line of greener alternative production methods, combined life cycle risk assessment analysis, and considering safe, eco-friendly, and environmentally sustainable products and processes.

4.   Reuse and Recycling of nanoproducts

4.1   Introduction to reusing and recycling: definition, benefits and their contribution to "sustainability"

Modern waste reduction is based on the three pillars, the 3Rs: "Reduce, Reuse, and Recycle", (in this precise hierarchical order). As such, in the recent past, efforts have been made to finally make a transition from a linear to a circular economy that implements these three pillars in our society. Also, when the concept of circular economy is applied to materials production, we have to keep in mind other 3Rs which are of utmost importance: "Recover, Redesign and Remanufacture" (as depicted in **Fig. 14**). Furthermore, other Rs are considered, such as "Repair" or "Remind" enlarging the vision of circular economy. European Commission defines circular economy as an economy 'where the value of products, materials, and resources is maintained in the economy for as long as possible, and the generation of waste is minimized.[178] In fact, there exist different ISO standards related to recycling, such as ISO 15270:2008 for plastics waste[179] and ISO 14001:2015 for environmental management control of recycling practice.[180] In particular, the concepts of 'reuse' and 'recycle' play a central role in the circular economy since they are both ways of extending a product's life.[181] Although there might seem pretty similar things, they are not, and it is worth introducing these terms and emphasizing their differences. On the one hand, the term **'reuse' refers to the use for a subsequent time of a product after this has fulfilled its original purpose**. Indeed, this concept implies that a product has not deteriorated, damaged, or irreversibly unimpaired after its first use, and thus that it is not yet wasted. Under these circumstances, a product can be collected, prepared, or " regenerated, " ideally with little effort, and reused. Worth to note is that such product can be reused for its original function or a different one (repurposing). On the other hand, the term **'recycle' refers to recovering material from waste and turning it into new products**. Thus, the genuine product is destroyed and processed or manufactured throughout this process. Some commonly used recycling processes are melting (this commonly occurs in plastics, glass, and metals), milling, and washing (i.e., paper). Other products that are more complex, like batteries, require an initial step to classify their different components, and in a subsequent step, these individual components are recycled, creating new products.

Figure 14. Basic concept of circular economy applied to materials production. Adapted from Ref 178.[178]

Extending a product's life to reach its maximum lifespan through reusing and recycling brings social, economic, and environmental benefits.[182] For example, the U.S. Environmental Protection Agency (EPA) highlights, among others, the following benefits:
- ***Social benefits:*** improved public health, preservation of natural resources, prevention of habitat destruction, increased caring for the planet

- *Economic benefits:* generation of jobs and business activity, reduction of the need to create new landfill space, reduce the consumption of fresh raw materials, energy recovery, and creation of a circular economy
- *Environmental benefits* include reducing greenhouse gas emissions and overall water and land pollution, lowering waste, and decreasing the ecosystem's destruction.

Recycling and reusing products pursue the societal goal of 'sustainability', which aims at guaranteed access of current and future generations to energy, food, clean water, health, and security, to name a few. Sustainability is a 'relatively old' term that can be traced back to 1600-1700, and it was applied to forestry[183], but it is still seen as a goal on the long term. This is because, currently, there are many barriers to achieving sustainability, which must be addressed for a "sustainability transition" to become possible. In addition, "sustainability" has a reputation as a "buzzword" and has received strong criticism. Some practices like "greenwashing",[184] which basically is a form of advertising that uses marketing strategies to persuade the public that an organization's products, aims, and policies are environmentally friendly, contribute to such criticism.

4.2   Recycling and reusing of materials in industry

If we attend to materials recycling, it must be pointed out that only part of the materials in industrial products can be recovered and recycled.[185] In this respect, industrial materials can be classified by paying attention to whether they are economically and technically compatible with recycling or not.[185] Considering the recent past prices and regulations, these groups are:

- *(1) Economically and technologically compatible:* this is the case of metals, steel, catalysts, paper, glasses, and some plastics (i.e., polyethylene terephthalate, PET, high-density polyethylene, HDPE).
- *(2) Economically not compatible*, but technically feasible, such as some structural and packing materials, most refrigerants and solvents.
- *(3) Technologically not compatible*, including coatings, pigments, pesticides, herbicides, preservatives, explosives, detergents, fertilizers, fuels, and other chemical products.

Many materials in group (1) have a high (>40%) recycling rate, i.e., concrete and steel used in construction,[186] aluminum and other alloying elements used in cans,[187] plastic like polyethylene terephthalate (PET),[188] paper,[189] cardboard and glass[190] and already have a clear solid strategy for it. In this respect, the same principles could be applied to material reuse in industry. However, technological progress has brought into play new materials whose reuse and recycling strategies are still pending to be developed and whose regulation is very recent and ever still under development. This is indeed the case with nanomaterials. There has been a tremendous "boom" in development and use of nanomaterials in industry, as seen in previous sections of this chapter. Indeed, such fast growth demands the creation of short-term reuse and recycling strategies, which has boosted the research in this precise matter, reaching a value, likely underestimated, of ca. 200 research articles about recycling and reusing nanomaterials. This, along with proper regulation in waste management, would promote the use of nanomaterials in the industry in a safe and efficient way, integrating this kind of new materials in the cycle of circular economy.

4.3   Recovery, reuse and recycle of nanoproducts

Fig. 15 summarizes key steps in the life cycle of a nanoproduct: its production, use, and reuse (if possible). If this is not possible, the nanomaterial will achieve its end-of-life, after which it is wasted. Finally, after a nanomaterial is wasted, it could be either recycled or the waste could serve as the precursor for subsequent synthesis of the nanomaterial (green synthesis-based). An important fact to bear in mind to reuse and recycling nanomaterials is that conventional waste recovery/separation processes, such as filtering and scrubbing, are not always suitable for nanoparticles. Moreover, incineration (the most common thermal treatment of waste to recover energy), does not always decompose nanoparticles. As such, nanomaterials waste that ends up in conventional plants is likely to be directly released into the environment. Thus, novel protocols must be developed to recover nanoparticles from industrial products at the end of their life.

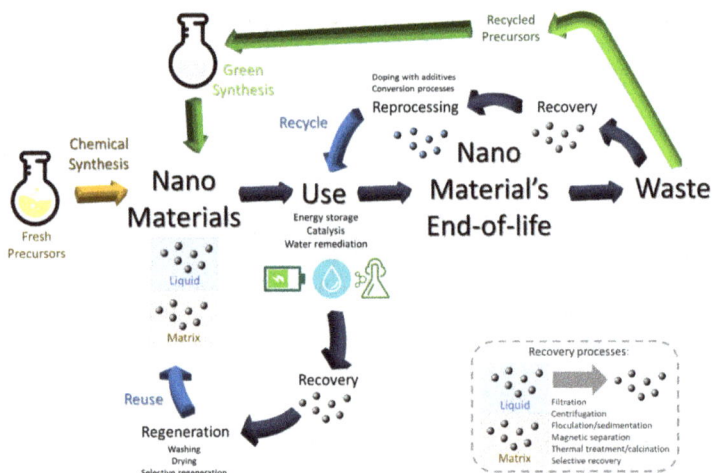

Figure 15. Life-cycle of nanomaterials, including recovery, reusing, and recycling processes, to expand their life and minimize waste. Worth noting is that just the main applications in which there have been generalized efforts to develop reusing and recycling strategies of nanomaterials appear: energy storage, catalysis, and water remediation.

Nanoparticles can be handled in powder, liquid, immobilized in a substrate, etc. When handled in liquid phase, **it is possible to separate and recover them from solution by filtration** (using submicrometer pores) **or centrifugation** (using a centrifugal force to separate the nanoparticles from the liquid through their sedimentation). Although these methods are widely used on a small scale, they present time- and energy-consuming drawbacks and problems related to the blockage of filtration units. Therefore, these methods are not always economically viable on a large scale.[191] Worth noting is that the magnetic nature of some nanoparticles can be exploited in order to tackle the problem of recovering and recycling the nanoparticles. This is the specific case of magnetic adsorbents,[192] which are a focus of special attention due to their straightforward separation and reusability in the field of water remediation and for many other applications that rely on magnetic separation. These nanocomposite particles frequently comprise Fe, Co, and Ni nanoparticles and have a **magnetic component, so they can be easily recovered by applying an external magnetic field**.[193] One good example of this is Invitrogen Dynabeads®, a commercial product used for magnetic separation-aided diagnosis technology,[194] which is done on a small scale. This commercial product can be reused several times using low pH or high salt concentration processes. However, other alternative methods for the separation and recycling of nanoparticles are needed for large-scale industrial applications. Other selective recovery processes have been recently developed. One good example is pH-controlled dynamic particle aggregation,[191] which avoids tedious centrifugation. The nanoparticle separation can be achieved by using reversible boronate ester and boroxine bonds formed between boronic acid-functionalized nanoparticles and poly(vinyl alcohol), or thiolated Poly(acrylic acid) coating,[195] whose aggregation state is sensitive to the adjustment of solvent pH. Thus, **nanoparticles functionalization may allow their recovery by adjusting the pH of the solution**. Other selective methods for the recovery of nanoparticles include nanoparticle recovery using a microemulsion[196] and cloud point extraction (CPE).[197] Finally, most nanomaterials are frequently embedded in a solid matrix, either inorganic, i.e., silica (SiO_2), or organic, like polystyrene (PS). Different methodologies such **as electroplating, thermal treatments, upcycling,** and other novel selective recovery processes have been reported in those cases where the recovery of nanomaterials from the selected matrix is needed.[198]

Next, this section provides specific examples where the reusing and recycling of the nanoparticles were assessed.

- Reusing of nanoparticles

The first examples are related to **water remediation application**s. In this application, it is essential to prepare environment-friendly materials that can reversibly adsorb heavy metals (Pb, Cd, Ni, and Cu) and other hazardous elements (i.e., Ag), degrade organic molecules (methylene blue, MB; Congo Red, Coomassie brilliant blue, Rhodamine B, RhB and 4-nitrophenol and many other), as well as to diminish bacterial concentration (to name a few, Escherichia coli, Staphylococcus aureus, and Pseudomonas aeruginosa), which have caused concerns due to the severe harm to the ecological environment. In addition, nanomaterials have been used to improve some **other characteristics of water, such as color and turbidity**, which are a sign of high concentrations of particulate matter, lowering the quality of water and having adverse effects in different industries with high usage of water (textile industry, paper industry, food and beverages industry, plastic industry, etc.).

In this area, **graphene oxide/magnetic lignin-based nanoparticles (GO-MLNPs)** were developed for the adsorption of Pb^{2+} and Ni^{2+} ions.[199] The advantage of using a material with a magnetic component is that it can be recycled easily by magnetic separation method. Furthermore, lignin is the second most abundant biopolymer of land-based biomass. Thus it has enormous potential for preparing various functional and sustainable materials as alternatives to plastics.[200] Finally, GO with hierarchical porous structures have abundant functional groups that can significantly improve the adsorption capacities of adsorbents.[201] To test the reusability performance of GO-MLNPs, the material was regenerated as follows: ions were desorbed by soaking in diluted acidic conditions for a few hours, then thoroughly washed and dried. After five adsorption-desorption cycles, the material maintained a performance above 85% of the initial value.[199] **Faujasite (FAU) zeolite nanocomposite decorated with cobalt ferrite nanoparticles** were developed and used to adsorb Pb^{2+} ions. Zeolite decoration in composites with magnetic ferrite nanoparticles has the advantage of high surface area maintenance, easy removal by the magnetic field, and low production cost. In this case, the adsorptive capacity was ca. 98% in the first two cycles.[202] However, the authors did not state how the composite was regenerated after magnetic separation and before the second cycle. **Silver nanoparticles decorated on an exfoliated graphitic carbon nitride/carbon sphere (AgNP/Eg-C$_3$N$_4$/CS)** were used to catalyze Cr^{4+} into Cr^{3+} with formic acid, generating hydrogen and carbon dioxide during the reaction. This material maintained its stability after six repeated cycles (98.5%). Once chromium was reduced, this material could be reused for the photocatalytic degradation of MB under visible light irradiation, showing degradation efficiencies of ca. 98%.[203] In this sense, carbon@g-C$_3$N$_4$ composite materials have shown the advantages of high surface area, high stability in physicochemical processes, low electron−hole recombination rate, good electronic transition band structure, and better visible (vis) light photocatalyst activity. On the other hand, Ag was integrated to enhance its catalytic activity and reduce the electron−hole recombination rate and it has shown low toxicity.[203] In addition, it was shown in similar materials that Ag can also enhance the antibacterial activity and generate reactive oxygen species, such as e^-, $\cdot O^{2-}$, and $\cdot OH$, which are accountable for the improved photocatalytic degradation of dyes.[204] **Au nanoparticles functionalized with thiolated Poly(acrylic acid) (AuNP-PAA)** have been studied for the reduction of a dye (4-nitrophenol) and their straightforward recovery thanks to their reversible aggregation and subsequent redispersion, which is induced by a change in the pH. The AuNP-PAA catalyst was highly active and reusable with 100% conversion, and minimal reductions in the reaction rate with up to four catalyst recycles. **Polyaniline (PANI) PANI/Fe$_3$O$_4$ nanofibers** were developed to treat wastewater containing Ag nanoparticles. Notably, the resulting product (PANI/Fe$_3$O$_4$/Ag composite) could be a catalysis for cleaning durable pollutant, 4-nitrophenol. After ten cycles, only a slight decrease in rate constant was found, indicating excellent reusability. Notably, this nanomaterial merges the ability to recover Ag NPs as pollutants and generates a recyclable material for environmental remediation.[205] Worth of note is that PANI is a well-known semi-flexible, low-cost, and environmentally stable rod-like conducting polymer, and it has also been exploited to remove many of the above-mentioned dyes (i.e., methyl orange,[206] Congo Red, Coomassie brilliant blue, Remazol brilliant Blue R, and MB) and heavy metals from the water waste.[207] Moreover, **ZnO–Ag–chitosan beads** were used for water disinfection and showed antimicrobial effects under visible light irradiation. Bacterial inactivation occurred due to the combined effects of $^{\bullet}O^{2-}$, $^{\bullet}OH$, and Ag species. The beads could provide clean and safe treated water for at least five cycles of operation,[208] between the beads must be regenerated by washing with distilled water and air-drying.

Respect to other characteristics necessary for the quality of water, i.e., color and turbidity, were improved using magnetic **iron oxide nanoparticles functionalized with Moringa oleifera (MO) compounds (Fe_3O_4@MO nanoparticles)**. MO has been widely used in water treatment processes to treat raw water for low-income locations due to its abundant availability, low cost, reduced by-product generation, biodegradability, non-toxicity, and multifunctional behavior.[209] This treatment is based on MO's ability to coagulate and flocculate particulate matter. Magnetic coagulants developed (Fe_3O_4@MO) could effectively remove ca. 90% of apparent color, ca. 90% of turbidity, and over 60% of UV_{254nm} absorption in river waters, after applying an external magnetic field for 30 min. On the other hand, conventional water treatment processes require several hours, which is a significant improvement. Further, iron oxide nanoparticles could be regenerated by washing with ethanol:water solution under stirring at room temperature, followed by magnetic separation. Fe_3O_4@MO can be reused without significant loss of efficiency. However, the washed regenerated nanoparticles required a subsequent functionalization with MO compounds.[210]

The next set of examples is focused on the **treatment and recovery of high-value elements from water**. Among others, oil, gas, and steel industries generate large volumes of water mixed with toxic hydrocarbons.[211] Worth to mention is that enhanced oil recovery (EOR) methods are applied to increase oil recovery rates, often using polymer flooding. However, **nanotechnology is being used in order to recover EOR polymers**,[212] since unwanted viscosity when discharging or reinjecting the water occurs.

Fe_3O_4 magnetic nanoparticles (Fe-MNP) were synthesized using the combustion synthesis method in the presence of glycine. They were used to recover oil from oil-in-water emulsions (O-in-W-emu). This simple nanomaterial showed a demulsification efficiency above 98% using a concentration of Fe-MNP of 10 mg/L. Furthermore, stability and reusability tests of the as-synthesized Fe-MNP exhibited an effective oil recovery of up to 90% after 7 cycles.[213] **Fe_3O_4 magnetic nanoparticles** prepared by the most common chemical coprecipitation route to treat stable oil-in-water emulsions to recover the oil. The cumulative percentage of oil recovery successively increased up to seven times of MNP's reuse, after which there was no significant recovery, and the the amount of recovered MNPs decreased with each successive reuse. The authors could reuse MNPs a maximum of 7 times. However, they did not report how MNP were regenerated. The final recovery of oil was ca. 70% (v/v), and of MNPs ca. 75% (w/w).[214] In line with this, **zinc imidazolate framework-8 (ZIF-8) nanoparticles are a suitable adsorbent to adsorb negatively charged oil droplets**. ZIF-8 has a sodalite-related zeolite-type structure comprising narrow six-membered ring pore windows of 3.4 Å and a much larger pores structure with thermal and chemical stabilities.[215] Its regeneration was reported by a simple ethanol-washing method, which exhibited more than 98% of regeneration efficiency over three cycles.[216] **Amine-functionalized magnetic nanoparticles** (MNPs) were used to remove an EOR polymer, partially hydrolyzed polyacrylamide (HPAM).[217] 100% **removal of HPAM** from water was achieved. The regeneration of MNPs relies on the adjustment of pH. After three cycles, 90% of the removal efficiency was maintained. This system has the advantage that MNPs-attached HPAM accelerates and enables efficient magnetic separation.

Nanomaterials for textile production are also being studied for their reusability. For instance, a system containing a **non-sulfur reducing agent, sodium borohydride, along with Ag nanoparticles** was designed to decolorize already spent dye baths after dyeing of nylon, silk and wool, followed by their reuse in repeated dyeing. Decolorized spent dye bath solutions were reused up to five times to dye fresh fabric samples, with negligible differences in all the dyed samples.[218]

The following example deals with reusing organocatalysts, which are important in the industry. Unlike traditional separation techniques (extraction, chromatography, filtration, etc.), **magnetic separation is an advantageous way to separate organocatalysts**, especially when solid-supported catalysts are used.[219]

The last set of examples focuses on the battery manufacturing industry. In particular, lead-acid battery (LAB) is commonly used in several devices, but the addition of carbon in its electrodes, forming lead-carbon batteries (LCB), has accounted for an improvement of reusability, among others. LABs mainly suffer from two problems: i) the negative plate cannot handle instantaneous high current charging, and ii) the negative plate undergoes rapid sulfuration, resulting in shorter battery life.[220] The addition of carbon in diverse forms has provided a longer life cycle.[221] For instance, **the incorporation of activated carbon and carbon nanotubes** has greatly increased the life span of the original LAB 93.5 %. At 70 % depth of discharge, the cycle life of the formed LCB was 7680 times.[220]

Nevertheless, the studies available indicate that, except in the case of the battery industry, the functionality of nanomaterials is retained only over a limited number of cycles (tens). When this functionality is lost, nanoproducts may be discarded as waste and possibly recycled.

Besides the reuse and recycling of nanoproducts, there is the possibility of reusing the precursors directly after being used to synthesize nanoparticles (after recovering the nanoparticles through a centrifugation step). This was the case with Se nanoparticles prepared in room-temperature ionic liquids (RTILs),[222] however, the nanoparticles obtained in subsequent "synthesis cycles" had altered size. Extending the use of a chemical brings economic savings and increases the overall efficiency of a process. However, the properties of the obtained nanomaterial should not be significantly changed in order to obtain similar functionalities.

- Recycling of nanoparticles

As explained above, 'recycle' refers to the process of **recovering material from waste and turning it into new products.** As such, the first relevant example employs water-containing Ag nanoparticles. In this case, using coagulants such as $TiCl_4$ allows to recover Ag nanoparticles from water and offers additional advantages since its flocculated sludge has been recycled to the valuable by-product TiO_2,[223] which has excellent catalytic properties improved by the merging of Ag and TiO_2 in a composite.[224] To recycle this nanomaterial, the sludge is dried and then calcinated for long periods. In addition, modified sodium alginate (MSA) was used in this study to improve the coagulation performance of $TiCl_4$. The second example deals with waste sintered Nd–Fe–B magnets, which can be recycled to obtain high-coercivity Nd–Fe–B powders or bonded magnets. The recycling process comprises different steps: first, the combination of hydrogen decrepitation (HD) process, which consists of the disintegration of the initial magnet, and its doping with NdH_x nanoparticles. Then the mixture is then degassed to remove the hydrogen, and the obtained powders are mixed with an epoxy resin. Finally, the obtained material is aligned under a 3 T magnetic field and pressed to obtain compact material, followed by heat treatment. Worth to note is that the coercivity could be restored to about 80% compared with the waste Nd–Fe–B scrap magnet.[225] Finally, the use of nanoparticles for coal liquefaction to maximize the oil yield and catalyst utilization using FeNi and FeMoNi catalysts (supported on carbon nanoparticles). Indeed, coal liquefaction is a method to convert a solid, in this case, coal, into liquid, in this case, hydrocarbons (liquid fuels or petrochemicals). The advantage of using carbon nanoparticles as support is that they provide high distillate yields and allow the separation from the solid residue by gravimetric separation for repeated use.[226] In addition, there is an urgent need to adopt green chemistry routes to **synthesize nanoparticles by recycling secondary waste resources**. Environmental wastes such as metallurgical slag, electronics (e-waste), and acid mine drainage (AMD) are rich sources of metals to produce nanoparticles.[227] This would have a positive economic and environmental impact, and the produced nanoparticles could be used for a wide range of applications. For example, CuO nanoparticles were obtained from recycling copper waste wires, and an electrochemical sensor was developed using these precise nanoparticles.[228] Likewise, large amounts of diamond-wire saw powder (DWSP) waste are generated in the photovoltaic industry. As such, a low-energy and simplified recycling method of the DWSP has been recently developed in order to prepare silica nanoparticles, which are a valuable material in the industry.[228]

Besides the recycling of nanomaterials, it was recently reported the possibility of applying a protocol, based on solvent distillation, for the recycling of colloidal nanoparticle synthesis surfactants and solvents for over 10 rounds of successive syntheses. The authors could produce a wide variety of mono- and bimetallic nanoparticles with reproducible sizes and compositions, leading to reproducible performance as heterogeneous catalysts. This technique greatly reduces the solvent-related costs of colloidal metal nanoparticle synthesis.[229]

4.4   Self and sustainable nanoproducts by design

Since the comprehension of how nanomaterials act after their littering and waste disposal in the environment is to date deficient, industry that uses nanoparticles-based products should create strategies for nanomaterial separation, recovery, reusing, and recycling, and in a cost-effective matter, as well as the assessment of the environmental sustainability of these processes. However, to save money, resources, and efforts, **products should be ideally designed so that their main components, including nanoparticles, cause no harm to the biocenosis and biotope of the ecosystem and so that they can be separated and reused/recycled as easily as possible.** All these concepts are integrated in the term 'Safe- and Sustainable-by-Design' (SbD) nanoproducts.[230] Indeed, this concept can be extended to other advanced materials, chemicals, and products in general.

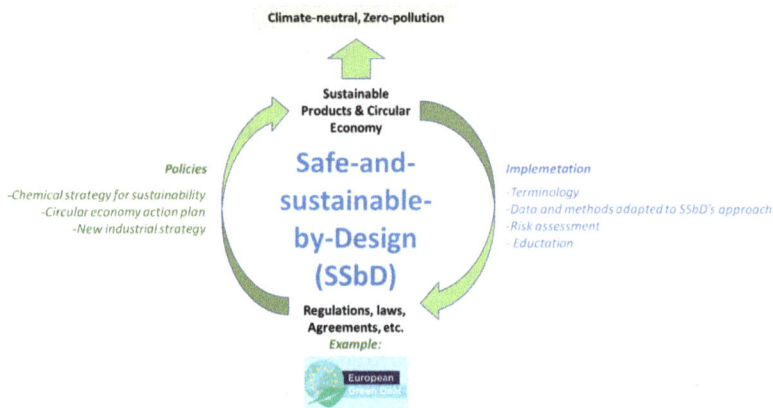

Figure 16. Safe-and-Sustainable-by-Design (SSbD) concept. The creation of regulations, laws, etc., have the final goals of fostering sustainable products and a circular economy and, finally, achieve a climate-neutral and zero-pollution situation. These goals are pursued through policies and the refinement/implementation of the SSbD approach. Below is an example of the 'European Green Deal'. Adapted from Ref. 230.[230]

The concept of SbD nano-products covers issues such as the identification of risks, uncertainty, and unpredictability associated with the use of nanomaterials and how these could compromise or harm humans and the environment and take these into account at an early phase of the innovation process of the nanomaterial so that the potential hazard(s) can be minimized. For this purpose, SbD's approach evaluates every step in the whole life cycle of nanomaterials, from the development phase to their waste, recovery, and recycling phases, as well as the risk of exposure.   In particular, SbD tackles the safety of a nanoproduct, its production, and its use. In addition, it tackles safer end-of-life. Indeed, SbD's approach favors a circular economy and sustainability. All these concepts are integrated in **Fig. 16**. In December 2019, the EU introduced its Green Deal, in which the ecological crisis is prioritized and SbD approach is highlighted. This has started a revolution in the cultural, economic, and political agenda of the EU. **Fig. 16** shows key aspects of the 'deal'. In particular, **some of the actions designed to achieve sustainable nanomaterials include i) the circumvention of properties** (structural, such as their volume; chemical, such as their composition, etc.) that may jeopardize the public's health; and **ii) the replacement of certain groups of chemicals likely to be** toxic (chemicals that do not degrade/react in a safe manner and that could (bio)accumulate or circulate, having a negative impact). Thus, SbD's approach works on these factors with the final goal of minimizing the environmental footprint and to lessen, among other, climate change, acidification (groundwater, streams, rivers, and lakes) and eutrophication, resource depletion (groundwater, minerals and other natural resources, deforestation, etc.), ozone depletion, photochemical ozone formation, (eco)toxicity, human toxicity and ionizing radiation.

SbD's approach aims at a new industrial strategy where, ideally, each step in the life cycle of a nanoproduct fulfills specific regulatory requirements. One example is existing environmental legislation from the European Union: REACH, which is the Registration, Evaluation, Authorization and restriction of Chemicals. A list of chemicals considered as 'substances of very high concern (SVHCs) emerged out of REACH. This list includes benzene, asbestos fibers, Arsenic, mercury and cadmium compounds, etc. This list is updated with new SVHCs every so often, and a general limit of 0.1% w/w (0.1% of total product weight) exists for each of the SVHCs. Nanoproducts are not included in this list to date, but REACH regulation already covers certain aspects related to the analysis of nanoparticles.[230]

Those nanomaterials presenting a multicomponent nature and displaying a stimuli-responsive behavior, i.e., (above-mentioned) 'Smart' NPs, are perhaps thought to be the solution to the SbD approach since these nanomaterials are required to comply with many requirements. However, the simplest designs often have several advantages, and choosing non-complex and already approved chemical compositions by the regulatory agencies, the use of bio-synthetic processes and inexpensive and widely available natural reagents definitely helps in the implementation of the SbD approach.

Concerning ecotoxicity, M. El-Shetehy et al.,[231] recently studied the replacement of currently used agrochemicals, such as plant biostimulants or pesticides, by SiO_2 nanoparticles. After applying the SiO_2 nanoparticles to the model plant Arabidopsis thaliana, the resistance against the bacterial pathogen Pseudomonas syringae was assessed. The authors demonstrated that SiO_2 nanoparticles induced disease resistance mechanisms

(SAR), achieving >90% bacterial inhibition.[231] The protective effect of SiO_2 NPs was demonstrated to be mediated by the activation of a salicylic acid (SA)-dependent plant immunity response and by the slow release of $Si(OH)_4$ from the nanoparticles, entering through the stomata of the plants (pores found in the epidermis of leaves) and clogging. Interestingly, the direct application of $Si(OH)_4$ (the reagent precursor of the SiO_2 NPs) induced a similar resistance toward pathogens. The authors studied the release rates of SiO_2 NPs in a continuously depleted ultrapure water system. They assessed the ecotoxicity of the nanoparticles and $Si(OH)_4$ through nematode tests, which are appropriate bioindicators of soil condition, and they are also suitable organisms for laboratory ecotoxicity testing.[232] The results obtained suggest that few (~10%) of the NPs could have dissolved within 48 h of SiO_2 NP exposure, and the ecotoxicity of SiO_2 NPs is 36-fold lower compared to $Si(OH)_4$. Thus, SiO_2 NPs proved to be safer for the plant and did not cause phytotoxicity even at concentrations 10-fold higher than the minimal dose needed for plant protection. Furthermore, the authors stated that nanomaterial has more than 1,000-fold material savings compared to solid bulk SiO_2 treatments. Finally, SiO_2 NPs have already been approved by the Food and Drug Administration (FDA), so they are considered safe and used as dietary additives (E551) in diverse foods. In particular, the established limit for the daily intake of nanoscale silica from food is ca. 2 $mg \cdot kg^{-1}$,[233] and $Si(OH)_4$ is generally used for plant nutrition.

In line with this, M. Delample et al., studied the replacement phosphine or carbene-palladium complexes catalysts, which are expensive and sensitive to oxygen, with glycerol in the presence of air-stable palladium nanoparticles. Glycerol is a cheap, safe, and sustainable solvent. The authors demonstrate that, in the presence of the NPs, it can be efficient for the catalytic and regioselective β,β-diarylation of acrylates.[234]

Concerning human toxicity, available cytotoxicity data (*in vitro* and *in vivo* assays in cell and animal models) on carbon-, metal-, and semiconductor-based NPs suggest that generally, cells can survive the short-term exposure and low concentrations ($<10 mg \cdot mL^{-1}$) of nanoparticles and that for increased doses cytotoxic effects appear, in a dose- and time-dependent manner.[49] While the causes for the increase in cell death observed at higher concentrations and longer exposure times are material specific, the generation of reactive oxygen species and the influence of cell internalization of nanoparticles are two common findings. Such possible adverse effects of nanoparticles should be continuously tested, and efforts should be made to improve the standardization of assays, including risk assessment studies of nanomaterials. This is crucial for the safe and sustainable development of emerging nanotechnologies.[235] In this respect, N. González-Ballesteros et al., developed a co-friendly, fast, one-pot synthetic route to synthesize Au NPs and assessed their potential for the biomedical application of cancer therapy as well as evaluating their safety in healthy cells.[236] The authors use brown macroalgae Cystoseira baccata (CB) extracts for the bio-synthesis of Au NPs (Au@CB). The synthesis occurs by slowly adding an aqueous solution containing the gold precursor ($HAuCl_4$) to the CB extract, and the NPs formed within a few minutes. While the NPs show a strong cytotoxic activity towards colorectal cancer cells, they display excellent biocompatibility on the healthy cell line tested. This example highlights one of the principles of the SbD concept since the use of toxic and expensive reagents has been avoided. Furthermore, CB algae used in this work is widely distributed worldwide and inhabits the low intertidal and subtidal levels of exposed rocky shores.[236]

Many efforts have been devoted to mitigate resource depletion, including renewable and non-renewable natural resources. In this sense, the world currently depends on fossil fuels as the major energy source, which is scarce. In addition to this, the use of fossil fuels is creating an environmental problem. Therefore, the demand for green energy will likely rise. Among all the developed green, safe and sustainable energies, hydrogen stands out for its inexpensiveness and no environmental pollution. The basic concept of this type of energy resource is that hydrogen is produced from water through water splitting using multiple processes, including photo-electro-chemical, photocatalytic, radiolysis, photo-biological, and thermal decomposition methods.[237] The water splitting research began using TiO_2 electrodes under UV radiation. The main advantage of using nanoparticles is the possibility of tuning the difference between valance and conduction bands. The NPs ideally absorb sunlight photons, and the electrons 'jump' from the valence to the conduction band creating positively charged holes. Both the electrons and holes migrate to the surface of the NPs and reduce water molecules to hydrogen. ZnO NPs can be prepared by green synthesis methods using *Moringaoleifera* natural extract. The photocatalytic generation of hydrogen was investigated under UV-Visible light irradiation. Upon the use of Na_2S and Na_2SO_3 sacrificial agents (electron donors added to the reaction system in order to consume the photogenerated holes and inhibit the oxidation of the NPs), the water splitting capability of the nanoparticles as well as their stability was demonstrated.[238]

Finally, SbD nanoremediation aims at innovative approaches for safe and sustainable remediation of persistent organic compounds, i.e., halogenated chemicals, perfluoroalkyl and polyfluoroalkyl substances (PFAS), and heavy metals. In particular, it was demonstrated the photocatalytic oxidation of hydrocarbons, phenols, aldehydes, halo-compounds, surfactants, dyes, drugs, or pesticides, while maintaining the health of environment recommendations.[239]In this area, magnetic nanoparticles can play a fundamental role. One-pot valorization process of different waste biomass containing, among others, polysaccharides, polyphenols, carbohydrates, lipids, enzymes, or proteins, has been used to develop biofunctionalized magnetite (Fe_3O_4) nanoparticles. Subsequently, sustainable, cost-effective, and eco-friendly Fe_3O_4 could act as: i) antimicrobial agent for water disinfection, ii) photo-degrader, and iii) adsorbent for diverse water pollutants. Other processes in which magnetic nanoparticles have demonstrated to be useful are biofuel production from lipids and lignocellulosic wastes and bio-upgrading of petroleum fractions. **Thus, these magnetic nanoparticles are attractive for a zero-waste green synthesis.**[131]

Conclusions and final remarks

This book chapter seeks to capture in a few words the current scenario of nanomaterials concerning one of the most important society's aims to create a healthy and sustainable planet for all. Nanomaterials are particular since all their regulations to ensure minimum or zero risks to both human health and the environment are still under development. However, in parallel, nanomaterials have already been introduced in the industry and, therefore, in our daily lives in products of very different nature, i.e., electronic devices, materials, cosmetics, drugs, and even foods. On top of that, the development of novel nanomaterials has intensified. Despite the high number of scientific papers devoted to the 'design' of novel nanomaterials, only a smaller fraction of scientific papers focus on the rest steps present in any product design process, like performing recovery/reusability/recycling testing, conducting risk assessment and (eco)toxicity tests, identify improvements, etc. This raises concerns about the real impact nanomaterials have on the environment.

Some research areas with high applicability in industry, like energy storage, catalysis, and water remediation, are already making efforts to make the life of a nanoproduct longer and to develop nanomaterials that have a minimized impact on the environment, evaluating the number of cycles that a nanomaterial can be reused for the same purpose with minimum regeneration treatment and without a significant loss of the efficiency of the nanomaterial. The studies available indicate that, except in the case of the battery industry, the functionality of nanomaterials is retained only over a limited number of cycles (tens). When this functionality is lost, nanoproducts may be discarded as waste and possibly recycled. In this sense, many nanomaterials lack a strategy for their recovery from the rest of the waste and apply the necessary treatments and remanufacturing processes for their recycling. In these cases, selective coagulation processes to sediment them or any selective physical method of separation, such as magnetic separation, is highly advantageous. Recycling nano-waste into precursors for subsequent synthesis of nanomaterials would also have enormous benefits, such as energy and material costs and reduction in waste generation. Indeed, we must consider the costs/efforts associated with the necessary treatments for the reuse/recycling of nanomaterials, which should be low enough to have an overall benefit.

The "safe- and sustainable-by-design" concept is vast, and it does not only consider the life cycle of the nanomaterial once built but also aims at optimizing, from a sustainability point of view, all the different processes involved in the fabrication of a nanomaterial, i.e., the synthetic approach used. In this sense, 'green chemistry' methods are associated as a base for sustainable processes, using natural resources that are safe to use, mild temperature conditions, saving energy, etc. Further, 'smart nanomaterials' are ideal candidates to implement the safe- and sustainable-by-design concept and to develop materials, which can respond to different stimuli, either physical or biological, so that they can be separated, degraded, etc., helping in their reuse and reducing their impact in the environment.

Indeed, the regulations related to nanomaterials are new-born and still under development. The regulations of both nanoproducts and nano-wastes will probably continue to appear and be refined. While these regulations are commonly implemented for chemicals and their disposal, a blank is still being filled for nanoproducts and nano-wastes. This book chapter gathers data about the nanomaterials' inherent toxicity and the toxicity associated with their small size and ever due to its shape. The data suggest that many nanoparticles induce oxidative stress, altering the redox potential and inducing protein denaturation or other cell injuries. Therefore, proper regulations are of utmost importance, ensuring a safe limit or safe dose of nanomaterials.

In summary, this book chapter highlights the need for material scientists and researchers, to keep in mind the concepts of 'green', 'safe', 'sustainable-by-design', 'eco-design', 'reuse', 'recycle', 'life-cycle', and 'eco/bio toxicity', and regularly apply and integrate them when developing nanoproducts. In this way, the ecological footprint that nanomaterials have in the environment and any hazards that nanomaterials could cause to humans could be minimized.

References

[1] ISO/TS 80004-4:2011 Nanotechnologies - Vocabulary - Part 4: Nanostructured materials, (2011).

[2] A. Nowak, K. Śliżewska, and A. Otlewska, Antigenotoxic activity of lactic acid bacteria, prebiotics, and products of their fermentation against selected mutagens, J. Regul. Toxicol. Pharmacol., vol. 73, no. 3 (2015) 938-946. https://doi.org/10.1016/j.yrtph.2015.09.021

[3] M. F. Hochella Jr, D. W. Mogk, J. Ranville, I. C. Allen, G. W. Luther, L. C. Marr, B. P. McGrail, M. Murayama, N. P. Qafoku, and K. M. Rosso, Natural, incidental, and engineered nanomaterials and their impacts on the Earth system, J. Sci., vol. 363, no. 6434 (2019) 8299. https://doi.org/10.1126/science.aau8299

[4] J. A. Dearing, K. L. Hay, S. M. Baban, A. S. Huddleston, E. M. Wellington, and P. Loveland, Magnetic susceptibility of soil: an evaluation of conflicting theories using a national data set, Geophys. J. Int., vol. 127, no. 3 (1996) 728-734. https://doi.org/10.1111/j.1365-246X.1996.tb04051.x

[5] F. J. Martin-Martinez, K. Jin, D. López Barreiro, and M. J. Buehler, The rise of hierarchical nanostructured materials from renewable sources: learning from nature, ACS Nano, vol. 12, no. 8 (2018) 7425-7433. https://doi.org/10.1021/acsnano.8b04379

[6] M. Loos, M. Loos, Ed. Nanoscience and Nanotechnology (2015). Oxford: William Andrew Publishing 1-36. https://doi.org/10.1016/B978-1-4557-3195-4.00001-1

[7] S. Philippe, A. H. Abbass, Ed. Nanoparticles in Ancient Materials: The Metallic Lustre Decorations of Medieval Ceramics (2012). Rijeka: IntechOpen 1-25.

[8] C. Ouellet-Plamondon, P. Aranda, A. Favier, G. Habert, H. Van Damme, and E. Ruiz-Hitzky, The Maya blue nanostructured material concept applied to colouring geopolymers, RSC Adv., vol. 5, no. 120 (2015) 98834-98841. https://doi.org/10.1039/C5RA14076E

[9] M. Sharon, M. Sharon, Ed. History of nanotechnology: from prehistoric to modern times (2019). John Wiley & Sons. https://doi.org/10.1002/9781119460534

[10] T. Pradell, J. Molera, A. D. Smith, A. Climent-Font, and M. S. Tite, Technology of Islamic lustre, J. Cult. Herit., vol. 9, (2008) 123-128. https://doi.org/10.1016/j.culher.2008.06.010

[11] J. Hulla, S. Sahu, and A. Hayes, Nanotechnology: History and future, J. Hum. Exp. Toxicol., vol. 34, no. 12 (2015) 1318-1321. https://doi.org/10.1177/0960327115603588

[12] J. Tersoff and D. R. Hamann, Theory and application for the scanning tunneling microscope, J. Phys. Rev. Lett., vol. 50, no. 25 (1983) 1998. https://doi.org/10.1103/PhysRevLett.50.1998

[13] S. Bayda, M. Adeel, T. Tuccinardi, M. Cordani, and F. Rizzolio, The history of nanoscience and nanotechnology: from chemical-physical applications to nanomedicine, J. Molecules, vol. 25, no. 1 (2019) 112. https://doi.org/10.3390/molecules25010112

[14] J. Hulla, S. Sahu, and A. Hayes, Nanotechnology: History and future, J. Hum. Exp. Toxicol., vol. 34, no. 12 (2015) 1318-1321. https://doi.org/10.1177/0960327115603588

[15] A. K. Yetisen, H. Qu, A. Manbachi, H. Butt, M. R. Dokmeci, J. P. Hinestroza, M. Skorobogatiy, A. Khademhosseini, and S. H. Yun, Nanotechnology in textiles, ACS Nano, vol. 10, no. 3 (2016) 3042-3068. https://doi.org/10.1021/acsnano.5b08176

[16] V. S. Saji, H. C. Choe, and K. W. Yeung, Nanotechnology in biomedical applications: a review, Int. J. Biomater., vol. 3, no. 2 (2010) 119-139. https://doi.org/10.1504/IJNBM.2010.037801

[17] B. S. Sekhon, Food nanotechnology-an overview, J. Nanotechnol. Sci. Appl. , vol. 3, (2010) 1-15.

[18] S. Manjunatha, D. Biradar, and Y. R. Aladakatti, Nanotechnology and its applications in agriculture: A review, Int. J. Farm Sci., vol. 29, no. 1 (2016) 1-13.

[19] M. T. Bohr, Nanotechnology goals and challenges for electronic applications, IEEE Trans. Nanotechnol., vol. 1, no. 1 (2002) 56-62. https://doi.org/10.1109/TNANO.2002.1005426

[20] E. Serrano, G. Rus, and J. Garcia-Martinez, Nanotechnology for sustainable energy, J. Renew. Sustain. Energy Rev., vol. 13, no. 9 (2009) 2373-2384. https://doi.org/10.1016/j.rser.2009.06.003

[21] R. K. Ibrahim, M. Hayyan, M. A. AlSaadi, A. Hayyan, and S. Ibrahim, Environmental application of nanotechnology: air, soil, and water, J. Environ. Sci. Pollut. Res. Int., vol. 23, no. 14 (2016) 13754-13788. https://doi.org/10.1007/s11356-016-6457-z

[22] V. Singh, P. Yadav, and V. Mishra, M. S. :Neha Srivastava, P. K. Mishra, Vijai Kumar Gupta, Ed. Recent advances on classification, properties, synthesis, and

characterization of nanomaterials (2020). Wiley Online Library 83-97. https://doi.org/10.1002/9781119576785.ch3

[23] A. Barhoum, M. L. García-Betancourt, J. Jeevanandam, E. A. Hussien, S. A. Mekkawy, M. Mostafa, M. M. Omran, M. S. Abdalla, and M. Bechelany, Review on natural, incidental, bioinspired, and engineered nanomaterials: history, definitions, classifications, synthesis, properties, market, toxicities, risks, and regulations, J. Nanomater., vol. 12, no. 2 (2022) 4418. https://doi.org/10.3390/nano12020177

[24] S. Saha, S. Bansal, and M. Khanuja, M. G. a. M. A. Shahid, Ed. Classification of nanomaterials and their physical and chemical nature (2022). Elsevier 7-34. https://doi.org/10.1016/B978-0-323-91009-5.00001-X

[25] W. J. Stark, P. R. Stoessel, W. Wohlleben, and A. Hafner, Industrial applications of nanoparticles, Chem. Soc. Rev., vol. 44, no. 16 (2015) 5793-5805. https://doi.org/10.1039/C4CS00362D

[26] S. D. Stranks and H. J. Snaith, Metal-halide perovskites for photovoltaic and light-emitting devices, Nat. Nanotechnol., vol. 10, no. 5 (2015) 391-402. https://doi.org/10.1038/nnano.2015.90

[27] Y.-B. Cheng, A. Pascoe, F. Huang, and Y. Peng, Print flexible solar cells, Nature, vol. 539, no. 7630 (2016) 488-489. https://doi.org/10.1038/539488a

[28] P.-C. Ma and Y. Zhang, Perspectives of carbon nanotubes/polymer nanocomposites for wind blade materials, J. Renew. Sust. Energ. Rev. , vol. 30, (2014) 651-660. https://doi.org/10.1016/j.rser.2013.11.008

[29] S. Das, A review on Carbon nano-tubes-A new era of nanotechnology, Int. J. Emerging Technol. Adv. Eng., vol. 3, no. 3 (2013) 774-783.

[30] Y. Wei, P. Zhang, R. A. Soomro, Q. Zhu, and B. Xu, Advances in the synthesis of 2D MXenes, Adv. Mater., vol. 33, no. 39 (2021) 2103148. https://doi.org/10.1002/adma.202103148

[31] M. Ajith, M. Aswathi, E. Priyadarshini, and P. Rajamani, Recent innovations of nanotechnology in water treatment: A comprehensive review, J. Bioresour. Technol., vol. 342, (2021) 126000. https://doi.org/10.1016/j.biortech.2021.126000

[32] H. Saleem and S. J. Zaidi, Nanoparticles in reverse osmosis membranes for desalination: A state of the art review, J. Desalination Water Treat., vol. 475, (2020) 114171. https://doi.org/10.1016/j.desal.2019.114171

[33] M. M. Abdullah, H. A. Al-Lohedan, and A. M. Atta, Novel magnetic iron oxide nanoparticles coated with sulfonated asphaltene as crude oil spill collectors, RSC Adv., vol. 6, no. 64 (2016) 59242-59249. https://doi.org/10.1039/C6RA09651D

[34] W. Sricharussin, P. Threepopnatkul, and N. Neamjan, Effect of various shapes of zinc oxide nanoparticles on cotton fabric for UV-blocking and anti-bacterial

properties, J. Fibers Polym., vol. 12, no. 8 (2011) 1037-1041.
https://doi.org/10.1007/s12221-011-1037-9

[35] M. Z. Khan, V. Baheti, M. Ashraf, T. Hussain, A. Ali, A. Javid, and A. Rehman, Development of UV protective, superhydrophobic and antibacterial textiles using ZnO and TiO_2 nanoparticles, J. Fibers Polym., vol. 19, no. 8 (2018) 1647-1654. https://doi.org/10.1007/s12221-018-7935-3

[36] S. B. Raut, D. Vasavada, and S. Chaudhari, Nano Particles-Application in Textile Finishing, J. Man-Made Text. India, vol. 53, no. 12 (2010) 7-12.

[37] M. Bilal and H. M. Iqbal, New insights on unique features and role of nanostructured materials in cosmetics, J. Cosmet. Dermatol. Sci. Appl., vol. 7, no. 2 (2020) 1-24. https://doi.org/10.3390/cosmetics7020024

[38] Y. Cui, S. Kumar, B. R. Kona, and D. van Houcke, Gas barrier properties of polymer/clay nanocomposites, RSC Adv., vol. 5, no. 78 (2015) 63669-63690. https://doi.org/10.1039/C5RA10333A

[39] A. P. Ramos, M. A. Cruz, C. B. Tovani, and P. Ciancaglini, Biomedical applications of nanotechnology, J. Biophys. Rev., vol. 9, no. 2 (2017) 79-89. https://doi.org/10.1007/s12551-016-0246-2

[40] H.-P. Phan, Implanted flexible electronics: Set device lifetime with smart nanomaterials, J. Micromachines, vol. 12, no. 2 (2021) 157. https://doi.org/10.3390/mi12020157

[41] H. Gavilán, S. K. Avugadda, T. Fernández-Cabada, N. Soni, M. Cassani, B. T. Mai, R. Chantrell, and T. Pellegrino, Magnetic nanoparticles and clusters for magnetic hyperthermia: Optimizing their heat performance and developing combinatorial therapies to tackle cancer, Chem. Soc. Rev., vol. 50, no. 20 (2021) 11614-11667. https://doi.org/10.1039/D1CS00427A

[42] M. Aflori, Smart nanomaterials for biomedical applications-a review, J. Nanomater., vol. 11, no. 2 (2021) 396. https://doi.org/10.3390/nano11020396

[43] P. Chandrasekharan, Z. W. Tay, D. Hensley, X. Y. Zhou, B. K. Fung, C. Colson, Y. Lu, B. D. Fellows, Q. Huynh, and C. Saayujya, Using magnetic particle imaging systems to localize and guide magnetic hyperthermia treatment: tracers, hardware, and future medical applications, J. Theranostics, vol. 10, no. 7 (2020) 2965. https://doi.org/10.7150/thno.40858

[44] E. Beltrán-Gracia, A. López-Camacho, I. Higuera-Ciapara, J. B. Velázquez-Fernández, and A. A. Vallejo-Cardona, Nanomedicine review: Clinical developments in liposomal applications, J. Cancer Nanotechnol., vol. 10, no. 1 (2019) 1-40. https://doi.org/10.1186/s12645-019-0055-y

[45] E. Boisselier and D. J. C. s. r. Astruc, Gold nanoparticles in nanomedicine: preparations, imaging, diagnostics, therapies and toxicity, Chem. Soc. Rev., vol. 38, no. 6 (2009) 1759-1782. https://doi.org/10.1039/b806051g

[46] S. Tran, P.-J. DeGiovanni, B. Piel, and P. Rai, Cancer nanomedicine: a review of recent success in drug delivery, J. Transl. Med., vol. 6, no. 1 (2017) 1-21. https://doi.org/10.1186/s40169-017-0175-0

[47] L. Xiao, K. Hong, C. Roberson, M. Ding, A. Fernandez, F. Shen, L. Jin, S. Sonkusare, and X. Li, Hydroxylated fullerene: a stellar nanomedicine to treat lumbar radiculopathy via antagonizing TNF-α-induced ion channel activation, calcium signaling, and neuropeptide production, ACS Biomater. Sci. Eng., vol. 4, no. 1 (2018) 266-277. https://doi.org/10.1021/acsbiomaterials.7b00735

[48] A. S. Chauhan, Dendrimers for drug delivery, J. Molecules, vol. 23, no. 4 (2018) 938. https://doi.org/10.3390/molecules23040938

[49] N. Lewinski, V. Colvin, and R. Drezek, Cytotoxicity of nanoparticles, Small, vol. 4, no. 1 (2008) 26-49. https://doi.org/10.1002/smll.200700595

[50] G. Yang, S. Z. F. Phua, A. K. Bindra, and Y. J. A. M. Zhao, Degradability and clearance of inorganic nanoparticles for biomedical applications, vol. 31, no. 10 (2019) 1805730. https://doi.org/10.1002/adma.201805730

[51] M. Li, K. T. Al-Jamal, K. Kostarelos, and J. Reineke, Physiologically based pharmacokinetic modeling of nanoparticles, ACS Nano, vol. 4, no. 11 (2010) 6303-6317. https://doi.org/10.1021/nn1018818

[52] S. A. M. Ealia and M. Saravanakumar, A review on the classification, characterisation, synthesis of nanoparticles and their application, in IOP conference series: materials science and engineering, 2017, vol. 263, no. 3, p. 032019: IOP Publishing. https://doi.org/10.1088/1757-899X/263/3/032019

[53] G. Gorrasi and A. Sorrentino, Mechanical milling as a technology to produce structural and functional bio-nanocomposites, J. Green Chem., vol. 17, no. 5 (2015) 2610-2625. https://doi.org/10.1039/C5GC00029G

[54] C. A. Charitidis, P. Georgiou, M. A. Koklioti, A.-F. Trompeta, and V. Markakis, Manufacturing nanomaterials: from research to industry, J. Manuf. Rev., vol. 1, (2014) 1-11. https://doi.org/10.1051/mfreview/2014009

[55] C. Spreafico, D. Russo, and R. Degl'Innocenti, Laser pyrolysis in papers and patents, J. Intell. Manuf., vol. 33, (2021) 353-385. https://doi.org/10.1007/s10845-021-01809-9

[56] M. Parashar, V. K. Shukla, and R. Singh, Metal oxides nanoparticles via sol-gel method: a review on synthesis, characterization and applications, J. Mater. Sci.: Mater. Electron., vol. 31, no. 5 (2020) 3729-3749. https://doi.org/10.1007/s10854-020-02994-8

[57] N. Baig, I. Kammakakam, and W. Falath, Nanomaterials: A review of synthesis methods, properties, recent progress, and challenges, Mater. Adv., vol. 2, no. 6 (2021) 1821-1871. https://doi.org/10.1039/D0MA00807A

[58] P. T. Anastas and J. C. Warner, Green chemistry, Acc. Chem. Res., vol. 640 (1998) 1998.

[59] M. K. Panda, Y. D. Singh, R. K. Behera, and N. K. Dhal, J. Patra, Fraceto, L., Das, G., Campos, E., Ed. Biosynthesis of nanoparticles and their potential application in food and agricultural sector (2020). Springer, Cham. 213-225. https://doi.org/10.1007/978-3-030-39246-8_10

[60] S. Hasan, A review on nanoparticles: their synthesis and types, J. Res. J. Recent Sci., vol. 2277, (2015) 2502.

[61] P. Mohanpuria, N. K. Rana, and S. K. Yadav, Biosynthesis of nanoparticles: technological concepts and future applications, J. Nanoparticle Res., vol. 10, no. 3 (2008) 507-517. https://doi.org/10.1007/s11051-007-9275-x

[62] J. Singh, T. Dutta, K.-H. Kim, M. Rawat, P. Samddar, and P. Kumar, 'Green'synthesis of metals and their oxide nanoparticles: applications for environmental remediation, J. Nanobiotechnology, vol. 16, no. 1 (2018) 1-24. https://doi.org/10.1186/s12951-018-0408-4

[63] N. Srivastava, M. Srivastava, P. Mishra, and V. K. Gupta, N. Srivastava, Ed. Green Synthesis of Nanomaterials for Bioenergy Applications (2021). Wiley & Sons Ltd. https://doi.org/10.1002/9781119576785

[64] J. M. Palomo and M. Filice, Biosynthesis of metal nanoparticles: novel efficient heterogeneous nanocatalysts, J. Nanomater., vol. 6, no. 5 (2016) 84. https://doi.org/10.3390/nano6050084

[65] Y. Choi and S. Y. Lee, Biosynthesis of inorganic nanomaterials using microbial cells and bacteriophages, Nat. Rev. Chem., vol. 4, no. 12 (2020) 638-656. https://doi.org/10.1038/s41570-020-00221-w

[66] J. Jeevanandam, S. F. Kiew, S. Boakye-Ansah, S. Y. Lau, A. Barhoum, M. K. Danquah, and J. Rodrigues, Green approaches for the synthesis of metal and metal oxide nanoparticles using microbial and plant extracts, Nanoscale, vol. 14, no. 7 (2022) 2534-2571. https://doi.org/10.1039/D1NR08144F

[67] J. Al-Haddad, F. Alzaabi, P. Pal, K. Rambabu, and F. Banat, Green synthesis of bimetallic copper-silver nanoparticles and their application in catalytic and antibacterial activities, J. Clean Technol. Environ. Policy, vol. 22, no. 1 (2020) 269-277. https://doi.org/10.1007/s10098-019-01765-2

[68] M. Srivastava, P. Mishra, V. K. Gupta, and N. Srivastava, M. S. Neha Srivastava, P. K. Mishra, Vijai Kumar Gupta, Ed. Green Synthesis of Nanomaterials for Bioenergy Applications (2020). John Wiley & Sons. https://doi.org/10.1002/9781119576785

[69] M. F. Al-Hakkani, Biogenic copper nanoparticles and their applications: A review, J. SN Appl. Sci., vol. 2, no. 3 (2020) 1-20. https://doi.org/10.1007/s42452-020-2279-1

[70] B. Malik, T. B. Pirzadah, M. Kumar, and R. U. Rehman, R. Prasad, Kumar, V., Kumar, M. , Ed. Biosynthesis of nanoparticles and their application in pharmaceutical industry (2017). Springer, Singapore 331-349. https://doi.org/10.1007/978-981-10-5511-9_16

[71] R. Chaudhary, K. Nawaz, A. K. Khan, C. Hano, B. H. Abbasi, and S. Anjum, An overview of the algae-mediated biosynthesis of nanoparticles and their biomedical applications, J. Biomolecules, vol. 10, no. 11 (2020) 1498. https://doi.org/10.3390/biom10111498

[72] P. Kurhade, S. Kodape, and R. Choudhury, Overview on green synthesis of metallic nanoparticles, J. Chem. Pap., vol. 75, no. 10 (2021) 5187-5222. https://doi.org/10.1007/s11696-021-01693-w

[73] S. Jadoun, R. Arif, N. K. Jangid, and R. K. Meena, Green synthesis of nanoparticles using plant extracts: A review, J. Environ. Chem. Lett. , vol. 19, no. 1 (2021) 355-374. https://doi.org/10.1007/s10311-020-01074-x

[74] S. Sinsinwar, M. K. Sarkar, K. R. Suriya, P. Nithyanand, and V. Vadivel, Use of agricultural waste (coconut shell) for the synthesis of silver nanoparticles and evaluation of their antibacterial activity against selected human pathogens, J. Microb. Pathog., vol. 124, (2018) 30-37. https://doi.org/10.1016/j.micpath.2018.08.025

[75] S. Busi and J. Rajkumari, A. M. Grumezescu, Ed. Microbially synthesized nanoparticles as next generation antimicrobials: scope and applications (2019). Elsevier 485-524. https://doi.org/10.1016/B978-0-12-816504-1.00008-9

[76] M. Ovais, A. T. Khalil, M. Ayaz, I. Ahmad, S. K. Nethi, and S. Mukherjee, Biosynthesis of metal nanoparticles via microbial enzymes: a mechanistic approach, Int. J. Mol. Sci., vol. 19, no. 12 (2018) 4100. https://doi.org/10.3390/ijms19124100

[77] A. Anandaradje, V. Meyappan, I. Kumar, and N. Sakthivel, A. Shukla, Ed. Microbial synthesis of silver nanoparticles and their biological potential (2020). Springer, Singapore. 99-133. https://doi.org/10.1007/978-981-13-8954-2_4

[78] R. K. Das, V. L. Pachapur, L. Lonappan, M. Naghdi, R. Pulicharla, S. Maiti, M. Cledon, L. M. A. Dalila, S. J. Sarma, and S. K. Brar, Biological synthesis of metallic nanoparticles: plants, animals and microbial aspects, J. Nanotechnol. Environ. Eng., vol. 2, no. 1 (2017) 1-21. https://doi.org/10.1007/s41204-017-0029-4

[79] T. Klaus-Joerger, R. Joerger, E. Olsson, and C.-G. Granqvist, Bacteria as workers in the living factory: metal-accumulating bacteria and their potential for materials science, Trends Biotechnol., vol. 19, no. 1 (2001) 15-20. https://doi.org/10.1016/S0167-7799(00)01514-6

[80] N. Durán, P. D. Marcato, O. L. Alves, G. I. De Souza, and E. Esposito, Mechanistic aspects of biosynthesis of silver nanoparticles by several Fusarium oxysporum strains, J. Nanobiotechnology, vol. 3, no. 1 (2005) 1-7. https://doi.org/10.1186/1477-3155-3-8

[81] A. Fariq, T. Khan, and A. Yasmin, Microbial synthesis of nanoparticles and their potential applications in biomedicine, J. Appl. Biomed., vol. 15, no. 4 (2017) 241-248. https://doi.org/10.1016/j.jab.2017.03.004

[82] I. Ocsoy, D. Tasdemir, S. Mazicioglu, C. Celik, A. Katı, and F. Ulgen, Biomolecules incorporated metallic nanoparticles synthesis and their biomedical applications, J. Mater. Lett., vol. 212, (2018) 45-50. https://doi.org/10.1016/j.matlet.2017.10.068

[83] M. Guilger-Casagrande and R. d. Lima, Synthesis of silver nanoparticles mediated by fungi: a review, J. Front. Bioeng. Biotechnol., vol. 7, (2019) 287. https://doi.org/10.3389/fbioe.2019.00287

[84] K. N. Thakkar, S. S. Mhatre, and R. Y. Parikh, Biological synthesis of metallic nanoparticles, J. Nanomed.: Nanotechnol. Biol. Med., vol. 6, no. 2 (2010) 257-262. https://doi.org/10.1016/j.nano.2009.07.002

[85] S. Mitra, A. Das, S. Sen, and B. Mahanty, Potential of metabolic engineering in bacterial nanosilver synthesis, World J. Microbiol. Biotechnol., vol. 34, no. 9 (2018) 1-10. https://doi.org/10.1007/s11274-018-2522-8

[86] G. Gahlawat and A. R. Choudhury, A review on the biosynthesis of metal and metal salt nanoparticles by microbes, RSC Adv., vol. 9, no. 23 (2019) 12944-12967. https://doi.org/10.1039/C8RA10483B

[87] N. A. Gow, J.-P. Latge, and C. A. Munro, The fungal cell wall: structure, biosynthesis, and function, J. Microbiol. Spectr., vol. 5, no. 3 (2017) 1-25. https://doi.org/10.1128/microbiolspec.FUNK-0035-2016

[88] A. Bhargava, N. Jain, M. A. Khan, V. Pareek, R. V. Dilip, and J. Panwar, Utilizing metal tolerance potential of soil fungus for efficient synthesis of gold nanoparticles with superior catalytic activity for degradation of rhodamine B, J. Environ. Manage., vol. 183, (2016) 22-32. https://doi.org/10.1016/j.jenvman.2016.08.021

[89] B. K. Salunke, S. S. Sawant, S.-I. Lee, and B. S. Kim, Comparative study of MnO_2 nanoparticle synthesis by marine bacterium Saccharophagus degradans and yeast Saccharomyces cerevisiae, J. Appl. Microbiol. Biotechnol., vol. 99, no. 13 (2015) 5419-5427. https://doi.org/10.1007/s00253-015-6559-4

[90] X. Zhang, Y. Qu, W. Shen, J. Wang, H. Li, Z. Zhang, S. Li, and J. Zhou, Biogenic synthesis of gold nanoparticles by yeast Magnusiomyces ingens LH-F1 for catalytic reduction of nitrophenols, J. Colloids Surf. A: Physicochem. Eng. Asp., vol. 497, (2016) 280-285. https://doi.org/10.1016/j.colsurfa.2016.02.033

[91] K. Chokshi, I. Pancha, T. Ghosh, C. Paliwal, R. Maurya, A. Ghosh, and S. Mishra, Green synthesis, characterization and antioxidant potential of silver nanoparticles

biosynthesized from de-oiled biomass of thermotolerant oleaginous microalgae Acutodesmus dimorphus, RSC Adv., vol. 6, no. 76 (2016) 72269-72274. https://doi.org/10.1039/C6RA15322D

[92] G. Singaravelu, J. Arockiamary, V. G. Kumar, and K. Govindaraju, A novel extracellular synthesis of monodisperse gold nanoparticles using marine alga, Sargassum wightii Greville, J. Colloids Surf. B: Biointerfaces, vol. 57, no. 1 (2007) 97-101. https://doi.org/10.1016/j.colsurfb.2007.01.010

[93] P. Vanathi, P. Rajiv, and R. Sivaraj, Synthesis and characterization of Eichhornia-mediated copper oxide nanoparticles and assessing their antifungal activity against plant pathogens, J. Mater. Sci., vol. 39, no. 5 (2016) 1165-1170. https://doi.org/10.1007/s12034-016-1276-x

[94] J. A. Lemire, J. J. Harrison, and R. J. J. N. R. M. Turner, Antimicrobial activity of metals: mechanisms, molecular targets and applications, Nat. Rev. Microbiol., vol. 11, no. 6 (2013) 371-384. https://doi.org/10.1038/nrmicro3028

[95] P. Mukherjee, S. Senapati, D. Mandal, A. Ahmad, M. I. Khan, R. Kumar, and M. Sastry, Extracellular synthesis of gold nanoparticles by the fungus Fusarium oxysporum, J. ChemBioChem., vol. 3, no. 5 (2002) 461-463. https://doi.org/10.1002/1439-7633(20020503)3:5<461::AID-CBIC461>3.0.CO;2-X

[96] M. Shahid, C. Dumat, S. Khalid, E. Schreck, T. Xiong, and N. K. Niazi, Foliar heavy metal uptake, toxicity and detoxification in plants: A comparison of foliar and root metal uptake, J. Hazard. Mater., vol. 325, (2017) 36-58. https://doi.org/10.1016/j.jhazmat.2016.11.063

[97] M. Ovais, A. T. Khalil, N. U. Islam, I. Ahmad, M. Ayaz, M. Saravanan, Z. K. Shinwari, and S. Mukherjee, Role of plant phytochemicals and microbial enzymes in biosynthesis of metallic nanoparticles, J. Appl. Microbiol. Biotechnol., vol. 102, no. 16 (2018) 6799-6814. https://doi.org/10.1007/s00253-018-9146-7

[98] P. Kuppusamy, M. M. Yusoff, G. P. Maniam, and N. Govindan, Biosynthesis of metallic nanoparticles using plant derivatives and their new avenues in pharmacological applications-An updated report, Saudi Pharm. J., vol. 24, no. 4 (2016) 473-484. https://doi.org/10.1016/j.jsps.2014.11.013

[99] C. Pandit, A. Roy, S. Ghotekar, A. Khusro, M. N. Islam, T. B. Emran, S. E. Lam, M. U. Khandaker, and D. A. Bradley, Biological agents for synthesis of nanoparticles and their applications, J. King Saud Univ. Sci., vol. 34, no. 3 (2022) 101869. https://doi.org/10.1016/j.jksus.2022.101869

[100] M. Rai and N. Duran, N. D. Mahendra Rai, Ed. Metal nanoparticles in microbiology (2011). Springer Berlin, Heidelberg. https://doi.org/10.1007/978-3-642-18312-6

[101] K. L. Kelly, E. Coronado, L. L. Zhao, and G. C. Schatz, The optical properties of metal nanoparticles: the influence of size, shape, and dielectric environment, Journal. Type of Article vol. 107, no. Issue (2003) 668-677. https://doi.org/10.1021/jp026731y

[102] N. Durán, P. D. Marcato, R. D. Conti, O. L. Alves, F. Costa, and M. J. J. o. t. B. C. S. Brocchi, Potential use of silver nanoparticles on pathogenic bacteria, their toxicity and possible mechanisms of action, J. Braz. Chem. Soc., vol. 21, (2010) 949-959. https://doi.org/10.1590/S0103-50532010000600002

[103] A. Bankar, B. Joshi, A. R. Kumar, and S. Zinjarde, Banana peel extract mediated novel route for the synthesis of silver nanoparticles, J. Colloids Surf. A: Physicochem. Eng. Asp., vol. 368, no. 1-3 (2010) 58-63. https://doi.org/10.1016/j.colsurfa.2010.07.024

[104] R. Ramanathan, A. P. O'Mullane, R. Y. Parikh, P. M. Smooker, S. K. Bhargava, and V. Bansal, Bacterial kinetics-controlled shape-directed biosynthesis of silver nanoplates using Morganella psychrotolerans, Langmuir, vol. 27, no. 2 (2011) 714-719. https://doi.org/10.1021/la1036162

[105] Y. K. Tak, S. Pal, P. K. Naoghare, S. Rangasamy, and J. M. Song, Shape-dependent skin penetration of silver nanoparticles: does it really matter?, Sci. Rep., vol. 5, no. 1 (2015) 1-11. https://doi.org/10.1038/srep16908

[106] S. Pal, Y. K. Tak, and J. M. Song, Does the antibacterial activity of silver nanoparticles depend on the shape of the nanoparticle? A study of the gram-negative bacterium Escherichia coli, J. Appl. Environ. Microbiol., vol. 73, no. 6 (2007) 1712-1720. https://doi.org/10.1128/AEM.02218-06

[107] S. Gurunathan, K. Kalishwaralal, R. Vaidyanathan, D. Venkataraman, S. R. K. Pandian, J. Muniyandi, N. Hariharan, and S. H. Eom, Biosynthesis, purification and characterization of silver nanoparticles using Escherichia coli, J. Colloids Surf. B: Biointerfaces, vol. 74, no. 1 (2009) 328-335. https://doi.org/10.1016/j.colsurfb.2009.07.048

[108] E. O. Mikhailova, Gold Nanoparticles: Biosynthesis and Potential of Biomedical Application, Funct. Biomater., vol. 12, no. 4 (2021) 70. https://doi.org/10.3390/jfb12040070

[109] T. Beveridge and R. Murray, Sites of metal deposition in the cell wall of Bacillus subtilis, J. Bacteriol., vol. 141, no. 2 (1980) 876-887. https://doi.org/10.1128/jb.141.2.876-887.1980

[110] B. Srinath, K. Namratha, and K. Byrappa, Eco-friendly synthesis of gold nanoparticles by Bacillus subtilis and their environmental applications, J. Adv. Sci. Lett., vol. 24, no. 8 (2018) 5942-5946. https://doi.org/10.1166/asl.2018.12224

[111] P. Mukherjee, A. Ahmad, D. Mandal, S. Senapati, S. R. Sainkar, M. I. Khan, R. Ramani, R. Parischa, P. Ajayakumar, and M. Alam, Bioreduction of AuCl4− ions by the fungus, Verticillium sp. and surface trapping of the gold nanoparticles formed,

Angew. Chem. Int., vol. 40, no. 19 (2001) 3585-3588. https://doi.org/10.1002/1521-3773(20011001)40:19<3585::AID-ANIE3585>3.0.CO;2-K

[112] S. K. Srivastava, R. Yamada, C. Ogino, and A. Kondo, Biogenic synthesis and characterization of gold nanoparticles by Escherichia coli K12 and its heterogeneous catalysis in degradation of 4-nitrophenol, J. Nanoscale Res. Lett., vol. 8, no. 1 (2013) 1-9. https://doi.org/10.1186/1556-276X-8-70

[113] P. Nagajyothi, K. Lee, and T. Sreekanth, Biogenic synthesis of gold nanoparticles (quasi-spherical, triangle, and hexagonal) using Lonicera japonica flower extract and its antimicrobial activity, J. Synth. React. Inorg. M., vol. 44, no. 7 (2014) 1011-1018. https://doi.org/10.1080/15533174.2013.797456

[114] S. Ahmed and S. Ikram, Biosynthesis of gold nanoparticles: a green approach, J. Photochem. Photobiol. B, Biol., vol. 161, (2016) 141-153. https://doi.org/10.1016/j.jphotobiol.2016.04.034

[115] K. S. Siddiqi, A. Husen, and R. A. Rao, A review on biosynthesis of silver nanoparticles and their biocidal properties, J. Nanobiotechnology, vol. 16, no. 1 (2018) 1-28. https://doi.org/10.1186/s12951-018-0334-5

[116] M. Saravanan, S. K. Barik, D. MubarakAli, P. Prakash, and A. Pugazhendhi, Synthesis of silver nanoparticles from Bacillus brevis (NCIM 2533) and their antibacterial activity against pathogenic bacteria, J. Microb. Pathog., vol. 116, (2018) 221-226. https://doi.org/10.1016/j.micpath.2018.01.038

[117] T. Klaus, R. Joerger, E. Olsson, and C.-G. Granqvist, Silver-based crystalline nanoparticles, microbially fabricated, Proc. Natl. Acad. Sci., vol. 96, no. 24 (1999) 13611-13614. https://doi.org/10.1073/pnas.96.24.13611

[118] S. A. Dahoumane, C. Jeffryes, M. Mechouet, and S. N. Agathos, Biosynthesis of inorganic nanoparticles: A fresh look at the control of shape, size and composition, J. Biomed. Eng., vol. 4, no. 1 (2017) 1-14. https://doi.org/10.3390/bioengineering4010014

[119] V. L. Das, R. Thomas, R. T. Varghese, E. Soniya, J. Mathew, and E. Radhakrishnan, Extracellular synthesis of silver nanoparticles by the Bacillus strain CS 11 isolated from industrialized area, J. 3 Biotech, vol. 4, no. 2 (2014) 121-126. https://doi.org/10.1007/s13205-013-0130-8

[120] M. S. Shekhawat, M. Manokari, N. Kannan, J. Revathi, and R. Latha, Synthesis of silver nanoparticles using Cardiospermum halicacabum L. leaf extract and their characterization, Int. J. Phytopharm., vol. 2, (2013) 15-20. https://doi.org/10.31254/phyto.2013.2503

[121] S. N. Kang, Y. M. Goo, M. R. Yang, R. I. H. Ibrahim, J. H. Cho, I. S. Kim, and O. H. Lee, Antioxidant and antimicrobial activities of ethanol extract from the stem and leaf of Impatiens balsamina L.(Balsaminaceae) at different harvest times, J. Molecules, vol. 18, no. 6 (2013) 6356-6365. https://doi.org/10.3390/molecules18066356

[122] V. Kathiravan, S. Ravi, S. Ashokkumar, S. Velmurugan, K. Elumalai, and C. P. Khatiwada, Green synthesis of silver nanoparticles using Croton sparsiflorus morong leaf extract and their antibacterial and antifungal activities, J. Spectrochim. Acta A Mol. Biomol. Spectrosc. , vol. 139, (2015) 200-205. https://doi.org/10.1016/j.saa.2014.12.022

[123] J. Kesharwani, K. Y. Yoon, J. Hwang, and M. Rai, Phytofabrication of silver nanoparticles by leaf extract of Datura metel: hypothetical mechanism involved in synthesis, J. Bionanoscience, vol. 3, no. 1 (2009) 39-44. https://doi.org/10.1166/jbns.2009.1008

[124] N. Vigneshwaran, N. Ashtaputre, P. Varadarajan, R. Nachane, K. Paralikar, and R. Balasubramanya, Biological synthesis of silver nanoparticles using the fungus Aspergillus flavus, J. Mater. Lett., vol. 61, no. 6 (2007) 1413-1418. https://doi.org/10.1016/j.matlet.2006.07.042

[125] N. Jain, A. Bhargava, S. Majumdar, J. Tarafdar, and J. Panwar, Extracellular biosynthesis and characterization of silver nanoparticles using Aspergillus flavus NJP08: a mechanism perspective, Nanoscale, vol. 3, no. 2 (2011) 635-641. https://doi.org/10.1039/C0NR00656D

[126] B. Ankamwar, C. Damle, A. Ahmad, and M. Sastry, Biosynthesis of gold and silver nanoparticles using Emblica officinalis fruit extract, their phase transfer and transmetallation in an organic solution, J. Nanotechnol., vol. 5, no. 10 (2005) 1665-1671. https://doi.org/10.1166/jnn.2005.184

[127] S. P. Dubey, M. Lahtinen, and M. Sillanpää, Tansy fruit mediated greener synthesis of silver and gold nanoparticles, J. Process Biochem., vol. 45, no. 7 (2010) 1065-1071. https://doi.org/10.1016/j.procbio.2010.03.024

[128] S. S. Shankar, A. Rai, A. Ahmad, and M. Sastry, Rapid synthesis of Au, Ag, and bimetallic Au core-Ag shell nanoparticles using Neem (Azadirachta indica) leaf broth, J. Colloid Interface Sci., vol. 275, no. 2 (2004) 496-502. https://doi.org/10.1016/j.jcis.2004.03.003

[129] S. Majidi, F. Zeinali Sehrig, S. M. Farkhani, M. Soleymani Goloujeh, and A. Akbarzadeh, Current methods for synthesis of magnetic nanoparticles, J. Artif. Cells Nanomed. Biotechnol., vol. 44, no. 2 (2016) 722-734. https://doi.org/10.3109/21691401.2014.982802

[130] L. S. Ganapathe, M. A. Mohamed, R. Mohamad Yunus, and D. D. Berhanuddin, Magnetite (Fe_3O_4) nanoparticles in biomedical application: From synthesis to surface functionalisation, J. Magnetochemistry, vol. 6, no. 4 (2020) 68. https://doi.org/10.3390/magnetochemistry6040068

[131] N. S. El-Gendy and H. N. Nassar, Biosynthesized magnetite nanoparticles as an environmental opulence and sustainable wastewater treatment, J. Sci. Total Environ., vol. 774, (2021) 145610. https://doi.org/10.1016/j.scitotenv.2021.145610

[132] M. H. Ehrampoush, M. Miria, M. H. Salmani, and A. H. Mahvi, Cadmium removal from aqueous solution by green synthesis iron oxide nanoparticles with tangerine peel extract, J. Environ. Health Sci. Eng. , vol. 13, no. 1 (2015) 1-7. https://doi.org/10.1186/s40201-015-0237-4

[133] Y. Mohamed, A. Azzam, B. Amin, and N. Safwat, Mycosynthesis of iron nanoparticles by Alternaria alternata and its antibacterial activity, Afr. J. Biotechnol., vol. 14, no. 14 (2015) 1234-1241. https://doi.org/10.5897/AJB2014.14286

[134] I. Kolinko, A. Lohße, S. Borg, O. Raschdorf, C. Jogler, Q. Tu, M. Pósfai, E. Tompa, J. M. Plitzko, and A. Brachmann, Biosynthesis of magnetic nanostructures in a foreign organism by transfer of bacterial magnetosome gene clusters, Nat. Nanotechnol., vol. 9, no. 3 (2014) 193-197. https://doi.org/10.1038/nnano.2014.13

[135] E. C. Descamps, J. B. Abbé, D. Pignol, and C. T. Lefèvre, D. Faivre, Ed. Iron oxides: from nature to applications: Controlled biomineralization of magnetite in bacteria (2016). Wiley Online Library. https://doi.org/10.1002/9783527691395.ch5

[136] L. Marcano, A. García-Prieto, D. Muñoz, L. F. Barquín, I. Orue, J. Alonso, A. Muela, and M. Fdez-Gubieda, Influence of the bacterial growth phase on the magnetic properties of magnetosomes synthesized by Magnetospirillum gryphiswaldense, J. Biochim. Biophys. Acta Rev. Cancer, vol. 1861, no. 6 (2017) 1507-1514. https://doi.org/10.1016/j.bbagen.2017.01.012

[137] P. Samaddar, Y. S. Ok, K. H. Kim, E. E. Kwon, and D. C. Tsang, Synthesis of nanomaterials from various wastes and their new age applications, J. Clean. Prod., vol. 197, (2018) 1190-1209. https://doi.org/10.1016/j.jclepro.2018.06.262

[138] S. K. Tiwari, M. Bystrzejewski, A. De Adhikari, A. Huczko, and N. Wang, Methods for the conversion of biomass waste into value-added carbon nanomaterials: Recent progress and applications, Prog. Energy Combust. Sci., vol. 92, (2022) 101023. https://doi.org/10.1016/j.pecs.2022.101023

[139] G. Zhao, Y. Li, G. Zhu, J. Shi, T. Lu, and L. Pan, Biomass-based N, P, and S self-doped porous carbon for high-performance supercapacitors, ACS Sustain. Chem. Eng., vol. 7, no. 14 (2019) 12052-12060. https://doi.org/10.1021/acssuschemeng.9b00725

[140] H. Zhao, Y. Cheng, Z. Zhang, B. Zhang, C. Pei, F. Fan, and G. Ji, Biomass-derived graphene-like porous carbon nanosheets towards ultralight microwave absorption and excellent thermal infrared properties, Carbon, vol. 173, (2021) 501-511. https://doi.org/10.1016/j.carbon.2020.11.035

[141] D. Mateo, A. Garcia-Mulero, J. Albero, and H. Garcia, N-doped defective graphene decorated by strontium titanate as efficient photocatalyst for overall water splitting, J. Appl. Catal. B: Environ., vol. 252, (2019) 111-119. https://doi.org/10.1016/j.apcatb.2019.04.011

[142] M. F. De Volder, S. H. Tawfick, R. H. Baughman, and A. J. Hart, Carbon nanotubes: present and future commercial applications, J. Sci., vol. 339, no. 6119 (2013) 535-539. https://doi.org/10.1126/science.1222453

[143] V. B. Mbayachi, E. Ndayiragije, T. Sammani, S. Taj, and E. R. Mbuta, Graphene synthesis, characterization and its applications: A review, J. Results Chem., vol. 3, (2021) 100163. https://doi.org/10.1016/j.rechem.2021.100163

[144] Y. Yang, H. Zhang, and Y. Yan, Synthesis of CNTs on stainless steel microfibrous composite by CVD: effect of synthesis condition on carbon nanotube growth and structure, J. Compos. B. Eng., vol. 160, (2019) 369-383. https://doi.org/10.1016/j.compositesb.2018.12.100

[145] M. Kumar, S. Yellampalli, Ed. Carbon nanotube synthesis and growth mechanism (2011) (Carbon nanotubes-synthesis, characterization, applications). Condens. Matter Phys. https://doi.org/10.5772/19331

[146] M. Kumar and Y. Ando, Carbon nanotubes from camphor: an environment-friendly nanotechnology, in Journal of Physics: Conference Series, 2007, vol. 61, no. 1, p. 129: IOP Publishing. https://doi.org/10.1088/1742-6596/61/1/129

[147] B. Makgabutlane, L. N. Nthunya, M. S. Maubane-Nkadimeng, and S. D. Mhlanga, Green synthesis of carbon nanotubes to address the water-energy-food nexus: A critical review, J. Environ. Chem. Eng., vol. 9, no. 1 (2021) 104736. https://doi.org/10.1016/j.jece.2020.104736

[148] A. A. Balandin, S. Ghosh, W. Bao, I. Calizo, D. Teweldebrhan, F. Miao, and C. N. Lau, Superior thermal conductivity of single-layer graphene, J. Nano Lett., vol. 8, no. 3 (2008) 902-907. https://doi.org/10.1021/nl0731872

[149] M. Orlita, C. Faugeras, P. Plochocka, P. Neugebauer, G. Martinez, D. K. Maude, A. L. Barra, M. Sprinkle, C. Berger, and W. A. de Heer, Approaching the Dirac point in high-mobility multilayer epitaxial graphene, J. Phys. Rev. Lett., vol. 101, no. 26 (2008) 267601. https://doi.org/10.1103/PhysRevLett.101.267601

[150] M. Kaur, H. Kaur, and D. Kukkar, Synthesis and characterization of graphene oxide using modified Hummer's method, in AIP conference proceedings, 2018, vol. 1953, no. 1, p. 030180: AIP Publishing LLC. https://doi.org/10.1063/1.5032515

[151] S. Pei and H.-M. Cheng, The reduction of graphene oxide, Carbon, vol. 50, no. 9 (2012) 3210-3228. https://doi.org/10.1016/j.carbon.2011.11.010

[152] B. Dai, L. Fu, L. Liao, N. Liu, K. Yan, Y. Chen, and Z. Liu, High-quality single-layer graphene via reparative reduction of graphene oxide, J. Nano Res., vol. 4, no. 5 (2011) 434-439. https://doi.org/10.1007/s12274-011-0099-8

[153] Q. Yu, J. Jiang, L. Jiang, Q. Yang, and N. Yan, Advances in green synthesis and applications of graphene, J. Nano Res., vol. 14, no. 11 (2021) 3724-3743. https://doi.org/10.1007/s12274-021-3336-9

[154] S. Malik, A. Vijayaraghavan, R. Erni, K. Ariga, I. Khalakhan, and J. P. Hill, High purity graphenes prepared by a chemical intercalation method, Nanoscale, vol. 2, no. 10 (2010) 2139-2143. https://doi.org/10.1039/c0nr00248h

[155] K. De Silva, H.-H. Huang, R. Joshi, and M. Yoshimura, Chemical reduction of graphene oxide using green reductants, Carbon, vol. 119, (2017) 190-199. https://doi.org/10.1016/j.carbon.2017.04.025

[156] J. Zhang, H. Yang, G. Shen, P. Cheng, J. Zhang, and S. Guo, Reduction of graphene oxide via L-ascorbic acid, J. Chem. Commun., vol. 46, no. 7 (2010) 1112-1114. https://doi.org/10.1039/B917705A

[157] S. Anju and P. Mohanan, Impact of Nanoparticles in Balancing the Ecosystem, Biointerface Res. Appl. Chem., vol. 11, no. 3 (2021) 10461-10481. https://doi.org/10.33263/BRIAC113.1046110481

[158] N. Musee, Nanowastes and the environment: Potential new waste management paradigm, J. Environ. Int., vol. 37, no. 1 (2011) 112-128. https://doi.org/10.1016/j.envint.2010.08.005

[159] S. A. Younis, E. M. El-Fawal, and P. Serp, C. Hussain, Ed. Nano-wastes and the environment: Potential challenges and opportunities of nano-waste management paradigm for greener nanotechnologies (2018). Springer, Cham. 1-72. https://doi.org/10.1007/978-3-319-58538-3_53-1

[160] Z. Zahra, Z. Habib, S. Hyun, and M. Sajid, Nanowaste: Another Future Waste, Its Sources, Release Mechanism, and Removal Strategies in the Environment, J. Sustainability, vol. 14, no. 4 (2022) 2041. https://doi.org/10.3390/su14042041

[161] M. S. Giroux, Z. Zahra, O. A. Salawu, R. M. Burgess, K. T. Ho, and A. S. Adeleye, Assessing the environmental effects related to quantum dot structure, function, synthesis and exposure, Environ. Sci. Nano, vol. 9, no. 3 (2022) 867-910. https://doi.org/10.1039/D1EN00712B

[162] O. V. Kharissova, L. M. T. Martínez, and B. I. Kharisov, L. M. T.-M. Oxana Vasilievna Kharissova, Boris Ildusovich Kharisov, Ed. Handbook of Nanomaterials and Nanocomposites for Energy and Environmental Applications (2021). Springer Cham. https://doi.org/10.1007/978-3-030-36268-3

[163] B. Nowack, J. F. Ranville, S. Diamond, J. A. Gallego-Urrea, C. Metcalfe, J. Rose, N. Horne, A. A. Koelmans, and S. J. Klaine, Potential scenarios for nanomaterial release and subsequent alteration in the environment, J. Environ. Toxicol. Chem., vol. 31, no. 1 (2012) 50-59. https://doi.org/10.1002/etc.726

[164] A. A. Keller, S. McFerran, A. Lazareva, and S. Suh, Global life cycle releases of engineered nanomaterials, J. Nanoparticle Res., vol. 15, no. 6 (2013) 1-17. https://doi.org/10.1007/s11051-013-1692-4

[165] M. Sajid, M. Ilyas, C. Basheer, M. Tariq, M. Daud, N. Baig, and F. Shehzad, Impact of nanoparticles on human and environment: review of toxicity factors, exposures, control strategies, and future prospects, J. Environ. Sci. Pollut. Res. Int., vol. 22, no. 6 (2015) 4122-4143. https://doi.org/10.1007/s11356-014-3994-1

[166] T. Takeuchi, M. Nakajima, and K. Morimoto, A human cell system for detecting asbestos cytogenotoxicity in vitro, J. Mutat. Res. Genet. Toxicol. Environ. Mutagen., vol. 438, no. 1 (1999) 63-70. https://doi.org/10.1016/S1383-5718(98)00163-6

[167] M. Barbarino and A. Giordano, Assessment of the carcinogenicity of carbon nanotubes in the respiratory system, J. Cancer, vol. 13, no. 6 (2021) 1318. https://doi.org/10.3390/cancers13061318

[168] S. Sharifi, S. Behzadi, S. Laurent, M. L. Forrest, P. Stroeve, and M. J. C. S. R. Mahmoudi, Toxicity of nanomaterials, Chem. Soc. Rev., vol. 41, no. 6 (2012) 2323-2343. https://doi.org/10.1039/C1CS15188F

[169] S. Medici, M. Peana, A. Pelucelli, and M. A. Zoroddu, An updated overview on metal nanoparticles toxicity, in Seminars in Cancer Biology, 2021, vol. 76, pp. 17-26: Elsevier. https://doi.org/10.1016/j.semcancer.2021.06.020

[170] E. Asmatulu, M. N. Andalib, B. Subeshan, and F. Abedin, Impact of nanomaterials on human health: a review, J. Environ. Chem. Lett., vol. 20, (2022) 2509-2529. https://doi.org/10.1007/s10311-022-01430-z

[171] O. A. Mohamed, S. Aysha Ajith, R. Sabouni, G. Husseini, A. Karami, and R. G. Bai, Toxicological impact of nanoparticles on human health: A review, J. Mater. Express, vol. 12, no. 3 (2022) 389-411. https://doi.org/10.1166/mex.2022.2161

[172] S. A. Younis, E. M. El-Fawal, and P. Serp, C. Hussain, Ed. Nano-wastes and the environment: Potential challenges and opportunities of nano-waste management paradigm for greener nanotechnologies (2018). Springer, Cham. 1-72. https://doi.org/10.1007/978-3-319-58538-3_53-1

[173] A. Surendranath and P. V. Mohanan, Impact of Nanoparticles in Balancing the Ecosystem, (2020).

[174] J. X. Bouillard, B. R'Mili, D. Moranviller, A. Vignes, O. Le Bihan, A. Ustache, J. A. Bomfim, E. Frejafon, and D. J. J. o. n. r. Fleury, Nanosafety by design: risks from nanocomposite/nanowaste combustion, J. Nanoparticle Res., vol. 15, no. 4 (2013) 1-11. https://doi.org/10.1007/s11051-013-1519-3

[175] B. Mrowiec, Kierunki i możliwości bezpiecznej gospodarki nanoodpadami, J. Chemik, vol. 70, no. 10 (2016).

[176] J. Wu, G. Zhu, and R. Yu, Fates and impacts of nanomaterial contaminants in biological wastewater treatment system: A review, J. Water Air Soil Pollut., vol. 229, no. 1 (2018) 1-21. https://doi.org/10.1007/s11270-017-3656-2

[177] ISO/TR 13121:2011 Nanotechnologies - Nanomaterial risk evaluation, (2011).

[178] I. M. de Waal, Coherence in law: A way to stimulate the transition towards a circular economy? A critical analysis of the European Commission's aspiration to achieve full coherence between chemicals legislation and waste legislation-and product legislation, Maastricht J. Eur. Comp., vol. 28, no. 6 (2021) 760-783. https://doi.org/10.1177/1023263X211048604

[179] ISO 15270: 2008. PlasticsGuidelines for the recovery and recycling of plastics waste, (2018).

[180] L. M. C. M. Da Fonseca, ISO 14001: 2015: An improved tool for sustainability, Int. J. Ind. Eng. Manag., vol. 8, no. 1 (2015) 37-50. https://doi.org/10.3926/jiem.1298

[181] S. F. Hansen, R. Arvidsson, M. B. Nielsen, O. F. H. Hansen, L. P. W. Clausen, A. Baun, and A. Boldrin, Nanotechnology meets circular economy, Nat. Nanotechnol., vol. 17, no. 7 (2022) 682-685. https://doi.org/10.1038/s41565-022-01157-6

[182] Available at: https://www.epa.gov/recycle/reducing-and-reusing-basics

[183] H. C. Von Carlowitz, Sylvicultura Oeconomica Oder Haußwirthliche Nachricht und Naturmäßige Anweisung zur Wilden Baum-Zucht Nebst Gründlicher Darstellung Wie... dem allenthalben und insgemein einreissenden Grossen Holtz-Mangel, Vermittelst Säe-Pflantz-und Versetzung vielerhand Bäume zu rathen... Worbey zugleich eine gründliche Nachricht von dem in Churfl. Sächß. Landen Gefundenen Turff... befindlich, vol. 1. Bey Johann Friedrich Brauns sel. Erben (1732).

[184] S. V. de Freitas Netto, M. F. F. Sobral, A. R. B. Ribeiro, and G. R. d. L. Soares, Concepts and forms of greenwashing: A systematic review, J. Environ. Sci. Eur., vol. 32, no. 1 (2020) 1-12. https://doi.org/10.1186/s12302-020-0300-3

[185] J.J. Cai, Z. W. Lu, and Q. Yue, Some problems of recycling industrial materials, J. Iron Steel Res. Int., vol. 15, no. 5 (2008) 37-41. https://doi.org/10.1016/S1006-706X(08)60246-0

[186] B. Björkman and C. Samuelsson, E. W. a. M. A. Reuter, Ed. Recycling of steel (2014). Elsevier 65-83. https://doi.org/10.1016/B978-0-12-396459-5.00006-4

[187] M. Niero and S. I. Olsen, Circular economy: To be or not to be in a closed product loop? A Life Cycle Assessment of aluminium cans with inclusion of alloying elements, J. Resour. Conserv. Recycl., vol. 114, (2016) 18-31. https://doi.org/10.1016/j.resconrec.2016.06.023

[188] T. Thiounn and R. C. Smith, Advances and approaches for chemical recycling of plastic waste, J. Polym. Sci., vol. 58, no. 10 (2020) 1347-1364. https://doi.org/10.1002/pol.20190261

[189] I. Ervasti, R. Miranda, and I. Kauranen, A global, comprehensive review of literature related to paper recycling: A pressing need for a uniform system of terms and definitions, J. Waste Manag., vol. 48, (2016) 64-71. https://doi.org/10.1016/j.wasman.2015.11.020

[190] R. Lebullenger and F. O. Mear, J. H. J. David Musgraves, Laurent Calvez, Ed. Glass recycling (2019). Springer Cham. 1355-1377. https://doi.org/10.1007/978-3-319-93728-1_39

[191] C. Liu, H. Gong, W. Liu, B. Lu, and L. Ye, Separation and recycling of functional nanoparticles using reversible boronate ester and boroxine bonds, J. Ind. Eng. Chem. Res., vol. 58, no. 11 (2019) 4695-4703. https://doi.org/10.1021/acs.iecr.9b00253

[192] M. Hassan, R. Naidu, J. Du, Y. Liu, and F. Qi, Critical review of magnetic biosorbents: Their preparation, application, and regeneration for wastewater treatment, J. Sci. Total Environ., vol. 702, (2020) 134893. https://doi.org/10.1016/j.scitotenv.2019.134893

[193] A. R. Mahdavian and M. A.-S. Mirrahimi, Efficient separation of heavy metal cations by anchoring polyacrylic acid on superparamagnetic magnetite nanoparticles through surface modification, J. Chem. Eng., vol. 159, no. 1-3 (2010) 264-271. https://doi.org/10.1016/j.cej.2010.02.041

[194] S. S. Leong, S. P. Yeap, S. C. Low, R. Mohamud, and J. Lim, N. T. K. Thanh, Ed. Current Progress in Magnetic Separation-Aided Biomedical Diagnosis Technology (2018). CRC Press, Taylor & Francis Group 175-200. https://doi.org/10.1201/9781315168258-10

[195] S. M. Ansar, B. Fellows, P. Mispireta, O. T. Mefford, and C. L. Kitchens, pH triggered recovery and reuse of thiolated poly (acrylic acid) functionalized gold nanoparticles with applications in colloidal catalysis, Langmuir, vol. 33, no. 31 (2017) 7642-7648. https://doi.org/10.1021/acs.langmuir.7b00870

[196] N. V. Mdlovu, C. L. Chiang, K.-S. Lin, and R. C. Jeng, Recycling copper nanoparticles from printed circuit board waste etchants via a microemulsion process, J. Clean. Prod., vol. 185, (2018) 781-796. https://doi.org/10.1016/j.jclepro.2018.03.087

[197] H. El Hadri and V. A. Hackley, Investigation of cloud point extraction for the analysis of metallic nanoparticles in a soil matrix, J. Environ. Sci. Nano, vol. 4, no. 1 (2017) 105-116. https://doi.org/10.1039/C6EN00322B

[198] M. Hezarkhani, A. A. Wis, Y. Menceloglu, and B. S. Okan, Nanomaterials recycling in industrial applications (2022). Elsevier 375-395. https://doi.org/10.1016/B978-0-323-90982-2.00017-2

[199] B. Du, L. Chai, W. Li, X. Wang, X. Chen, J. Zhou, and R. C. Sun, Preparation of functionalized magnetic graphene oxide/lignin composite nanoparticles for adsorption of heavy metal ions and reuse as electromagnetic wave absorbers, J. Sep. Purif. Technol., vol. 297, (2022) 121509. https://doi.org/10.1016/j.seppur.2022.121509

[200] H.-M. Wang, T. Q. Yuan, G. Y. Song, and R. C. Sun, Advanced and versatile lignin-derived biodegradable composite film materials toward a sustainable world, J. Green Chem., vol. 23, no. 11 (2021) 3790-3817. https://doi.org/10.1039/D1GC00790D

[201] Z. Yan, T. Wu, G. Fang, M. Ran, K. Shen, and G. Liao, Self-assembly preparation of lignin-graphene oxide composite nanospheres for highly efficient Cr (VI) removal, RSC Adv., vol. 11, no. 8 (2021) 4713-4722. https://doi.org/10.1039/D0RA09190A

[202] E. C. Paris, J. O. Malafatti, H. C. Musetti, A. Manzoli, A. Zenatti, and M. T. Escote, Faujasite zeolite decorated with cobalt ferrite nanoparticles for improving removal and reuse in Pb^{2+} ions adsorption, Chin. J. Chem. Eng., vol. 28, no. 7 (2020) 1884-1890. https://doi.org/10.1016/j.cjche.2020.04.019

[203] E. Prabakaran and K. Pillay, Self-Assembled Silver Nanoparticles Decorated on Exfoliated Graphitic Carbon Nitride/Carbon Sphere Nanocomposites as a Novel Catalyst for Catalytic Reduction of Cr (VI) to Cr (III) from Wastewater and Reuse for Photocatalytic Applications, ACS Omega, vol. 6, no. 51 (2021) 35221-35243. https://doi.org/10.1021/acsomega.1c00866

[204] M. E. Khan, T. H. Han, M. M. Khan, M. R. Karim, and M. H. Cho, Environmentally sustainable fabrication of $Ag@g\text{-}C_3N_4$ nanostructures and their multifunctional efficacy as antibacterial agents and photocatalysts, ACS Appl. Nano Mater., vol. 1, no. 6 (2018) 2912-2922. https://doi.org/10.1021/acsanm.8b00548

[205] Q. Yang, Removal and reuse of Ag nanoparticles by magnetic polyaniline/Fe_3O_4 nanofibers, J. Mater. Sci., vol. 53, no. 12 (2018) 8901-8908. https://doi.org/10.1007/s10853-018-2181-z

[206] X. Guo, G. T. Fei, H. Su, and L. De Zhang, Synthesis of polyaniline micro/nanospheres by a copper (II)-catalyzed self-assembly method with superior adsorption capacity of organic dye from aqueous solution, J. Mater. Chem., vol. 21, no. 24 (2011) 8618-8625. https://doi.org/10.1039/c0jm04489j

[207] V. Janaki, B.-T. Oh, K. Shanthi, K.-J. Lee, A. Ramasamy, and S. Kamala-Kannan, Polyaniline/chitosan composite: an eco-friendly polymer for enhanced removal of dyes from aqueous solution, J. Synth. Met., vol. 162, no. 11-12 (2012) 974-980. https://doi.org/10.1016/j.synthmet.2012.04.015

[208] P. Chatterjee, M. M. Ghangrekar, and S. Rao, Disinfection of secondary treated sewage using chitosan beads coated with ZnO-Ag nanoparticles to facilitate reuse of treated water, J. Chem. Technol. Biotechnol., vol. 92, no. 9 (2017) 2334-2341. https://doi.org/10.1002/jctb.5235

[209] N. U. Yamaguchi, L. F. Cusioli, H. B. Quesada, M. E. C. Ferreira, M. R. Fagundes-Klen, A. M. S. Vieira, R. G. Gomes, M. F. Vieira, and R. Bergamasco, A review of Moringa oleifera seeds in water treatment: Trends and future challenges, J. Process Saf. Environ., vol. 147, (2021) 405-420. https://doi.org/10.1016/j.psep.2020.09.044

[210] T. R. T. dos Santos, M. F. Silva, M. B. de Andrade, M. F. Vieira, and R. Bergamasco, Magnetic coagulant based on Moringa oleifera seeds extract and super paramagnetic nanoparticles: optimization of operational conditions and reuse

evaluation, J. Desalin. Water Treat., vol. 106, (2018) 226-237. https://doi.org/10.5004/dwt.2018.22065

[211] M. A. Miranda, A. Ghosh, G. Mahmodi, S. Xie, M. Shaw, S. Kim, M. J. Krzmarzick, D. J. Lampert, and C. P. Aichele, Treatment and Recovery of High-Value Elements from Produced Water, J. Water, vol. 14, no. 6 (2022) 880. https://doi.org/10.3390/w14060880

[212] J. G. Speight, Heavy oil production processes (2013). Gulf Professional Publishing. https://doi.org/10.1016/B978-0-12-404570-5.00006-5

[213] W. F. Elmobarak and F. Almomani, Application of magnetic nanoparticles for the removal of oil from oil-in-water emulsion: Regeneration/reuse of spent particles, J. Pet. Sci. Eng., vol. 203, (2021) 108591. https://doi.org/10.1016/j.petrol.2021.108591

[214] P. Verma and A. N. Bhaskarwar, S. T. Ange Nzihou, Nandakumar Kalarikkal, K.P. Jibin, Ed. Reuse of Magnetite (Fe$_3$O$_4$) Nanoparticles in De-Emulsification of Emulsion Effluents of Steel-rolling Mills (2022). Taylor & Francis 183. https://doi.org/10.1201/9781003339304-12

[215] P. Pillai, S. Dharaskar, S. Sasikumar, and M. Khalid, Zeolitic imidazolate framework-8 nanoparticle: a promising adsorbent for effective fluoride removal from aqueous solution, J. Appl. Water Sci., vol. 9, no. 7 (2019) 1-12. https://doi.org/10.1007/s13201-019-1030-9

[216] M. G. F. Rodrigues, T. L. A. Barbosa, and D. P. A. Rodrigues, Zinc imidazolate framework-8 nanoparticle application in oil removal from oil/water emulsion and reuse, J. Nanoparticle Res., vol. 22, no. 11 (2020) 1-15. https://doi.org/10.1007/s11051-020-05036-w

[217] S. Ko, H. Lee, and C. Huh, Efficient removal of enhanced-oil-recovery polymer from produced water with magnetic nanoparticles and regeneration/reuse of spent particles, J. SPE Prod. Oper, vol. 32, no. 03 (2017) 374-381. https://doi.org/10.2118/179576-PA

[218] A. Nautiyal and S. R. Shukla, Silver nanoparticles catalyzed reductive decolorization of spent dye bath containing acid dye and its reuse in dyeing, J. Water Process. Eng. , vol. 22, (2018) 276-285. https://doi.org/10.1016/j.jwpe.2018.02.014

[219] R. Mrówczyński, A. Nan, and J. Liebscher, Magnetic nanoparticle-supported organocatalysts-an efficient way of recycling and reuse, RSC Adv., vol. 4, no. 12 (2014) 5927-5952. https://doi.org/10.1039/c3ra46984k

[220] Z. Wang, X. Tuo, J. Zhou, and G. Xiao, Performance study of large capacity industrial lead-carbon battery for energy storage, J. Energy Storage, vol. 55, (2022) 105398. https://doi.org/10.1016/j.est.2022.105398

[221] S. Mandal, S. Thangarasu, P. T. Thong, S.-C. Kim, J.-Y. Shim, and H.-Y. Jung, Positive electrode active material development opportunities through carbon addition

in the lead-acid batteries: A recent progress, J. Power Sources, vol. 485, (2021) 229336. https://doi.org/10.1016/j.jpowsour.2020.229336

[222] A. Guleria, C. M. Baby, A. Tomy, D. K. Maurya, S. Neogy, A. K. Debnath, and S. Adhikari, Size Tuning, Phase Stabilization, and Anticancer Efficacy of Amorphous Selenium Nanoparticles: Effect of Ion-Pair Interaction,−OH Functionalization, and Reuse of RTILs as Host Matrix, J. Phys. Chem. C., vol. 125, no. 25 (2021) 13933-13945. https://doi.org/10.1021/acs.jpcc.1c02894

[223] H. Shon, S. Vigneswaran, J. Kandasamy, M. Zareie, J. Kim, D. Cho, J.-H. J. S. S. Kim, and Technology, Preparation and characterization of titanium dioxide (TiO2) from sludge produced by TiCl4 flocculation with FeCl3, Al2(SO4)3 and Ca(OH)2 coagulant aids in wastewater, J. Sep. Sci. Technol., vol. 44, no. 7 (2009) 1525-1543. https://doi.org/10.1080/01496390902775810

[224] Z. Wang, Y. Wang, C. Yu, Y. Zhao, M. Fan, and B. Gao, The removal of silver nanoparticle by titanium tetrachloride and modified sodium alginate composite coagulants: floc properties, membrane fouling, and floc recycle, J. Environ. Sci. Pollut. Res. Int., vol. 25, no. 21 (2018) 21058-21069. https://doi.org/10.1007/s11356-018-2240-7

[225] X. Li, M. Yue, W. Liu, and D. Zhang, Recycle of waste Nd-Fe-B sintered magnets via NdH x nanoparticles modification, J. IEEE Trans. Magn., vol. 51, no. 11 (2015) 1-3. https://doi.org/10.1109/TMAG.2015.2438073

[226] U. Priyanto, K. Sakanishi, O. Okuma, and I. Mochida, Liquefaction of Tanito Harum coal with bottom recycle using FeNi and FeMoNi catalysts supported on carbon nanoparticles, J. Fuel Process. Technol., vol. 79, no. 1 (2002) 51-62. https://doi.org/10.1016/S0378-3820(02)00101-7

[227] K. K. Brar, S. Magdouli, A. Othmani, J. Ghanei, V. Narisetty, R. Sindhu, P. Binod, A. Pugazhendhi, M. K. Awasthi, and A. Pandey, Green route for recycling of low-cost waste resources for the biosynthesis of nanoparticles (NPs) and nanomaterials (NMs)-A review, J. Environ. Res., vol. 207, (2022) 112202. https://doi.org/10.1016/j.envres.2021.112202

[228] A. Nassar, H. Salah, N. Hashem, M. Khodari, and H. Assaf, Electrochemical Sensor Based on CuO Nanoparticles Fabricated From Copper Wire Recycling-loaded Carbon Paste Electrode for Excellent Detection of Theophylline in Pharmaceutical Formulations, J. Electrocatalysis, vol. 13, no. 2 (2022) 154-164. https://doi.org/10.1007/s12678-021-00698-z

[229] C. J. Wrasman, C. Zhou, A. Aitbekova, E. D. Goodman, and M. Cargnello, Recycling of Solvent Allows for Multiple Rounds of Reproducible Nanoparticle Synthesis, J. Am. Chem. Soc., vol. 144, no. 26 (2022) 11646-11655. https://doi.org/10.1021/jacs.2c02837

[230] A. Mech, S. Gottardo, V. Amenta, A. Amodio, S. Belz, S. Bøwadt, J. Drbohlavová, L. Farcal, P. Jantunen, and A. Małyska, Safe-and sustainable-by-design: The case of Smart Nanomaterials. A perspective based on a European workshop, J. Regul. Toxicol. Pharmacol., vol. 128, (2022) 105093. https://doi.org/10.1016/j.yrtph.2021.105093

[231] M. El-Shetehy, A. Moradi, M. Maceroni, D. Reinhardt, A. Petri-Fink, B. Rothen-Rutishauser, F. Mauch, and F. Schwab, Silica nanoparticles enhance disease resistance in Arabidopsis plants, Nat. Nanotechnol., vol. 16, no. 3 (2021) 344-353. https://doi.org/10.1038/s41565-020-00812-0

[232] I. Sochová, J. Hofman, and I. Holoubek, Using nematodes in soil ecotoxicology, J. Environ. Int., vol. 32, no. 3 (2006) 374-383. https://doi.org/10.1016/j.envint.2005.08.031

[233] A. M. Mebert, C. J. Baglole, M. F. Desimone, and D. Maysinger, Nanoengineered silica: Properties, applications and toxicity, J. Food Chem. Toxicol., vol. 109, (2017) 753-770. https://doi.org/10.1016/j.fct.2017.05.054

[234] M. Delample, N. Villandier, J.-P. Douliez, S. Camy, J.-S. Condoret, Y. Pouilloux, J. Barrault, and F. Jérôme, Glycerol as a cheap, safe and sustainable solvent for the catalytic and regioselective β, β-diarylation of acrylates over palladium nanoparticles, J. Green Chem., vol. 12, no. 5 (2010) 804-808. https://doi.org/10.1039/b925021b

[235] B. Fadeel and A. E. Garcia-Bennett, Better safe than sorry: Understanding the toxicological properties of inorganic nanoparticles manufactured for biomedical applications, Adv. Drug Deliv. Rev., vol. 62, no. 3 (2010) 362-374. https://doi.org/10.1016/j.addr.2009.11.008

[236] N. González-Ballesteros, S. Prado-López, J. Rodríguez-González, M. Lastra, and M. Rodríguez-Argüelles, Green synthesis of gold nanoparticles using brown algae Cystoseira baccata: Its activity in colon cancer cells, J. Colloids Surf. B: Biointerfaces, vol. 153, (2017) 190-198. https://doi.org/10.1016/j.colsurfb.2017.02.020

[237] I. Ali, Water photo splitting for green hydrogen energy by green nanoparticles, Int. J. Hydrog. Energy, vol. 44, no. 23 (2019) 11564-11573. https://doi.org/10.1016/j.ijhydene.2019.03.040

[238] B. Archana, K. Manjunath, G. Nagaraju, K. C. Sekhar, and N. Kottam, Enhanced photocatalytic hydrogen generation and photostability of ZnO nanoparticles obtained via green synthesis, Int. J. Hydrog. Energy, vol. 42, no. 8 (2017) 5125-5131. https://doi.org/10.1016/j.ijhydene.2016.11.099

[239] A. Charanpahari, N. Gupta, V. Devthade, S. Ghugal, and J. Bhatt, L. Martínez, Kharissova, O., Kharisov, B., Ed. Ecofriendly nanomaterials for sustainable photocatalytic decontamination of organics and bacteria (2019). Springer, Cham. 1777-1805. https://doi.org/10.1007/978-3-319-68255-6_179

New Materials for a Circular Economy
Materials Research Foundations 149 (2023) 311-367

Materials Research Forum LLC
https://doi.org/10.21741/9781644902639-9

Chapter 9

Advances in the Treatment of Waste Derived from Electronic Components: The Future of Cars: An Assessment through Raw Materials

A. Ortego[1], A. Valero[1], I. García-Díaz[2], F.A. López[2], M. Iglesias-Émbil[3,4]

[1]CIRCE Institute, Universidad de Zaragoza, Mariano Esquillor,15, Zaragoza, 50018, Spain

[2]National Center for Metallurgical Research (CENIM). Spanish National Research Council (CSIC), Avda. Gregorio del Amo,8, Madrid, 28040, Spain

[3]SEAT S.A. Autovía A-2 Km. 585, 08760 Martorell, Spain

[4]Sostenipra Research Group (2017 SGR 1683), Institut de Ciència i Tecnologia Ambientals ICTA-UAB, (CEX2019-0940-M), Z Building, Universitat Autònoma de Barcelona (UAB), Campus UAB, Bellaterra, Barcelona, 08193 Spain

Abstract

A conventional car needs more than 50 different types of metals, being critical most of them. In addition, the renovation of current cars for cleaner, safer and more comfortable ones will increase the critical metal demand to manufacture some components: batteries, LEDs for lighting, permanent magnets for motors, electronic units and different kinds of sensors. As a result, the availability of enough raw materials is a matter of concern for the automobile industry so new approaches from resource efficiency must be urgently applied. This chapter presents a review of the car´s resource efficiency from those metals needed to be manufactured with special attention to those used in electronic components. To accomplish with this purpose, three main goals have been established in this study: (1) analyzing car compositions and the most strategical metals for the automotive industry, (2) examining possible raw material bottlenecks before 2050, and ultimately (3) studying the effectiveness of current recycling policies and alternatives from eco-design.

Contents

1. Metal composition of passenger cars

Advances towards cleaner passenger cars (cars) are encouraging the continuous renovation of car fleets so it is expected that in the following decades a complete renovation will take place. This renovation seems irreversible considering the European parliament has recently voted to ban combustion engine cars from 2035.

This new generation of cars will reduce its fossil fuel dependency. But in contrast, it will demand a huge quantity of other kinds of natural resources being some of them even scarcer than oil. Some of these resources will be necessary to manufacture the following components: batteries (Co, Ni, Mn or Li), LEDs for lighting (Ga, Ge, Y), permanent magnets for motors (Nd, Dy, Pr), electronic control units (Au, Ag, Sn, Ta, Yb), different kinds of sensors (Ce, Tb, Se, La), infotainment screens (In), automotive high-performance steel or aluminum alloys (Nb, Mo, Cr, Ti, V, Sc, W).

Unfortunately, these resources are also finite, some of them are very scarce being even considered as critical for the European commission, and other institutions from several perspectives such as economic importance, supply risk, ecological risks or geological availability.

In order to know which specific resources are highly critical to the automotive industry, the starting point consists in determining the material composition of a car, with special attention to electronics and batteries.

In Table 1, four different car compositions are compared. In all cases, the compared vehicles are hatchback models among which are one ICEV[1] (Internal Combustion Engine Vehicle) and three BEV[2]s (Battery Electric Vehicle) (BEV_333; BEV_622; BEV_811)[3]. The reason behind analyzing three BEVs responds to the different chemistries used in batteries over time, that require different amounts of raw materials like cobalt, nickel, lithium or manganese.

Derived from the results, it can be concluded that BEVs will have a metal weight around 50 % higher than ICEV ones (from around 800 kg to more than 1.200 kg). In particular, Fe (+ 249 kg), Al (+ 45 kg), Cu (+ 44 kg), Ni (+ 32 kg), Mn (+ 14 kg), Co (+ 13 kg) and Li (+ 7 kg) are the main contributors to this increase.

It is also highlighted that other metals will be in more demand. As an example, this is the case of rare earth elements Dy, Pr and Nd that are used to manufacture permanent magnets for electrical engines. Specifically, the demand of Dy, Pr and Nd will by multiplied by 147, 126 and 65 respectively with respect to the ICEV content.

---

[1] ICEV: Internal Combustion Engine Vehicle.
[2] BEV: Battery Electric Vehicle.
[3] BEV_333 = battery electric vehicle with battery Li-ion 333; BEV_622 = battery electric vehicle with battery Li-ion 622; BEV_811 = battery electric vehicle with battery Li-ion 811.

Table 1.   *Metal composition of four different types of cars. Source [1].*

Metal	Type of vehicle			
	ICEV	BEV_333	BEV_622	BEV_811
Ag	23.572	34.669	34.669	34.669
Al	116,917.516	161,464.682	161,464.682	161,464.682
As	2.347	0.879	0.879	0.879
Au	0.612	2.411	2.411	2.411
B	27.137	57.238	57.238	57.238
Ba	771.345	1,634.146	1,634.146	1,634.146
Be	0.020	0.053	0.053	0.053
Bi	85.373	34.500	34.500	34.500
Cd	0.119	0.242	0.242	0.242
Ce	0.060	0.137	0.137	0.137
Co	21.691	22,620.363	12,324.363	5,460.363
Cr	6,318.717	6,534.836	6,534.836	6,534.836
Cu	14,939.189	59,517.256	59,517.256	59,517.256
Dy	0.438	59.727	59.727	59.727
Fe	648,195.108	894,671.166	894,671.166	894,671.166
Ga	0.319	0.495	0.495	0.495
Gd	0.000	0.002	0.002	0.002
Ge	0.006	0.006	0.006	0.006
Hf	0.016	0.000	0.000	0.000
In	0.408	0.089	0.089	0.089
La	0.452	3.671	3.671	3.671
Li	4.367	7,955.167	7,211.567	6,353.567
Mg	2,751.794	2,339.420	2,339.420	2,339.420
Mn	4,698.876	27,511.743	17,959.343	11,552.943
Mo	216.697	559.189	559.189	559.189
Nb	163.412	198.234	198.234	198.234
Nd	15.369	972.896	972.896	972.896
Ni	2,974.248	23,357.810	37,600.610	43,835.410
Pb	11,299.782	11,561.676	11,561.676	11,561.676
Pd	0.704	0.416	0.416	0.416
Pr	0.073	9.290	9.290	9.290
Pt	0.050	0.058	0.058	0.058
Ru	0.013	0.042	0.042	0.042
Sb	51.706	129.988	129.988	129.988
Se	0.014	0.580	0.580	0.580
Sm	0.356	0.000	0.000	0.000
Sn	246.760	392.525	392.525	392.525
Sr	124.114	326.283	326.283	326.283
Ta	5.813	11.514	11.514	11.514
Tb	0.048	2.240	2.240	2.240
Te	0.013	0.040	0.040	0.040
Ti	2,577.262	1,359.722	1,359.722	1,359.722

V	98.454	146.421	146.421	146.421
W	1.642	5.239	5.239	5.239
Y	0.002	0.100	0.100	0.100
Zn	6,419.453	5,921.346	5,921.346	5,921.346
Zr	81.722	6.227	6.227	6.227
Metal weight (kg)	819.037	1,229.405	1,223.056	1,215.162
Metal weight over the total (%)	67.7	70.4	70.3	70.2
Other mat. (kg)	389.777	516.055	515.595	515.595
Total weight (kg)	1,208.814	1,745.460	1,738.651	1,730.757
Note: "Other mat." refers to rubbers, plastics, glasses, and fluids.				

Due to the electrification of mobility will make more sophisticated cars from electrical and electronic points of view, a more detailed analysis focused on electrical and electronic components is presented as follows.

Table 2 compares the metal demand of an ICEV and a BEV (it is considered only the case of BEV_333) to manufacture the electrical and electronics components in both cases. It must be considered that this subsystem (electrical and electronics) includes the contribution of batteries.

Table 2.   Metal needed to manufacture electrical and electronic components. ICEVs and BEVs. [1]. In red those which demand grow in BEV with respect to ICEV and are also considered as critical raw material (CRM) by the European Commission (EC).

	Mass (g) (ICEV)	Mass (g) (BEV 333)	Difference (g) (BEV 333 – ICEV)	Is it considered CRM by the EC?
Ag	2.5900	20.5345	17.9445	
Al	2,114.3303	80,710.9797	78,596.6494	
As	2.3367	0.1320	-2.2048	
Au	0.2148	2.0370	1.8223	
B	0.7236	0.0000	-0.7236	X
Ba	18.3722	48.3467	29.9745	X
Be	0.0039	0.0325	0.0286	X
Bi	0.1439	0.4372	0.2932	X
Cd	0.0651	0.1938	0.1287	
Ce	0.0602	0.0002	-0.0601	X
Co	2.6378	22,585.7512	22,583.1134	X
Cr	57.9853	5,091.3693	5,033.3840	
Cu	9,695.8173	54,308.9773	44,613.1601	
Dy	0.0615	55.6052	55.5437	X
Fe	9,356.0333	95,618.0713	86,262.0380	
Ga	0.0010	0.1119	0.1110	X
Gd	0.0000	0.0013	0.0013	
Ge	0.0019	0.0057	0.0037	X

Hf	0.0054	0.0000	-0.0054	X
In	0.0145	0.0802	0.0656	X
La	0.0000	0.3204	0.3204	X
Li	0.3251	7,952.8326	7,952.5075	X
Mg	48.8900	391.9370	343.0470	X
Mn	63.1531	21,802.3485	21,739.1955	
Mo	1.7777	21.0179	19.2402	
Nb	0.4804	5.1740	4.6936	
Nd	1.2983	854.9973	853.6990	X
Ni	39.7377	22,762.9312	22,723.1935	
Pb	11,199.7492	11,511.3155	311,5664	
Pd	0.0768	0.3469	0.2701	X
Pr	0.0000	0.9916	0.9916	X
Pt	0.0000	0.0218	0.0218	X
Ru	0.0046	0.0319	0.0273	
Sb	41.4622	110.0660	68.6037	X
Se	0.0099	0.5751	0.5652	
Sm	0.1161	0.0000	-0.1161	X
Sn	72.8275	280.6893	207.8618	
Sr	14.8717	37.0552	22.1835	X
Ta	2.0911	9.0986	7.0074	X
Tb	0.0215	0.0000	-0.0215	X
Te	0,0096	0,0319	0,0223	
Ti	6.8975	0.0000	-6.8975	X
V	0.3296	0.9975	0.6679	X
W	0.6153	0.1040	-0.5113	X
Y	0.0002	0.0000	-0.0002	X
Zn	637.2904	4.1794	-633.1110	
Zr	0.0361	0.0000	-0.0361	
TOTAL	**33,383**	**325,950**	**292,566**	

As a result, it can be concluded that the weight of electrical and electronic components in the case of BEV is 292 kg higher than in the ICEV case. The main contributing metals to this increment are Fe (+86 kg), Al (+78 Kg), Cu (+44 kg), Co (+22 kg), Ni (+22 kg), Mn (+21 kg) and Li (+8 kg). Note that the highest number of metals, in which the demand grows up from BEV to ICEV are also considered critical by the European commission. Therefore, it can be stated that cars are becoming more critical from a raw material point of view, being the electronic components largely responsible for this fact.

2. How to measure resource efficiency in a car? The problem of mixing oranges with apples

Although at the end of the previous chapter, it was stated that cars are becoming more critical, this statement is based on the critical criteria applied by the EC. These criteria comprise both the supply risk and the economic importance of the material in question. In both cases, these parameters are not constant but depend on cyclical issues. However, a criticality assessment requires universal units based on physical and non-changing parameters.

On the other hand, metals are contained in a car in very different proportions. While there are kilograms of Fe or Al, other metals, such as rare earths, were found in milligrams. Can these minor metals be more critical than others, which are used in kilograms? How can we compare the criticality of metals whose contribution to the car's weight has different orders of magnitude?

Taking into consideration the current European ELV recycling (End of Life Vehicle) legislation (Directive 2000/53/EC), the accounting of metals is made on a mass basis. This means that, to accomplish the recycling target, 1 gram of Co is as important as 1 gram of Fe. That said, it is well known that their scarcity in the earth crust and hence physical value are different.

For a better understating of the weakness of the mass approach, we will continue explaining the weighting method used by ELV Directive. Given that it defines a recycling target of 85% over the total car weight, for an ICEV, like the example shown in Table 1 with a total weight of 1.208 kg, the recycling target could be achieved with a non-recycled fraction of 181 kg (85% of 1.208 kg).

Following with the same example, the weight of all metals in semiconductors (Cd, Ga, B, In, Ge, As, Sb, Se, Te) sums around 80 g (0,007 % over the total car weight). It means that all metallic mass in semiconductors could be deposited in a landfill in 2.200 treated ICEVs and, at the same time, accomplishing with the legislation. Nevertheless, the semiconductor crisis suffered by the automotive sector since 2020 has demonstrated the importance and dependency of these materials that are used in very small mass fractions, but from the perspective of recycling are as valuable as others more common like iron or copper.

For better quantification, economics could suggest a monetary approach. However, the economic accounting has its own weaknesses. One of them consists in monetary depreciation (the purchasing potential of 1 € of 2022 is different than that of 1 € of 2003). Another problem of the economic approach relies on its fluctuations. For example, cobalt price fell from 90,000 $/ton to 80,000 $/ton in less than one month [2] or most recently London Metal Exchange (LME) was forced to halt nickel trading after prices doubled in from one day to another [3].

Therefore, a more appropriate indicator to assess resource use in a car must be sought. In this regard, the authors propose the thermodynamic rarity indicator, which allows to account for the quality of each raw material. By means of this thermodynamic approach, those metals, which are used in very small quantities (even smaller than 1 gram), will be higher rated because of their greater physical value (measured through their thermodynamic rarity). Thus, this methodology enables the identification of those car parts, which are the most critical ones from a resource efficiency viewpoint, without mixing oranges with apples.

New Materials for a Circular Economy   Materials Research Forum LLC
Materials Research Foundations 149 (2023) 311-367   https://doi.org/10.21741/9781644902639-9

2.1   Thermodynamic rarity

Thermodynamic rarity can be thus defined as the amount of exergy resources needed to obtain a mineral commodity from ordinary rock (a hypothetical state called Thanatia where are minerals are dispersed), using the prevailing technology. Hence, it allows taking nature into account as it apprehends both ideas: 1) conservation, because it advices to preserve those minerals that are scarce through exergy replacement costs or natural bonus and 2) efficiency, because embodied exergies indicate real energy expenditures that should be decreased in order to be cost-effective from mine to market. Thermodynamic rarity (or simply rarity) varies from mineral to mineral, as a function of absolute scarcity in nature and the state of technological development.

Figure 1 represents the exergy required to extract a commodity from the mine and to process it up to market applications depending on the ore grade of the mines. As it can be observed, the smaller is the ore grade the higher is the exergy required to obtain a commodity. Moreover, this dependency is exponential, so small reductions in ore grades make big increments of exergy requirements.

Note that, if technology does not change, rarity can be assumed constant. Before effective mining appeared, mineral deposits were abundant and highly concentrated, (the natural bonus was high and so the exergy replacement costs). In turn, it was very easy to mine and beneficiate minerals (embodied exergy costs were low). Throughout history, the low hanging fruits have been already mined, steadily resulting in exploitation of lower ore grade mines. Nowadays, the mining industry needs to go deeper, further and more energy intensively than before (the natural bonus has been decreased), i.e., exergy replacement costs have been converted into embodied exergies.

In short, the authors propose the evaluation of resource efficiency in cars through the thermodynamic rarity indicator. This method was also recommended by the German advisory council [5] due to current ELV policies based on mass targets are ineffective to recover minor metals.

It should be additionally mentioned that measuring resource efficiency of products by means of the European commission criticality assessment is not possible, given that both indicators supply risk and economic importance are dimensionless.

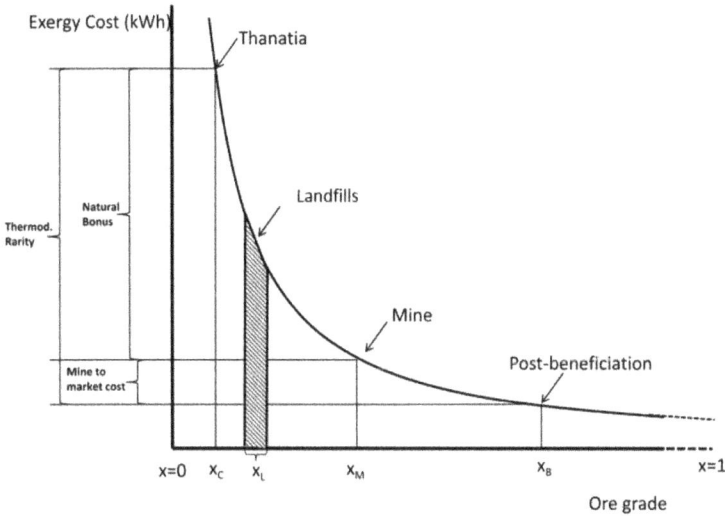

Figure 1.   Thermodynamic rarity represents the exergy cost (kWh) needed for producing a given mineral commodity from bare rock to market, i.e., from Thanatia to the mine and then to post beneficiation. Source [4].

2.2   Mass and rarity approaches

In this section, a comparison of the metal composition of cars detailed in Table 1 is carried out. This assessment is performed both in mass and rarity terms, comparing metal content per whole car (Figures 2a, 2b) as well as per subsystem (Figures 2c, 2d).

As before mentioned, the metal weight of BEV is around 400 kg higher than in ICEV (Figure 2a). However, when the comparison is made through the rarity approach (Figure 2b), the rarity of BEVs varies between 3,6; 2,8; and 2,2 times higher than the one in ICEV, in the cases of BEV_333, BEV_622 and BEV_811, respectively. The reason behind such different rarity values in BEVs resides in the battery composition, being Co the main contributor for this difference. It becomes thus clear that changes in cobalt demand have an important impact in the overall car rarity.

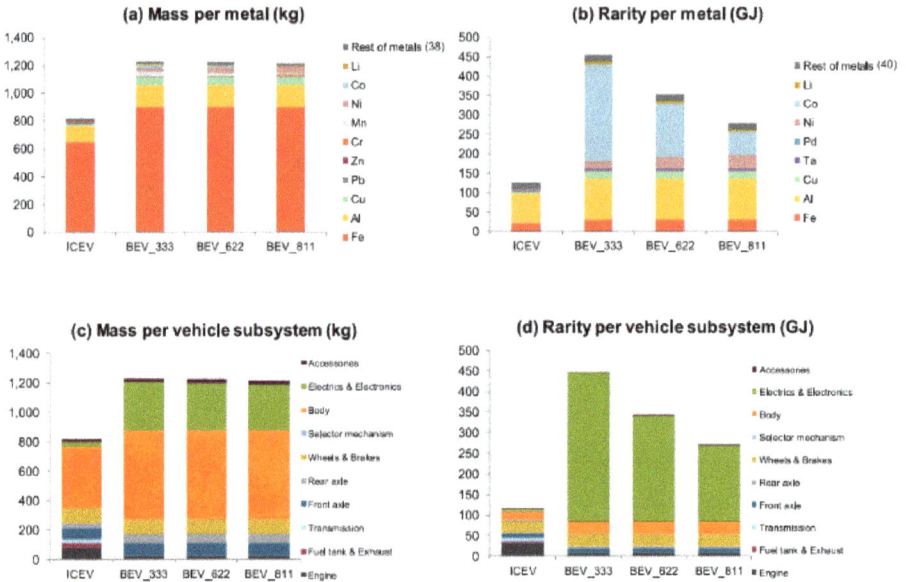

Figure 2. *Mass and Rarity comparison of the vehicles represented in Table 1.*

Focusing on the car's subsystems analysis, electrics and electronics represent around 25 % of the total metal weight in BEVs. Nevertheless, when applying rarity, the contribution of electrics and electronics varies between 77 %, 71 % and 64 % for BEV_333, BEV_622 and BEV_811, respectively.

The comparison in mass and rarity terms allows the identification of those metals more valuable because of their mineral capital or those subsystems, which are more critical due to the metals, used to manufacture them.

As can be observed in Figure 2, cars are becoming growing consumers of raw materials, given that BEV's rarity is more than double that in the ICEV. Precisely, Figure 3 represents this weakness as the *Achilles heel* of automotive manufacturers. In this case, the evolution of energy efficiency in cars over time is compared with the evolution of resources for manufacturing cars using thermodynamic rarity.

New Materials for a Circular Economy   Materials Research Forum LLC
Materials Research Foundations 149 (2023) 311-367   https://doi.org/10.21741/9781644902639-9

From Figure 3 can be concluded that, as more efforts are carried out for developing more efficient cars from an energy efficiency perspective, the more inefficient they are from a resource efficiency viewpoint. Furthermore, it is also remarkable to mention that the manufacturing of electric cars exponentially increases the resource dependency of the automotive industry. In fact, small steps towards improving energy efficiency imply large increases in the use of critical raw materials.

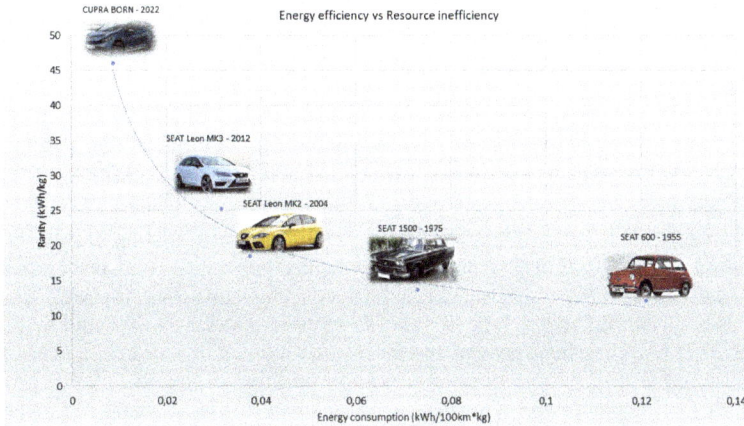

Figure 3.   Energy efficiency vs resource inefficiency over time. Own elaboration.

3.   What are the most strategic metals for the automotive industry?

Thermodynamic rarity can be considered as a new dimension to assess the physical value for any metal. Nevertheless, and as before mentioned, there are other non-physical but relevant factors for the supply of raw materials. Some of these parameters are the economic importance or supply risk, which are considered by the European commission to define the metal criticality.

In this section, an index named Strategic Metal Index (SMI) is presented, with the aim of estimating a holistic approach about criticality applied to car manufacturers as well as checking the reliability of the rarity indicator with respect to non-physical point of view (markets, geopolitics, vulnerability, substitution, dependency).

3.1   Physical and non-physical approach: Strategic metal index

The SMI was defined by [6] as an indicator, whose value ranges from 0 to 100 and it is calculated considering the following variables:

A= Automotive manufacturing demand of each metal with respect to total production.

B= Metal known resources with respect to total cumulative demand

C= Supply risk indicator from the European Union

Next, a detailed description of the parameters and its calculation is shown.

3.1.1 Automotive metal demand

The first factor needed for the SMI calculation, A, represents the automotive manufacturing sector demand of each metal with respect to total production of each commodity. To make these estimations, the average metal content of different types of cars is required to assess the cumulative demand of each metal from 2018 to 2050. On the other hand, the total cumulative demand for all sectors for each studied metal during that same period is also needed. Dividing the cumulative demand of the automotive sector by the total cumulative demand from 2018 to 2050 gives an idea of the importance of the automotive sector in the world capacity production of each metal. On the other hand, projections presented in this study go up to 2050 and the car's lifetime is considerably lower than the period analyzed. Accordingly, metal demand from fleet renovation is considered as well.

The demand projection for each type of car is presented in Figure 4. With respect to annual world sales (left) it must be underlined that the fleet's renovation effect is already included. In all cases, the considered average lifetime is 17 years, as stated by data published by the Spanish ELV recycling management system as, after a literature research, it is among the worst case scenarios [7]. Projections of sales and world fleet evolution are estimated using data from the International Energy Agency technology roadmaps and the International Organization of Automobile Manufacturers [8], [9].

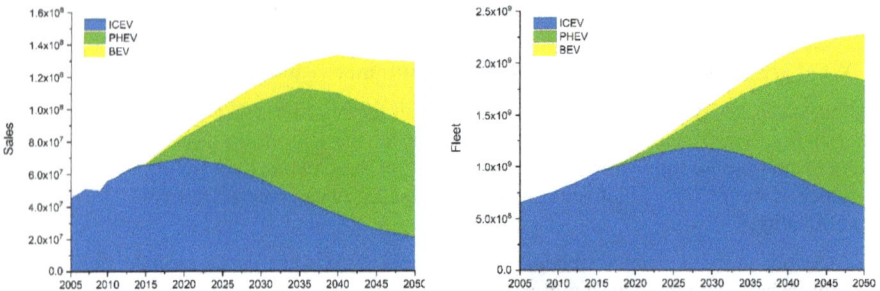

Figure 4.   World vehicle sales projection (left) and world fleet evolution (right). Scale: millions of units. [10, 11].

As stated by these reports, world fleet will increase to 2,200 million vehicles in 2050, where an average of 26, 53 and 19% will correspond to ICEV, PHEV and BEV, respectively. In addition, PHEV sales could surpass ICEV sales in 2032, whereas BEV sales in 2044. Additionally, it can be observed that beyond 2020, ICEV world sales will start to decrease.

New Materials for a Circular Economy   Materials Research Forum LLC
Materials Research Foundations 149 (2023) 311-367   https://doi.org/10.21741/9781644902639-9

3.1.2 Demand and known mineral resources

The second factor, B, is defined as the metal known resources with respect to total cumulative demand of each metal. This factor is calculated by dividing the total cumulative demand of each metal, from 2018 to 2050, by the current known mineral resources. Specifically, mineral resources are based on current geological knowledge. They are concentrations of solid materials of economic interest which are not economically extractable at the time of determination but that have a grade or quality that could make them extractable in the future.

There is some uncertainty associated with resources information as, obviously, the Earth's composition and mineral concentrations cannot be fully studied in every part of the planet due to economic limitations. The available information depends on the exploration carried out by mining companies, which usually tend to focus on more profitable minerals and large deposits. Still, resources information remains approximately constant over time compared with reserves or reserves base information, that strongly depend on the market price of each element each year and can be used for our calculations [12]. Using resources data might provide an overestimation of what can be really mined, as some mineral deposits might be impossible to put in operation, for instance, resources located in natural protected areas, under cities, etc.

3.1.3 Supply risk indicator

Since 1975, the European Union has been concerned regarding raw material supply and availability. In 2008, the Raw Materials Initiative (RMI) was created to meet three main goals for ensuring (1) fair and sustainable supply of raw materials from global markets, (2) sustainable supply inside the EU as well as (3) resource efficiency and supply of raw materials coming from secondary sources (recycling) [13]. Related to this RMI, the European Commission periodically publishes a list of critical raw materials in the EU, being the last list available in the 2020 report [14]. This critical raw material list has been used in literature in many analysis of mineral trade, compared with other criticality assessment methods and applied to analyze the impact of the elements of this list on emerging technologies [15–18].

As before stated, there are two main parameters considered in these reports, used to determine the criticality of each material: economic importance and supply risk. In the first case, economic importance tries to provide an insight on the importance that a selected material has for the EU economy in terms of end-use applications and the value added of each corresponding sector. In this section, we have already incorporated this information using the A variable, which considers the amount of each element applied in the automotive sector compared to the rest of the sectors. On the other hand, supply risk (SR) refers to the countries that produce these elements. In other words, whether the material is produced spread between many countries or concentrated only in a few ones. In this regard, the governance performance of the producing countries as well as the import reliance of the EU are included in the calculation of the index.

In this section, we have included the SR parameter in our calculations (C), given that it reflects not only the concentration of production of each selected element, but also the substitutability index. This SR parameter ranges from 0 to 10, so that an extrapolation has been done to a 0 to 100 scale for applying it in the SMI index formula.

3.2   Strategic metal index (SMI)

Each of the variables (A to C) reflects a particular situation for each element. Still, to obtain a global and unique value that can better assess the three factors as a whole, weighting coefficients are needed to standardize the data. The SMI for each individual element can be therefore calculated as the sum of the three variables by means of using weighting coefficients for each one, as follows (Eq. 1):

$$SMI = \alpha \cdot A + \beta \cdot B + \gamma \cdot C \tag{1}$$

Each coefficient's value (α, β and γ) ranges from 0 to 1, depending on the importance that the constraints could have in each variable. Once the values of the coefficients are assigned for calculating the SMI, they must be weighted, so that the sum of α, β and γ equals to 1. The ranked results of the SMI for each element analyzed in this section can be found in Figure 5.

Metals, which present a SMI value higher than 40 (in red), are considered as strategic, meaning that they have a high criticality. Values between 20 and 40 (in orange) indicate metals with a medium criticality. Whereas values lower than 20 (in green) reflect metals with low criticality. These boundaries have been selected to better represent the changes in tendency of the SMI values, corresponding a quarter of the total to metals labelled as strategic, while the remaining three quarters are more or less equally divided between medium and low categories.

As a result, the obtained SMI values indicate as strategic for the automotive sector the following twelve metals: tellurium, terbium, antimony, dysprosium, nickel, lithium, silver, bismuth, indium, neodymium, gold and boron. Of the twelve elements that have a SMI value higher than 40, only those belonging to the rare earth element group (terbium, neodymium and dysprosium), along with antimony, indium and boron (which are semiconductors) were labelled as critical by the European Commission. Noteworthy to highlight is the condition of nickel being more strategical than cobalt, because of the evolution of the battery chemistry to less cobalt demanding cathodes at the expense of nickel.

Applying the SMI assessment can help to put the focus on the specific elements that are used in the automotive manufacturing industry instead of using a more generalist approach. It provides useful insights on which metals should be considered strategic in the sector, particularly: tellurium, terbium, antimony, dysprosium, nickel, lithium, silver, bismuth, indium, neodymium, gold and boron. Thus, each automotive manufacturer can carry out studies to identify the different parts of the car where these elements are contained.

New Materials for a Circular Economy   Materials Research Forum LLC
Materials Research Foundations 149 (2023) 311-367   https://doi.org/10.21741/9781644902639-9

Accordingly, specific strategies can be implemented regarding the substitution of such metals or establishing eco-design measures to improve recyclability or even reusability of certain parts.

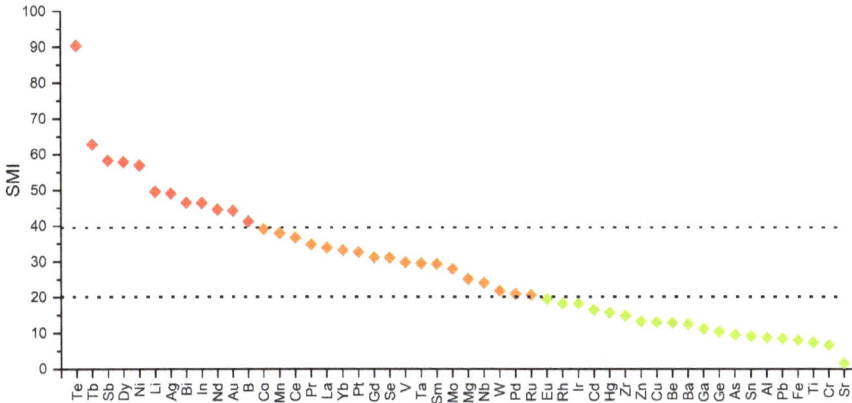

Figure 5.   *Ranked SMI values for each element. Source: Ortego, A et al (2018)*

4. Resource efficiency of recycling processes

At this point, it is time to analyze what is happening with the materials contained in a car when achieving its End of Life (EoL). The general belief is that ELV treatment is a solved problem, because recycling rates for whole car are greater than 85 % in weight terms, thus accomplishing with the European legislation. Yet, what is really taking place regarding all metals involved in the recycling process? Are these functionally recycled? Are minor but scarce metals recovered to be introduced again in new cars or at least in other sectors?

In the automotive manufacturing many resources and thus efforts (time of designers and developers, an enormous mixology of scarce metals, complex manufacturing processes, advance logistics…) are intended to manufacture sophisticate devices such as lightings, sensors or electronics. Accordingly, many resources are spent to produce a high-tech component resulting from a complex mixture of materials. On the contrary, when this component ends up in a recycling plant, the applied processes are designed to spend as less resources as possible. These typical processes comprise shredding as well as magnetic, eddy current and gravity separation technologies, specifically designed to consume small quantities of energy while processing an enormous quantity of cars each day. In conclusion, it can be stated that much fewer resources are put into play to separate than to mix.

For a better comprehension of the different efforts employed to manufacture or recycle, one can analyze the time spent in both operations. A conventional car requires around 20 hours in a production chain to be completely produced, whereas it is recycled in a shredding plant in less than 40 seconds. It means that the time used to manufacture a car is 1.800 times the time to recycle it.

At this point, the question is… Is it really easier to separate than to mix? Everybody can check how easy it is to mix water with sugar (it is only necessary to turn the spoon several times in the glass) and how difficult it is to separate them (we need to spend much more energy than before for evaporating the water). In thermodynamics, this phenomenon is well known: the irreversibility created in any mixing process. Indeed, mixing processes are highly entropic, so that an enormous quantity of exergy must be spent to separate. Moreover, in this separation process wastes will be always produced.

If thermodynamics already demonstrates that segregation of flows is far more resource intensive than their mixing, why in current separation (recycling) processes much less resources are spent than for mixing (manufacturing)? Are current recycling processes effective in terms of the recovery of scarce metals contained in a car? Is it still effective the ELV legislation published more than 20 years ago for the next generation of cars full of electric and electronic components?

4.1   ELV recycling processes

Figure 6 illustrates the ELV recycling process, where red arrows represent the material flow of a recycling operation, blue boxes show the destination in a landfill, green boxes show the output of recycled material, the yellow boxes show the output to an energy recovery process and the grey box shows reusing. According to Vidovic and coauthors [19], there are mainly five entities involved: (1) users who must deliver their car to collection facilities; (2) collection facilities, i.e., dealers or repair garages, which collect the ELVs; (3) authorized ELV treatment companies, which remove hazardous parts of cars that cannot be depolluted in landfills such as fuel, oil, tires, batteries, or air conditioning cooling gas, and disassemble reusable parts such as starters, suspensions or engines; (4) shredding plants, which receive decontaminated ELVs and shred them to separate them into three fractions: ferrous metals, non-ferrous metals, and the rest (a mix of rubber, foam, and plastics called ASR) using magnetic processes; (5) post-shredding plants, which receive the non-ferrous metal fraction from shredding plants and apply eddy current and density processes to separate mainly aluminum, zinc, and copper.

The scrap resulting from (4) and (5) is sent to smelters to produce steel or aluminum from secondary sources. The ASR resulting from (4) and (5) usually ends up in landfill or energy recovery plants, although the latter option is more complex owing to the heterogeneity of ASR composition, which contains a small fraction of metals that often hinders energy recovery.

Consequently, in a conventional vehicle recycling process, no specific operations to recycle minor but valuable metals are present. According to Ohno et al. [20], this fact entails the loss of most alloy elements either because they are downcycled or because they end up in the automobile shredder residue (ASR), ultimately becoming landfilled.

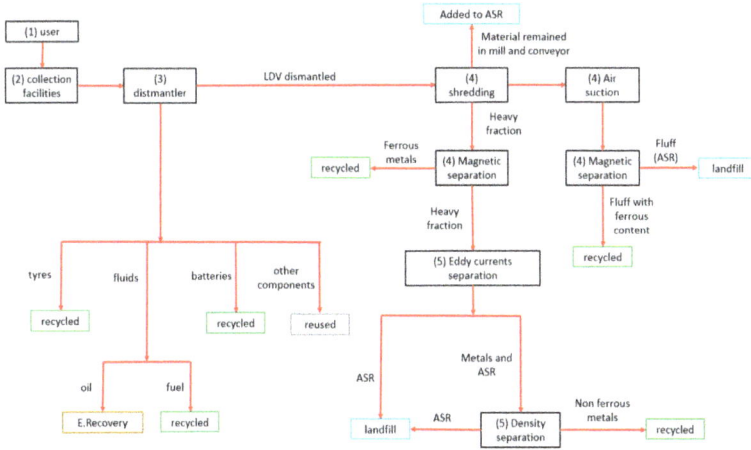

Figure 6.   Vehicle recycling scheme. (1) operations performed by users; (2) operations performed by collection facilities; (3) operations performed by dismantlers; (4) operations performed by shredding plants, and (5) operations performed by post-shredding plants. Source: own elaboration.

4.2   The problem of downcycling

Downcycling can be calculated as the additional quantity of virgin metals that would be required to manufacture a whole car if the starting point is scrap coming from ELV recycling facilities [21]. According to this definition, it is assumed that all metals become diluted in one of the different scrap types from ELV recycling facilities. This is an idealization since a portion of metals (usually below 5%) ends also in landfills.

Yet, the main goal of this section is to identify those components that are more critical from a downcycling point of view and thus provide automotive manufacturers and policy makers with valuable information to improve resource efficiency. It should be stated that this hypothesis is not very far removed from current approaches used by automotive manufacturers. In the homologation process of cars, manufacturers need to assess the recyclability degree of the new car for ensuring that they meet the ELV Directive. To that end, it is considered that all metals incorporated in a car are recycled while minor metals are downcycled and cannot be used again for their intrinsic properties.

Downcycling assessment presented in this section has always a negative value as it represents a loss incurred from an initial situation to a final situation, where quality decreases. A high negative value of downcycling indicates that an important quantity of metals contained in the given car part are not functionally recycled. On the contrary, a small value of downcycling means that these metals are not only recycled but also to a higher degree functionally recycled. It may occur that a heavy car part denoted as "A" has

New Materials for a Circular Economy   Materials Research Forum LLC
Materials Research Foundations 149 (2023) 311-367   https://doi.org/10.21741/9781644902639-9

a lower downcyclability degree than a lighter one denoted as "B". The reason for this can come from two factors: 1) because A, as opposed to B, is composed of mainly fully recyclable materials (such as steel), and/or 2) because B incorporates non-recyclable materials with a much higher rarity than A.

Table 3 presents the results. Metal downcycling is equal to 32.8 kg, accounting for 4.5 % of the total analyzed weight of the car, meaning that 32.8 kg of metals have not been functionally recycled. This indicates that the ELV Directive is satisfied, as more than 85 % of the vehicle would be recycled. However, a completely different situation occurs when one assigns a quality value to metals using the thermodynamic rarity indicator. In such a case, the loss increases to 21,647 MJ, or equivalently, −26.95 rt% of the total thermodynamic rarity of the analyzed metals in the vehicle. It means that the mineral capital of one in four recycled ELVs is lost.

If the ELV Directive targets were expressed in terms of rarity, current recycling systems would not be enforcing the law. In order, the most downcycled subsystems are the following: accessories, electrical and electronic equipment, fuel tank, exhaust system, and engine.

Table 3.   Downcycling by vehicle subsystem in mass and thermodynamic rarity

	Mass (g/veh)		Thermodynamic Rarity (MJ/veh)	
Engine	−6,024.39	−5.34%	−10,905.50	−35.64%
Fuel tank and exhaust system	−3,480.60	−19.95%	−762.76	−58.79%
Transmission	−696.55	−1.77%	−168.27	−5.27%
Front axle	−1,729.59	−3.75%	−678.24	−8.23%
Rear axle	−827.91	−3.03%	−313.54	−25.40%
Wheels and brakes	−2,907.38	−5.85%	−1,246.27	−13.97%
Selector mechanism	−114.48	−3.22%	−43.23	−26.82%
Body	−11,602.71	−2.85%	−2,369.72	−13.11%
Electrical and electronic	−5,169.93	−20.85%	−4,461.13	−57.36%
Accessories	−285.73	−10.90%	−698.52	−87.30%
Total	**−32,839.28**	**−4.50%**	**−21,647.22**	**−26.95%**

Finally, Table 4 presents the top 3 most downcycled car parts for each subsystem given in Table 3, with the three most downcycled metals presented for each case. This type of analysis allows identifying the parts where eco-design efforts should be made, in order to recover the most from valuable raw materials. The list has been made by ordering downcycling in percentage with respect to the initial thermodynamic rarity value for each subsystem.

Table 4. The top 3 most downcycled components in each category

Description	Vehicle subsystem	Thermodynamic Rarity (MJ/veh)	Downcycling (%)	Metal 1	Metal 2	Metal 3
Particle filter	Engine	8,613.48	−99.72%	Pt	Pd	Cu
Turbo distributor	Engine	260.71	−97.86%	Nb	Cr	-----
Turbo guide vane housing	Engine	185.32	−88.77%	W	Nb	Cr
Clamp	Fuel tank and exhaust system	12.35	−75.88%	Ni	Mo	Cr
Front pipe	Fuel tank and exhaust system	379.57	−74.89%	Ni	Al	Cr
Closing cap	Fuel tank and exhaust system	0.91	−73.38%	Ni	Cr	Zn
Shaft	Transmission	11.68	−33.74%	Ni	Mo	Cr
Reverse shaft	Transmission	35.35	−28.92%	Ni	Sn	Mo
5th gear shaft	Transmission	32.74	−28.44%	Ni	Mo	Cu
Input pinion	Front axle	13.11	−27.87%	Ni	Cr	-----
Track control arm	Front axle	260.76	−20.07%	Mo	Nb	V
Steering wheel	Front axle	147.94	−19.70%	Ta	Au	Cu
Axle dampers	Rear axle	206.55	−80.74%	Al	Cu	Zn
Feathering	Rear axle	0.30	−17.13%	Zn	-----	-----
Rear axle	Rear axle	840.17	−16.31%	Al	V	Ni
Break assistance pump	Wheels and brakes	352	−82.10%	Al	Ga	Ta
Brake distributor	Wheels and brakes	9.05	−58.76%	Al	Cu	Ni
Brake sensor	Wheels and brakes	7.02	−57.80%	Al	Cu	Ni
Foot adjusting	Selector mechanism	12.93	−76.22%	Ta	Al	Pd
Clear level	Selector mechanism	0.78	−71.85%	Ni	Cr	Zn
Ventilator	Selector mechanism	0.11	−58.41%	Ni	Cr	Zn
Seat belt	Body	84.02	−83.90%	Al	Zn	Ni
Bracket	Body	9.74	−76.30%	Ni	Cr	-----
Inner rear mirror	Body	185.02	−74.90%	Au	Fe	Mg
Airbag circuit	Electrical and electronic	118.51	−98.81%	Ta	Au	Pd
Rain sensor	Electrical and electronic	7.23	−98.51%	Ta	Au	Pd
Temperature sensor	Electrical and electronic	45.01	−98.39%	Pt	Ni	Cu
Speed sensor	Accessories	5.16	−98.46%	Au	Pd	Cu
Control unit	Accessories	600.52	−95.04%	Ta	Au	Al
Aerial amplifier	Accessories	37.41	−83.53%	Au	Ta	Pt

In the case of the engine, the application of metals Pt and Pd (particle filter), Nb and Cr (turbo distributor), and W and Nb (turbo guide vane housing), is notable. With respect to the fuel tank and exhaust subsystems, the most downcycled metals are Ni and Cr, which are contained in alloys for exhaust pipe, closing cap, and clamp. Regarding the transmission, Ni and Mo, applied to manufacture shafts, are the most downcycled metals. In the front axle, the steering wheel was highlighted, owing to the use of Au, Ta, and Cu in the electricity transmission system located in the steering rod. Further, in the rear axle, Al is the most downcycled metal, which is also present in axle dampers. In addition, in the breaking system, the metals Al, Cu, Ga, Ta, and Ni used in sensor, distributor, and pump are the most downcycled. In the selector mechanism category, the application of Ni, Cr, and Zn as steel alloys in the clear level and ventilator together with Ta, Al, and Pd in foot-adjusting is notable. Moreover, the inner mirror is the most downcycled part in the body owing to its gold content for the anti-glare system. Also highlighted should be the seat belts that use Al, Zn, and Ni. In electric and electronic equipment, the metals used in sensors such as Ta, Au, Ni, Pd, or Pt are the most downcycled. Finally, with respect to accessories, which also include electric and electronic equipment, the most downcycled parts are those including Au and Ta, such as the control units or aerial amplifier as well as those presenting Au and Pd such as the speed sensor.

4.3 Downcycling reduction recommendations

Several recommendations are suggested to reduce downcycling in ELV recycling processes. It must be taken into consideration that the recommendations listed below do not consider the car design phase.

- Disassembly of electric and electronic components containing valuable metals such as Au, Ag, REEs, platinum group metals (PGMs), Sn, Ta, or Te before shredding. Some of the most identified critical parts are the following: panel instrument, lighting switchers, LED lamps, power window motors, windscreen cleaner motors, electronic control units, rain sensors, electric mirrors, aerial amplifiers, and infotainment devices. According to Li et al. [22] this operation could be carried out even using automation by means of robots and would allow that specific recycling processes are implemented at a later stage.
- Application of hybrid recycling processes for the aforementioned parts, in the case of waste electric and electronic equipment (WEEEs), as it was suggested in previous studies by Arda et al. [23], Awasthi and Li [24] and Cui and Zhang (J. Cui & Zhang, 2008). These recycling processes should comprise mechanical, hydrometallurgical and biometallurgical technologies. This is because, as Ardente et al. [26] mentioned, the sole application of mechanical techniques is not compatible with the recycling of other materials such as indium in the displays or REEs in LEDs. Using this approach, valuable metals such as Au, Ag, Cr, Cu, Ga, In, Mg, Mo, Nb, Ni, Pd, Pt, Sn, Ta, V, or W would be recycled. This operation could be even developed in WEEE recycling plants, given that some of them are already implementing these processes.
- Disassembly of engine and gearbox components made of special steel alloys (high content of Cr, Mo, Nb, Ni, Ti, or W). Some of these components include the exhaust pipe, o-rings, turbos, pinions, and gear shafts. Owing to the excessive time required for removing some of these parts (i.e., removing o-ring from an engine requires the cylinder head cap, cylinder head, connecting rods, and pistons to be disassembled before and this operation requires at least 1 h), an intermediate situation could be implemented. This situation could imply that engines and gearbox are disassembled from the rest of the vehicle before shredding.

New Materials for a Circular Economy   Materials Research Forum LLC
Materials Research Foundations 149 (2023) 311-367   https://doi.org/10.21741/9781644902639-9

Notably, in a car manufacturing plant, the entire powertrain (including front axle, gearbox, engine, and rear axle) is joined to the body in less than 30 second and hence, this operation could be implemented via a reverse approach.

- Application of specific shredding processes for different car parts mainly made of steel and aluminum alloys (i.e., engines, gearboxes, and bodies) to produce different scrap qualities with the aim of manufacturing different qualities of alloys using them. Moreover, this measure would avoid the contamination of steel alloys due to high levels of some metals. The elements that have lower oxygen affinity than iron, such as Cu, Sn, Co and Ni, remain in the final alloy. As it was published by Daehn, Cabrera and Allwood [27] and Harsco Minerals [28] the use of low quality scraps provokes the production of off-specification steel and, in addition to the direct impact of this on the steelmaking process, it could be also considered to be a loss of these valuable elements.

- Because of the mentioned reasons, it is important for ELV authorized treatment centers to be equipped with information systems showing the location of the components that must be disassembled beforehand and indicating the proper disassembly procedure. This recommendation could be implemented using the International Dismantling Information System [29], which currently shows information related to batteries, fluids, or airbags and in which the automotive manufacturers are involved.

- To implement novel post-shredder treatments to recycle critical metals from the scrap obtained as output after the application of conventional recycling methods and from the ASR.

5. Recycling of electrical and electronic components

Electrical and electronic waste (E-Waste) has a very complex nature containing various components such as plastics, glass, ceramics and scarce metals (e.g. precious metals and rare earths) [30][31]. So, selective recovery of metals of interest can be complicated. To increase the technological and economic viability of recovery processes it is necessary to eliminate unnecessary parts through pre-treatments [32].

Therefore, the recovery of rare earths and precious metals from electronic and electrical waste generally consists of three steps: collection, pre-processing and processing (pyrometallurgy, hydrometallurgy and biohydrometallurgy) [33]. Taking into consideration the downcycling recommendation, an adequate collection and disassembly of electronic and electrical vehicle components is necessary before the pre-processing and processing treatment of them.

5.1   Pre-processed

A schematic summary of the possible paths that can be followed in the recycling of E-waste is shown in Figure 7. The composition of E-waste is heterogeneous and consists of a large number of materials, including plastics, glass and metals. In order to achieve a circular economy, all these materials need to be collected and recycled.

After the collection of electronic waste, the pre-processing stage takes place. These processes of release of metals of interest from other components are an indispensable stage for the recovery of important fractions of E-Waste. The loss of valuable and critical items in the recycling chain is usually because these elements end up in erroneous output streams, which implies the optimization of the stage of each of the stages of pre-treatment. [34].

In general, the pre-processing phases consist of:

• Manually disassembly/dismantling

• Mechanical treatment processes

• Combination of manual and mechanical pre-processing.

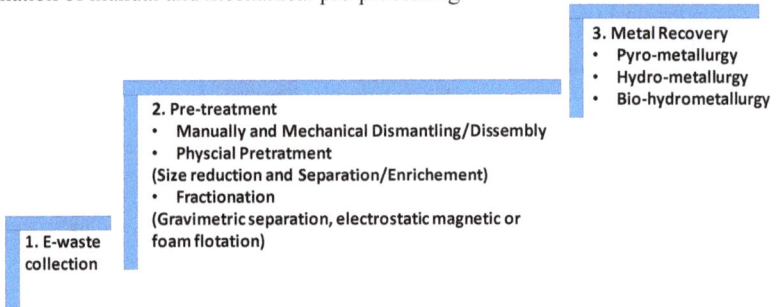

3. Metal Recovery
• Pyro-metallurgy
• Hydro-metallurgy
• Bio-hydrometallurgy

2. Pre-treatment
• Manually and Mechanical Dismantling/Dissembly
• Physcial Pretratment
(Size reduction and Separation/Enrichement)
• Fractionation
(Gravimetric separation, electrostatic magnetic or foam flotation)

1. E-waste collection

Figure 7.   Three stages of E-waste recycling. Based on [35]

This begins with the disassembly of the E-waste, which can be carried out manually or mechanically, depending on the characteristics of the equipment treated. The process of disassembling the E-waste can be a simple process, being necessary only to remove the fasteners such as screws or bolts or become a more complicated task due to the fusion of the coatings by welding or encapsulation, which will require an additional mechanical process of separation of the materials. In these stages of disassembly, faces and wiring boards are removed [36]. The main objective of manual disassembly is to achieve the elimination of hazardous components or with negative effects on the environment. This type of dismantling is more specific, managing to separate the most valuable parts or units from the E-Waste [37].

The components that will subsequently be recycled are subjected to a reduction in size using crushers, hammer mills, rotary crushers, disc crushers and ball mills [38][39]. The reduction in size may also involve heating the electrical components above the melting point used in the welds (240-250ºC), as well as subjecting the material to external forces such as impact, shear and vibration. From this approach, it is possible to achieve disassembly values of 94% [40]. Most of the machinery used in the E-waste dismantling process is equipped with sieves to collect the different particle size fractions that are generated [39]. The required granulometry depends on the next stage, sizes less than 200 µm is a requirement for leaching in stirred tank, which is the main stage of the hydrometallurgical process. The thickest material is fed to pyrometallurgical processes. The efficiency of pre-treatments depends on the degree of release, which is usually a function of particle size. In PCBs (Printed Circuit Boards) it is necessary to reduce the particle size to 5mm to achieve a degree of release of metals greater than 97% [41].

After size reduction, the material can be incorporated into the recycling processes or be subjected to an additional pre-processing to achieve a greater separation between the metal and non-metallic components. This additional separation process may include gravimetric separation, electrostatic magnetic or foam flotation [42] [32].

Gravimetric separation achieves the concentration of particles according to their specific gravity. This depends on the density of the particles, as well as the size [43]. Water, air, heavy media and sieves have been used to separate metals from plastic and ceramic materials [39].

Magnetic separation can be used to separate metal components from non-metallic ones. This method is excellent at separating ferrous from non-ferrous materials using high-intensity magnets. Size reduction is a critical factor for magnetic separation to prevent agglomeration of particles that could impede the efficient separation of ferrous and non-ferrous materials. Electrostatic separation can be useful for classifying materials of different electrical conductivity. They can be divided into electrostatic corona separation, triboelectric separation and eddy current separation. Foam flotation uses the natural hydrophobicity of the particles to separate the metal and non-metallic components. Reducing the particle size of PCBs have been shown to result in a base metal recovery rate of 95.6%. Although this process is efficient in recovering Cu, it can result in gold losses of up to 24.5% [43].

Any of the above processes or combination of processes, is essential to concentrate rare earths and metals prices and remove other components such as ceramics and plastics from the E-waste for subsequent recycling. The improvement of the metal content is crucial for the subsequent processing of waste by hydrometallurgical and biometallurgical methods, not being so essential for pyrometallurgical treatments [42].

5.2   Metal recovery

After the pre-processing stage, the E-waste will move on to the next stage, the processing. Currently, there are different methods for the recovery of metals from E-waste, among them pyrometallurgy and hydrometallurgy are processes frequently used for the recovery of non-ferrous metals, rare earths, and precious metals. The use of microorganisms and their metabolites through biohydrometallurgy currently is a very promising way.

5.2.1 Pyrometallurgical methods

The recovery of metals only by physical methods is not usually feasible because these metals form complex matrices with non-metals and with other materials such as ceramics or plastics. To achieve an efficient recovery, it is necessary to use other ways. A good option may be pyrometallurgical methods [44].

New Materials for a Circular Economy   Materials Research Forum LLC
Materials Research Foundations 149 (2023) 311-367   https://doi.org/10.21741/9781644902639-9

Pyrometallurgical methods are the route traditionally used for the separation and recovery of non-ferrous metals from metallurgical materials at high temperatures. It is a process that takes place at high temperature in oxidative or reducing conditions where physical and chemical transformations occur allowing the recovery of the metals of interest [45]. Pyrometallurgical treatments of E-waste involve smelting or incineration in blast furnace or plasma arc furnace, sintering, melting processes and gaseous phase reactions at high temperature [25, 33, 39].

Figure 8 illustrates the fundamental stages of pyrometallurgy, the selection of one or other operations depends fundamentally on the nature of the electronic waste and their requirements on smelting operations [46].

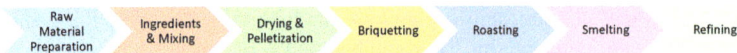

| Raw Material Preparation | Ingredients & Mixing | Drying & Pelletization | Briquetting | Roasting | Smelting | Refining |

Figure 8.   Pyrometallugical process. Based on [35]

Pyrometallurgical methods are implemented at an industrial level. In recent times, these processes have improved significantly increasing maximum recovery and decreasing their environmental impact. The recycling of E-waste is carried out in different copper and lead smelters. More than 70% of PCBs are treated by pyrometallurgical methods [25]. Once crushed, the PCBs are incorporated as raw material in the foundries for the separation and recovery of copper and precious metals such as gold and silver.

The recovery of the metals was carried out in a smelting process, where the raw materials are fed to a molten metal bath at 1250°C that is stirred with a mixture of supercharged air (up to 39% oxygen). In this process the precious metals will become together with copper or iron sulphide and others will oxidize and become part of the slag.

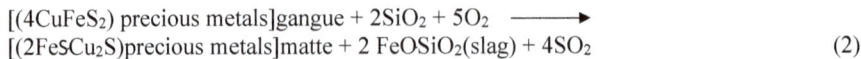

$$[(4CuFeS_2) \text{ precious metals}]\text{gangue} + 2SiO_2 + 5O_2 \longrightarrow$$
$$[(2FeSCu_2S)\text{precious metals}]\text{matte} + 2\ FeOSiO_2(\text{slag}) + 4SO_2 \quad\quad (2)$$

The matte is oxidized to a greater extent producing liquid copper blister

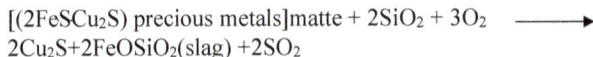

$$[(2FeSCu_2S) \text{ precious metals}]\text{matte} + 2SiO_2 + 3O_2 \longrightarrow$$
$$2Cu_2S + 2FeOSiO_2(\text{slag}) + 2SO_2$$

$$2Cu_2S + O_2 \longrightarrow 2Cu_{blister} + SO_2 \qquad (3)$$

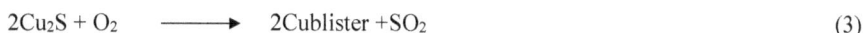

The blister copper is enriched in the converters and in an anode furnace where anodes of 99.1% purity are obtained, the remaining 0.9% corresponds to precious metals, Au, Ag, Pd, Pt along with metals such as Se, Te and Ni, which are recovered in the electrorefining stage.

Some companies using pyrometallurgical processes such as Aurubis in Germany [47], the Noranda foundry in Quebec [25], the Ronnskar foundry in Sweden [25], the Umicore foundry in Belgium [48] and the Dowa foundry in Japan are optimized industrial processes of state-of-the-art foundries [49]. Next, we will discuss some of these processes in more depth. The details of these foundries can be seen in the Table 5.

The Noranda process from Quebec, Canada, recycles E-waste (electronic scrap, PCB, etc.) together with extraction copper concentrates. The materials fed into the reactor are immersed in a molten metal bath at 1250°C that is stirred with a mixture of supercharged air. A reduction in the energy cost of the process is associated with the combustion of plastics and other flammable materials that are fed together with the E-waste. The impurities of iron, lead and zinc are converted into oxides that will be fixed in the silicic slag. The slag is cooled and ground to recover valuable metals before disposal. The copper matte which contains the valuable metals is transferred to the converter. After upgrading in the converters, the liquid blister copper is refined in anode furnaces achieving a purity of the 99.1%. The remaining 0.9% corresponds among others to precious metals, for their recovery they will be treated in the electrorefining process [25], [50].

The largest pyrometallurgical facility that processes e-waste is the Umicore smelter in Belgium, which is primarily focused on recovering precious metals from E-waste [49]. Various industrial wastes and by-products of non-ferrous industries (e.g., slag, matte, speiss, anodic sludge), precious metal ingots, spent industrial catalysts, as well as recyclable materials from consumption, such as automotive catalysts or printed circuit boards are used as raw materials. Umicore processes 350,000 tons of e-waste/year and recovers more than 100 tons of gold and 2400 tons of silver per year [33][48]. The material fed in the foundry plant is dismantled and processed to remove large pieces of plastic, iron and aluminium. The first step in precious metal extraction (PMO) takes place in the ISASMELT™ submerged lance combustion furnace [48]. This involves injecting air enriched with oxygen and fuel into a molten bath and the reduction of metals is achieved by incorporating coke as a reducing agent. Plastics and organic substances that may be contained in food replace fuel as an energy source and partially coke as a reducing agent. In the smelting stage, precious metals and copper are concentrated in ingots and the rest in lead slag [33][48]. The copper ingot will be subjected to electrowinning processes thus recovering the precious metals. Lead slag will be treated in base metal operations (BMO) subjected to different processes in which a lead blast furnace, a lead refinery (Harris process), a plant for the recovery of base metals, price metals and special metals such as indium, selenium and telluride will be used [25]. It should be noted that in addition to pyrometallurgical techniques, the plant uses hydrometallurgical and electrochemical methods [25, 48].

Table 5. Examples of pyrometallurgical companies that recovery metals form E-waste. Based on [42, 25].

Facility	Location	Metals Recovery	Process Overview
Umicore	Hoboken, Belgium	Ag, As, Au, Bi, Cu, In, Ir, Ni, Pb, Pd, Pt, Rh, Ru, Sb, Se, Sn	ISA SMELT™ smelting, copper leaching and electrowinning, precious metal refining. E-Waste cover up to 10% of the feed. Plastics partially substitute the coke as reducing agent and fuel.
Aurubis	Lünen, Germany	Ag, Au, Cu, Pb, Sn, Zn	Top submerged lance bath smelting, black copper processing and electrorefining precious metal refining
Noranda	Quebec, Canada	Ag, Au, Cu, Ni, Pd, Pt, Se, Te	Smelting and electrorefining, precious metal refining. Feeding to copper smelter with copper concentration (14% of the total throughput)
Rönskä	Boliden, Sweden	Ag, Au, As, Bi, Cu, In, Ir, Ni, Pb, Pd, Pt, Rh, Ru, Sb, Se, Sn	Kaldo reactor smelting, copper refining and purification, precious metal refining. Feeding to Kaldo reactor with lead concentrates. Plastic was tested as reducing agent and fuel.
Dowa smelter	Kosaka, Japan	Ag, Au, Bi, Cu, Ni, Pb, Sb, Sn, Te	Top submerged lance bath smelting and copper refining, precious metal refining.

In order to minimize the emissions in Unicore's IsaSmelt process, an emission control system was installed. The gases are cooled with energy recovery, these gases are cleaned using some techniques, such as sleeve filters, electro filters and scrubbers. Sulphur is converted into SO_2, which is transformed into sulphuric acid in the "contact" plant. In the chimney, SO_2 and NO_x are continuously monitored, to avoid any kind of environmental problem. According to Hageluken's report, the plant's emissions are well below European limits.

One of the main advantages of pyrometallurgical methods is that any form of electronic scrap can be used as a raw material for the recovery of metals. However, these processes also have some drawbacks, such as high slag production, non-selective operations and risk of formation and dioxins and release of toxic compounds [52].

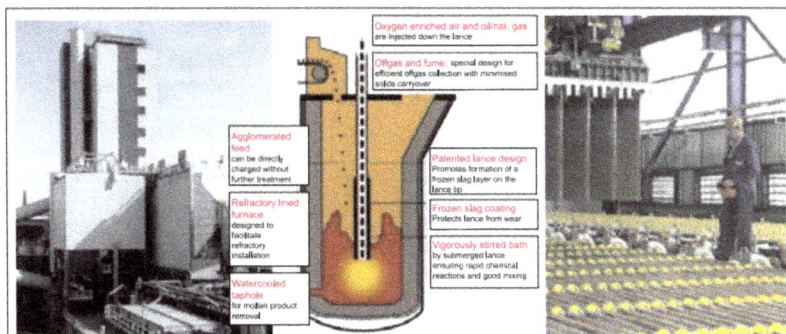

Figure 9.   Umicore's IsaSmelt furnace (left and centre) and electrowinning plant (right) [48].

5.2.2 Hydrometallurgical method

Hydrometallurgy is a branch of extractive metallurgy that involves the use of aqueous solutions to extract metals from solid raw material. Hydrometallurgical methods have certain advantages over pyrometallurgical methods, as they are more accurate [25], more easily controllable [53], have lower environmental impact [39, 33], lower infrastructure requirements and be more predictable [25, 54].

The main steps of hydrometallurgical methods consist of a leaching stage, with the aim of solving the metals of interest, the solid material will be leached with different leaching agents, such as acids, alkalis and ionic liquids, etc. The solutions are then processed using separation and purification methods such as precipitation, cementation, solvent extraction, adsorption and ion exchange, with the aim of separating and concentrating the metals of interest. The final recovery of metals is done using electrorefine, chemical reduction or crystallization techniques [55].

Figure 10 shows a summary of the hydrometallurgical methods used for the recovery of metals in E-waste. E-waste contains base metals (Cu, Fe, Co, Ni, Zn, Sn, Pb and Al), rare earths and precious metals (Eu, Y, Au, Ag Pt, Pd, etc.), even though the former have lower economic value than the latter, base metals are present in higher quantities representing most of the metal content in electronic waste.

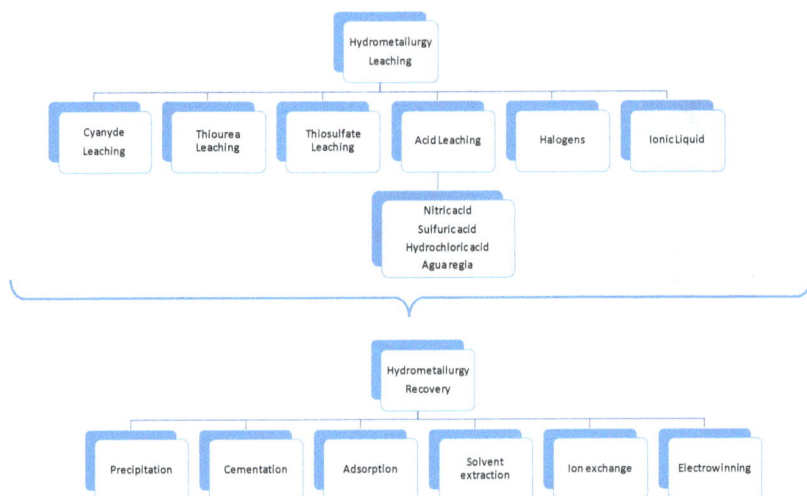

Figure 9.   Summary of the hydrometallurgical methods used for the recovery of metals. Based on [52]

5.2.2.1 Leaching

Particle size, leaching agent type, leaching agent concentration, temperature, pH, solid/liquid ratio, agitation, and redox potential are parameters that control the degree of leaching and its kinetics. Sometimes the use of oxidizing or reducing agents improves oxidation efficiency [56].

PCBs are composed of base metals (Cu, Ni and Fe) in addition to precious metals such as Ag, Au and Pd. The concentration of the latter is much higher than what can be found in natural deposits, hence the importance of their recycling. As will be seen below, one of the limitations for the recovery of PCBs from precious metals is the high content of Cu. This causes the consumption of the reagents and decreases the recovery of precious metals. Therefore, the recovery processes of precious metals take place after the removal of copper, which means an improvement in the performance of the process.

The recovery of gold and precious metals from PCBs can be carried out with different leaching agents such as cyanide, thiosulfate, thiourea, mineral acids and halides. Next, the reaction mechanisms will be presented, as well as pros and cons of each leaching agent.

The leaching of gold with cyanide has been employed by mining industry for more than a century [57]. It is the most used reagent for the extraction of gold from minerals and secondary sources, being 90% of Au and Ag extracted with this leaching agent. The leaching mechanism of Au with cyanide corresponds to the electrochemical process, according to the following reactions:

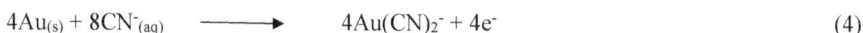

$$4Au_{(s)} + 8CN^-_{(aq)} \longrightarrow 4Au(CN)_2^- + 4e^- \tag{4}$$

$$O_{2(g)} + 2H_2O_{(aq)} + 4e^- \longrightarrow 4\,OH^-_{(aq)} \tag{5}$$

The reactivity order is Au>Ag>Pd>Pt. The study of the pH effect on the leaching of gold, silver, palladium and platinum metals showed that the maximum leaching is reached for pH values between 10-10.5 [58]. In this pH range the cyanidation process is safe, economical and has a lower environmental impact. At this pH the cyanide anion is formed which prevents the formation of HCN. Which has a greater volatility and high toxicity.

Although cyanidation is very effective for the leaching of these metals, in recent years different accidents have taken place in the Au mines that have caused the contamination of rivers and groundwater [52, 59, 60]. This hinders its future implementation, which is why alternative systems are being studied, among which thiourea, thiosulfate and halides have been proposed [61, 62, 63].

Thiourea, $(NH_2)_2CS$, is a potential agent leaching. It is soluble in water which with the addition of an oxidant reacts rapidly with gold or silver to form a stable cationic complex [64], equation 6. Leaching of precious metals with thiourea can achieve efficiency of up to 90%, and recoveries of 99% for Au [65]. Leaching studies suggest that the electron pairs of the nitrogen atom have a greater ability to create a coordination bond with Au and Ag than with cyanide [66, 67].

$$Au_{(s)} + 2CS(NH_2)_{2(aq)} \longrightarrow Au[CS(NH_2)_2]_2^+ + e^- \tag{6}$$

The most effective oxidant for the leaching of gold is Fe(III) in sulfuric acid [64].

$$Au_{(s)} + 2CS(NH_2)_{2(aq)} + Fe^{3+}_{(aq)} \longrightarrow Au[CS(NH_2)_2]_2^+ + Fe^{2+} \tag{7}$$

Apart from the leaching reaction, other competitive reactions take place in the reaction medium. It hinders the dissolution of gold and consumes thiourea or encourages its oxidation to irreversible products, which causes the consumption of large quantities of thiourea and makes it an expensive process. Control of these competitive reactions is achieved with low iron concentrations [65].

$$2CS(NH_2)_2 + 2\,Fe^{3+} \longrightarrow [CS(NH_2)(NH)]_2 + 2H^+ + 2Fe^{2+} \tag{8}$$

$$[CS(NH_2)(NH)]_2 \longrightarrow CS(NH_2)_2 + NH_2CN + S^0 \tag{9}$$

$$Fe^{3+} + SO^{2-}_4 + CS(NH_2)_2 \longrightarrow [FeSO_4(CS(NH_2)_2)]^+ \tag{10}$$

$$Fe^{3+} + CS(NH_2)_2 \longrightarrow [FeCS(NH_2)_2]^{3+} \tag{11}$$

$$Fe^{3+} + 2CS(NH_2)_2 \longrightarrow [Fe(CS(NH_2)_2)_2]^{3+} \tag{12}$$

The leaching percentage of gold depends on different factors such as the concentration of thiourea and oxidant, in addition to the pH of the solution. Leaching is normally carried out in a pH range between 1.0 and 2.0, in which the thiourea is stable. It decomposes rapidly at basic pH. If the thiourea content in the solution is too high, it is easily oxidized with the ferric ion in acid solution to the formation of formamidine disulfide.

A recovery of 90% Au and 50% Ag was achieved with a thiourea concentration of 24 g/L and 0.6% Fe(III) at room temperature from waste mobile phones PCBs. The complete recovery of Au and Ag was studied by Lee et al. [68], they proposed two leaching stage with thiourea and H_2SO_4 and ferric sulfate at room temperature. Several studies propose methods of leaching in two stages. In a first stage the leaching of Cu takes place and then in a second stage with acid thiourea the recovery of 80 and 70% of the Au and the Ag presents [64, 68, 69].

Despite the leaching efficacy of gold in thiourea, it is a method that does not have great applications at an industrial level, mainly due to the high price of cyanide and its high consumption associated with secondary reactions [25].

Thiosulfate ($S_2O_3^{2-}$) has been proposed as an alternative to cyanide for gold leaching by having a lower environmental impact and a low cost. However, it has lower leaching efficiency than cyanide and its consumption is higher [70].

$$4\,Au_{(s)} + 8\,S_2O_3^{2-} + O_{2(g)} + 2H_2O_{(aq)} \longrightarrow 4[Au(S_2O_3)_2]^{3-}{}_{(aq)} + 4OH^-{}_{(aq)} \tag{13}$$

Increased reactivity is achieved by channelling the reaction with cupric ions, as an oxidant and the presence of ammonia [71].

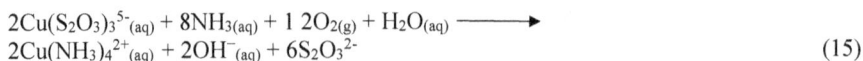

$$Au_{(s)} + 5S_2O_3^{2-} + Cu(NH_3)_4^{2+}{}_{(aq)} \longrightarrow AuS_2O_3^{2-}{}_{(aq)} + 4NH_{3(aq)} + Cu(S_2O_3)_3^{5-}{}_{(aq)} \tag{14}$$

$$2Cu(S_2O_3)_3^{5-}{}_{(aq)} + 8NH_{3(aq)} + 1\,2O_{2(g)} + H_2O_{(aq)} \longrightarrow$$
$$2Cu(NH_3)_4^{2+}{}_{(aq)} + 2OH^-{}_{(aq)} + 6S_2O_3^{2-} \tag{15}$$

The stability of the thiosulfate complex with gold depends on the conditions of the medium. It is necessary to maintain a basic means to prevent decomposition and this also favouring the solubilization of the Cu(II)-ammonia complex [72]. It is necessary to indicate that the Cu present in PCBs can affect the leaching process of gold, due to the decomposition of thiosulfate. Another parameter that affects the stability of the $Au(S_2O_3)_3^{2-}$-complex is

New Materials for a Circular Economy   Materials Research Forum LLC
Materials Research Foundations 149 (2023) 311-367   https://doi.org/10.21741/9781644902639-9

temperature, where the increase in temperature has destabilizing effects of the complex. Therefore, the best reaction conditions are room temperature and pH values between 9-10, value which is determined by ammonia [25, 56].

Tests carried out by Ficeriova et al. [73] showed that the 98 and 93% of Au and Ag was leaching from the reaction of thiosulfate with PCBs which has a particle size of less than 0.80 mm. Petter et al. [74] demonstrated an increase in leaching efficacy by adding a $CuSO_4$ concentration between 0.015 and 0.030 M with sodium thiosulfate. A two-stage leaching process has been developed for the recovery of metals from PCBs, in the first stage the base metals are leached with a mixture of H_2SO_4 and H_2O_2, in the second stage a thiosulfate solution was used. The highest yields are obtained for a concentration of 0.2 M of thiosulfate during 48 hours of reaction [75]. Işildar et al. [76] demonstrated that leaching percentages greater than 90% of Au could be achieved, for a concentration of 0.038 $CuSO_4$ and NH_4OH (0.3 to 0.38M) for time reaction of 6.73 hours

As in the case of thiourea, one of the main problems of thiosulfate leaching is the high consumption of reagent. So, it makes economically unviable leaching, despite its environmental benefits.

Although the most common leaching agents are not selective for the leaching of precious metals, a wide range of inorganic acids such as (H_2SO_4, HCl, agua regia o H_2SO_4 y HNO_3) can be used to leach Ag along with other base metals present in E-Waste [77, 78, 79]. HNO_3 is a particularly researched leaching agent for the leaching of Cu, Pb and Sn in E-waste [80, 81]. The 82.7% of Ag together with 94% of copper was leached with a 4 M HNO_3 solution at 65°C [82]. Neto et al. [83] achieved a leaching percentage of 97% Ag and 78% Cu for a 2M solution of HNO_3 at 50°C, Au values were less than 3%. Serpe et al. [84] achieved base metals leaching with citric acid, in addition study the effect of the NH_3 and IO^{3-}/I^-mixture. It allows the oxidation of Ag and Cu and the selective separation of both by the precipitation of AgI. The leaching extraction of Ag and Pd from PCBs was studied by Yazici et al. [85], they achieved the complete recovery of Cu as well as more than 90% of the Ag together with 58% of the Pd presents, for this they used an H_2SO_4-$CuSO_4$-NaCl solution, at a temperature of 80°C, where the Cl^-/Cu^{2+} ratio was 21.

The combination of HNO_3 with HCl (aqua regia) is a leaching agent that has been extensively studied in the leaching of precious metals, focusing mainly on Au [86][87]. Research results show that the percentage of leaching increases with acid concentration and reaction time. A sequential leaching, where the first two stages are used as a leaching agent HNO_3 and a third with aqua regia increases the recovery of Au and Ag [87]. Park and Fray [88] demonstrated that it is possible to leach Ag, Pd and Au from PCBs with an aqua regia solution at the same time. Studies carried out to achieve the simultaneous extraction and electrodeposition of Cu in HNO_3 or aqua regia showed that the highest percentages of recovery are obtained for HNO_3, however, this solution cannot be used directly in electrorefine processes. Copper deposition was inhibited by high acid contents [89].

The leaching reaction of copper with acid:

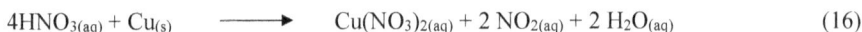

$$4HNO_{3(aq)} + Cu_{(s)} \longrightarrow Cu(NO_3)_{2(aq)} + 2\ NO_{2(aq)} + 2\ H_2O_{(aq)} \tag{16}$$

An effective and promising alternative to HNO_3 is H_2SO_4 combined with an oxidizing agent, such as H_2O_2. Piranha solution ($2M\ H_2SO_4$: $0.2\ M\ H_2O_2$) at a temperature of 85°C during 12 h of reaction, achieves leaching 95% of copper, the results showed that the particle size of PCBs is a crucial factor in copper recovery, the best leaching results are obtained for particle size less than 1 mm [75]. Below is the leaching reaction in $H_2SO_4 + H_2O_2$

$$Cu_{(s)} + 2\ H^+_{(aq)} + H_2O_{2(aq)} \longrightarrow Cu^{2+}_{(aq)} + 2\ H_2O_{(aq)} \tag{17}$$

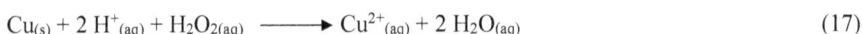

One of the problems associated with acid leaching agents is corrosion; aqua regia is a highly effective leaching agent for leaching Cu, Au and Ag. However, its use at an industrial level is very limited by its high corrosive power, being necessary the construction of rubber-coated steel reactors. A similar case is what happens with the piranha solution, although its corrosion power is lower than that of aqua regia. Sulfuric acid has a low corrosion power and is also easy to regenerate which makes it more viable for use from an industrial point of view.

Sometimes the recovery of precious metals is combined with the recovery of Cu, through sequential leaching processes, where the first stage corresponds to leaching with an acid followed by a leaching stage in cyanide, thiosulfate or thiourea to extract the precious metals.

Previously the use of cyanide for the dissolution of gold, halogens (fluorine, chlorine, bromine, iodine and astatine) were used, although there is no evidence of this dissolving capacity for fluorine and astatine [57, 90]. Depending on the chemical conditions both gold (I) and (III) is able to form complexes with chloride, bromide and iodide. Of all of them, the most studied is chlorine [25, 91, 92]. The efficiency of chlorination is greater for low pH values, high concentration of chlorides, high temperatures and surfaces. As indicated above, a traditional means for the leaching of precious metals is aqua regia. Leaching with chlorides takes place in the presence of a strong oxidant H_2O_2 o $NaClO_3$, under these conditions Cl_2 is produced. It is extremely toxic and corrosive, so its handling is very complicated, as well as its industrial use as seen previously. Studies of the leaching of Pd have been carried out using HCl and NaCl as a leaching agent with two oxidants HNO_3 and H_2O. Both oxidants in HCl achieved Pd recovery percentages of 93 and 95% respectively. [93]. The patent filed by Zhou [94] leaching 92% of Au and Pd using HCl and $NaClO_3$. The study of the effect of Cu^{2+} and Cl⁻ as an oxidant and ligand for silver recovery showed that at low concentrations of Cl silver precipitated as AgCl, an increase in concentration increases the recovery of Ag. The recovery percentage of 92% is reached for a concentration of Cu^{2+} of 4g/L and 46.6 g/L of Cl⁻ at 80°C. The Au extraction of PCBs was studied using H_2O_2-HCl as a leaching medium followed by an electro-obtaining stage. The highest leaching percentages are obtained for a mixture of 1 M H_2O_2 and 5M HCl. Lu et al. [95] described a two-stage chlorination leaching process to achieve the selective recovery of Cu and Au from PCB waste by controlling the redox potential of the solution with $NaClO_3$. Under optimal conditions, a recovery of 96.5% of Cu and 93% of Au is achieved at potential redox of 0.4 and 1.1 V respectively.

Iodide also has the ability to leach precious metals. The iodine-iodide system has certain advantages over chlorine for the leaching of Au, as it is not considered toxic or corrosive and it has a high selectivity for Au recovery. Another advantage is the recovery and reuse of iodine and iodide. Iodine dissolves in the presence of iodide forming the triiodide that will oxidize the elemental Au and form the gold iodide complex.

Despite the good results of iodization, the two major disadvantages are the high consumption and cost of the reagents. One of the ways to minimize iodine consumption is the addition of an oxidant in the systems that also improves the leaching percentages, making it a more economical process [96, 97]. The leaching of PCBs with 1% H_2O_2 and iodine concentrations between 1-2% manage to leach 95% of the Au [96, 97]. A leaching in two stages was also studied, in the first stage Cu and base metals are leached with H_2SO_4 2M and H_2O_2 0.2M at 80°C for 120 minutes. The second stage consisted of a leaching with iodine in the presence of 2% of H_2O_2 achieving the leaching of 93% of the Au present [62].

Ionic liquids have been a breakthrough in hydrometallurgical methods. They are made up of organic cations and anions that can be organic or inorganic. Figure 11 shows the most common cations and anions. Due to their physical and chemical properties such as low vapour pressure, low volatility, high thermal stability, low melting point (100°C), wide range of temperatures in which they are liquid, as well as being friendly to the environment have emerged as an alternative in leaching. They are also compounds that can be designed to selectively leach metals of interest in E-waste [98]. Metals such as Cu, Zn, Pb, Fe, Au and Ag can be digested with ionic liquids [99]. Zhang et al. [100] Leached 98.3% of Cu from PCB residues using an acidic ionic liquid with H2O2. Several literature reviews address this topic in depth [44, 101, 102, 103, 104].

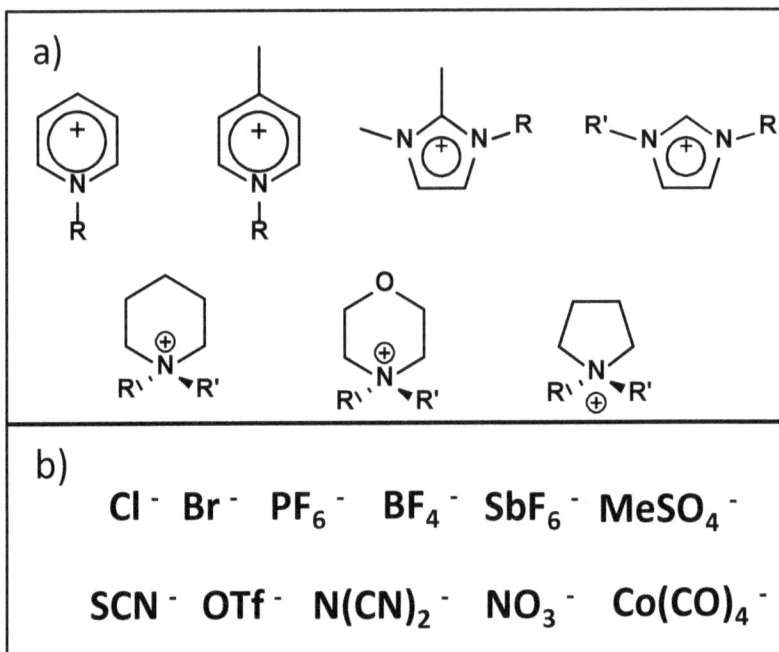

Figure 101. Different organic cations (a) and inorganic anions (b) [105]

5.2.2.2 Metal recovery

The leaching solutions obtained are multielemental, being necessary to develop selective strategies for the recovery of metals. The next stage in hydrometallurgical methods consists of the selective recovery of metals from leaching solutions. Recovery methods include precipitation, cementation [106], solvent extraction [107, 108, 109], adsorption [110, 111], ion exchange [112, 113] and electrowinning [114, 115].

Precipitation:

Precipitation is a widely established method for the recovery of metals in multielemental solutions. The incorporation into the solution of chemical compounds such as sulfides, hydroxides, carbonates or oxalates can change the ionic balance of the reaction and precipitate the metals in their salt form.

Precipitation processes consist of three phases, nucleation, core growth, and aggregation and crystallization [116]. The pH and concentration of the metals in the solution are factors that affect its precipitation. The main drawbacks of precipitation are the co-precipitation of the unwanted metals present in the solution, the amount of reagents needed to adjust the pH values and the generation of sludge [117, 56].

The precipitation of metal hydroxides is achieved using strong bases such as NaOH, $Ca(OH)_2$, the general equation of the process:

$$M^{2+}_{(aq)} + OH_{(aq)}^{-} \longrightarrow M(OH)_{2(sol)} \tag{18}$$

The precipitation of metal hydroxides has low cost and easy application [118]. However, one of its great disadvantages is the pH control, the modification of the pH involves the solubilization of hydroxides form, and the soluble complex $M(OH)^{+}$ is formed.

Research has shown that selective precipitation of the In ion as $In(OH)_3$ with NH_4OH is possible in a pH range of 5.0 and 9.0. At pH 7.4 maximum precipitation is reached, 99.8% [119]. Lee et al. [120] demonstrated the precipitation of Nd as hydroxide by modifying the pH of the solution with NaOH. The pH of the H_2SO_4 leaching solution was $-0.13<pH<0.02$, the increase in pH to 0.6 when adding NaOH means the recovery of 95% of the Nd in the solution. Another way to obtain metal hydroxides is the precipitation of metals as carbonates and subsequent conversion to hydroxide.

Precipitation of metal sulphides is another alternative for precipitation of metals in solution. The main precipitation reagents used can be FeS, CaS, Na_2S, NaHS, $(NH_4)_2S$, H_2S. Sulphide precipitation has several advantages, is more selective than other methods and can precipitate metals with low concentrations, on the order of ppb [121]. In addition, the reaction rate and stability of precipitates is high [116].

Same as in the hydroxide precipitation, pH plays a fundamental role in precipitation; a change in pH can lead to increased sulfide solubility [116]. Another key parameter in precipitation reactions is the concentration of sulphide. If it is in excess or depleted, sulfides or metals will remain in the solution. Precipitation with sulfides has other disadvantages such as the formation of polysulfides. This implies the excess consumption of reagent and a low precipitation of the metal. It is also observed a supersaturation effect, which results in very fine particles, which hinders filtration processes [122]. Selective recovery of In from an In/Sn solution was carried out by precipitating the Sn as SnS with gaseous H_2S [122]. Precipitation of 98% Ga from HNO_3 solutions is achieved by the addition of drop by drop (5 ml/min) of Na_2S [123].

Oxalate is another precipitation agent for dissolving metals. Studies carried out demonstrate the precipitation of Y with oxalate [64]. The selective recovery of metals from HNO3 solutions of magnetic sludge have also been researched, this solution was formed by Nd, Dy, Fe and B. Fe was precipitated as $Fe(OH)_3$ with a solution of NaOH, for pH values 2.0-3.0, under these conditions the co-precipitation of 25% of the Dy and the Nd in solution is observed. More than 70% of Nd remaining in solution was precipitated with oxalic acid [124].

Cementation:

Zinc cementation has been used commercially for the recovery of gold from cyanide solutions since 1890. This process is known as Merill-Crowe [125]. The main cementation reactions consist of a cathodic deposition of gold and anodic corrosion of zinc:

$$2Au(CN)^{2-} + 2e^- \longrightarrow 2Au + 4\ CN^- \tag{19}$$

$$Zn + 4CN^- \longrightarrow Zn(CN)_4^{2-} + 2e^- \tag{20}$$

The yields in the cementation of gold are practically constant for pH values e 8 to 11. However, the presence of impurities of lead, copper, nickel, arsenic, antimony and sulphur has a negative effect on the cementation of Au. The recovery of gold was studied from solutions of thiourea, thiosulfate or thiocyanth by means of a reduction-precipitation using as a stabilizing agent a solution of 12% by weight of $NaBH_4$ and 40% of NaOH. The gold ion can be reduced to metallic gold from thiourea solutions with the addiction of sodium borohydride, therefore it can be used as a leaching agent and as a gold reextraction agent in solvent extraction and in resins [126].

Solvent extraction:

Solvent extraction is also known as liquid-liquid extraction. It consists of mixing two immiscible liquid phases, between which the metal or metals of interest will be distributed; it is therefore an equilibrium process that can be represented in a general way as:

$$M^{n+}_{(aq)} + n\ HR \longrightarrow MR_n + n\ H^+_{(aq)} \tag{21}$$

HR is the organic liquid phase, MRn, organometallic spice (extracted), H^+ proton released by the extraction organic compound to be able to extract the cationic metal.

The efficiency of the extraction is determined by the pH of the aqueous phase, the counter-ion of the leaching agent, the selection of the organic extraction agent and the diluent, the reaction time, the temperature, the ratio aqueous phase organic phase. Some of the extraction agents used for the recovery of metals from the leaching solutions of E-waste are tributyl phosphate TBP, bis(2,4,4-trimetahylpentyl), Bis(2-ethylhexyl) phophoric acid D2EHPA, phosphinic acid Cyanex 272 or a mixture of different phosphine oxides.

Multiple solvent extraction systems have been studied for gold recovery [127, 128, 129]. Cyanex 921 is used to extract gold cyanide at all pH values, the presence of lithium ion improves efficiency in extraction. Cyanex 921 is also a suitable extraction agent in chloride medium, requiring lower amounts of extraction agent than in cyanide solutions. Re-extraction is efficiently achieved with water [130]. The recovery of In from leaching H_2SO_4 or HCl solution from LCD using D2EHPA showed that the best extraction conditions are achieved with a concentration of D2EHPA 0.25 M and a pH value<1.0. The exothermic nature of D2EHPA is reflected in the improvement of extraction percentages at temperatures below 20°C [131].

The selective recovery and separation of Nd, Dy, Gd and Pr from out-of-use magnets with D2EHPA was studied. Previously, NdFeBs were sulfated, roasted and leached with water. Different concentrations of the extraction agent and different eluents were evaluated, Solvent 70, hexane, octane, toluene, 1-octanol, cyclohexanone and chloroform. The results demonstrated greater efficacy in aliphatic organic diluents than in polar ones. The best conditions for separation between heavy rare earths (Nd, Dy and Pr) and light Gd is achieved for a concentration of 0.3 M D2EHPA in hexane. The 100% re-extraction is achieved with a concentration equal or greater than 2M HCl [132]. Ionic liquids can also be efficient extraction agents for the recovery of metals from leaching solutions [56, 133, 134].

Adsorption:

Adsorption is considered one of the simplest and most efficient methods for the recovery of metals in solution. The two main fields of application of adsorption are the removal of toxic elements or contaminants from industrial effluents prior to disposal [135] and the recovery of precious metals from aqueous solutions [91]. The parameters that influence adsorption are pH, temperature, initial metal concentration, reaction time, stirring speed, adsorbent used [56, 91].

One of the first methods for the recovery of Au and Ag cyanide complexes was on activated carbon [136]. Currently, there is a wide variety of different adsorbents for the recovery of metals [137, 138, 139].

Ion exchange:

Synthetic resins are generally used in ion exchange. These resins are capable of exchanging the ion of their functional group for the metal ions that are to be recovered from the leaching solution.

Strong-based anion exchange resins such as amberlite have demonstrated a high gold and silver capacity recovery [140]. The selective complexion of copper by Amberlite XAD 7HP resin offers a high separation between gold and copper from the dissolution of mobile phone PCBs. Aminophosphoric IRC-747 and aminomethylphosphonic TP260 show selective adsorption of rare earths in concentrated solutions of phosphoric acid (4mol/L) having great potential for application at an industrial level [141]. Resins for ion exchange are a versatile method for selective metal recovery; an increase in recovery percentages is achieved by adjusting the elution solution as well as the elution parameters. Ion exchange is an easy method to operate where reagent losses are minimal. The main obstacles to its commercialization are the amount of secondary waste that is generated after elution.

New Materials for a Circular Economy   Materials Research Forum LLC
Materials Research Foundations 149 (2023) 311-367   https://doi.org/10.21741/9781644902639-9

Electrowinning:

Electrowinning or electrorefining is a highly selective method for the recovery of metals in solution. In addition, its high selectivity presents other advantages such as a low generation of secondary waste and low investment cost.

Electrowinning methods have been widely used for the selective recovery of base metals such as Cu or Pb [142, 143, 114]. Currently the application of these methods for the recovery of critical and precious metals is in its early stages of development. One of the great challenges of electrowinning methods is to achieve the selective recovery of precious metals, Au, in the presence of Cu. The selective separation of Cu, Ag, Au and Pd from PCBs leached in aqua regia with electrowinning was carried out using four sequential electrowinning cuvettes. The selective recovery of each metal was achieved, firstly, Cu, then the Au, followed by the Pd and finally the Ag [56]. Recent research has evaluated direct recovery of Cu from PCB leaching and subsequent electrowinning [144, 145, 79]. The researchers studied the effect of $CuSO_4 5H_2O$, NaCl and H_2SO_4 concentration, current density and electrolysis time on recovery percentages and the size of recovered copper powder [146].

5.2.3 Biohydrometalurgical methods

The recovery of metals by biohydrometalurgical methods is one of the most promising technologies in recent years [147, 148, 149]. The great potential of these methods represents a technological advance, which entails the interest of the materials industry and the mining industry [51]. Biohydrometallurgical methods were already adopted in pre-Roman times at the Rio Tinto mine in southwestern Spain to recover Cu, Ag and Al.

Biohydrometalurgical methods present two main areas of bioleaching processes in which microorganisms solubilize the elements of solid substances to be later recovered. And biosorption where metals interact physicochemically with the surface of living or dead microorganisms [51, 150]. Biohydrometalurgical treatments have great advantages such as a low investment cost, low energy consumption and low environmental impact [52, 35].

Biological methods involve microorganisms for metals recovery such as Thiobacilli sp, Pseudomonas sp, Aspergillus sp., Penicillium sp., etc. All microorganisms use metal species for structural functions and/or catalytic functions. These methods instead of using chemical reagents for leaching use various microbes such as iron-oxidizing bacteria (Acidithiobacillus ferrooxidans) or sulfur-oxidizing bacteria. Different microorganisms into their removable or soluble form naturally transform solid metal components. A clear example of this is the extraction of metals from microorganisms in sulfide ores, such as the recovery of gold and silver, bacteria remove metal sulfides that interfere with minerals containing precious metals before cyanidation treatment [151].

The recovery of metals such as Co, Mo, Ni, Pb, Zn from sulphurous minerals, minerals generally carriers of base and precious metals, has been successfully carried out by bioleaching [152, 153]. However, the only industrial process in significant proportions is the recovery of copper and gold [152, 154].Brandl et al. [155] conducted a study to recover

Cu, Zn, Ni and Al from electronic waste. The study was carried out on the dust of the scrap crushing processes. Leaching tests were carried out with fungal cultures of *A. niger, P. simplicissimum* and the mixed consortium of *Thiobacillus thiooxidans* and *T. rerroxidans*. The results showed that adding high amounts of scrap increases the initial pH due to the alkalinity of the material. To minimize the toxic effect on microorganisms the process was carried out in two steps. First, biomass was produced and at a later stage, electronic scrap dust was added. For scrap concentrations of 5 and 10 g/L thiobacilli leaching of more than 90% of the available Al, Cu, Ni and Zn. In the solutions the presence of Pb and Sn is not observed, this is due to the precipitation of both compounds such as $PbSO_4$ and SnO. Higher concentrations of scrap decrease mobilization percentages. The analysis of the results of the mushroom culture showed that for 100 g/L of scrap the almost complete leaching of Ni, Pb, Sn and Zn is achieved, in a two-stage leaching process [155]. Recovery percentages greater than 80% of Ni, Cu and Zn are achieved by treating electronic waste with thermophilic microorganisms, Sulfobacillus thermosulficooxidans [147]. Liang et al. [156] recovered more than 90% of Cu, Ni, Zn and Pb using a mixture of T thiooxidans and T. ferroxidans. The 94.8% of Cu was recovered from electronic circuit boards, PCBs, in an incubation period of 28 days with T. ferroxidans [157].

Therefore, different studies have shown the resilience of precious metals (e.g., Ag, Au, Co), as well as base metals (e.g., Cu, Ni, Fe, Zn) by biohydrometallurgy (Table 6). The microorganism A. niger MXPE6 of or Acid ithiobacillus thiooxidans are suitable for the recovery of Au and Aspergillus nomius is more promising for the mobilization of Cu.

Table 6.   Microorganism used for bioleaching of E-waste. Based on [158]

Microorganism	E-Waste	Target metal	References
A. niger MX5 and A. niger MXPE6	PCBs	Cu, Au	[159]
Aspergillus niger MXPE6	PCBs	Au	[160]
Acidithiobacillus caldus Sulfobacillus thermosulphidooxans	PCBs	Cu, Zn	[161]
Aspergilllus niger	PCBs	Cu, Sn	[162]
Pseudomonas balearica SAE1	PCBs	Au, Ag	[163]
P. fluorescens	PCBs	Ag	[164]
Acidithiobacillus thiooxidans	PCBs	Au	[165]
Purpureocillium lilacinum, Aspergillusniger	PCBs	Cu, Al, Zn, Pb	[166]
Acidithiobacillus thiooxidans, Acidithiobacillus ferrooxidan	PCBs	Fe	[167]

One of the disadvantages of biohydrometallurgy compared to traditional hydrometallurgical processes is the low efficiency and long bioleaching time. However, under optimal conditions, the suitable microorganism culture could improve the efficiency of these processes [158].

6.   Recommendations through eco-design

The main goal of this section is to define eco-design strategies in cars, which must go beyond "design for recycling" as it is understood now. Instead, the strategy should be oriented towards "design for functional recycling", with the aim that scarce and valuable metals do not become lost at the EoL and form part of the technosphere as much time as possible.

6.1   Eco-design examples

Hereafter different eco-design examples will be presented based on cases researched previously by Ortego and coauthors [168, 169]. The measures are proposed in several car parts, which can be considered as critical from its rarity value in a conventional car. The measures are proposed in four different eco-design categories: (1) facilitating disassembly, (2) critical metal substitutability, (3) retrofitting and (4) new approaches.

6.1.1   Facilitating disassembly

The generator, which is currently placed at the bottom part of the engine, should be preferably placed in the upper part, as this would facilitate disassembly and the recovery of its valuable metals. The generator is moved by a multifunction belt, meaning that this measure could be easily implemented.

Other example in this section is the exterior mirrors, which have electrical engines, LEDs and even an anti-frozen mirror system. All these systems require scarce metals. The mirror should be designed to be removed from the exterior part of the door. This would avoid the disassembly of the inner door cover (see Figure 12). As for the wires that connect the battery, they have a high thickness (and copper content) due to the maximum power demand required to move the starter. With a redesign, they could be also disassembled when the batteries are removed in ELV authorized centers.

Figure 112.   Disassemblability process of exterior mirrors [168]

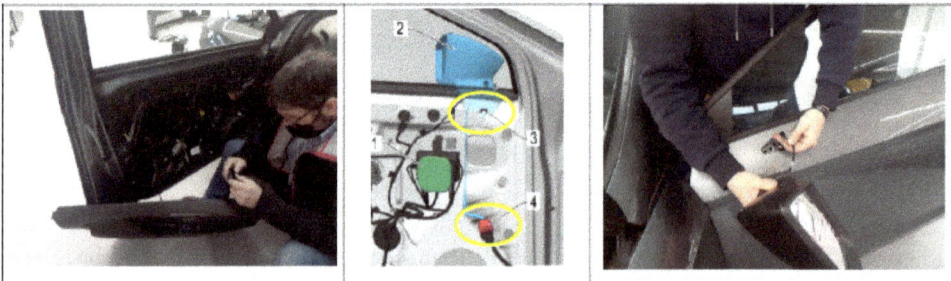

6.1.2 Critical metal substitutability

Now several substitutability recommendations based on copper, gold, tantalum and palladium are presented. The recommendations are focused on these metals, because they are the most commonly applied in parts identified as critical in an ICEV [169]. Below we list some of the possible alternatives:

- Aluminum instead of copper in certain wirings: as it is well known, aluminum is an alternative to copper for wiring.
- Silver plating instead of gold plating in electronic contacts: gold, as a native element, has very useful properties such as high conductivity, corrosion resistance, high melting point or reflectivity.
- Ceramic capacitors instead of tantalum capacitors: Ceramic capacitors have the highest market share but tantalum capacitors provide a feasible alternative if higher breakdown strengths are required. The reduced costs, smaller size (suitable for space-constrained electronic circuits), high-frequency characteristics, higher reliability, ripple control and longevity are driving the market to replace tantalum capacitors with ceramic capacitors wherever possible.

6.1.3 Retrofitting

From a retrofitting point of view, it is key the availability of reusable parts. Such parts can only come from models with high production values such as those in segment C. The destination of such parts would be in turn vehicles with smaller sales figures and where the focus is not placed on newest designs. This is why a clear destination of retrofitted parts would be in industrial vehicles. Particularly, retrofitting would be a plausible option for the combi instrument, lighting switcher, rain sensor, air quality sensor or exhausts gases temperature sensor (see Figure 13 with examples about different components that could be retrofitted).

Retrofitting of engines could be also considered mainly in the future case of electric vehicles due to retrofitting is much easier in electric engines than in combustion ones.

The gearbox is manufactured with some valuable metals like magnesium, nickel, chromium and molybdenum. In a conventional car the gearbox contains 80 % of the total magnesium used by the vehicle. Moreover, the gearbox is a very unfailing and robust component and is therefore a perfect candidate to be retrofitted. Manual gearbox design has not substantially changed throughout the years and this operation is a common technique applied in specialist gearbox repair garages.

New Materials for a Circular Economy   Materials Research Forum LLC
Materials Research Foundations 149 (2023) 311-367   https://doi.org/10.21741/9781644902639-9

13.a lighting switcher

13.b rain sensor

13.c rear screen cleaner motor

Figure 123. Different components that could be retrofitted

6.1.4 New approaches

In the field of new approaches, it is recommended to assess the possibility to centralize all (or at least the most critical) electronic control units (i.e., on board supply control unit, door unit, airbag unit, electronic control unit, combi instrument electronic or electronic from infotainment) in a common unit. In this way, this central control unit could be easily disassembled and sent from ELV authorized centers to specific recycling plants, as it is happening with batteries or tires. Further, it is also recommended to assess the impact of changing the vehicle voltage from 12 V to 24 V or 48 V. This measure is being proposed to improve the engine efficiency and the performance of hybrid systems. Note that this measure could also reduce the section of wiring. In this sense, the use of Integrated Starter and Generator Technology, as it is used by some manufacturers like VOLVO [170], may well reduce the demand of copper and permanent magnets containing rare earths. Finally, it is proposed to assess the impact that would have the inclusion of combi instrument information and switchers in the screen of infotainment unit. This measure would not only avoid the use of such devices but also the associated wiring.

Figure 134. Different electronic control units used in an ICEV. From left to right: onboard supply unit, airbag unit, electronic control unit and door unit.

Conclusions

As it has been shown, vehicles are evolving to complex machines more similar to a computer like a typical car. This evolution results in more electronic needed and as consequence more scarce, critical and valuable materials. Moreover, this growing trend is being exponentially so circular economy measures must be urgently applied to guarantee the future of the sector.

To this end, it is necessary to advance in eco-design, extend the lifetime of electronic components, advance in the development of advanced and automated disassembly technologies, reduce the production of E-waste and develop and deploy a complete ecosystem allowing European Electronics to be green with a low environmental impact in targeting very low electronic waste.

Acknowledgments

This chapter has received funding from the Spanish Ministry of Science and Innovation (PID2020-116851RB-I00 and PIE-RTC-202260E098. Irene García-Díaz was supported by contract PTA2020-018866-I from the Ministry of Science and Innovation

References

[1] M. Iglesias-Émbil, A. Valero, A. Ortego, M. Villacampa, J. Vilaró, and G. Villalba, Raw material use in a battery electric car - a thermodynamic rarity assessment, Resour Conserv Recycl. 158 (2020) https://doi.org/10.1016/j.resconrec.2020.104820

[2] Information on: https://www.lme.com/Metals/Minor-metals/Cobalt#tabIndex=2

[3] Information on: https://www.reuters.com/business/lme-suspends-nickel-trading-day-after-prices-see-record-run-2022-03-08/

[4] A. Valero, A. Valero, Thanatia, the destiny of the Earth's mineral resources. 2014.

[5] K. Sperlich, Open Public Consultation on the ELV Evaluation (Directive 2000/53/EC) Recommendations of German Advisory Council on the Environment (SRU-Sachverständigenrat für Umweltfragen), 2019.

[6] A. Ortego, A. Valero, A. Valero, M. Iglesias, and M. Villacampa, Strategic metal ranking for the automobile sector, in: Sustainable Development of Energy, Water and Environment Systems, 2018.

[7] SIGRAUTO, Age of ELV treatments, 2018.

[8] International Energy Agency, Energy Technology Perspectives: Scenarios & Strategies To 2050, 2010.

[9] International Organization of Motor Vehicle Manufacturers, Sales statistics, 2016.

[10] International Energy Agency, Energy Technology Perspectives: Scenarios & Strategies to 2050. 2010.

[11] International Organization of Motor Vehicle Manufacturers, Sales statistics, 2016.

[12] G. Calvo, G. Mudd, A. Valero, A. Valero, Decreasing ore grades in global metallic mining: A theoretical issue or a global reality? Resources. 5 (2016) https://doi.org/10.3390/resources5040036

[13] European Commission, The raw materials initiative - meeting our critical needs for growth and jobs in Europe. Communication from the Commission to the European Parliamente and the Council. COM (2008) 699 final. 2008.

[14] European Commission, Study on the EU's list of Critical Raw Materials (2020) Final Report, 2020.

[15] G. Calvo, A. Valero, L. G. Carmona, K. Whiting, Physical Assessment of the Mineral Capital of a Nation: The Case of an Importing and an Exporting Country, Resources, 4 (2015) 857-870. https://doi.org/10.3390/resources4040857

[16] R. L. Moss, E. Tzimas, P. Willis, J. Arendorf, L. T. Espinoza, Critical metals in the path towards the decarbonisation of the EU energy sector. Assessing rare metals as supply-chain bottlenecks in low-carbon energy technologies. European Commission Joint Research Centre, 2013.

[17] D. G. Angerer, Raw materials for emerging technologies. A report commissioned by the German Federal Ministry of Economics and Technology, 2009.

[18] G. Calvo, A. Valero, A. Valero, Thermodynamic approach to evaluate the criticality of raw materials and its application through a material flow analysis in Europe, J. Ind. Ecol. 22 (2017) 839-852. https://doi.org/10.1111/jiec.12624

[19] M. Vidovic, B. Dimitrijevic, B. Ratkovic, V. Simic, A novel covering approach to positioning ELV collection points, Resour. Conserv. Recycl. 57, (2011) 1-9. https://doi.org/10.1016/j.resconrec.2011.09.013

[20] H. Ohno, K. Matsubae, K. Nakajima, Y. Kondo, S. Nakamura, T. Nagasaka, Toward the efficient recycling of alloying elements from end of life vehicle steel scrap, Resour. Conserv. Recycl. 100 (2015) 11-20. https://doi.org/10.1016/j.resconrec.2015.04.001

[21] A. Ortego, A. Valero, A. Valero, M. Iglesias, Downcycling in automobile recycling process: A thermodynamic assessment, Resour. Conserv. Recycl. 136 (2018) https://doi.org/10.1016/j.resconrec.2018.04.006

[22] J. Li, M. Barwood, S. Rahimifard, Robotic disassembly for increased recovery of strategically important materials from electrical vehicles, Robot. Comput. Integr. Manuf. 50 (2018) 203-212. https://doi.org/10.1016/j.rcim.2017.09.013

[23] I. Arda, E. Rene, E. D. van Hullebusch, P. Lens, Electronic waste as a secondary source of critical metals: Management and recovery technologies, Resour. Conserv. Recycl. 135, (2018) 296-312. https://doi.org/10.1016/j.resconrec.2017.07.031

[24] A. K. Awasthi, J. Li, An overview of the potential of eco-friendly hybrid strategy for metal recycling from WEEE, Resour. Conserv. Recycl. 126 (2017) 228-239. https://doi.org/10.1016/j.resconrec.2017.07.014

[25] J. Cui, L. Zhang, Metallurgical recovery of metals from electronic waste: A review, J. Hazard. Mater. 158 (2008) 228-256. https://doi.org/10.1016/j.jhazmat.2008.02.001

[26] F. Ardente, F. Mathieux, M. Recchioni, Recycling of electronic displays: Analysis of pre-processing and potential ecodesign improvements, Resour. Conserv. Recycl. 92 (2014) 158-171. https://doi.org/10.1016/j.resconrec.2014.09.005

[27] K. E. Daehn, A. Cabrera Serrenho, J. M. Allwood, How Will Copper Contamination Constrain Future Global Steel Recycling? Environ. Sci. Technol. 51 (2017) 6599-6606. https://doi.org/10.1021/acs.est.7b00997

[28] H. M. & Minerals, 2018. Scrap contaminants. Retrieved from: http://www.harscom.com/217/Scrap-Contaminants.aspx

[29] IDIS, 2016. International Dismantling Information System. Retrieved from: http://www.idis2.com/

[30] A. B. Patil, M. Tarik, R. P. W. J. Struis, C. Ludwig, Exploiting end-of-life lamps fluorescent powder e-waste as a secondary resource for critical rare earth metals, Resour. Conserv. Recycl. 164 (2021) 105153. https://doi.org/10.1016/j.resconrec.2020.105153

[31] A. B. Botelho Junior, D. C. R. Espinosa, and J. A. S. Tenório, The use of computational thermodynamic for yttrium recovery from rare earth elements-bearing residue, J. Rare Earths. 39 (2021) 201-207. https://doi.org/10.1016/j.jre.2020.02.019

[32] J. Cui, E. Forssberg, Mechanical recycling of waste electric and electronic equipment: a review, J Hazard Mater, 99 (2003) 243-263. https://doi.org/10.1016/S0304-3894(03)00061-X

[33] A. Khaliq, M. Rhamdhani, G. Brooks, S. Masood, Metal extraction processes for electronic waste and existing industrial routes: A review and australian perspective, Resources, 3 (2014) 152-179. https://doi.org/10.3390/resources3010152

[34] A. Marra, A. Cesaro, and V. Belgiorno, Separation efficiency of valuable and critical metals in WEEE mechanical treatments, J. Clean. Prod. 186 (2018) 490-498. https://doi.org/10.1016/j.jclepro.2018.03.112

[35] R. Nithya, C. Sivasankari, A. Thirunavukkarasu, Electronic waste generation, regulation and metal recovery: a review, Environ. Chem. Lett. 19 (2021) 1347-1368. https://doi.org/10.1007/s10311-020-01111-9

[36] S. Zhang E. Forssberg, Mechanical separation-oriented characterization of electronic scrap, Resour Conserv Recycl. 21 (1997) 247-269. https://doi.org/10.1016/S0921-3449(97)00039-6

[37] C. Lee, C. Chang, K. Fan, T. Chang, An overview of recycling and treatment of scrap computers, J Hazard Mater. 114 (2004) 93-100. https://doi.org/10.1016/j.jhazmat.2004.07.013

[38] I. Dalrymple, N. Wright, R. Kellner, N. Bains, K. Geraghty, M. Goosey, L. Lightfoot, An integrated approach to electronic waste (WEEE) recycling, Circuit World, 33 (2007) 52-58. https://doi.org/10.1108/03056120710750256

[39] M. Kaya, Recovery of metals and nonmetals from electronic waste by physical and chemical recycling processes, J. Waste Manag. 57 (2016) 64-90. https://doi.org/10.1016/j.wasman.2016.08.004

[40] S. Zhang, E. Forssberg, Mechanical separation-oriented characterization of electronic scrap, Resour. Conserv. Recycl. 21 (1997) 247-269. https://doi.org/10.1016/S0921-3449(97)00039-6

[41] Shunli Zhang, E. Forssberg, Mechanical recycling of electronics scrap - the current status and prospects, Waste. Manag. Res. 16 (1998) 119-128. https://doi.org/10.1177/0734242X9801600204

[42] J. Van Yken, N. J. Boxall, K. Y. Cheng, A. N. Nikoloski, N. R. Moheimani, A. H. Kaksonen, E-waste recycling and resource recovery: A review on technologies, barriers and enablers with a focus on oceania, Metals. 11 (2021) https://doi.org/10.3390/met11081313

[43] M. Sarvar, M. M. Salarirad, M. A. Shabani, Characterization and mechanical separation of metals from computer Printed Circuit Boards (PCBs) based on mineral processing methods, Waste Manage. 45 (2015) 246-257. https://doi.org/10.1016/j.wasman.2015.06.020

[44] S. C. Chakraborty, M. W. U. Zaman, M. Hoque, M. Qamruzzaman, J. U. Zaman, D. Hossain, B. K. Pramanik, L. N. Nguyen, L. D. Nghiem, M. Mofijur, M. I. H. Mondal, J. A. Sithi, S. M. S. Shahriar, M. A. H. Johir, M. B. Ahmed, Metals extraction processes from electronic waste: constraints and opportunities, Environ. Sci. Pollut. Res. 29 (2022) 32651-32669. https://doi.org/10.1007/s11356-022-19322-8

[45] B. Ebin, M. I. Isik, Pyrometallurgical processes for the recovery of metals from WEEE, in WEEE Recycling, Elsevier, 2016, pp. 107-137. https://doi.org/10.1016/B978-0-12-803363-0.00005-5

[46] E. Ma, Recovery of waste printed circuit boards through pyrometallurgy, in Electronic waste management and treatment technology, Elsevier, 2019, pp. 247-267. https://doi.org/10.1016/B978-0-12-816190-6.00011-X

[47] R. Kahhat, E. Williams, Product or waste? Importation and end-of-life processing of computers in Peru, Environ. Sci. Technol. 43 (2009) 6010-6016. https://doi.org/10.1021/es8035835

[48] C. Hagelüken, Recycling of electronic scrap at Umicore's integrated metals smelter and refinery, Erzmetall. 59 (2006) 152-161.

[49] M. Kaya, Electronic Waste and Printed Circuit Board Recycling Technologies, Springer, 2019. https://doi.org/10.1007/978-3-030-26593-9

[50] B. S. H. Veldbuizen, Mining discarded electronics, Industrial Environment. 17 (1994) 7.

[51] J. Cui. L. Zhang, Metallurgical recovery of metals from electronic waste: A review, J. Hazard. Mater. 158 (2008) 228-256. https://doi.org/10.1016/j.jhazmat.2008.02.001

[52] E. Hsu, K. Barmak, A. C. West, A.-H. A. Park, Advancements in the treatment and processing of electronic waste with sustainability: a review of metal extraction and

recovery technologies, Green Chem. 21 (2019) 919-936.
https://doi.org/10.1039/C8GC03688H

[53] J. K. M. V. Ahamed Ashiq, Hydrometallurgical recovery of metals from E-waste, in Electronic waste management and treatment technology, Elsevier I., 2019, pp. 225-246. https://doi.org/10.1016/B978-0-12-816190-6.00010-8

[54] D. Andrews, A. Raychaudhuri, C. Frias, Environmentally sound technologies for recycling secondary lead, J. Power. Sources. 88 (2000) 124-129. https://doi.org/10.1016/S0378-7753(99)00520-0

[55] M. Sadegh Safarzadeh, M. S. Bafghi, D. Moradkhani, M. Ojaghi Ilkhchi, A review on hydrometallurgical extraction and recovery of cadmium from various resources, Miner. Eng. 20 (2007) 211-220. https://doi.org/10.1016/j.mineng.2006.07.001

[56] M. Sethurajan, E. D. van Hullebuscha , D. Fontana, A Akcil, H. Deveci, B. Batinic, J P. Leal, T. A. Gasche, M. A. Kucuker, K. Kuchta, I. F. F. Neto, H. M. V. M. Soares, A. Chmielarz, Recent advances on hydrometallurgical recovery of critical and precious elements from end of life electronic wastes - a review, Crit. Rev. Environ. Sci. Technol. 49 (2019) 212-275. https://doi.org/10.1080/10643389.2018.1540760

[57] G. Hilson, A. J. Monhemius, Alternatives to cyanide in the gold mining industry: what prospects for the future? J. Clean. Prod. 14 (2006) 1158-1167. https://doi.org/10.1016/j.jclepro.2004.09.005

[58] R. W. R. Dorin, Determination of leaching rates of precious metals by electrochemical techniques, J. Appl. Electrochem. 21 (1991) 419. https://doi.org/10.1007/BF01024578

[59] F. Korte, M. Spiteller, F. Coulston, The cyanide leaching gold recovery process is a nonsustainable technology with unacceptable impacts on ecosystems and humans: the disaster in Romania, Ecotoxicol. Environ. Saf. 46(2000) 241-245. https://doi.org/10.1006/eesa.2000.1938

[60] D. Pant, D. Joshi, M. K. Upreti, R. K. Kotnala, Chemical and biological extraction of metals present in E waste: A hybrid technology, J. Waste. Manag. 32 (2012) 979-990. https://doi.org/10.1016/j.wasman.2011.12.002

[61] M. Gökelma, A. Birich, S. Stopic, B. Friedrich, A review on alternative gold recovery re-agents to cyanide, J. mater. sci. chem. Eng. 04 (2016) 8-17. https://doi.org/10.4236/msce.2016.48002

[62] M. Sahin, A. Akcil, C. Erust, S. Altynbek, C. S. Gahan, A. Tuncuk, A potential alternative for precious metal recovery from E-waste: Iodine leaching, Sep Sci Technol, 50 (2015) 2587-2595 . https://doi.org/10.1080/01496395.2015.1061005

[63] H. Cui, C. Anderson, Hydrometallurgical treatment of waste printed circuit boards: bromine leaching, Metals. 10 (2020) 462. https://doi.org/10.3390/met10040462

[64] I. Birloaga, I. De Michelis, F. Ferella, M. Buzatu, F. Vegliò, Study on the influence of various factors in the hydrometallurgical processing of waste printed circuit boards for copper and gold recovery, Waste. Manag. 33 (2013) 935-941. https://doi.org/10.1016/j.wasman.2013.01.003

[65] D. A. Ray, M. Baniasadi, J. E. Graves, A. Greenwood, S. Farnaud, Thiourea leaching: an update on a sustainable approach for gold recovery from E-waste, J. Sustain. Metall. 8 (2022) 597-612. https://doi.org/10.1007/s40831-022-00499-8

[66] M. Gurung, B. B. Adhikari, H. Kawakita, K. Ohto, K. Inoue, S. Alam, Recovery of gold and silver from spent mobile phones by means of acidothiourea leaching followed by adsorption using biosorbent prepared from persimmon tannin, Hydrometallurgy. 133 (2013) 84-93. https://doi.org/10.1016/j.hydromet.2012.12.003

[67] S. S. Konyratbekova, A. Baikonurova, A. Akcil, Non-cyanide leaching processes in gold hydrometallurgy and iodine-iodide applications: a review, Miner. Process. Extr. Metall. Rev. 36 (2015) 198-212. https://doi.org/10.1080/08827508.2014.942813

[68] C.-H. Lee, L.-W. Tang, S. R. Popuri, A study on the recycling of scrap integrated circuits by leaching, Waste. Manag. Res. 29 (2011) 677-685. https://doi.org/10.1177/0734242X10380995

[69] I. Birloaga, F. Vegliò, Study of multi-step hydrometallurgical methods to extract the valuable content of gold, silver and copper from waste printed circuit boards, J. Environ. Chem. Eng. vol. 4 (2016) 20-29. https://doi.org/10.1016/j.jece.2015.11.021

[70] F. Z. Zhang XY, Chen L, Review on gold leaching from PCB with non-cyanide leach reagents, Non-ferr. Met. 61 (2009) 72-76.

[71] M. G. Aylmore, D. M. Muir, Thermodynamic analysis of gold leaching by ammoniacal thiosulfate using Eh/pH and speciation diagrams, Miner. Metall. Process. 18 (2001) 221-227. https://doi.org/10.1007/BF03403254

[72] F. Arslan, B. Sayiner, Extraction of gold and silver from turkish gold ore by ammoniacal thiosulphate leaching, Miner. Process. Extr. Metall. Rev. 29 (2007) 68-82. https://doi.org/10.1080/08827500601141784

[73] E. Ficeriova, J. Balaz, P.Gock, Leaching of gold, silver and accompanying metals from circuit boards (PCBs) waste, Acta. Montan. Slovaca 16 (2011) 128-131.

[74] P. M. H. Petter, H. M. Veit, A. M. Bernardes, Evaluation of gold and silver leaching from printed circuit board of cellphones, J. Waste Manag. 34 (2014) 475-482. https://doi.org/10.1016/j.wasman.2013.10.032

[75] C. J. Oh, S. O. Lee, H. S. Yang, T. J. Ha, M. J. Kim, Selective leaching of valuable metals from waste printed circuit boards, J. Air. Waste. Manage. Assoc. 53 (2003) 897-902. https://doi.org/10.1080/10473289.2003.10466230

[76] P. N. Işildar, A. van de Vossenberg, J. Rene, E. R. van Hullebusch, E. D. Lens, Biorecovery of metals from electronic waste in Sustainable Heavy Metal Remediation, Springer, 2017, pp. 241-278. https://doi.org/10.1007/978-3-319-61146-4_8

[77] H. Yang, J. Liu, J. Yang, Leaching copper from shredded particles of waste printed circuit boards, J. Hazard. Mater. 187 (2011) 393-400. https://doi.org/10.1016/j.jhazmat.2011.01.051

[78] D. Das, S. Mukherjee, M. G. Chaudhuri, Studies on leaching characteristics of electronic waste for metal recovery using inorganic and organic acids and base, Waste. Manag. Res. 39 (2021) 242-249. https://doi.org/10.1177/0734242X20931929

[79] T. E. Lister, P. Wang, A. Anderko, Recovery of critical and value metals from mobile electronics enabled by electrochemical processing, Hydrometallurgy. 149 (2014) 228-237. https://doi.org/10.1016/j.hydromet.2014.08.011

[80] I. F. F. Neto, H. M. V. M. Soares, Sequential separation of Ag, Al, Cu and Pb from a multi-metal leached solution using a zero waste technology, Sep. Sci. Technol. 53 (2018) 2961-2970. https://doi.org/10.1080/01496395.2018.1482342

[81] Ž. Kamberović, M. Ranitović, M. Korać, Z. Andjić, N. Gajić, J. Djokić, S. Jevtić, Hydrometallurgical process for selective metals recovery from waste-printed circuit boards, Metals, 8 (2018) 441. https://doi.org/10.3390/met8060441

[82] N. Naseri Joda F. Rashchi, Recovery of ultra-fine grained silver and copper from PC board scraps, Sep. Purif. Techno. 92 (2012) 36-42. https://doi.org/10.1016/j.seppur.2012.03.022

[83] I. F. F. Neto, C. A. Sousa, M. S. C. A. Brito, A. M. Futuro, H. M. V. M. Soares, A simple and nearly-closed cycle process for recycling copper with high purity from end life printed circuit boards, Sep. Purif. Technol. 164 (2016) 19-27. https://doi.org/10.1016/j.seppur.2016.03.007

[84] A. Serpe, A. Rigoldi, C. Marras, F. Artizzu, M. Laura Mercuri, P. Deplano, Chameleon behaviour of iodine in recovering noble-metals from WEEE: towards sustainability and 'zero' waste, Green Chem. 17 (2015) 2208-2216. https://doi.org/10.1039/C4GC02237H

[85] E. Y. Yazici, H. Deveci, Extraction of metals from waste printed circuit boards (WPCBs) in H2SO4-CuSO4-NaCl solutions, Hydrometallurgy, 139 (2013) 30-38. https://doi.org/10.1016/j.hydromet.2013.06.018

[86] P. Cyganowski, K. Garbera, A. Leśniewicz, J. Wolska, P. Pohl, D. Jermakowicz-Bartkowiak, The recovery of gold from the aqua regia leachate of electronic parts using a core-shell type anion exchange resin, J. Saudi Chem. Soc. 21 (2017) 741-750. https://doi.org/10.1016/j.jscs.2017.03.007

[87] P. P. Sheng T. H. Etsell, Recovery of gold from computer circuit board scrap using aqua regia, Waste. Manag. Res. 25 (2007) 380-383. https://doi.org/10.1177/0734242X07076946

[88] Y. J. Park, D. J. Fray, Recovery of high purity precious metals from printed circuit boards, J. Hazard. Mater. 164 (2009) 1152-1158. https://doi.org/10.1016/j.jhazmat.2008.09.043

[89] C. M. Maguyon, M. C. C., Alfafara, C. G., Migo, V. P., Movillon, J. L. Rebancos, Recovery of copper from spent solid printed-circuit-board (PCB) wastes of a PCB manufacturing facility by two-step sequential acid extraction and electrochemical deposition, J. Environ. Sci. Manag. 15 (2012) 17-27.

[90] S. R. La Brooy, H. G. Linge, G. S. Walker, Review of gold extraction from ores, Miner Eng, 7 (1994) 1213-1241. https://doi.org/10.1016/0892-6875(94)90114-7

[91] S. Syed, Recovery of gold from secondary sources-A review, Hydrometallurgy. 115-116 (2012) 30-51. https://doi.org/10.1016/j.hydromet.2011.12.012

[92] J. Cui, L. Zhang, Metallurgical recovery of metals from electronic waste: A review, J. Hazard. Mater. 158 (2008) 228-256. https://doi.org/10.1016/j.jhazmat.2008.02.001

[93] A. Quinet, P., Proost, J., Van Lierde, Recovery of precious metals from lectronic scrap by hydrometallurgical processing routes, Miner. Metall. 8Process. 22 (2005) 17-22. https://doi.org/10.1007/BF03403191

[94] J. Zhou, P. Zheng, Z. Tie, Technological process for extracting gold, silver and palladium from electronic industry waste, CN Patent 1603432A (C22B 11/00). (2005)

[95] Y. Lu, Q. Song, Z. Xu, Integrated technology for recovering Au from waste memory module by chlorination process: Selective leaching, extraction, and distillation, J. Clean. Prod. 161 (2017) 30-39. https://doi.org/10.1016/j.jclepro.2017.05.033

[96] H. M. Xu Qu, Chen Donghui, Chen Liang, Iodine leaching process for recovery of gold from waste PCB, Chin. J. Environ. Eng. 3 (2009) 911-914

[97] M. H. Xu, Q., Chen, D. H., Chen, L., Huang, Gold leaching from waste printed circuit board by iodine process, Non-ferr. Met. 62 (2010) 88-90.

[98] J. G. Huddleston, A. E. Visser, W. M. Reichert, H. D. Willauer, G. A. Broker, R. D. Rogers, Characterization and comparison of hydrophilic and hydrophobic room temperature ionic liquids incorporating the imidazolium cation Green Chem. 3 (2001) 156-164. https://doi.org/10.1039/b103275p

[99] T. Makanyire, S. Sanchez-Segado, A. Jha, Separation and recovery of critical metal ions using ionic liquids, Adv. Manuf. 4 (2016) 33-46. https://doi.org/10.1007/s40436-015-0132-3

[100] D. Zhang, L. Dong, Y. Li, Y. Wu, Y. Ma, B. Yang, Copper leaching from waste printed circuit boards using typical acidic ionic liquids recovery of e-wastes' surplus

value, Waste Manage. 78 (2018) 191-197.
https://doi.org/10.1016/j.wasman.2018.05.036

[101] G. Inman, I. C. Nlebedim, D. Prodius, Application of ionic liquids for the recycling
and recovery of technologically critical and valuable metals, Energies. 15 (2022) 628.
https://doi.org/10.3390/en15020628

[102] C. Vallejos-Michea, Y. Barrueto, Y. P. Jimenez, Life cycle analysis of the ionic
liquid leaching process of valuable metals from electronic wastes, J. Clean. Prod. 348
(2022) 131357. https://doi.org/10.1016/j.jclepro.2022.131357

[103] Y. Barrueto, P. Hernández, Y. P. Jiménez, J. Morales, Properties and application of
ionic liquids in leaching base/precious metals from e-waste. A review.
Hydrometallurgy. 212 (2022) 105895. https://doi.org/10.1016/j.hydromet.2022.105895

[104] K. Kurniawan, S. Kim, J. Lee, Ionic liquids-assisted extraction of metals from
electronic waste, in Ionic Liquid-Based Technologies for Environmental
Sustainability, Elsevier, 2022, pp. 295-329. https://doi.org/10.1016/B978-0-12-
824545-3.00019-2

[105] A. R. Salvador, Líquidos iónicos a temperatura ambiente: un nuevo medio para las
reacciones químicas, Rev. Real Acad. Cienc. Exactas Fis. Nat. 102 (2008) 79-90.

[106] W. L. Choo M. I. Jeffrey, An electrochemical study of copper cementation of
gold(I) thiosulfate, Hydrometallurgy. 71 (2004) 351-362.
https://doi.org/10.1016/S0304-386X(03)00087-2

[107] M. D. Rao, K. K. Singh, C. A. Morrison, J. B. Love, Recycling copper and gold
from e-waste by a two-stage leaching and solvent extraction process, Sep. Purif.
Technol. 263 (2021) 118400. https://doi.org/10.1016/j.seppur.2021.118400

[108] F. J. Alguacil, P. Navarro, Non-dispersive solvent extraction of Cu(II) by LIX
973N from ammoniacal/ammonium carbonate aqueous solutions, Hydrometallurgy. 65
(2002) 77-82. https://doi.org/10.1016/S0304-386X(02)00093-2

[109] F. J. Alguacil, I. Garcia-Diaz, F. Lopez, and O. Rodriguez, Recycling of copper
flue dust via leaching-solvent extraction processing, Desalination. Water. Treat. 56
(2015) 1202-1207. https://doi.org/10.1080/19443994.2014.963148

[110] I. García-Díaz, F. A. López, F. J. Alguacil, Carbon nanofibers: A new adsorbent for
copper removal from wastewater, Metals. 8 (2018) 1-13.
https://doi.org/10.3390/met8110914

[111] F. J. Alguacil, I. Garcia-Diaz, F. Lopez, O. Rodriguez, Removal of Cr(VI) and
Au(III) from aqueous streams by the use of carbon nanoadsorption technology,
Desalination. Water. Treat. 63 (2017) 351-356. https://doi.org/10.5004/dwt.2017.0264

[112] F. J. Alguacil, La eliminación de metales tóxicos presentes en efluentes líquidos
mediante resinas de cambio iónico. Parte XIII: Zinc(II)/H+/Lewatit OC-1026, Rev. de
Metal. 56 (2020) 172. https://doi.org/10.3989/revmetalm.172

New Materials for a Circular Economy   Materials Research Forum LLC
Materials Research Foundations 149 (2023) 311-367   https://doi.org/10.21741/9781644902639-9

[113] F. J. Alguacil, P. Adeva, M. Alonso, Processing of residual gold (III) solutions via ion exchange, Gold Bull. 38 (2005) 9-13. https://doi.org/10.1007/BF03215222

[114] P. R. Jadhao, A. Pandey, K. K. Pant, K. D. P. Nigam, Efficient recovery of Cu and Ni from WPCB via alkali leaching approach, J. Environ. Manage. 296 (2021) 113154. https://doi.org/10.1016/j.jenvman.2021.113154

[115] V. Rai, D. Liu, D. Xia, Y. Jayaraman, J.-C. P. Gabriel, Electrochemical approaches for the recovery of metals from electronic waste: a critical review, Recycling. 6, (2021) 53. https://doi.org/10.3390/recycling6030053

[116] A. E. Lewis, Review of metal sulphide precipitation, Hydrometallurgy. 104 (2010) 222-234. https://doi.org/10.1016/j.hydromet.2010.06.010

[117] P. Paranjape, M. D. Yadav, Recent advances in the approaches to recover rare earths and precious metals from E-waste: A mini-review, Can. J. Chem. Eng. 101 (2023) 1043. https://doi.org/10.1002/cjce.24435

[118] J. Huisman, C. B. Boks, A. L. N. Stevels, Quotes for environmentally weighted recyclability (QWERTY): Concept of describing product recyclability in terms of environmental value, Int. J. Prod. Res. 41 (2003) 3649-3665. https://doi.org/10.1080/0020754031000120069

[119] A. V. M. Silveira, M. S. Fuchs, D. K. Pinheiro, E. H. Tanabe, D. A. Bertuol, Recovery of indium from LCD screens of discarded cell phones, Waste Manage. 45 (2015) 334-342. https://doi.org/10.1016/j.wasman.2015.04.007

[120] C.-H. Lee, M.-K. Jeong, M. Fatih Kilicaslan, J.-H. Lee, H.-S. Hong, S.-J. Hong, Recovery of indium from used LCD panel by a time efficient and environmentally sound method assisted HEBM, Waste Manage. 33 (2013) 730-734. https://doi.org/10.1016/j.wasman.2012.10.002

[121] E. Kim, M. Kim, J. Lee, B. D. Pandey, Selective recovery of gold from waste mobile phone PCBs by hydrometallurgical process, J. Hazard. Mater. 198, (2011) 206-215. https://doi.org/10.1016/j.jhazmat.2011.10.034

[122] A. Lewis, R. van Hille, An exploration into the sulphide precipitation method and its effect on metal sulphide removal, Hydrometallurgy, 81 (2006) 197-204. https://doi.org/10.1016/j.hydromet.2005.12.009

[123] S.-H. Hu, M.-Y. Xie, Y.-M. Hsieh, Y.-S. Liou, W.-S. Chen, Resource recycling of gallium arsenide scrap using leaching-selective precipitation, Environ. Prog. Sustain. Energy. 34 (2015) 471-475. https://doi.org/10.1002/ep.12019

[124] J. P. Rabatho, W. Tongamp, Y. Takasaki, K. Haga, A. Shibayama, Recovery of Nd and Dy from rare earth magnetic waste sludge by hydrometallurgical process, J. Mater. Cycles. Waste. Manag. 15 (2013) 171-178. https://doi.org/10.1007/s10163-012-0105-6

[125] G. Chi, M. C. Fuerstenau, J. O. Marsden, Study of Merrill-Crowe processing. Part I: solubility of zinc in alkaline cyanide solution, Int. J. Miner. Process. 49 (1997) 171-183. https://doi.org/10.1016/S0301-7516(96)00043-9

[126] F. T. Awadalla, G. M. Ritcey, Recovery of gold from thiourea, thiocyanate, or thiosulfate solutions by reduction-precipitation with a stabilized form of sodium borohydride, Sep. Sci. Technol. 26 (1991) 1207-1228. https://doi.org/10.1080/01496399108050525

[127] Y.-F. Huang, S.-L. Chou, S.-L. Lo, Gold recovery from waste printed circuit boards of mobile phones by using microwave pyrolysis and hydrometallurgical methods, Sustain. Environ. Res. 32 (2022) 6. https://doi.org/10.1186/s42834-022-00118-x

[128] H. Mahandra, F. Faraji, A. Azizitorghabeh, A. Ghahreman, Selective extraction and recovery of gold from complex thiosulfate pregnant leach liquor using cyphos IL 101, Ind. Eng. Chem. Res. 61 (2022) 5612-5619. https://doi.org/10.1021/acs.iecr.2c00388

[129] M. Wang, Q. Wang, J. Wang, R. Liu, G. Zhang, Y. Yang, Homogenous liquid-liquid extraction of Au(III) from acidic medium by ionic liquid thermomorphic systems, ACS Sustain. Chem. Eng. 9 (2021) 4894-4902. https://doi.org/10.1021/acssuschemeng.1c00497

[130] F. J. Alguacil, C. Caravaca, A. Cobo, S. Martinez, The extraction of gold(I) from cyanide solutions by the phosphine oxide Cyanex 921, Hydrometallurgy. 35 (1994) 41-52. https://doi.org/10.1016/0304-386X(94)90016-7

[131] F. Yang, F. Kubota, Y. Baba, N. Kamiya, M. Goto, Selective extraction and recovery of rare earth metals from phosphor powders in waste fluorescent lamps using an ionic liquid system, J. Hazard. Mater. 254-255 (2013) 79-88. https://doi.org/10.1016/j.jhazmat.2013.03.026

[132] M. Gergoric, C. Ekberg, B.-M. Steenari, T. Retegan, Separation of heavy rare-earth elements from light rare-earth elements via solvent extraction from a neodymium magnet leachate and the effects of diluents, J. Sustain. Metall. 3 (2017) 601-610. https://doi.org/10.1007/s40831-017-0117-5

[133] X. Sun, J. R. Bell, H. Luo, S. Dai, Extraction separation of rare-earth ions via competitive ligand complexations between aqueous and ionic-liquid phases, Dalton Trans. 40 (2011) 8019. https://doi.org/10.1039/c1dt10873e

[134] J. Park, Y. Jung, P. Kusumah, J. Lee, K. Kwon, C. Lee, Application of Ionic Liquids in Hydrometallurgy, Int. J. Mol. Sci. 15 (2014) 15320-15343. https://doi.org/10.3390/ijms150915320

[135] B. Volesky, Biosorption and me, Water Res. 41 (2007) 4017-4029. https://doi.org/10.1016/j.watres.2007.05.062

[136] T. G. Chapman, F.W. McQuiston, U.S. Patent 2545239 (C01G 5/00), (1951)

[137] L. M. T. Rosa, W. G. Botero, J. Braga do Carmo, G. V. M. Gabriel, W. R. Waldman, A. D.M. Cavagis, D. Goveia, L. Camargo de Oliveira, Application of natural organic residues in the remediation of metals from e-waste, Environ. Technol. Innov. 27 (2022) 102452. https://doi.org/10.1016/j.eti.2022.102452

[138] X. Zhang, H. Li, M. Ye, H. Zhang, G. Wang, Y. Zhang, Bacterial cellulose hybrid membrane grafted with high ratio of adipic dihydrazide for highly efficient and selective recovery of gold from E-waste, Sep. Purif. Technol. 292 (2022) 121021. https://doi.org/10.1016/j.seppur.2022.121021

[139] S. C. R. Santos, H. A. M. Bacelo, R. A. R. Boaventura, C. M. S. Botelho, Tannin-adsorbents for water decontamination and for the recovery of critical metals: current state and future perspectives, Biotechnol. J. 14 (2019) 1900060. https://doi.org/10.1002/biot.201900060

[140] Z. Dong, T. Jiang, B. Xu, Y. Yang, Q. Li, recovery of gold from pregnant thiosulfate solutions by the resin adsorption technique, Metals. 7 (2017) 555. https://doi.org/10.3390/met7120555

[141] X. Hérès, V. Blet, P. Di Natale, A. Ouaattou, H. Mazouz, D. Dhiba, F. Cuer, Selective extraction of rare earth elements from phosphoric acid by ion exchange resins, Metals. 8 (2018) 682. https://doi.org/10.3390/met8090682

[142] A. Mecucci, K. Scott, Leaching and electrochemical recovery of copper, lead and tin from scrap printed circuit boards, J. Chem. Technol. Biotechnol. 77 (2002) 449-457. https://doi.org/10.1002/jctb.575

[143] A. Fathima, J. Y. B. Tang, A. Giannis, I. M. S. K. Ilankoon, M. N. Chong, Catalysing electrowinning of copper from E-waste: A critical review, Chemosphere. 298 (2022) 134340. https://doi.org/10.1016/j.chemosphere.2022.134340

[144] Y. Chu, M. Chen, S. Chen, B. Wang, K. Fu, H. Chen, Micro-copper powders recovered from waste printed circuit boards by electrolysis, Hydrometallurgy. 156 (2015) 152-157. https://doi.org/10.1016/j.hydromet.2015.06.006

[145] E. Rudnik, E. Bayaraa, Electrochemical dissolution of smelted low-grade electronic scraps in acid sulfate-chloride solutions, Hydrometallurgy. 159 (2016) 110-119. https://doi.org/10.1016/j.hydromet.2015.11.008

[146] S. Fogarasi, F. Imre-Lucaci, Á. Imre-Lucaci, P. Ilea, Copper recovery and gold enrichment from waste printed circuit boards by mediated electrochemical oxidation, J. Hazard. Mater. 273 (2014) 215-221. https://doi.org/10.1016/j.jhazmat.2014.03.043

[147] S. Ilyas, M. A. Anwar, S. B. Niazi, M. Afzal Ghauri, Bioleaching of metals from electronic scrap by moderately thermophilic acidophilic bacteria, Hydrometallurgy. 88 (2007) 180-188. https://doi.org/10.1016/j.hydromet.2007.04.007

[148] L. E. Macaskie, N. J. Creamer, A. M. M. Essa, N. L. Brown, A new approach for the recovery of precious metals from solution and from leachates derived from

electronic scrap, Biotechnol. Bioeng. 96 (2007) 631-639.
https://doi.org/10.1002/bit.21108

[149] A. N. Mabbett, D. Sanyahumbi, P. Yong, L. E. Macaskie, Biorecovered Precious Metals from Industrial Wastes: Single-Step Conversion of a Mixed Metal Liquid Waste to a Bioinorganic Catalyst with Environmental Application, Environ Sci Technol. 40 (2006) 1015-1021. https://doi.org/10.1021/es0509836

[150] A. Marra, A. Cesaro, E. R. Rene, V. Belgiorno, P. N. L. Lens, Bioleaching of metals from WEEE shredding dust, J. Environ. Manage. 210 (2018) 180-190. https://doi.org/10.1016/j.jenvman.2017.12.066

[151] M. Vera, A. Schippers, W. Sand, Progress in bioleaching: fundamentals and mechanisms of bacterial metal sulfide oxidation-part A, Appl. Microbiol. Biotechnol., 97 (2013) 7529-7541. https://doi.org/10.1007/s00253-013-4954-2

[152] D. Morin, A. Lips, T. Pinches, J. Huisman, C. Frias, A. Norberg, E. Forssberg, BioMinE - Integrated project for the development of biotechnology for metal-bearing materials in Europe, Hydrometallurgy. 83 (2006) 69-76. https://doi.org/10.1016/j.hydromet.2006.03.047

[153] H. L. Ehrlich, Microbes and metals, Appl. Microbiol. Biotechnol, 48 (1997) 687-692. https://doi.org/10.1007/s002530051116

[154] G. J. Olson, J. A. Brierley, C. L. Brierley, Bioleaching review part B:, Appl. Microbiol. Biotechnol. 63 (2003) 249-257. https://doi.org/10.1007/s00253-003-1404-6

[155] H. Brandl, S. Lehmann, M. A. Faramarzi, D. Martinelli, Biomobilization of silver, gold, and platinum from solid waste materials by HCN-forming microorganisms, Hydrometallurgy. 94 (2008) 14-17. https://doi.org/10.1016/j.hydromet.2008.05.016

[156] G. Liang, Y. Mo, Q. Zhou, Novel strategies of bioleaching metals from printed circuit boards (PCBs) in mixed cultivation of two acidophiles, Enzyme. Microb. Technol. 47 (2010) 322-326. https://doi.org/10.1016/j.enzmictec.2010.08.002

[157] S. Chen, Y. Yang, C. Liu, F. Dong, B. Liu, Column bioleaching copper and its kinetics of waste printed circuit boards (WPCBs) by Acidithiobacillus ferrooxidans, Chemosphere. 141 (2015) 162-168. https://doi.org/10.1016/j.chemosphere.2015.06.082

[158] A. Islam, T. Ahmed, Md. R. Awual, A. Rahman, M. Sultan, A. Abd Aziz, M. Uddin Monir, S. Hwa Teo, M. Hasan Advances in sustainable approaches to recover metals from e-waste-A review, J. Clean. Prod. 244 (2020) 118815. https://doi.org/10.1016/j.jclepro.2019.118815

[159] M. E. Díaz-Martínez, R. Argumedo-Delira, G. Sánchez-Viveros, A. Alarcón, Ma. R. Mendoza-López, microbial bioleaching of Ag, Au and Cu from printed circuit boards of mobile phones, Curr. Microbiol. 76 (2019) 536-544. https://doi.org/10.1007/s00284-019-01646-3

[160] R. Argumedo-Delira, M. J. Gómez-Martínez, B. J. Soto, Gold bioleaching from printed circuit boards of mobile phones by aspergillus niger in a culture without agitation and with glucose as a carbon source, Metals. 9 (2019) 521. https://doi.org/10.3390/met9050521

[161] S. Akbari, A. Ahmadi, Recovery of copper from a mixture of printed circuit boards (PCBs) and sulphidic tailings using bioleaching and solvent extraction processes, Chem. Eng. Process. 142 (2019) 107584. https://doi.org/10.1016/j.cep.2019.107584

[162] K. Brandl G. Plitas, C. N. Mihu, C. Ubeda, T. Jia, M. Fleisher, B. Schnabl, R. P. DeMatteo, E.G. Pamer., Vancomycin-resistant enterococci exploit antibiotic-induced innate immune deficits, Nature, 455 (2008) 804-807. https://doi.org/10.1038/nature07250

[163] A. Kumar, H. S. Saini, S. Kumar, Bioleaching of gold and silver from waste printed circuit boards by pseudomonas balearica SAE1 isolated from an e-waste recycling facility, Curr Microbiol, 75 (2018) 194-201. https://doi.org/10.1007/s00284-017-1365-0

[164] Z. Yuan, Z. Huang, J. Ruan, Y. Li, J. Hu, R. Qiu, Contact behavior between cells and particles in bioleaching of precious metals from waste printed circuit boards, ACS Sustain. Chem. Eng. 6 (2018) 11570-11577. https://doi.org/10.1021/acssuschemeng.8b01742

[165] A. Luyima, H. Shi, L. Zhang, Leaching studies for metals recovery from waste printed wiring boards, JOM, 63 (2011) 38-41. https://doi.org/10.1007/s11837-011-0135-x

[166] M. Xia, P. Bao, A. Liu, M.Wang, L. Shen, R. Yu, Y. Liu, M. Chen, J. Li, X. Wu, G. Qiu, W. Zeng, Bioleaching of low-grade waste printed circuit boards by mixed fungal culture and its community structure analysis, Resour. Conserv. Recycl. 136 (2018) 267-275. https://doi.org/10.1016/j.resconrec.2018.05.001

[167] S. Wang, L. Chen, X. Zhou, W. Yan, R. Ding, B. Chen, C. Wang, F. Zhao, Enhanced bioleaching efficiency of copper from printed circuit boards without iron loss, Hydrometallurgy. 180 (2018) 65-71. https://doi.org/10.1016/j.hydromet.2018.07.010

[168] A. Ortego, A. Valero, R. Magdalena, M. Iglesias-Embil, TREASURE project - D3.2: Report on disassemblability analysis, Zaragoza, 2022.

[169] A. Ortego, A. Valero, A. Valero, M. Iglesias, Toward material efficient vehicles: ecodesign recommendations based on metal sustainability assessments, SAE Int. J. Mater. 11 (2018) 213-228. https://doi.org/10.4271/05-11-03-0021

[170] VOLVO, "Integrated starter generator," 2001. Retrieved from: https://www.media.volvocars.com/global/en-gb/media/pressreleases/5278

New Materials for a Circular Economy
Materials Research Foundations 149 (2023) 368-392

Materials Research Forum LLC
https://doi.org/10.21741/9781644902639-10

Chapter 10

Biomaterials with Bioactive features Developed using Ulvan as Key Component within a Bioeconomy Approach

M.D. Torres[1], H. Domínguez[1,*]

[1]CINBIO, Department of Chemical Engineering, Universidade de Vigo (Campus Ourense), Edificio Politécnico, As Lagoas, 32004, Ourense, Spain

herminia@uvigo.es

Abstract

Ulvan is a bioactive marine sulphated polysacharide that can be extracted from green algae. This biopolymer can form thermoreversible gels depending on the chemical composition and structure, showing a great potential. The cold-set gelling properties and unique shear-thickening fluid properties of this biopolymer could be valuable for the exploration of ulvans as a new source of water-soluble biodegradable gelling polysaccharides. The lack of cytotoxicity and specific properties make these gels also useful for novel biomedical applications. This chapter presents an overview on the extraction conditions of ulvans, which are key since they can affect the macromolecular distribution, rheological or textural, and bioactive properties of the isolated biopolymer, and offer a wide application field. Their chemical properties, atypical gelling performance and the corresponding mechanical characteristics are addressed. Their potential to formulate biodegradable hydrogels, to be used in 3D printing, and some representative applications will be covered.

Keywords

Green Seaweeds, Sulphated Polysaccharides, Bioactive, Biopolymers, Printable Hydrogels

Contents

New Materials for a Circular Economy   Materials Research Forum LLC
Materials Research Foundations 149 (2023) 368-392   https://doi.org/10.21741/9781644902639-10

1.   Introduction

Ulva is a widely distributed cosmopolitan, green macroalgal genus, with well established cultivation and usually destined for food uses. These species can also be collected from blooms, which are increasingly frequent and undesirable events that could represent an underutilized feedstock to obtain useful and valuable products [1-3]. There is a need to provide the tools for the effective utilization of macroalgal blooms, particularly Ulva sp. and other chlorophyta, which are predominantly problematic [2]. The major components of this group of seaweeds are carbohydrates with storage function, such as starch, with structural function, such as ulvan, and other minor sulfated polysaccharides, and it also contains proteins, minerals, chlorophylls, carotenoids [1], polyunsaturated fatty acids with high ω-3/ω-6 ratio [4] and other interesting compounds such as squalene, α-tocopherol, and phytosterols [5].

One of the exclusive and most attractive compounds found in this genus is ulvan, which is a polyanionic sulfated polysaccharide consisting of D-glucuronic acid, L-iduronic acid, L-rhamnose-3-sulfate and xylose (Figure 1), accounting for 9-40 % of green seaweed dry weight. Ulvan has similar structural features as glycosaminoglycans, such as chondroitin sulfate, hyaluronan, dermatan sulfate, and heparan sulfate [6].

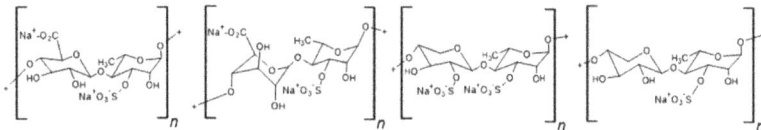

Figure 1. Basic structure of ulvans [7].

The chemical composition and structural properties of ulvans, such as charge density and molecular weight, are determined by species, environmental factors and extraction procedures, and these characteristics define the biological properties. Ulvan presents a variety of activities including radical scavenging and reducing power, antimicrobial, antiviral, immunomodulatory, anti-inflammatory, antitumoral, anticoagulation, prebiotic, antidiabetic, antihypertensive, blood lipid level reduction and regenerative activities [8-24].

Since the cultivation technology of Ulva sp. is well established, the interest in diversifying its commercial uses has increased [1]. Ulvans can form thermoreversible gels and can be used as thickeners, stabilizers, and materials useful for food, chemical, cosmetic [19] and

New Materials for a Circular Economy   Materials Research Forum LLC
Materials Research Foundations 149 (2023) 368-392   https://doi.org/10.21741/9781644902639-10

biomedical applications, including drug delivery, tissue engineering, and wound healing [6, 25, 26]. They exhibit high biocompatibility and biodegradability when compared with synthetic ones and can be combined with other polymers and with active substances to prepare various forms of dressings. The major difficulties for their incorporation in biomedical uses are related to the difficulties in standardizing their microstructure and physicochemical properties due to batch-to-batch variations. In addition, since the evaluation criteria for these biomaterials have not been well established [22]. Ulvan gels present a great potential, but have currently fewer commercial applications than other algal hydrocolloids. Therefore, the advances in their extraction and formulation processes are needed to attain a sustainable utilization of this resource.

2.   Extraction strategies

Extraction has a significant impact on the chemical composition and properties of ulvan [11, 20, 24, 27], i.e., a HCl-based extraction led to lower molecular weight ulvan with less charge and antiviral properties than an ammonium oxalate-based extraction [28].

The conventional industrial methods for the extraction of seaweed hydrocolloids following traditional protocols are well optimized and established, but improvements regarding the high consumption of solvent and energy and the impact on the environment could be improved. There is a growing interest to include innovative technologies into greener and more sustainable processes with enhanced efficiency over conventional processes by increasing the yields of hydrocolloids extraction, decreasing the time and energy consumption, and maintaining the products properties [23, 29].

The most promising innovative technologies described in literature involve the use of chemical- and organic solvents-free processes based on intensification techniques, such as ultrasound, microwave, enzyme, pulsed electric field assistance and subcritical water extraction. They are aimed at improving the yield of polysaccharides, and also to achieve targeted hydrocolloids for specific purposes and functionalities. Process intensification has emerged as a tool to develop equipment and techniques, leading to compact, smaller and cleaner plants with increased efficiency, products quality, safety and improved control and automation, and also with decreased byproducts formation, capital cost and energy consumption [30, 31].

Compared to conventional extraction methods, innovative assisted technologies, applied either alone or in combination [32, 33], enhanced yields maintaining sustainability [29, 34] during the extraction of biopolymers [34] and phenolics [35]. Even when most of these novel technologies offer lower energy consumption and greener character, the sustainability of the overall process should be assessed.

The pre-treatments and extraction methods employed to extract hydrocolloids are mainly aimed to enhance extraction yields. Frequently, a previous extraction with solvents is required, i.e. de-pigmentation with hexane, soaking in 80-95% ethanol to remove soluble impurities, such as free sugars, amino acids, phenolics, and low molecular weight compounds [11, 20, 36, 37].

New Materials for a Circular Economy   Materials Research Forum LLC
Materials Research Foundations 149 (2023) 368-392   https://doi.org/10.21741/9781644902639-10

Ulvan is usually extracted in hot water (65-90 °C) for 1-20 h [10, 11, 36, 38, 39], and sometimes in media with chelating agents added to breakdown the ionic bonds between Ca^{2+} and the compounds of the cell wall [37]. Alternatively, acid or enzymatic assistance have been described [20, 40, 41]. The possibility of using seawater has been reported [42]. This alternative can be attractive for the direct processing of seaweed biomass. When the feedstock is stored before use, low drying temperatures or times are required to maintain the quality of other bioactives such as carotenoids or phenolic compounds [43], although are less relevant for polysaccharides.

The soluble ulvan can be recovered from the extract by precipitation with ethanol [11, 37, 38, 44]. Either HCl or NaOH have been combined with ethanol precipitation to recover the polysaccharides. However, this stage is not specific and the crude ulvan precipitated also contains proteins and inorganic material [40]. Alternatively, membrane technology has been proposed for the recovery of solubilized ulvan fractions [20]. Further fractionation by chromatographic separation has been proposed for characterization purposes and to obtain more active fractions [11]. The water-soluble sulfated polysaccharides contain neutral sugars (rhamnose, glucose, galactose, xylose and arabinose), uronic acids and proteins [10, 11, 37, 45]. The resulting product shows sulphate content in the range 6-20% and molecular weight 50-1800 kDa depending on the extraction and purification procedure [10, 27, 37].

Different innovative strategies to enhance the performance of the extraction process have been proposed (Table 1), the operational conditions are species dependent and should be optimized case by case.

Table 1. Examples of innovative intensification techniques for the extraction of ulvan and sulphated polysaccharides from Ulva~~nulvan~~ sp.

Assisted technique	Solvent	Biological activity	References
Microwave	Distilled water 80% ethanol 50-600 W 80-180 °C, 10 min-2 h 0.05-0.1 M HCl	Functional properties Antioxidant, Pancreatic lipase inhibition	44, 46, 47, 48
Ultrasound	53 kHz, 180 W, 66 °C, 40-60 min	Antioxidant Immunostimulant	9, 49
Enzyme assistance	Protease, cellulases, pectinases and mixtures 6% enzyme, 50 °C, 2-3 h	Antiviral Antioxidant Anti-inflammatory Anti-ageing	8, 17, 20, 32
Pulsed electric fields	1-18 kV, 1.2-3 Hz, 0.96-5000 µs 0.04% NaCl solution, conductivity of 1250 µS/cm	Antioxidant Skin whitening Anti-aging	32, 50, 51, 52
Subcritical water	Distilled or seawater 130-180 °C, 40-80 min	Antioxidant	4, 53

Microwave-assisted extraction can simplify the extraction and recovery process [54], because ulvan is easily released into the solvent without requiring chelating agents to disrupt the ulvan interactions with the cell-wall [46]. Microwave-assistance could highly accelerate the reaction rate during depolymerization of ulvan, cleaving glycosidic linkages without breaking significant structural units [55]. Yuan et al. [47] reported that during microwave assisted acid hydrothermal extraction the sulfate content increased with temperature and acid concentration, but the molecular weight decreased, therefore affecting both functional (water and oil holding capacities, foaming capacity and stability) and biological properties.

Ultrasound-assisted extraction required relatively short times during water extraction [9], but also combination with chemical processes, i.e., after alkaline extraction, have been proposed [49]. Limitations derived from the possible degradation of the extracted compounds at extreme conditions that could lead to depolymerisation of hydrocolloids [56] should be considered.

Enzyme-assisted extraction, using hydrolytic enzymes, mainly proteases and carbohydrases, and exhibit improved extraction yields compared to conventional techniques positively affect the quality of the extracted hydrocolloids avoiding chemical residues. Enzyme treatment has been frequently applied with commercial formulations including cellulases, pectinases and both neutral and alkaline endo-proteases provided significant enhancement of extraction yields, neutral sugar and protein contents compared to the control [8, 32, 20]. Among different carbohydrates and sulfatases a polysaccharide lyase family and an ulvan lyase were particularly effective to degrade *U. lactuca* cell walls and to release mono- and oligosaccharides [57]. In addition, marine microbial enzymes have been proposed [58]. Enzyme-assistance can be valid both for extraction and for depolymerization of protein and carbohydrate [19, 32]. Hung et al. [21] proposed hot water extraction assisted by multiple hydrolases, including ulvanolytic enzymes, amylase, cellulase, and xylanase, produced by marine bacteria, to obtain ulvan oligosaccharides from *Ulva lactuca*, further recovered by ultrafiltration. The influence of hydrolytic treatments on both the biological and functional properties should be optimized; the viscosity values were lower for the ulvan extracted from the enzyme assisted process than for those biopolymers from ultrasound and from hot water extraction, but the antiradical properties were improved [33].

Subcritical water refers to water at temperatures under 374 °C, usually in the range 110 to 230 °C, and proved suitable for the extraction of compounds of lower polarity than those extracted with water at lower temperatures, and also favor depolymerization of polysaccharide, protein and polyphenolics fractions, thus enhancing the biological properties of the final product [4]. The yields attained are usually higher than with other technologies, being almost quantitative after 2 h at 130 °C [53]. The type of compounds regarding different polarity, and the molecular weights and viscosity of polysaccharides can be controlled by varying the extraction temperature [44], which is a relevant variable [48]. The influence of temperature in this range is important also during further processing,

New Materials for a Circular Economy   Materials Research Forum LLC
Materials Research Foundations 149 (2023) 368-392   https://doi.org/10.21741/9781644902639-10

after sterilization for 15 and 30 min at 121 °C the extracted ulvan completely lost its activity, indicating that it is thermolabile and susceptible to very high temperatures [39].

Pulsed electric field assisted extraction could cut the energy costs associated to drying and deashing and allowed the extraction of different organic compounds, but one of the associated challenges is in relation to the high salt content of marine macroalgal biomass. Levkov et al. [51] combined the pulse generator with the sliding electrodes to provide a continuous liquid phase extraction during electroporation. This technology has been proposed for inducing cell permeabilization of the fresh biomass and to allow deashing from Ulvan sp. [51], removing five of the major elements (K, Mg, Na, P and S) [59]. It was also proved to be suitable for macroalgal biomass fractionation and valorization of other Ulva fractions, such as protein [59, 60], phenolics [52] and starch [50]. Besides, this procedure was adequate to remove salts, proteins and other small molecules concentrating other fractions in the solid residue remaining after pressing [50]. This approach provided similar results to the traditional hot water extraction although allowed shortening by more than four the extraction time [52]. High shear homogenization for up to 35 °C and 40 min provided the highest protein and carbohydrate yields from *Ulva lactuca* and an energy reduction compared to osmotic shock, enzymatic incubation and pulsed electric field [32].

Combination of technologies

The sequential or simultaneous application of conventional and innovative technologies can overcome the extraction limitations, aiming at enhanced yields with minimal degradation. In addition, the synergistic effects can provide additional benefits on the process performance and efficiency, such as those resulting from the combination of microwaves and ultrasound with enzyme assistance. The use of enzymes improved the yields of extraction, which were further improved by the application of ultrasonic-enzyme treatment, and those treatments also showed radical scavenging ability over that extracted by hot water extraction and ultrasonic-assisted treatment [33]. Higher ulvan yield was obtained after combining enzymatic and chemical extraction procedures [61]. Microwave-assisted hydrothermal extraction was employed for the isolation of sulfated polysaccharides from green seaweeds, in only 10 min after reaching 160 °C [44]. The further saccharification of hydrothermally treated *Ulva pertusa* with cellulase, alfa-amylase, and beta-glucosidase could be performed in 3 h whereas longer processing time (around 24 h) were required for the untreated seaweed [48].

In addition to the methods of extraction, molecular modification, such as degradation, carboxymethylation, and sulfonation could alter the biological activities of polysaccharides [23, 25]. Depolymerization is the most widely used [62]. Different biological properties of ulvans can be enhanced after depolymerization to a molecular weight in the range 7-670 kDa [7, 19, 20, 22, 25]. Oligosaccharides can be obtained by acid and by enzymatic hydrolysis, by radical depolymerization with hydrogen peroxide [7, 22], or by chemical- and solvent-free processes by ion exchange resin depolymerization [17, 25].

Apart from ulvan, other high valuable products can be obtained from Ulva sp. green seaweed, such as protein [63], phenolics [35], chlorophylls and carotenoids (lutein, β-

New Materials for a Circular Economy   Materials Research Forum LLC
Materials Research Foundations 149 (2023) 368-392   https://doi.org/10.21741/9781644902639-10

carotene, neoxanthin, β-cryptoxanthin, violaxanthin, antheraxanthin and zeaxanthin) [1, 50, 64-66]. The joint valorization of the different fractions in the framework of a seaweed biorefinery is desirable [41, 37], particularly with cleaner, efficient and environmentally friendly technologies. Most of the biomass produced today from *Ulva sp.* is lost, but it could be the feedstock for the biorefinery either alone [51] or in combination with microalgae [67]. Seaweeds are advantageous feedstocks because polysaccharides can be obtained under milder, more economic processing conditions than those reported for lignocellulosic materials [68]. However, integration of macroalgae into biorefinery network requires new processing sustainable technologies that will lead to energy efficient and zero waste [26, 50, 64]. Different successful examples can be recently found for producing a salt-rich product, ulvan and protein fractions [36, 41, 69].

3.   Mechanical properties

Rheological features of ulvans in terms of viscosity and viscoelasticity exhibited high variability depending on the green seaweed species used as source of this biopolymer [14], indicating that further understanding of the structure-activity relationships are dramatically relevant. The bead conformation of ulvan decreases its intermolecular interactions resulting in the low viscosity of its solutions, as well as influencing gel strength and activity. The influence of pH over the solution properties of ulvans is a convenient mechanism by which to fine-tune its rheology to suit specific applications, since in high pH solutions (~13) ulvan presents a more open conformation rising the intermolecular interactions that give higher viscosities and stronger gel properties [7].

Table 2 summarizes representative current examples of ulvans extracted from different green seaweeds sources, with the corresponding biopolymer-based formulations developed, the mechanical properties analysed and the selected matrices for food and therapeutic applications.

Morelli et al. [76] studied the exploitation of ulvan form *Ulva* spp. for potential biomedical applications, due to the chemical versatility and widely ascertained bioactivity of this biomaterial. This work was the first successful attempt of development of ulvan-based hydrogels displaying thermogelling behaviour. Ulvan was provided with thermogelling properties by grafting poly(N-isopropylacrylamide) chains onto its backbone as thermosensitive component. They identified sol-gel transition of the copolymer at 30-31 °C, thus indicating the feasibility of ulvan for being used as in-situ hydrogel forming systems for biomedical applications.

Tsubaki et al. [44] assessed the impact of microwave hydrothermal treatment on the properties of extracted ulvans from different *Ulva* ssp. species. The findings of this work indicated that molecular weights and viscosity values of the extracted biopolymers could be controlled by varying the extraction temperature. They observed that viscosity decreased with increasing processing temperature. Ulvan from *Ulva meridionalis* exhibited 1.42 ± 0.08 mPa s (100 °C), while that from *Ulva ohnoi* exhibited 2.18 ± 0.09 mPa s (120 °C).

Table 2. Examples of the mechanical features of matrices of biomaterials formulated with ulvans from different green seaweeds.

Ulvan source	Formulations	Mechanical properties	Matrices	References
Ulva fenestrata	Ulvan/chitosan	η: $1\text{-}10^4$ Pa s, 25°C G'_0, G_0'': 10^3, 100 Pa	Hydrogel	70
Ulva lactuca	Ulvan (5-10%); Glycerol (2%)	η: 200-700 cP, Room temperature	Wound dressing	6
Ulva genus	Ulvan (2%)	Blade, G': 0.1–6.6 Pa Filament,G':22.7-74.2Pa	Food solution	14
Ulva fasciata	Ulvan (2.4%) / PVA (1:2)	TS: 11.6–20.4 MPa	Nanofiber	71
Ulva rigida	Ulvan (10%) / PCA (1:10)	CME: 1.6 -4.9 MPa	Scaffold	72
Ulva lactuca	Ulvan (3%); Sorbitol or Glycerol (1-2%)	TS:0.5-3.5 MPa E: 10-45%	Smart film	73
Ulva linza	Ulvan (1%)	η: $0.01\text{-}10^3$ Pa s, 20°C G'_0, G_0'': 20, 1 Pa	Food solution	45
Ulva lactuca	Ulvan (1-4%)	η: 18 mPa, 25°C	Yogur	74
Ulva fasciata	Ulvan; Carrageenan; Glycerol (30%)	TS: 36.78-49.12 MPa G'_0, G_0'': 150, 50 Pa E: 9.3-11%	Edible film	75
Ulva spp.	Ulvan (4%)	η: $1\text{-}10^4$ Pa s, 15-40°C η*: $1\text{-}10^4$ Pa s,15-40°C	Biomedical	76
Ulva meridionalis Ulva ohnoi	Ulvan (1%)	η:1-2.5 mPa s, 100-180°C	Solution	44

CME: Compressive modulus of elasticity; E: Elongation at break; G': Elastic modulus; G'': Viscous modulus; η: Apparent viscosity; η*: Complex viscosity; PVA: Polyvinyl alcohol; PCA: Polycaprolactone; TS: Tensile strengh

Ganesan et al. [75] developed edible films with ulvan polysaccharides from *Ulva fasciata*, and also combinations with semirefined carragenans from *Kappaphycus alvarezii*. The molecular weights of tested biopolymers ranged between 210 kDa and 72 kDa in the presence of glycerol. These authors found that low molecular weight films presented better antioxidant activity, whereas high molecular weight films possessed adequate mechanical features.

Rheological studies conducted on ulvan from Ulva linza indicate that this biopolymer exhibited shear-thinning pseudoplastic behavior for a concentration range between 0.5 and 3% et al. [45]. Ulvan-based gelled matrices prepared at 1% of biopolymer content showed weak gels behaviour based on the measured storage and loss moduli. These authors also proposed a hydrophobically modified uronamide ulvan to increase the dynamic viscosity and improved the rheological properties with the introduction of octyl fatty chains by creating hydrophobic associations in the solution of the modified ulvan.

Shalaby & Amin [74] prepared symbiotic yogurt by using ulvan polysaccharide by different percentage as a prebiotic ingredient and using probiotic starter bacteria in the manufacture of set symbiotic yogurt. From all the results of physical and viscosity of symbiotic yogurt, they stated that the addition of ulvan (1-2%) as prebiotics improved the rheological properties of the resultant symbiotic yogurt by increasing the viscosity and firmness and reducing the syneresis rate. The viscous properties of the yogur increased adequately in the presence of ulvan due to the interactions between the biopolymer and milk protein and reduction of syneresis.

Guidara et al. [73] proposed the formulation of smart ulvan films from *Ulva lactuca* responsive to stimuli of plasticizer type and concentration as well as extraction conditions. They also reported that extracted ulvans displayed the presence of the peculiar chemical properties and self-aggregation molecular behavior, so it was proved the feasibility of these advantageous ulvans for being used as in-situ films forming systems. This strategy could hold promise to be utilized as a part of smart/intelligent food packaging or biomedical technologies.

Kidgell et al. [14] compared ulvans from blade (*Ulva australis*, *Ulva rigida*, *Ulva* sp. B or *Ulva* spp.) and filamentous (*Ulva flexuosa*, *Ulva compressa*, *Ulva prolifera* or *Ulva ralfsii*) *Ulva* species reporting high variability in the chemical, structural and mechanical features. Those biopolymers from blade species had higher yields (14.0–19.3%) and iduronic acid content (7-18 mol%), but lower molecular weight (Mw: 190-254 kDa) and elastic moduli (G': 0.1–6.6 Pa) when compared with filamentous species (yield: 7.2–14.6 %; iduronic acid: 4-7 mol%; molecular weight: 260-406 kDa; G': 22.7–74.2 Pa).

Madany et al. [71] reported the use of ulvan/polyvinyl alcohol for the fabrication of successful nanofibers by electrospining, as a promising candidate for biomedical applications. The ulvan-based nanofibers exhibited a range of specific tensile stress of 11.6–20.4 MPa. The elongation fracture elongation at fracture ranged from 15.0–27.0% and the range of stress was 24.5–26.5 MPa. These authors stated that the nano-size effect of the ulvan increased the mechanical properties of the generated nanofibrous hybrid mat and reduced the range of mechanical stress that leads to stable mechanical characteristics.

Kikinois et al. [72] found that ulvan materials can be an attractive natural choice for bone tissue engineering. They synthesized polycaprolactone/ulvan-based hybrid scaffolds, observing that all fabricated biomaterials contained 10% biopolymer exhibited sesired mechanical features. Hybrid scaffolds incorporating only ulvan in their polymer matrix, presented a faster differentiation rate, confirming the osteoinductive potential of ulvan towards the development of biomedical scaffolds for bone tissue regeneration applications.

Sulastri et al. [77] showed the potential of ulvan-based hydrogel films as wound dressing biomaterials. Hydrogels with ulvan content ranging from 5 to 10% were prepared, showing that higher biopolymer content involved higher viscosity values (200 - 700 cps), swelling degree (82 - 130% at 1 h), moisture content (24 - 18.4%), and the water vapour transmission (1856–2590 $g/m^2/24$ h). Authors stated that the swelling behavior and water vapor transmission of these films were great, indicating potential as a wound dressing biomaterial, supported by their antimicrobial and antioxidant properties.

Ren et al. [70] proposed the preparation of a natural polysaccharide-based hydrogel matrix made using ulvan dialdehyde, chitosan, dopamine and silver nanoparticles for chronic diabetic wound healing, with suitable mechanical properties. The rheological results indicated biomaterials with shear-thinning behaviour, whose rigidity and viscosity increased in the presence of different used additives, which allows better protection to the injured tissue. This smart hydrogel involves an easy and effective way for diabetic chronic wound management. Besides, it provides a new route for using *Ulva* as a valuable biomaterial for the global and large-scale manufacturing of valued added biomaterials.

4.   Printability possibilities

In recent years, sulfated polysaccharides from marine sources such as alginates or carrageenans have been rapidly picked up by scientists in the biomedical field as printable biomaterials. They are being studied alone or in combination with other biopolymers to facilitate the manufacture of tailored scaffolds promoting better regeneration [78]. From the 3D printing point of view, there are scarce articles in the literature which has utilized ulvan as a bioink. The enhancement of the rheological properties of ulvans can pave way to introduce this biodegradable biopolymer as a new kind of bioink for novel 3D printing applications [79]. Figure 2 shows an overview of the potential of ulvan for functional biomaterials and different sectors of applications, including bioinks.

Figure 2. Scheme of the potential applicability of printable ulvan biomaterials.

Kogelenberg [80] focused on the fabrication of ulvan based structures for cell culture in wound healing. Author explored several different printing strategies, implementing Poloxamer 407, a thermosensitive block copolymer with favourable mechanical properties,

New Materials for a Circular Economy   Materials Research Forum LLC
Materials Research Foundations 149 (2023) 368-392   https://doi.org/10.21741/9781644902639-10

as a support for the printing of modified ulvan, a biologically favourable material. Promising results for further study were obtained combining ulvan and Poloxamer 407 for skin regeneration.

Chen et al. [33] worked on a traditional gelatin methacryloyl based bioink (3.6%), tuned by addition of an ulvan type polysaccharide (<10%), isolated from a cultivated source of a specific Australian Ulvacean macroalga for dermal-like structures using a 3D bioplotter (Envision TEC, Germany). Printing processing conditions included 4-layers, printing temperature of 25 °C, a pressure of 3.5 bar and printing speed of 8 mm/s using a nozzle of 0.2 mm. The incorporation of ulvan in the bioinks favoured the extrusion printing process by reducing yield stress, improved the mechanical strength and regulated the rate of scaffold degradation. This study demonstrated progression towards a biocompatible and biofunctional ink that simultaneously delivered improved mechanical, structural and stability traits that are important in promoting real world applications in skin tissue regeneration.

5. Applications

Different uses can be proposed based on the different technological and biological properties of ulvans; the most common are summarized in this section (Table 3). The whole seaweed can be destined for food and feed purposes, whereas the extracts and fractions are promising for cosmetics and pharmaceuticals. The richness in macro- and micronutrients, and the presence of bioactive components converts Ulva sp. biomass in a valuable food and feed feedstock [87], and the one health approach, which considers the conexion among human, animal, and environmental health can be also applied to their cultivation [88]. The consumption of *Ulva* sp. algae can provide essential elements, but depending on the cultivation area, some toxic compounds (such as Cu, Ni, Mn, and Pb) could be present and their concentrations should be controlled to protect the consumer against potential adverse health risks [89, 90].

Ulvan is a promising potential ingredient for functional food, with immunomodulation, anti-aging, apoptotic, hepatoprotective and prebiotic actions [24]. Dietary macroalgae offers a complete set of antioxidants, macro- and micronutrients that converts them in adequate agents for involuntary intervention in noncommunicable diseases, such as obesity, hypertension, diabetes, and diseases that can be ameliorated with diet strategies [91].

Polysaccharides and oligosaccharides also show desirable physiochemical and gelling properties to improve the quality and shelf-life of food [24], and are interesting for their biological properties as new functional foods [3]. Alipour et al. [81] proposed the use of Ulva intestinalis sulfated polysaccharides (more than 65% carbohydrate, 8% protein, 18% sulfate and 6% uronic acid, with average molecular weight of 64.2 kDa) for surimi formulation, and observed enhanced emulsifying stability maintaining the emulsification capacity and textural characteristics and water holding capacity of surimi gels. The incorporation of the seaweed powder proved also suitable for surimi fortification, although

New Materials for a Circular Economy   Materials Research Forum LLC
Materials Research Foundations 149 (2023) 368-392   https://doi.org/10.21741/9781644902639-10

whiteness of gels decreased significantly, while emulsifying stability of surimi pastes were notably improved [92]. Thunyawanichnondh et al. [93] proposed the addition of this polysaccharide at 0.25 % (w/v) to facilitate the bonding and sheet formation and reducing product fracture in a dried crispy high fiber and protein healthy snack. Utilization of the whole seaweed for food or use of some components, i.e. oil in microemulsions [94], enrichment of bread crumbs and crusts showed a darker and greener color, with higher phenolic compounds, chlorophylls and antioxidant capacity [95].

Abundant examples related to the incorporation of the whole seaweed in different animal diets can be found. Peña-Rodríguez et al. [83] confirmed that a plant-based diet in the presence of live *Ulva clathrata* available ad libitum, allowed a sustainable brown shrimp *Farfantepenaeus californiensis* production improving the feed conversion ratio and protein efficiency ratio. Dietary inclusion of *Ulva* sp. seaweed meal had no adverse effect on visceral organ size, carcass and meat quality traits, and meat stability of Boschveld indigenous hens and resulted in similar shelf life as a control diet [96]. *Ulva* sp. represent an inexpensive diet for amphipods, which are used as an alternative for fishes and cephalopods in aquaculture [97], for the formulation of a settlement substrate and in the initial post-metamorphosis period for echiniculture [98], and *U. ohnoi* for salmon diets without causing any major effect in growth and feed efficiency [99].

Pharmacological properties have been reported for different *Ulva* sp. extracts. The water soluble polysaccharides from *U. pertusa* significantly decreased the atherogenic index in mice [38] and *U. rigida* extracts in zebrafish diets promoted an immunomodulatory action after 1 week by upregulating cytokine expression without compromising intestinal integrity [82]. The ulvan extract as well as its vitamin-enriched formulation reduced the load of Marek's disease virus in infected chicken fibroblasts at concentrations that are innocuous for the other cells [85].

Ulvans can be used in cosmetics for their thickening, moisturizing and antioxidant activities properties and may induce cell proliferation and collagen biosynthesis due to the high percentage of rhamnose and the moisturizing properties provided by glucuronic acid. Fournière et al. [19] confirmed the beneficial action of ulvan poly- and oligosaccharide fractions obtained by enzyme-assisted extraction and depolymerization for dermo-cosmetic applications based on the action against inflammatory acneic and non-acneic *Cutibacterium acnes* strains on keratinocytes. Ulvans enhanced the synthesis of glycosaminoglycans, total collagen, and MMP-1 (Matrix Metalloproteinase-1) regulation without showing cytotoxic effect on fibroblasts [17; 19].

New Materials for a Circular Economy   Materials Research Forum LLC
Materials Research Foundations 149 (2023) 368-392   https://doi.org/10.21741/9781644902639-10

Table 3. Some examples of applications of ulvan and green seaweed polysaccharides for food, feed and therapeutics applications.

Application	Species/application	Properties	References
Food and feed	*Ulva intestinalis* Surimi formulations	Enhanced emulsifying stability Maintained texture and water holding	81
	U. rigida/ diet with seaweed extracts	Enhanced immunomodulatory action zebrafish juveniles	82
	live *Ulva clathtrata* and plant diet	Improved the feed conversion ratio and protein efficiency of brown shrimp juveniles	83
	Ulva rigida	Facilitated the bonding in a dried crispy high fiber and protein healthy snack	84
Therapeutic	*Ulva pertusa* polysaccharides	Significantly lowered the atherogenic index	38
	Ulva sp.	Increase in hyaluronan production	25
	Ulva sp.	Active against inflammatory acneic and non-acneic *Cutibacterium acnes* strains	18, 19
	Ulva armoricana extract	Antiviral against Marek's disease virus in chicken	85
	Ulva lactuca gold and silver nanoparticles	Apoptotic activity against colorectal cancer cells	86

Ulvan show extraordinary potential based on its unique features, biological activities and the excellent biocompatibility, biodegradability, low cost, and abundance, can favour its integration into novel biomaterials. They could be proposed in a variety of applications [22]. Some illustrative examples of these novel biomaterials could be mentioned. Ben Amor et al. [37] used a high molecular weight ulvan fraction for the synthesis of a polyelectrolyte- ulvan/chitosan biomembrane with anticoagulant properties. Ibrahim et al. [39] reported antimicrobial activity of mixtures of ulvan-chitosan hydrogels against different pathogens, in some cases being more potent than ulvan.

Ulvan can form thermoreversible gels in the presence of calcium ions at pH in the range 7.5-8.0 and has also been used for 3D printing development of novel scaffolds [79]. Ulvan has been used to produce ulvan/polyvinyl alcohol nanofibers with adequate mechanical properties [71] and has been proposed for encapsulating other bioactives, such as nisin [100] or curcumin [101, 102]. Sari-Chmaysssem et al. [45] prepared a hydrophobically

New Materials for a Circular Economy   Materials Research Forum LLC
Materials Research Foundations 149 (2023) 368-392   https://doi.org/10.21741/9781644902639-10

modified uronamide ulvan in aqueous medium in the presence of octylamine to improve the rheological properties by creating hydrophobic associations in the solution. Guidara et al. [103] produced active films with ulvan from *U. lactuca*, using as plasticizer glycerol to provide a more compact structure, more negative temperature of transition, and higher antiradical properties, or with sorbitol, showing light color and higher chelating ability. Don et al. [61] prepared bioactive films based on ionic crosslinking between chitosan and ulvan and adding tripolyphosphate and glycerol to enhance the tensile strength compared to chitosan-based films, and also showed antioxidant and whitening activities, being biocompatible to normal cells but toxic to melanoma cancer cells.

The natural polysaccharide is a great alternative to the toxic chemical reagents for the synthesis of nanomaterials. They can serve as stabilizing/capping and reducing agents for the synthesis of stable nanoparticles, and the nanomaterials could enhance the bioactivity and biocompatibility of the polysaccharides. Chemical constituents from *Ulva* sp. aqueous extracts [86], from purified ulvan [104] and from ulvan and other Ulva polysaccharides acted as reducing and stabilizing agents for silver, gold and gold-silver nanoparticles formation [105].

6.   Future perspectives

The utilization of exclusive compounds from marine macroalgae continues growing since these feedstocks are abundant and easily available, within a biorrefinery and circular economy approach. The relationship between activities and structural features of these compounds with heterogeneous composition is not known and requires further investigation, preferably in a multidisciplinary and multi-scale approach [22]. The commercial utilization of sulfated polysaccharides from green seaweeds requires further research on their structure-activity relationships, the mechanisms controlling the atypical gelling performance and efficient preparation of novel products and applications [12; 22]. The extraction conditions are key since can affect the macromolecular distribution, rheological or textural, and bioactive properties of the biopolymer, and then to the pallet of applications for the integral valorization of the seaweed. For this purpose, the incorporation of innovative technologies at different stages of their extraction and modification to reduce or eliminate the use of multiple solvents in improved processes within a circula bioeconomy concept is being studied [29]. Innovative extraction technologies, including those assisted by ultrasound, microwaves, pulsed electric fields, high pressures and enzymes, are considered as clean and efficient alternatives to conventional ones. In addition, developments involving ulvan combination with other materials using advanced tools such as nanotechnology and three-dimensional printing are promising [22].

References

[1] A.I. Eismann, R. Perpetuo Reis, A. Ferreira da Silva, D. Negrão Cavalcanti. *Ulva* spp. carotenoids: Responses to environmental conditions. Algal Res. 48 (2020) 101916. https://doi.org/10.1016/j.algal.2020.101916

[2] C.F.H. Joniver, A. Photiades, P.J. Moore, A.L. Winters, A. Woolmer, J.M.M. Adams. The global problem of nuisance macroalgal blooms and pathways to its use in the circular economy. Algal Res. 58 (2021) 102407. https://doi.org/10.1016/j.algal.2021.102407

[3] L. Ning, Z. Yao, B. Zhu. Ulva (Enteromorpha) Polysaccharides and Oligosaccharides: A Potential Functional Food Source from Green-Tide-Forming Macroalgae. Marine Drugs 20 (2022) 202. https://doi.org/10.3390/md20030202

[4] R. Pangestuti. Nutritional value and biofunctionalities of two edible green seaweeds (*Ulva lactuca* and *Caulerpa racemosa*) from Indonesia by subcritical water hydrolysis. Marine Drugs 19 (2021) 578. https://doi.org/10.3390/md19100578

[5] M. Kendel. Lipid composition, fatty acids and sterols in the seaweeds *Ulva armoricana*, and *Solieria chordalis* from brittany (France): An analysis from nutritional, chemotaxonomic, and antiproliferative activity perspectives. Mar. Drugs 13 (2015) 5606 - 5628. https://doi.org/10.3390/md13095606

[6] E. Sulastri, R. Lesmana, M.S. Zubair, K.M. Elamin, N. Wathoni. A comprehensive review on ulvan based hydrogel and its biomedical applications. Chem. Pharm. Bull. 69 (2021) 432 - 443. https://doi.org/10.1248/cpb.c20-00763

[7] J.T. Kidgell, C.R.K. Glasson, M. Magnusson, G. Vamvounis, I.M. Sims, S.M. Carnachan, S.F.R. Hinkley, A.L. Lopata, R. Nys, A.C. Taki. The molecular weight of ulvan affects the *in vitro* inflammatory response of a murine macrophage. Int. J. Biol. Macromol. 150 (2020) 839 - 8481. https://doi.org/10.1016/j.ijbiomac.2020.02.071

[8] K. Hardouin, G. Bedoux, A.-S. Burlot, C. Donnay-Moreno, J.-P. Bergé, P. Nyvall-Collén, N. Bourgougnon. Enzyme-assisted extraction (EAE) for the production of antiviral and antioxidant extracts from the green seaweed *Ulva armoricana* (Ulvales, Ulvophyceae). Algal Res. 16 (2016) 233 - 239. https://doi.org/10.1016/j.algal.2016.03.013

[9] F. Rahimi, M. Tabarsa, M. Rezaei. Ulvan from green algae *Ulva intestinalis*: optimization of ultrasound-assisted extraction and antioxidant activity. J. Appl. Phycol. 28 (2016) 2979 - 2990. https://doi.org/10.1007/s10811-016-0824-5

[10] T.T.T. Thanh, T.M.T. Quach, T.U. Nguyen, D. Vu Luong, M.L. Bui, T.T.V. Tran. Structure and cytotoxic activity of ulvan extracted from green seaweed *Ulva lactuca*. Int. J. Biol. Macromol. 93 (2016) 695-702. https://doi.org/10.1016/j.ijbiomac.2016.09.040

[11] M. Tabarsa, S. G. You, E. H. Dabaghian, U. Surayot. Water-soluble polysaccharides from *Ulva intestinalis*: Molecular properties, structural elucidation and immunomodulatory activities. J. Food Drug Anal. 26 (2018) 599 - 608. https://doi.org/10.1016/j.jfda.2017.07.016

[12] T.T.V. Tran, H.B. Truong, N.H.V. Tran, T.M.T. Quach, T.N. Nguyen, M.L. Bui, Y. Yuguchi, T.T.T. Thanh. Structure, conformation in aqueous solution and antimicrobial

activity of ulvan extracted from green seaweed *Ulva reticulata*. Natural Prod. Res. 32 (2018) 2291-2296. https://doi.org/10.1080/14786419.2017.1408098

[13] X.-Y. Liu, D. Liu, G.-P. Lin, Y.-J. Wu, L.-Y. Gao, C. Ai, Y.-F. Huang, M.-F. Wang, H.R. El-Seedi, X.-H. Chen, C. Zhao. Anti-ageing and antioxidant effects of sulfate oligosaccharides from green algae *Ulva lactuca* and *Enteromorpha prolifera* in SAMP8 mice. Int. J. Biol. Macromol. 139 (2019) 342 - 351. https://doi.org/10.1016/j.ijbiomac.2019.07.195

[14] J.T. Kidgell, S.M. Carnachan, M. Magnusson, R.J. Lawton, I.M. Sims, S.F.R. Hinkley, R. de Nys, C.R.K. Glasson. Are all ulvans equal? A comparative assessment of the chemical and gelling properties of ulvan from blade and filamentous *Ulva*. Carbohydr. Polym. 264 (2021) 118010. https://doi.org/10.1016/j.carbpol.2021.118010

[15] K.P. Anjali, B.M. Snageetha, G. Devi, R. Raghunathan, S. Dutta. Bioprospecting of seaweeds (*Ulva lactuca* and *Stoechospermum marginatum*): The compound characterization and functional applications in medicine-a comparative study. J. Photochem. Photobiology B: Biol. 200 (2019) 111622. https://doi.org/10.1016/j.jphotobiol.2019.111622

[16] N.F. Ardita, L. Mithasari, D. Untoro, S.I. Oktavia-Salasia. Potential antimicrobial properties of the *Ulva lactuca* extract against methicillin-resistant *Staphylococcus aureus*-infected wounds: A review. Veterinary World 14 (2021) 1116 - 1123. https://doi.org/10.14202/vetworld.2021.1116-1123

[17] M. Fournière, T. Latire, M. Lang, N. Terme, N. Bourgougnon, G. Bedoux. Production of active poly- and oligosaccharidic fractions from *Ulva* sp. by combining enzyme-assisted extraction (EAE) and depolymerization. Metabolites 9 (2019) 182. https://doi.org/10.3390/metabo9090182

[18] M. Fournière, G. Bedoux, N. Lebonvallet, R. Leschiera, C. Goff-Pain, N. Bourgougnon, T. Latire. Poly-and oligosaccharide Ulva sp. fractions from enzyme-assisted extraction modulate the metabolism of extracellular matrix in human skin fibroblasts: Potential in anti-aging dermo-cosmetic applications. Marine Drugs 19 (2021) 156. https://doi.org/10.3390/md19030156

[19] M. Fournière, G. Bedoux, D. Souak, N. Bourgougnon, M.G.J. Feuilloleyv, T. Latire. Effects of *Ulva* sp. Extracts on the growth, biofilm production, and virulence of skin bacteria microbiota: *Staphylococcus aureus*, *Staphylococcus epidermidis*, and *Cutibacterium acnes* strains. Molecules 26 (2021) 4763. https://doi.org/10.3390/molecules26164763

[20] M. Guidara, H. Yaich, I. B. Amor, J. Fakhfakh, J. Gargouri, S. Lassoued, C. Blecker, A. Richel, H. Attia, H. Garna. Effect of extraction procedures on the chemical structure, antitumor and anticoagulant properties of ulvan from *Ulva lactuca* of Tunisia coast. Carbohydrate Polym. 253 (2021) 117283. https://doi.org/10.1016/j.carbpol.2020.117283

[21] Y.-H.R. Hung, G.-W. Chen, C.-L. Pan, H.-T.V. Lin. Production of ulvan oligosaccharides with antioxidant and angiotensin-converting enzyme-inhibitory activities by microbial enzymatic hydrolysis. Fermentation 7 (2021) 160. https://doi.org/10.3390/fermentation7030160

[22] M.-C. Wan, W. Qin, C. Lei, Q.-H. Li, M. Meng, M. Fang, W. Song, J.-H. Chen, F. Tay, L.-N. Niu. Biomaterials from the sea: Future building blocks for biomedical applications. Bioactive Mat. 6 (2021) 4255 - 4285. https://doi.org/10.1016/j.bioactmat.2021.04.028

[23] T. Wassie, K. Niu, C. Xie, H. Wang, W. Xin. Extraction Techniques, Biological Activities and Health Benefits of Marine Algae *Enteromorpha prolifera* Polysaccharide. Frontiers Nutri., 87 (2021) 747928. https://doi.org/10.3389/fnut.2021.747928

[24] D. Liu, Y. Ouyang, R. Chen, M. Wang, C. Ai, H.R. El-Seedi, Md.M.R. Sarker, X. Chen, C. Zhao. Nutraceutical potentials of algal ulvan for healthy aging. Int. J. Biol. Macromol. 194 (2021) 422 - 434. https://doi.org/10.1016/j.ijbiomac.2021.11.084

[25] S. Shen, X. Chen, Z. Shen, H. Chen. Marine polysaccharides for wound dressings application: An overview. Pharmaceutics 13 (2021) 1666. https://doi.org/10.3390/pharmaceutics13101666

[26] F. Menaa, U. Wijesinghe, G. Thiripuranathar, N.A. Althobaiti, A.E. Albalawi, B.A. Khan, B. Menaa. Marine algae-derived bioactive compounds: A new wave of nanodrugs? Mar. Drugs 19 (2021) 484. https://doi.org/10.3390/md19090484

[27] C. Costa, A. Alves, P.R. Pinto, R.A. Sousa, E.A. Borges Da Silva, R.L. Reis. Rodrigues A.E. Characterization of ulvan extracts to assess the effect of different steps in the extraction procedure. Carbohydr. Polym. 88 (2012) 537 - 5462. https://doi.org/10.1016/j.carbpol.2011.12.041

[28] S. Shefer, A. Robin, A. Chemodanov, M. Lebendiker, R. Bostwick, L. Rasmussen, M. Lishner, M. Gozin, A. Golberg. Fighting SARS-CoV-2 with green seaweed Ulva sp. extract: Extraction protocol predetermines crude ulvan extract anti-SARS-CoV-2 inhibition properties in *in vitro* Vero-E6 cells assay. Peer J. 9 (2021) 12398. https://doi.org/10.7717/peerj.12398

[29] L.P. Gomez, C. Alvarez, M. Zhao, U. Tiwari, J. Curtin, M. Garcia-Vaquero, B.K. Tiwari. Innovative processing strategies and technologies to obtain hydrocolloids from macroalgae for food applications. Carbohydr. Polym. 24815 (2020) 116784. https://doi.org/10.1016/j.carbpol.2020.116784

[30] H. Vaghari, M. Eskandari, V. Sobhani, A. Berenjian, Y. Song, H. Jafarizadeh-Malmiri. Process Intensification for Production and Recovery of Biological Products, American Journal of Biochemistry and Biotechnology, 11, (2015), 37-43. https://doi.org/10.3844/ajbbsp.2015.37.43

[31] F. Tian, R.-H. Zhang, X. Wang. A positive feedback onto ENSO due to tropical instability wave (TIW)-induced chlorophyll effects in the Pacific. Geophysical Research Letters, 46, 2019, 889-897. https://doi.org/10.1029/2018GL081275

[32] P.R. Postma, O. Cerezo-Chinarro, R.J. Akkerman, G. Olivieri, R.H. Wijffels, W.A. Brandenburg, M.H.M. Eppink. Biorefinery of the macroalgae *Ulva lactuca*: extraction of proteins and carbohydrates by mild disintegration. J. Appl. Phycol. 30 (2018) 1281 - 1293. https://doi.org/10.1007/s10811-017-1319-8

[33] J. Chen, W. Zeng, J. Gan, Y. Li, Y. Pan, J. Li, H. Chen. Physicochemical properties and anti-oxidation activities of ulvan from *Ulva pertusa* Kjellm. Algal Res. 55 (2021) 102269. https://doi.org/10.1016/j.algal.2021.102269

[34] A. Kartik, D. Akhil, D. Lakshmi, K. Panchamoorthy Gopinath, J. Arun, R. Sivaramakrishnan, A. Pugazhendhi. A critical review on production of biopolymers from algae biomass and their applications. Biores. Tech. 329 (2021) 124868. https://doi.org/10.1016/j.biortech.2021.124868

[35] Y. Kumar, S. Singhal, A. Tarafdar, A. Pharande, M. Ganesan, P.C. Badgujar Ultrasound assisted extraction of selected edible macroalgae: Effect on antioxidant activity and quantitative assessment of polyphenols by liquid chromatography with tandem mass spectrometry (LC-MS/MS). Algal Res. 5 (2020) 102114. https://doi.org/10.1016/j.algal.2020.102114

[36] C.R.K. Glasson, L. Donnet, A. Angell, M.J. Vucko, A.J. Lorbeer, G. Vamvounis, R. de Nys, M. Magnusson. Multiple response optimisation of the aqueous extraction of high quality ulvan from *Ulva ohnoi*. Biores. Technol Rep. 7 (2019) 100262. https://doi.org/10.1016/j.biteb.2019.100262

[37] C. Ben Amor, M.A. Jmel, P. Chevallier, D. Mantovani, I. Smaali. Efficient extraction of a high molecular weight ulvan from stranded *Ulva* sp. biomass: application on the active biomembrane synthesis. Biomass Conversion Biorefinery 10.1007/s13399-021-01426-9.

[38] Y. Pengzhan, Z. Quanbin, L. Ning, X. Zuhong, W. Yanmei, L. Zhi'en. Polysaccharides from *Ulva pertusa* (Chlorophyta) and preliminary studies on their antihyperlipidemia activity. J. Appl. Phycol. 15 (2003) 21 - 27. https://doi.org/10.1023/A:1022997622334

[39] M.I.A. Ibrahim, M.S. Amer, H.A.H. Ibrahim, E.H. Zaghloul. Considerable Production of Ulvan from *Ulva lactuca* with Special Emphasis on Its Antimicrobial and Anti-fouling Properties. Applied Biochem. Biotechnol. 194 (2022) 3097 - 3118. https://doi.org/10.1007/s12010-022-03867-y

[40] H. Yaich, H. Garna, S. Besbes, M. Paquot, C. Blecker, H. Attia. Effect of extraction conditions on the yield and purity of ulvan extracted from *Ulva lactuca*. Food Hydrocoll. 31 (2013) 375-382. https://doi.org/10.1016/j.foodhyd.2012.11.013

[41] C.R.K. Glasson, I.M. Sims, S.M. Carnachan, R.de Nys, M. Magnusson. A cascading biorefinery process targeting sulfated polysaccharides (ulvan) from *Ulva ohnoi*. Algal Res. 27 (2017) 383-391. https://doi.org/10.1016/j.algal.2017.07.001

[42] M. Polikovsky, A. Gillis, E. Steinbruch, A. Robin, M. Epstein, A. Kribus, A. Golberg. Biorefinery for the co-production of protein, hydrochar and additional co-products from a green seaweed *Ulva* sp. with subcritical water hydrolysis. Ener. Conv. Manag. 225 (2020) 113380. https://doi.org/10.1016/j.enconman.2020.113380

[43] A.F.R. Silva, H. Abreu, A.M.S. Silva, S.M. Cardoso. Effect of oven-drying on the recovery of valuable compounds from *Ulva rigida*, *Gracilaria* sp. and *Fucus vesiculosus.* Marine Drugs 17 (2019) 90. https://doi.org/10.3390/md17020090

[44] S. Tsubaki, O. Kiriyoa, H. Masanoric, O. Ayumub, M. Tomohiko. Microwave-assisted hydrothermal extraction of sulfated polysaccharides from *Ulva* spp. and *Monostroma latissimum*. Food Chem. 210 (2016) 311 - 316. https://doi.org/10.1016/j.foodchem.2016.04.121

[45] N. Sari-Chmayssem, S. Taha, H. Mawlawi, J.-P. Guégan, J. Jeftić, T. Benvegnu. Extacted ulvans from green algae *Ulva linza* of Lebanese origin and amphiphilic derivatives: evaluation of their physico-chemical and rheological properties. J. Applied Phycol. 31 (2019) 1931 - 19461. https://doi.org/10.1007/s10811-018-1668-y

[46] B. Le, K.S. Golokhvast, S.H. Yang, S. Sun. Optimization of microwave-assisted extraction of polysaccharides from *Ulva pertusa* and evaluation of their antioxidant activity. Antioxidants 8 (2019) 129. https://doi.org/10.3390/antiox8050129

[47] Y. Yuan, X. Xu, C. Jing., P. Zou, C. Zhang, Y. Li. Microwave assisted hydrothermal extraction of polysaccharides from *Ulva prolifera*: Functional properties and bioactivities. Carbohyd. Polym. 181 (2018) 902 - 910. https://doi.org/10.1016/j.carbpol.2017.11.061

[48] J. Kim, S.H. Ha. Hydrothermal pretreatment of *Ulva pertusa* Kjellman using microwave irradiation for enhanced enzymatic hydrolysis. Korean Chem. Eng. Res. 53 (2015) 570 - 575. https://doi.org/10.9713/kcer.2015.53.5.570

[49] H. Tian, X. Yin, Q. Zeng, L. Zhu, J. Chen. Isolation, structure, and surfactant properties of polysaccharides from *Ulva lactuca* L. from South China Sea. Int. J. Biol. Macromol. 79 (2015) 577 - 582. https://doi.org/10.1016/j.ijbiomac.2015.05.031

[50] M. Prabhu, K. Levkov, Y.D. Livney, A. Israel, A. Golberg. High-Voltage Pulsed electric field preprocessing enhances extraction of starch, proteins, and ash from marine macroalgae *Ulva ohnoi*. ACS Sust. Chem. Eng. 7 (2019) 17453-17463. https://doi.org/10.1021/acssuschemeng.9b04669

[51] K. Levkov, Y. Linzon, B. Mercadal, A. Ivorra, C. A. González, A. Golberg. High-voltage pulsed electric field laboratory device with asymmetric voltage multiplier for marine macroalgae electroporation. Innov. Food Sci. Emerg. Technol. 60 (2020) 102288. https://doi.org/10.1016/j.ifset.2020.102288

[52] N. Castejón, K.A. Thorarinsdottir, R. Einarsdótti, K. Kristbergsson, G. Marteinsdóttir. Exploring the potential of icelandic seaweeds extracts produced by aqueous pulsed electric fields-assisted extraction for cosmetic applications. Marine Drugs 19 (2021) 662. https://doi.org/10.3390/md19120662

[53] R. Pezoa-Conte, A. Leyton, A. Baccini, M. C. Ravanal, P. Mäki-Arvela, H. Grénman, C. Xu, S. Willför, M. E. Lienqueo, J.-P. Mikkola. Aqueous Extraction of the Sulfated Polysaccharide Ulvan from the Green Alga *Ulva rigida*-Kinetics and Modeling. Bioenergy Res. 10 (2017) 915 - 9281. https://doi.org/10.1007/s12155-017-9853-4

[54] N. Flórez-Fernández, M.D. Torres, M.J. González-Muñoz, H. Domínguez. Potential of intensification techniques for the extraction and depolymerization of fucoidan. Algal Res. 30 (2018) 128 - 148. https://doi.org/10.1016/j.algal.2018.01.002

[55] B. Li, S. Liu, R. Xing, K. Li, R. Li, Y. Qin, X. Wang, Z. Wei, P. Li. Degradation of sulfated polysaccharides from *Enteromorpha prolifera* and their antioxidant activities. Carbohydr. Polym. 92 (2013) 1991 - 1996. https://doi.org/10.1016/j.carbpol.2012.11.088

[56] F. Chemat, N. Rombaut, A.G. Sicaire, A. Meullemiestre, A.S. Fabiano-Tixier, M. Abert-Vian. Ultrasound assisted extraction of food and natural products. mechanisms, techniques, combinations, protocols and applications. a review. Ultrasonics Sonochemistry, 34, (2017), 540-560. https://doi.org/10.1016/j.ultsonch.2016.06.035

[57] M.M. Costa, L.B. Pio, P. Bule, V.A. Cardoso, M. Duarte, C.M. Alfaia, D.F. Coelho, J.A. Brás, C.M.G.A. Fontes, J.A.M. Prates. Recalcitrant cell wall of *Ulva lactuca* seaweed is degraded by a single ulvan lyase from family 25 of polysaccharide lyases. Animal Nutrition 9 (2022) 184 - 192. https://doi.org/10.1016/j.aninu.2022.01.004

[58] A.S. Jagtap, N.P.V. Sankar, R.I. Ghori, C.S. Manohar. Marine microbial enzymes for the production of algal oligosaccharides and its bioactive potential for application as nutritional supplements. Folia Microbiologica 67 (2022) 175 - 191. https://doi.org/10.1007/s12223-021-00943-4

[59] A. Robin, M. Kazir, M. Sack, A. Israel, W. Frey, G. Mueller, Y.D. A. Livney, Golberg. Functional Protein Concentrates Extracted from the Green Marine Macroalga *Ulva* sp., by High Voltage Pulsed Electric Fields and Mechanical Press. ACS Sust. Chem. Engi. 6 (2018) 13696 - 137055. https://doi.org/10.1021/acssuschemeng.8b01089

[60] M. Polikovsky, F. Fernand, M. Sack, W. Frey, G. Müller, A. Golberg. Towards marine biorefineries: Selective proteins extractions from marine macroalgae *Ulva* with pulsed electric fields. Innov. Food Sci. Emerg. Technol. 37 (2016) 194 - 200. https://doi.org/10.1016/j.ifset.2016.03.013

[61] T.-M. Don, L.-M. Liu, M. Chen, Y.-C. Huang. Crosslinked complex films based on chitosan and ulvan with antioxidant and whitening activities. Algal Res. 58 (2021) 102423. https://doi.org/10.1016/j.algal.2021.102423

[62] R. Zhong, W. Xuzhi, D-Q. Wang, C. Zhao, D. Liu, L. Gao, M. Wang, C. Wu, S.M. Nabavid, M. Daglia, J. Xiao, H. Cao. Polysaccharides from Marine *Enteromorpha*: Structure and function. Trends Food Sci. Tech. 99 (2020) 11 - 20. https://doi.org/10.1016/j.tifs.2020.02.030

[63] L. Juul. Ulva fenestrata protein - Comparison of three extraction methods with respect to protein yield and protein quality. Algal Res. 60 (2021) 102496. https://doi.org/10.1016/j.algal.2021.102496

[64] M. Prabhu, A. Israel, R.R. Palatnik, D. Zilberman, A. Golberg. Integrated biorefinery process for sustainable fractionation of *Ulva ohnoi* (Chlorophyta): process optimization and revenue analysis. J. Appl. Phycol. 32 (2020) 2271 - 2282. https://doi.org/10.1007/s10811-020-02044-0

[65] V.A. Mantri, M.A. Kazi, N.B. Balar, V. Gupta, T. Gajaria. Concise review of green algal genus *Ulva* Linnaeus. J. Appl. Phycol. 32 (2020) 2725 - 2741. https://doi.org/10.1007/s10811-020-02148-7

[66] M. Martins, R. Oliveira, J.A.P. Coutinho, M.A.F. Faustino, M.G.M.S. Neves, D.C.G.A. Pinto, S.P.M. Ventura. Recovery of pigments from *Ulva rigida*. Sep. Pur. Technol. 2551 (2021) 117723. https://doi.org/10.1016/j.seppur.2020.117723

[67] A. Agarwal, A. Mhtre, R. Pandit, A.M. Lali. Synergistic biorefinery of Scenedesmus obliquus and *Ulva lactuca* in poultry manure towards sustainable bioproduct generation. Biores. Tech. 29 (2020) 122462. https://doi.org/10.1016/j.biortech.2019.122462

[68] C. Andrade, P.L. Martins, L.C. Duarte, A.C. Oliveira, F. Carvalheiro,. Development of an innovative macroalgae biorefinery: Oligosaccharides as pivotal compounds. Fuel 32015 (2022) 123780. https://doi.org/10.1016/j.fuel.2022.123780

[69] M. Magnusson, C.R.K. Glasson, M.J. Vucko, A. Angell, T.L. Neoh, R. Nysbet. Enrichment processes for the production of high-protein feed from the green seaweed Ulva ohnoi. Algal 4 (2019) 101555. https://doi.org/10.1016/j.algal.2019.101555

[70] Y. Ren, A. Aierken, L. Zhao, Z. Lin, J. Jiang, B. Li, J. Wang, J. Hua, Q. Tu. hUC-MSCs lyophilized powder loaded polysaccharide ulvan driven fuctional hydrogel for chronic diabetic wound healing. Carbohydr. Polym. 288 (2022) 119404. https://doi.org/10.1016/j.carbpol.2022.119404

[71] M.A. Madany, M.S. Abdel-Kareem, A.K. Al-Oufy, M. Haroun, S. Sheweita. The biopolymer ulvan from *Ulva fasciata*: Extraction towards nanofibers fabrication. Int. J. Biol. Macromol. 177 (2021) 401-412. https://doi.org/10.1016/j.ijbiomac.2021.02.047

[72] S. Kikinois, E. Ioannou, E. Aggelidou, L.A. Tziveleka, E. Demiri, A. Bakopoulou, S. Zinelis, A. Kritis, V. Roussis. The marine polysaccharide ulvan confers potent osteoinductive capacity to PCL-based scaffolds for bone tissue engineering applications. Int. J. Mol. Sci. 22 (2021) 3086. https://doi.org/10.3390/ijms22063086

[73] M. Guidara, H. Yaich, S. Benelhadj, Y.D. Adjouman, A. Richel, C. Blecker, M. Sindic, S. Boufi, H. Attia, H. Garna. Smart ulvan films responsive to stimuli of plasticizer and extraction condition in physico-chemical, optical, barrier and mechanical properties. Int. J. Biol. Macromol. 150 (2020) 714 - 7261. https://doi.org/10.1016/j.ijbiomac.2020.02.111

[74] S. Shalaby & H. Amin. Potential Using of Ulvan Polysaccharide from *Ulva lactuca* as a Prebiotic in Synbiotic Yogurt Production. Journal of Probiotics & Health. (2019) 07. 10.35248/2329-8901.19.7.208. https://doi.org/10.35248/2329-8901.19.7.208

[75] A.R. Ganesan, S. Munisamy, R. Bhat. Producing edible films from semi refined carrageenan (SRC) and ulvan polysaccharides for potential food applications. Int. J. Biol. Macromol. 112 (2018) 1164-1170. https://doi.org/10.1016/j.ijbiomac.2018.02.089

[76] A. Morelli, M. Betti, D. Puppi, F. Chiellini. Design, preparation and characterization of ulvan based thermosensitive hydrogels. Carbohydr. Polym. 136 (2016) 1108-1117. https://doi.org/10.1016/j.carbpol.2015.09.068

[77] E. Sulastri, M.S. Zubair, R. Lesmana, M.N. Wathoni. Development and characterization of ulvan polysaccharides-based hydrogel films for potential wound dressing applications. Drug design, development and therapy 15 (2021) 4213 - 4226. https://doi.org/10.2147/DDDT.S331120

[78] J. Dinoro, M. Maher, S. Talebian, M. Jafarkhani, M. Meharli, G. Orive, J. Foroughi, M.S. Lord, A. Dolatshahi-Pirouz. Sulfated polysaccharide-based scaffolds for orthopaedic tissue engineering. Biomat. 214 (2016) 119214. https://doi.org/10.1016/j.biomaterials.2019.05.025

[79] B. Mahendiran, S. Muthusamy, S. Sampath, S.N. Jaisankar, K.C. Popat, R. Selvakumar, G.S. Krishnakumar. Recent trends in natural polysaccharide based bioinks for multiscale 3D printing in tissue regeneration: A review. Int. J. Biol. Macromol. 183 (2021) 564 - 588. https://doi.org/10.1016/j.ijbiomac.2021.04.179

[80] S. Kogelenberg (2017) Fabrication of ulvan based structures for cell culture in wound healing. Conference proceedings of University of Wollongong Australia.

[81] H.J. Alipour, M. Rezaei, B. Shabanpour, M. Tabarsa. Effects of sulfated polysaccharides from green alga *Ulva intestinalis* on physicochemical properties and microstructure of silver carp surimi. Food Hydrocoll. 74 (2018) 87 - 96. https://doi.org/10.1016/j.foodhyd.2017.07.038

[82] M. Monteiro, A.S. Lavrador, A. Oliva-Teles, A. Couto, A.P. Carvalho, P. Enes, P. Díaz-Rosales. Macro- and microalgal extracts as functional feed additives in diets for

zebrafish juveniles. Aquaculture Res. 52 (2021) 6420 - 6433.
https://doi.org/10.1111/are.15507

[83] A. Peña-Rodríguez, R. Elizondo-González, M.G. Nieto-López, D. Ricque-Marie, L.E. Cruz-Suárez. Practical diets for the sustainable production of brown shrimp, *Farfantepenaeus californiensis*, juveniles in presence of the green macroalga *Ulva clathrata* as natural food. J. Appl. Phycol. 29 (2017) 413 - 421.
https://doi.org/10.1007/s10811-016-0846-z

[84] J. Thunyawanichnondh, N. Suebsiri, S. Leartamonchaikul, W. Pimolsri, W. Jittanit, S. Charoensiddhi. Potential of green seaweed *Ulva rigida* in Thailand for healthy snacks. J. Fisheries Environ. 44 (2015) 29 - 39.

[85] F. Bussy, S. Rémy, M. Le Goff, P.N. Collén, L. Trapp-Fragnet. The sulphated polysaccharides extract ulvans from *Ulva armoricana* limits Marek's disease virus dissemination in vitro and promotes viral reactivation in lymphoid cells. BMC Veterinary Res. 18 (2022) 155. https://doi.org/10.1186/s12917-022-03247-y

[86] N. González-Ballesteros, M.C. Rodríguez-Argüelles, S. Prado-López, M. Lastra, M. Grimaldi, A. Cavazza, L. Nasi, G. Salviati, F. Bigi. Macroalgae to nanoparticles: Study of Ulva lactuca L. role in biosynthesis of gold and silver nanoparticles and of their cytotoxicity on colon cancer cell lines. Materials Sci. Eng. C 97 (2019) 498 - 509.
https://doi.org/10.1016/j.msec.2018.12.066

[87] R. Radulovich, S. Umanzor, R. Cabrera, R. Mata. Tropical seaweeds for human food, their cultivation and its effect on biodiversity enrichment. Aquaculture 436 (2015) 40 - 46. https://doi.org/10.1016/j.aquaculture.2014.10.032

[88] G. Bizzaro, A.K. Vatland, D.M. Pampanin. The One-Health approach in seaweed food production. Environment International 158 (2022) 106948.
https://doi.org/10.1016/j.envint.2021.106948

[89] D. Desideri, C. Cantaluppi, F. Ceccotto, M.A. Meli, C. Roselli, L. Feduzi. Essential and toxic elements in seaweeds for human consumption. Journal of Toxicology and Environmental Health - Part A, 79 (2016) 112 - 122.
https://doi.org/10.1080/15287394.2015.1113598

[90] A.R. Ganesan, K. Subramani, B. Balasubramanian, W.C. Liu, M.V. Arasu, N.A. Al-Dhabi, V. Duraipandiyan. Evaluation of *in vivo* sub-chronic and heavy metal toxicity of under-exploited seaweeds for food application. J. King Saud University - Sci. 32 (2020) 1088 - 1095. https://doi.org/10.1016/j.jksus.2019.10.005

[91] M.L. Cornish, A.T. Critchley, O.G. Mouritsen. A role for dietary macroalgae in the amelioration of certain risk factors associated with cardiovascular disease. Phycologia 54 (2015) 649 - 666. https://doi.org/10.2216/15-77.1

[92] H. Jannat-Alipour, M. Rezaei, B. Shabanpour, M. Tabarsa. Edible green seaweed, *Ulva intestinalis* as an ingredient in surimi-based product: chemical composition and

New Materials for a Circular Economy   Materials Research Forum LLC
Materials Research Foundations 149 (2023) 368-392   https://doi.org/10.21741/9781644902639-10

physicochemical properties. J. Applied Phycol. 31 (2019) 2529 - 2539.
https://doi.org/10.1007/s10811-019-1744-y

[93] J. Thunyawanichnondh, N. Suebsiri, S. Leartamonchaikul, W. Pimolsri, W. Jittanit,
S. Charoensiddhi. Potential of green seaweed *Ulva rigida* in Thailand for healthy
snacks. J Fish Environ. 44(1), (2020), 29-39

[94] N. Aprilianti, R.D. Saraswati, S.A. Budhiyanti. The quality of *Ulva lactuca* fatty
acid microemulsion with ascorbic acid antioxidant during storage. IOP Conference
Series: Earth and Environmental Science. 4th International Symposium on Marine and
Fisheries Research, ISMFR, 919 (2021) 012035. https://doi.org/10.1088/1755-
1315/919/1/012035

[95] T. Amoriello, F. Mellara, M. Amoriello, D. Ceccarelli, R. Ciccoritti.Powdered
seaweeds as a valuable ingredient for functional breads. European Food Res. Technol.
247 (2021) 2431 - 2443. https://doi.org/10.1007/s00217-021-03804-z

[96] L.T. Nhlane, C.M. Mnisi, V. Mlambo, M.J. Madibana. Effect of seaweed-containing
diets on visceral organ sizes, carcass characteristics, and meat quality and stability of
Boschveld indigenous hens. Poultry Sci. 100 (2021) 949 - 956.
https://doi.org/10.1016/j.psj.2020.11.038

[97] P. Jiménez-Prada, I. Hachero-Cruzado, J.M. Guerra-García. Aquaculture waste as
food for amphipods: the case of *Gammarus insensibilis* in marsh ponds from southern
Spain. Aquaculture Int. 29 (2021) 139 - 153. https://doi.org/10.1007/s10499-020-
00615-z

[98] S. Carbonara, R. D'Adamo, A. Novelli, S. Pelosi, A. Fabbrocini 2018. Ground Ulva
solution (GUS): A promising metamorphosis cue for *Paracentrotus lividus*
larviculture. Aquaculture 491 (2018) 289 - 2941.
https://doi.org/10.1016/j.aquaculture.2018.03.044

[99] F. Norambuena, K. Hermon, V. Skrzypczyk, J.A. Emery, Y. Sharon, A. Beard,
Turchini G.M. Algae in fish feed: Performances and fatty acid metabolism in juvenile
Atlantic Salmon. PLoS ONE 10 (2015) 0124042.
https://doi.org/10.1371/journal.pone.0124042

[100] R. Gruskiene, T. Kavleiskaja, R. Staneviciene, S. Kikionis, E. Ioannou, E.
Serviene, V. Roussis, J. Sereikaite. Nisin-loaded ulvan particles: Preparation and
characterization. Food 10 (2021) 1007. https://doi.org/10.3390/foods10051007

[101]. T.H. Bang, T.T.T. Van, L.X. Hung, N.D. Nhut, T.T.T. Thuy, B.T. Huy. Nanogels
of acetylated ulvan enhance the solubility of hydrophobic drug curcumin. Bull. Mat.
Sci. 42 (2017) 1. https://doi.org/10.1007/s12034-018-1682-3

[102] S. Coiai, B. Campanella, R. Paulert, F. Cicogna, E. Bramanti, A. Lazzeri, L.
Pistelli, M.-B. Coltelli. Rosmarinic acid and Ulvan from terrestrial and marine sources
in anti-microbial bionanosystems and biomaterials. Applied Sci. 11 (2021) 9249.
https://doi.org/10.3390/app11199249

[103] M. Guidara, H. Yaich, A. Richel, C. Blecker, S. Boufi, H. Attia, H. Garna. Effects of extraction procedures and plasticizer concentration on the optical, thermal, structural and antioxidant properties of novel ulvan films. Int. J. Biol. Macromol. 135 (2019) 647 - 658. https://doi.org/10.1016/j.ijbiomac.2019.05.196

[104] A. Massironi, A. Morelli, L. Grassi, D. Puppi, S. Braccini, G. Maisetta, S. Esin, G. Batoni, C.D. Pina, F. Chiellini. Ulvan as novel reducing and stabilizing agent from renewable algal biomass: Application to green synthesis of silver nanoparticles. Carbohydr. Polym. 203, (2019), 310-321. https://doi.org/10.1016/j.carbpol.2018.09.066

[105] T. Ashokkumar, K. Vijayaraghavan. Mono- and Bimetallic Au(Core)-Ag(Shell) nanoparticles mediated by *Ulva reticulata* extracts. Chem. Select, 37 (2019) 11009 - 110149. https://doi.org/10.1002/slct.201903202

New Materials for a Circular Economy
Materials Research Foundations 149 (2023) 393-445

Materials Research Forum LLC
https://doi.org/10.21741/9781644902639-11

Chapter 11

Lignocellulosic resources: A Key Player in the Transition to a Circular Bioeconomy

E. Espinosa[1,a,*], E. Rincón[1,b], L.M. Reyes Mendez[2,c], R. Morcillo-Martín[1,3,d],
L. Rabasco-Vílchez[3,e], J. Moreno-García[4,f], J. de Haro[1,3,g], A. Lucena[1,3,h], A. Rodríguez[1,i]

[1]Biopren Group (RNM 940), Chemical Engineering Department, Instituto Químico para la Energía y el Medioambiente (IQUEMA), Facultad de Ciencias, Universidad de Córdoba, 14014, Córdoba, Spain

[2]Universidad Nacional Abierta y a Distancia-UNAD, Escuela de Ciencias Básicas, Tecnología e Ingeniería, 111511, Bogotá DC, Colombia

[3]Department of Food Science and Technology, Faculty of Veterinary, Universidad de Córdoba, 14014, Córdoba, Spain

[4]Department of Agricultural Chemistry, Edaphology, and Microbiology, Universidad de Córdoba, 14014, Córdoba, Spain

[a]eduardo.espinosa@uco.es, [b]b32rirue@uco.es, [c]laura.reyes@unad.edu.co, [d]t62momar@uco.es, [e]t62ravil@uco.es, [f]b62mogaj@uco.es, [g]q42hanij@uco.es, [h]b52luopa@uco.es, [i]a.rodriguez@uco.es

*eduardo.espinosa@uco.es

Abstract

The trend towards sustainable development and the twin goals of reducing the carbon footprint using natural resources have put lignocellulosic residues in the spotlight as a feedstock for the development of biorefinery systems. These facts are compounded by the massive increase in the production of waste derived from the food supply chain, both agricultural and agri-food residues. Biorefineries are sustainable processing systems, encompassing a network of different processes and equipment, for the fractionation of biomass main components. These main fractions are then used to obtain new materials and high-added value products such as fuels and chemicals. This chapter presents the main industrial uses of agricultural and agri-food residues as part of the biorefinery. These industrial applications range from the most classical and widely studied, such as the papermaking industry and the production of fine chemicals, to the most novel applications in the food industry, the development of composite materials, clinical applications and for energy production, etc.

Keywords

Biorefinery, Lignocellulosic Residues, Circular Bioeconomy, Valorization, Bioproducts

Contents

1.   Introduction

The current trend, at last, is to evolve towards sustainable development, hand in hand with the circularity of processes and the management of natural resources. As our parents and grandparents would say, *nihil novum sub sole*, that is, "nothing new under the sun". Indeed, previous generations had been doing this in their daily lives, reusing packaging, buying local products, etc. Until technological development and new oil refining techniques led to linear models of raw material consumption and product production. As mentioned, it seems that society, not individually, has become aware of the imperative need to properly manage the natural resources at our disposal, on the one hand because they are not infinite, and on the other because their management and treatment should not lead to environmental problems.

New Materials for a Circular Economy   Materials Research Forum LLC
Materials Research Foundations 149 (2023) 393-445   https://doi.org/10.21741/9781644902639-11

The world population is growing at a dizzying pace. It was estimated that 2050 would be the year in which the world's population would reach 10,000 million inhabitants, and there are already many authors who affirm that this figure will be reached before then. This population increase is accompanied, of course, by an increase in food production and in the generation of associated waste. This type of waste, lignocellulosic residue from agri-food activities, is currently scarcely valorized and used, and even less in value-added products. The majority composition of these residues, also called vegetable biomass or lignocellulosic biomass, in cellulose, hemicelluloses and lignin, makes them susceptible to be valorized by biorefinery processes.

The valorization of these residues is therefore a challenge that will allow a real approach to the concept of sustainable development and circular bioeconomy. This chapter shows the wide application potential of the lignocellulosic fractions of these residues, reviewing the latest related work.

Table 1. List of the most produced crops susceptible to generate large quantities of lignocellulosic residues [1].

Raw material	Production (million tons)
Sugar, Sugarcane	1869.71
Corn	1162.35
Wheat	760.93
Rice	756.74
Palm oil	418.44
Soya	353.46
Sugarbeet	252.97
Tomato	186.82
Barley	157.03
Banana	119.83
Grape	78.03
Orange	75.45
Rapeseed	72.38
Sorghum	58.71
Eggplant	56.62

2. Papermaking industry

Agricultural waste are residues, mostly of lignocellulosic origin, accumulated after the harvesting of annual plants during the summer or autumn seasons. In the world, more than

9.8 billion tons of food products are produced annually, a large part of which are susceptible to generate a huge amount of lignocellulosic residues in the form of pruning, leaves, skin, bone, etc. Among the most outstanding crops in terms of waste production are cereal straws, fruit trees, vegetables, and oil extraction products. The 15 most cultivated crops of this nature alone account for more than 50% (5321 million tons) of total annual food production (*Table 1*) [1].

According to literature estimates, this type of waste generates a waste:fruit ratio of 0.8 - 1.2 kg [2]. Considering this, it can be estimated that 4250 - 6385 million tons of lignocellulosic waste are produced annually from these sources, which can be recovered in various industrial applications, one of the most important is the paper and cardboard industry.

Nowadays, the dominant raw material in the pulp production industry for paper and board is wood. However, any lignocellulosic material containing a reasonable amount of fibers can be used as a raw material to produce cellulosic pulp. This possibility, together with the increase in consumer awareness of environmental preservation, has increased the interest in the search for new paper production processes. These new processes must be environmentally friendly, both in the use of raw materials and in the manufacturing processes. In addition, government agencies are devoting increasing financial and human resources to research into alternative raw materials to conventional ones. Moreover, non-wood plant sources have shown different types of advantages in the production of pulp and paper compared to wood resources, as presented in *Table 2*.

In recent decades, countless potential sources of fibers for papermaking have been studied to replace or complement wood. These raw materials can be classified into two broad categories: i) Common non-wood raw materials or hardwood alternatives, such as straws, bagasse, bamboo, reeds, and grasses, kenaf esparto, corn stalks, sorghum stalks, etc., and ii) Special non-wood or softwood substitutes, such as cotton linters; fax, hemp and kenaf, bamboo, etc. [5]. One of the key factors to take into account when introducing non-wood raw materials in pulp and paper production is their availability. The availability of raw materials is very important when considering the possibility of an industrial facility for pulp and paper production. Availability is related to the production and location of the various lignocellulosic materials that can be used for the intended purpose. Availability provides information about the volume of raw materials available and the most suitable location for a future industrial plant.

To isolate the cellulose fibers from the lignocellulosic matrix of these wastes, it is necessary to subject the raw material to a delignification process, also known as a pulping process, where the lignin is separated from the raw material, thus concentrating the cellulosic fraction in the matrix.

Table 2. Comparison of non-wood resources for pulp and papermaking [3,4].

Description	Fiber Resources	
	Wood	**Non-wood**
Cycle growth	Long growth cycles [X]	Short cycle growth [V)
Cellulose content	Higher cellulose content [V]	Lower cellulose content depends on the types of non-wood [X]
Lignin content	Contain higher lignin content [X]	Contain lower lignin content [V]
Chemical uses	Use a large volume of chemical during pulping process [X]	Use of a small amount of the chemical in pulping process [V]
Time pulping	Need long time for pulping process [X]	Shorten time for pulping process [V]
Cost operation	Expensive due to the limitation resource [X]	Cheaper cost because the abundance resources [V]
Environmental impact	Increase environmental problem: global warming and soil erosion responsible for about 17% of global carbon budget [X]	Reduce environmental impact which reduce the deforestation problem and improve sustainable forestry [V]
Handling and storage	Easy [V]	Difficult [X]
Washing after process	Easy [V]	Difficult [X]
Chemical recovery	Possible [V]	Difficult [X]
Paper machine speed	High [V]	Low [X]

[X]: Disadvantage; [V]: Advantage.

One of the simplest pulping processes is the so-called "alkaline pulping process". It uses a solution of sodium hydroxide (NaOH) as a cooking lye and is mainly used to obtain semi-chemical pulps. In addition, it is considered environmentally favorable since, unlike the Kraft process, it does not use sulfur compounds as a delignifying reagent [6]. This process can be used with agricultural residues such as rice straw [7], wheat straw [8], vine shoots [9], barley straw [10], among others, to obtain cellulosic pulps. The most industrially used process for wood raw materials is Kraft pulping. This process uses an aqueous solution of NaOH and Na_2S at high temperatures. The lignin contained in the lignocellulosic material reacts stoichiometrically with the hydroxide and hydrosulfide anions, being depolymerized in small fragments that dissolve in the basic solution. As a drawback, this type of process generates large amounts of residual lye with high polluting power due to its sulfur content

New Materials for a Circular Economy   Materials Research Forum LLC
Materials Research Foundations 149 (2023) 393-445   https://doi.org/10.21741/9781644902639-11

[11]. It is also possible to delignify the raw material using organic solvents. The organosolv pulping process was born from the search for a pulping process that avoids odor emissions, high water consumption and energy costs, as well as the high use of reagents in the traditional paper industry [6]. In this process, a distinction is made between low-boiling solvents (methanol, ethanol, acetone, ethyl acetate, etc.), which can be easily recovered by distillation, and high-boiling solvents (triethylene glycol, ethylene glycol, etc.), which require less expensive installations since they withstand less pressure, although they are difficult to recover [12].

With the increase in demand for paper worldwide, the incorporation of non-wood raw materials to meet the potential shortage of wood for papermaking is becoming increasingly interesting. However, it is still necessary to develop cleaner production technology, reduce pulp manufacturing costs, improve product quality, and carry out industrial modernization to enable the incorporation of these raw materials.

3. Fine chemicals

Biorefinery aims a complete valorization of the biomass with a minimum energy cost that increases the value of the production chain. Large-scale biorefinery systems that are already operational generate mostly biofuels (i.e., ethanol, methanol, biodiesel, biogas, etc.) and chemicals isolated directly from biomass (i.e., tannins, pigments, polysaccharides, etc.). However, new processes can be developed to obtain high-value chemical products by converting these compounds extracted directly from the biomass. This is known as secondary biorefinery, where the main fractions of lignocellulosic residues (cellulose, hemicellulose, and lignin) are subjected to conversion and synthesis processes that give rise to platform chemicals such as furfural, levulinic acid, vanillin, and other chemical derivatives and additives [13].

Regardless of the products to be obtained from the biorefinery system, the lignocellulose must first be depolymerized and partially deoxygenated. To achieve this, there are mainly two pathways, thermochemical and hydrolytic. The thermochemical process involves obtaining a mixture of charcoal and pyrolysis oil, or syngas (mixture of carbon monoxide and hydrogen), depending on whether pyrolysis or gasification is carried out, respectively. However, this syngas must compete economically with syngas from natural gas. Hydrolytic conversion of lignocellulose, catalyzed by acids or enzymes, results in C5 and C6 monosaccharides, fatty acids plus glycerol and amino acids, from hemicellulose, cellulose, and triglycerides, respectively. Lignin is generally used to produce energy. This recalcitrant polymer could gain further added value through its selective conversion to chemically useful aromatic components, but this remains a challenge. In any case, the key to obtaining chemicals from lignocellulosic biomass lies in the right choice of feedstock and utilizing each of its components to their maximum value [14].

There are basically two approaches to convert cellulosic biomass and its constituent in building blocks. The first is a complete deoxygenation to petroleum hydrocarbons and subsequent processing into commodity chemicals. There are numerous strategies for this

New Materials for a Circular Economy   Materials Research Forum LLC
Materials Research Foundations 149 (2023) 393-445   https://doi.org/10.21741/9781644902639-11

approach encompassing the production of lower alcohols and their subsequent dehydration into olefins, production of mixtures of gasoline, diesel and kerosene range alkenes or the standard mixture of benzene, toluene, and xylenes (BTX) [15,16]. In recent years, agri-food residues such as sugarcane bagasse, almond shells or olive pomace have been used for this purpose, yielding in 22-26% BTX by means of fast catalytic pyrolysis [17,18].

The second approach involves the direct conversion of oxygenates into platform molecules by biotechnological or chemical processes. The former results in lower alcohols, diols, mono- and di-carboxylic acids by fermentation, lactic acid, citric acid, and succinic acid [19,20]. Chemical processes result in the production of furfural and 5-hydroxymethyl furfural (5-HMF), subsequently converted to levulinic acid (LA) and γ-valerolactone or furan-2,5-dicarboxylic acid (FDCA), in the production of different organic acids such as gluconic acid, formic acid, acetic acid and oxalic acid, and isosorbide production [21–24].

Both furfural and 5-HMF are two of the most promising and important platform chemicals from lignocellulose biorefinery as they are precursors for a variety of high added-value products such as bioplastics, liquid fuels, and pharmaceuticals [25]. Similarly, LA has also been ranked in the top 12 building blocks by the U.S. Department of Energy. Both LA and levulinate esters, also obtained from lignocellulose, are promising intermediates in the synthesis of numerous chemicals such as fragrance additives, solvents, pharmaceuticals, and plasticizers [26].

In this context, this section aims to summarize the latest advances in obtaining important platform molecules such as furfural, 5-HMF, LA and levulinate esters from lignocellulosic biomass and, more specifically, agri-food residues.

3.1   Furfural and 5-hydroxymethyl furfural

Among all chemicals derived from lignocellulosic biomass, furfural and 5-HMF are two of the most important building block components for the industry to produce a wide variety of organic motifs. In fact, these two chemicals are in the top 10 of versatile platform chemicals from lignocellulose. Typically, furfural is obtained from hemicelluloses, and 5-HMF by catalytic conversion of cellulose [27]. Generally, their production is the result of effective pretreatment of biomass followed by hydrolysis to the corresponding hexose sugar units and subsequent catalytic dehydration to the corresponding 5-HMF or furfural product (*Fig. 1*) [28].

Furfural is used to obtain other high value-added chemicals such as pharmaceuticals, industrial solvents, food additives, solvents, and monomers, highlighting its versatility [29]. For its part, 5-HMF is considered a feedstock for bioenergy, construction chemistry and medicine [30]. Therefore, developing strategies for obtaining these chemicals from waste biomass is of particular interest.

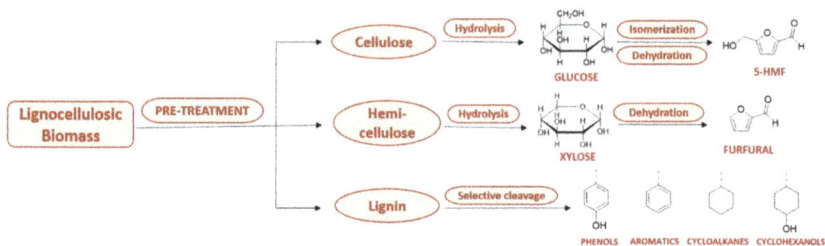

Figure 1. Lignocellulose biomass to furan/aryl based commercially viable platform molecules. Taken with permission from [28].

Over the years, numerous procedures have been developed for the synthesis of furfural and 5-HMF from sugar derivatives, but these procedures are mainly dependent on edible monosaccharides substrates. This is not the case if direct transformation of lignocellulosic biomass is performed. In this sense, sugarcane bagasse, one of the most abundant agri-food residues, can be used for its direct transformation in 5-HMF. This synthesis can be carried out in a batch type reactor at 270 °C for 10 min obtaining a maximum yield of 5-HMF of 3.09% wt. without the need to use a reaction catalyst [31]. However, more efficient methodologies such as microwave (MW) are already used for the same purpose. 5-HMF synthesis can be carried out under MW-assisted heating using single or combined metal chloride catalysts in DMA-LiCl solvent. Specifically, the combined catalyst $Zr(O)Cl_2/CrCl_3$ allows obtaining 42% of 5-HMF from sugarcane bagasse, being an easy, efficient, and low-cost methodology. Moreover, these catalysts can be reused up to four cycles for the same process without significant loss in 5-HMF yield, thus increasing the efficiency of the process [32]. The most recent studies for the synthesis of 5-HMF from sugarcane bagasse have reported that 5-HMF yield can reach 8.1% wt. if hydrothermal MW liquefaction is carried out under acidic seawater conditions. This efficient and sustainable methodology allows transforming biomass into valuable compounds using seawater or brine wastewater [33]. A very interesting approach that allows increasing the 5-HMF yield to more than 65% is by pretreating sugarcane bagasse with ultrasound-ionic liquid and subsequent synthesis in a solid microwave-solid acid-ionic liquid system. These conditions suggest that the acoustic cavitation phenomenon to which the feedstock is subjected promotes the transformation of bagasse into 5-HMF [34]. Finally, the most recent studies for obtaining 5-HMF from this raw material report high yields of 5-HMF (>55%) if a pretreatment with deep eutectic solvent is performed. Specially, the deep eutectic solvent composed of chlorine chloride and carboxylic acids significantly decreases the lignin content of sugarcane bagasse and partially hydrolyzes cellulose and hemicelluloses, increasing the yield in the subsequent synthesis of 5-HMF [35].

New Materials for a Circular Economy   Materials Research Forum LLC
Materials Research Foundations 149 (2023) 393-445   https://doi.org/10.21741/9781644902639-11

Another of the most used agri-food residues to produce furfural and 5-HMF are those derived from corn crops (corn stalks, corn stover, corncob and corn straw). Continuing with the MW heating synthesis methodology, corn stover results in moderate yields of 5-HMF (20-35%) but high yields for furfural (51-66%) using the $AlCl_3 \cdot 6H_2O$ catalyst in a two-phase water/tetrahydrofuran medium with fast kinetics (10 min) [36]. This MW-assisted synthesis from corn stover can also be catalyzed by niobium phosphate in aqueous medium reaching furfural yields of 23 mol% with respect to the initial raw biomass [37]. Going a step further, corn stalks can be subjected to hydrothermal carbonization followed by sulfonation resulting in a solid carbonaceous acid catalyst. This catalyst promotes the synthesis of 5-HMF from corn stalks in the ionic liquid (IL)1-butyl-3-methyl imidazolium chloride yielding 44.1% 5-HMF [38]. Using corn stalks, furfural yields of 83.5% have been achieved at the expense of a lower yield of 5-HMF (19.5%) with slightly more severe reaction conditions (190 °C for 100 min), employing an acid solid catalyst prepared by copolymerization of p-toluenesulfonic acid and paraformaldehyde [39]. As mentioned above, the use of ionic liquids (IL) seems to favor the synthesis of 5-HMF. When corn stalks are converted with an IL such as 1-allyl-3-methylimidazolium chloride and an organic solvent such as isopropanol, 5-HMF yields of more than 60% are obtained, using magnesium and tin ions loaded biochar as catalyst [40]. Another methodology that has allowed obtaining 5-HMF from corn stalks is fast pyrolysis. The use of this technique for corn stalks has reported an optimum 5-HMF yield of 4.87% wt. at 300 °C mainly from the degradation of fructose and sucrose if the stalk pulp is used in the ripening stage [41]. It is also possible to simultaneously obtain 5-HMF and furfural from corncob using the solid acid catalyst polytriphenylamine-SO$_3$H. The use of this catalyst for the reaction carried out by conventional heating in an oil bath at 175 °C gives yields of 73.9 and 32.3% of furfural and 5-HMF, respectively, by using a safe and environmentally friendly reaction medium such as γ-valerolactone [27]. As previously mentioned, the use of IL to produce 5-HMF is a promising research approach as it allows for a significant increase in yields. The use of IL in the case of corncob results in a 5-HMF yield of 66.58% using a mixture of catalysts (MCMP-Al, Cr and Mg) with infrared radiation heating during distillation under reduced pressure. Interestingly, imidazolium IL and this catalyst mixture can be reused up to 40 times for the same reaction with very low loss of catalytic activity [42]. In the same line, the use of IL 1-butyl-3-methylimidazolium hydrogen sulfate results in a high efficiency of obtaining 5-HMF for corn straw, using the Amberlyst-15 catalyst. This system is applicable to other raw material such as sorghum and allows the additional obtaining of furfural and LA [43]. As can be seen, research for obtaining 5-HMF or furfural tends towards the use of IL or environmentally favorable reaction media such as γ-valerolactone. In addition, in recent years, agri-food residues that are not as common as those mentioned so far are beginning to be used. This is the case, for example, of residues such as the outer part of melon rind. This waste can be used as a substrate for obtaining 5-HMF. It has been demonstrated that the use of sulfuric acid and clay Montmorillonite KSF as catalyst under MW irradiation in the two-phase water/THF system provides an efficient extraction of the reaction products [44].

Materials Research Forum LLC
https://doi.org/10.21741/9781644902639-11

3.2   Levulinic acid and levulinate esters

Another of the most important building blocks is LA, which can be produced from deep hydrolysis of biomass. The importance of this compound lies in its dual functional groups of ketone and carboxylic acid that make it an ideal platform molecule for various synthesis reactions of high added-value compounds such as γ-valerolactone or valeric acid. The reaction mechanism for LA production is very similar to that of 5-HMF and it is widely accepted that 5-HMF is a key intermediate for LA formation in the hydrolysis process. When rehydration of 5-HMF occurs after dehydration of the hexoses that produce it, LA is obtained (*Fig. 2*) [45].

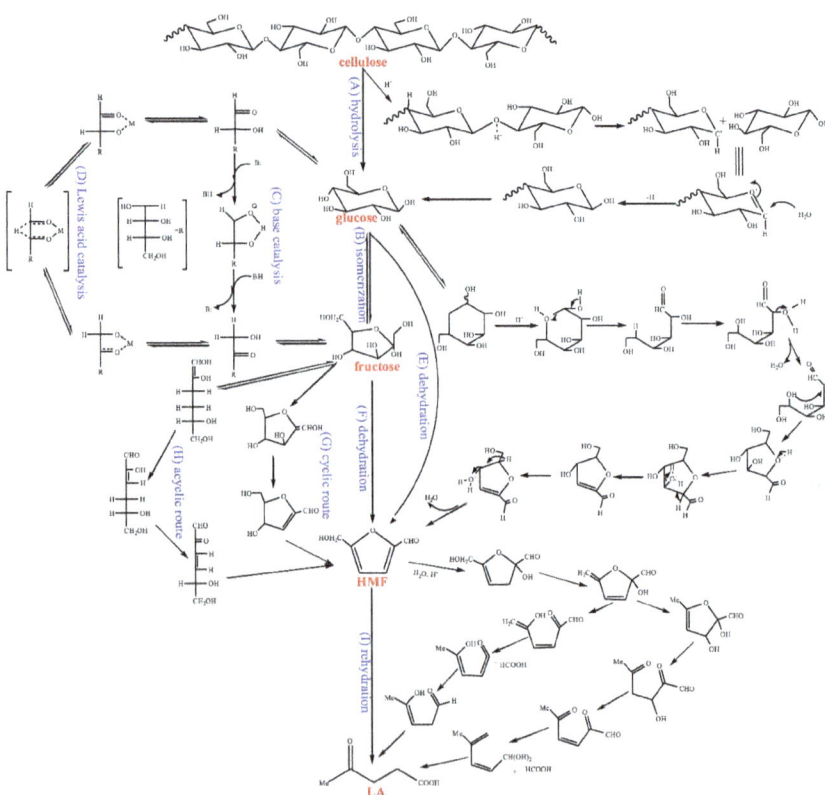

Figure 2. Cellulose hydrolysis to levulinic acid. Taken with permission from [45].

Table 3. Production of levulinic acid and its esters from agri-food residues.

Agri-food residue	Pretreatment	Hydrolysis technique	Catalyst	Main product	Product yield	Reference
Rice husks	Soxhlet aqueous extraction	Batch-pressurized reactor	HCl 4.5% (v/v)	LA	59.4%, w/w	[49]
Empty fruit bunch (EFB)	-	Batch-pressurized reactor	$CrCl_3$-HY zeolite	LA	53.2%	[50]
Hazelnut shells	-	Autoclave-MW reactor	HCl 37 wt.%	LA	9-12 wt.%	[51]
Corn stalk	-	Batch-pressurized reactor	$FeCl_3$	LA	48.9%	[52]
Corncob	-	Batch reaction incorporating distillation	p-toluenesulfonic acid 1.18M	LA	61.6%	[53]
Corn stover	Ball milling	Liquefaction under MW heating	H_2SO_4	ELA	31.2%	[54]
Potato peel waste	-	MW irradiation	H_2SO_4, $CrCl_3 \cdot 6H_2O$; $AlCl_3 \cdot 6H_2O$	LA	49%	[55]
Wheat straw	Hemicellulose and lignin removal	Pressurized reactor	$CuSO_4$	MLA	20.2 wt.%	[56]
Corn stover	-	MW irradiation	H_2SO_4 1-4 wt.%	ELA	58.1 mol%	[57]
Corn stover	-	Autoclave	SAPO-18 zeolites	LA	70.2%	[58]
Wheat straw	-	Liquefaction with IL	1-methyl-3-(4-sulfobutyl) imidazole bisulphate	ELA	85.5%	[59]
Corn stover	Ball milling	Alcoholysis under MW	$Al_2(SO_4)_3$	MLA	64.9 mol%	[60]
Rice Husk	Delignification	Conventional heating	Mn_3O_4	LA	39.8%	[61]
Bay tree pruning	Autohydrolysis	MW irradiation	$Al_2(SO_4)_3$	MLA	40 wt.%	[62]

On the other hand, levulinate esters are produced after esterification of LA with alcohol. The best known levulinate esters are methyl levulinate (MLA) and ethyl levulinate (ELA). MLA is a gasoline additive while ELA is more soluble in diesel-type fuels. In general, levulinate esters are also used as solvents, plasticizers, or precursors in the production of various synthetic materials. Ultimately, both LA and levulinate esters are important platform molecules for current and future biorefineries [46]. Many studies have been conducted in the production of these compounds from lignocellulosic biomass, such as eucalyptus wood [47], rubber wood, palm oil frond, and bamboo [48]. In the context of this chapter, *Table 3* lists the studies of recent years on the production of LA or levulinate esters from agri-food residues.

As displayed, the direct conversion of biomass by means of chemical routes with technologies ranging from conventional heating or autoclave to liquefaction and MW-assisted irradiation make it possible to obtain platform molecules and high added-value products such as furfural, 5-HMF, LA or levulinate esters, among others. These technologies, together with the continuous study and innovation in catalytic systems, allow increasing the range of biorefinery systems that use agri-food residues.

4.   Food industry

4.1   Packaging

Plastic materials have been the most extensively used due to its versatility and low cost. In recent years, the excessive use of this synthetic materials has led to their massive accumulation in different environments e.g., marine, lands, foods, etc., creating a serious pollution problem for aquatic life, environment, and humans. This scenario has promoted the research for sustainable, renewable, biodegradable, and compostable materials that possess the required features for the preservation of food products. Lignocellulosic biomass has been proposed as a renewable feedstock for the complete or partial replacement of petroleum-based materials. In this subsection the different packaging applications studied for lignocellulosic biomass obtained from agri-food industry will be discussed.

4.1.1   Molded fiber and pulp products

One of the most used forms of storage in the food industry are rigid containers such as rigid trays, tetrabricks, single-ply boxes, bags, corrugated boxes, etc., used for the protection, storage, and conservation of the product prior and after their commercialization. Many research articles have described ways to develop rigid packaging from different lignocellulosic feedstocks. In this regard, extensive research has been carried out to obtain bio-composites from lignocellulosic biomass blended with biobased polymers as well as biopolymer obtention by microbial fermentation [63–65]. However, to obtain these types of biopolymers, several transformation stages are necessary, some of which are too complex, costly, and not environment friendly. Therefore, there is a bias towards simpler and more sustainable methodologies that will have a real beneficial impact on the population.

As one of sustainable and green packaging materials, molded fiber and its products have gained attention due to its renewability, recyclability, and biodegradability [66]. Molded fiber products are obtained by subjecting lignocellulosic biomass to chemical and/or

New Materials for a Circular Economy   Materials Research Forum LLC
Materials Research Foundations 149 (2023) 393-445   https://doi.org/10.21741/9781644902639-11

mechanical defibrillation, obtaining chemical, mechanical or semi-chemical pulps [67]. These are used for the manufacture of molded fiber/pulps following a series of processes, including pulp preparation, forming, pressing, and drying, creating different types of three-dimensional products [66,68]. Traditionally, virgin fibers from woody sources (coniferous tree species) and secondary fibers (books, newspapers, etc.). have been used for cellulosic pulp obtention. Nevertheless, its reliance with the wood sector leads to deforestation issues while residual chemicals in re-pulps pose a critical risk to food packaging safety [69]. For this reason, in recent years, cellulose fibers from non-woody species have been used to obtain cellulosic pulps for food packaging. Due to the large-scale production of agricultural crops and the considerable amounts of biomass generated from their cultivation, it has been proposed as a renewable feedstock for the total or partial substitution of wood-derived pulps. Since mechanical and barrier properties of these pulps are not always suitable, they are often mixed with other synthetic or bio-based polymers such as expanded polystyrene (EPS), polypropylene (PP), polylactic acid (PLA), polyhydroxyalkanoates (PHA), polyhydroxybutyrate (PHB), etc. [64,65,70,71]. Water and oil-proofing agents can be added to the polymer matrix such as lignin, chitosan, glycerol, or alkyl-ketene-dimer (AKD) as an alternative for polymer blending [72–74]. Blends of pulps obtained from different raw materials have also been used to improve mechanical properties of the final material. Molded-fiber packages from banana stem (B), pineapple leaf (P) and rice straw (R) have been obtained. The mixing ratio of 30:70 (%) for P:B and R:B mixtures have proven to be sufficient to improve both tensile strength and Young`s modulus values of B properties in the range of 63–167% and 55–117%, respectively [75]. Oats, maize, rapeseed, and barley straws have also been evaluated for pulp obtention by ecological processes obtaining promising results to produce ecological containers [76]. Recently, good results have been obtained with a bio-pulping process of rice straw fibers in the formation of corrugated paper for packaging, reaching tensile strength and ring pressure values of 8 N and 2.46 N·m/g, respectively [77]. Therefore, the production of rigid packaging from non-woody resources could be a feasible and sustainable alternative for the packaging industry in the next few years.

4.1.2 Flexible films

Another widely used packaging technology, especially in the food industry, are film materials. Different fractions of the lignocellulosic biomass have been used in this type of packaging. First, some cellulose derivatives, such as cellulose nanofibers (CNF) and cellulose nanocrystals (CNC), have been synthesized from agricultural wastes to be used as polymeric matrix for films [78–80]. These types of nanomaterials can be applied in two different forms, in the formation of nanopapers and as reinforcement of different polymeric matrices such as chitosan, alginate and starch among many others [81].

Nanopapers are a type of nanomaterial comprised from both nanocellulose and other nano-scale particles, all of which add desirable properties to the package, such as high strength and low weight, making nanopaper useful in packaging applications [82]. However, water resistance and durability values are not comparable to those for plastic materials. To avoid these problems, several additives and coating materials are added [83,84]. Secondly, the use of cellulose nanocrystals (CNC) and cellulose nanofibers (CNF) as reinforcers has a positive effect on the improvement of the physical, mechanical, barrier and optical properties of the films. This effect normally occurs at concentrations of around 5% wt. on the dry weight of the polymer, being more stable with the use of CNC instead of CNF [81]. Considering the economic cost, the performance of CNF should be further studied as well as the safety issues in terms of migration characteristics.

Hemicelluloses are other interesting fraction for the development of film packages due to their biocompatibility, biodegradability, non-toxicity, and film-forming potential [85]. Different types of agri-food wastes such as wheat, barley and oat brans, cotton stalks, sugarcane bagasse, or bamboo, have been used for the extraction of hemicelluloses and their application in films [85–88]. Hemicellulose derived films have shown good oxygen properties compared to other polymer matrices. They also present good flexibility, low gas permeability and high-water resistance [89]. However, their hygroscopic nature and low mechanical properties limit their usage as a starting feedstock. Due to their excellent properties as reinforcing agents, many authors have developed hemicellulose blends with cellulose nanoparticles and/or lignin to confer rigidity, hydrophobicity, and antioxidant capacity to the final material [72,90–94].

4.1.3 Coatings

Another packaging technology used for the protection of food matrices are coating materials. These are composed of a thin layer made of different polymers such as polysaccharides, proteins and/or lipids, usually applied by dipping or spraying in a liquid form, on the food surface [95]. As mentioned for film materials, different formulations of food coatings have been developed from lignocellulosic biomass fractions, being cellulose fraction the most used. They are usually in the form of cellulose derivatives, CNF and/or CNC, combined with each other, with other biopolymers or even with other biomass fractions such as hemicelluloses or lignin [96]. Tian *et al.*, (2021) carried out the extraction of hemicelluloses from sugar bagasse and subsequent blending with CNC, adding AKD as a hydrophobic agent and crosslinker. This formulation prolonged the shelf-life of green asparagus from 5 to 7 days [97]. Similarly, Mugwagwa & Chimphango (2022) isolated hemicelluloses and pectins from wheat straw and mango peels, adding CNC as a reinforcing agent [92]. Studying their mechanical, physical and antioxidant properties they concluded that biocomposites with > 50% pectin were appropriate for application as edible films and coatings for the prevention of lipid oxidation in fatty foods. For lignin fraction, extensive research has been made on its application as a polymeric matrix in films, active agent, reinforcing agent as well as a coating for packaging materials, but few research has been done regarding its use in food coating materials.

4.2 Bioactive compounds

In addition to being a source of natural polymers, lignocellulosic residues derived from the agri-food industry are characterized by being rich in nutrients such as vitamins, minerals,

New Materials for a Circular Economy   Materials Research Forum LLC
Materials Research Foundations 149 (2023) 393-445   https://doi.org/10.21741/9781644902639-11

carbohydrates, lipids, and proteins, which makes these substrates ideal for obtaining natural active compounds (NAs) and biomaterials through different techniques such as hydrolysis, fermentation, biocatalysis, among others [98–100].

Different methodologies have been explored to obtain these active compounds, e.g., ultrasound-assisted extraction in sugarcane bagasse [101] and potato waste [102,103]; liquid-liquid extraction of grape pomace of *Cabernet sauvignon* variety [104]; ohmic heating assisted of potato peel [105]; MW-assisted extraction of cacao (*Theobroma cacao* L.) pod husk [106], sea buckthorn pomace and seed extracts [107]; fermentation of peanut waste [108], wheat waste [109,110], rice waste [109,111]; and others. *Table 4* presents some research on obtaining active compounds from lignocellulosic agro-industrial waste and its primary biological function.

Table 4. Active compounds obtained from lignocellulosic materials from agroindustry.

Raw material	Active compounds	Principal function	Reference
Banana peels	Flavonols, hydroxycinnamic acids, flavan-3- ols, and catecholamines.	Antioxidant capacity, antimicrobial activity, and dietary fiber	[112,113]
Avocado (peels and seeds)	Catechin, chlorogenic acid, caffeic acid, p-coumaric acid and ferulic acid.	Antioxidant capacity	[114]
Calendula officinalis (leaves and stem)	Vanillin and vanillic acid	Antioxidant capacity, essential flavoring and chemical additive	[115]
Yellow-fleshed and white cassava (peels and stems)	Phytate, polyphenols, tannins, oxalate, alkaloids and saponins	Antioxidant capacity	[116]
Rice bran	Ferulic, ellagic and cinnamic acids	Antioxidant capacity	[117]
	Coumarins, flavonoids, saponin, steroid and alkaloids	Antioxidant capacity	[118]
Sugarcane bagasse	Tricin 4-*O*-guaiacylglyceryl ether-7-*O*-glucopyranoside, genistin, *p-coumaric* acid, quercetin and genistein	Antioxidant capacity	[119]

The use of NAs as substitutes for synthetic additives (SAs) in foods, with antimicrobial, antioxidant, and anti-browning characteristics and properties, as well as to improve nutritional and sensory properties [120,121], are current market trends, which increasingly demand the use of NAs. They are also being studied in the development of biodegradable and edible packaging and coatings as a method of food preservation [122–124]. Incorporating NAs with antimicrobial activity, antioxidant capacity, stabilizer, emulsifier, preservative, and colorant, in biodegradable and edible films and coatings is a new trend in active and intelligent packaging [125,126]. The biopolymers used for developing these materials have shown a high potential to transport active ingredients, which must be biocompatible, non-toxic, and environmentally friendly for the food industry and biomedical applications [127,128].

The incorporation of NAs obtained from lignocellulosic residues such as corn stigma extract [129], *Ficus carica* Linn leaves extract [130], *Eriobotrya japonica* leaves [131] and date palm pits [132], for the development of films and coatings based on biopolymers (gelatin, corn starch, chitosan, ripe banana peel, and others) have potentiated their application in food preservation due to their antimicrobial activity and antioxidant capacity, in addition to improving the mechanical and physical properties. Some examples of the use of these materials for food preservation are presented in *Table 5*.

Table 5. Active films and coatings used in food preservation.

Raw material	Polymeric matrix	Food application	Principal function	Reference
Tea ground waste	Potato starch, Carboxymethyl cellulose, gelatin and cellulose	salmon fillets	Antioxidant capacity and antimicrobial activity	[133]
Feijoa peel flour	Pinhão starch and citric pectin	Apple	Antioxidant capacity and antimicrobial activity	[134]
Shallot onion wastes	Sodium alginate and carboxymethyl cellulose	Fresh-cut apple and potato	Anti-browning	[135]
Akebia trifoliata (Thunb.) Koidz. peel extracts	Chitosan	A. trifoliata fruits	Delaying crack and mature	[136]
Ppricot (*Prunus armeniaca*) kernel essential oil	Chitosan	Spiced beef	Antimicrobial activity (*Listeria monocytogenes*), decrease in lipid oxidation and better sensory attributes	[137]

These active packaging and coatings with active compounds obtained from lignocellulosic waste from the agro-industry have been used to preserve foods, mainly to increase the shelf life of the products and maintain their physicochemical and sensory characteristics and quality.

New Materials for a Circular Economy   Materials Research Forum LLC
Materials Research Foundations 149 (2023) 393-445   https://doi.org/10.21741/9781644902639-11

4.3   Functional foods: Prebiotics

Functional food science is a rapidly growing field, which has developed due to the public's desire for healthier lifestyles. Functional food is a term used for the first time in Japan in 1984, where there is a specific legislative procedure to approve these types of foods which are called as Foods for Specificied Health Use (FOSHU) and defined as *"Food products fortified with special constituents that possess advantageous physiological effects"*. Nevertheless, in most countries these foods are not legally recognized so there is not an agreed definition [138].

In the European Union there is no formal legislative definition for functional food, but nutritional and health claims must be listed in the Annex of Regulation (EC) No 1924/2006 [139]. These claims must be endorsed by scientific evidence and four steps are required: (1) Identification of the active food or component, (2) fulfil clinical studies and meta-analysis, (3) measure health endpoint either directly or through effective biomarkers, and (4) statistically significant health benefits.

According to Granato *et al.* (2020), functional foods can be defined as natural or processed foods that have a positive effect on health beyond nutritional uses when they are regularly consumed within a diverse diet at efficacious levels [140]. Nowadays the most demanded functional foods by consumers are yogurts with digestive benefits, cereals, margarines, and butters that control cholesterol metabolisms, and energy or protein bars and beverage [141]. Among the most widespread functional ingredients, prebiotics compounds industry produced in 2018 US$ 3.4 billion and will reach US$ 8.34 billion in 2026 [142].

Prebiotics are defined by the International Scientific Association for Probiotics and Prebiotics (ISAPP) as *"a substrate that is selectively utilized by host microorganisms conferring a health benefit"*. This definition involves any microbial community present in the human body, not only gut, but also skin and vaginal tract. In addition, the beneficial effect(s) of the prebiotic must be confirmed through studies conducted in the target animal and mediated by the microbiota. However, molecular-based methods have determined that other genera, which are different among hosts and ecosystems, are also able to exploit prebiotic substrates by fermentation and other metabolic pathways [143]. In this sense, in two human studies beyond bifidobacteria, other microbial groups such as *Faecalibacterium prausnitzii* and *Anaerostipes* spp. increased. In addition to selectivity, prebiotics must also produce a beneficial effect on health [144,145]. Prebiotics have a positive effect on boosting the immune system by regulating the action of lipogenic enzymes in the liver and increasing the synthesis of short-chain fatty acids [145]. In humans, fructooligosaccharides (FOS) and inulin are metabolized specifically by several strains of bifidobacterial. This process alters the gut ecosystem, stimulating the production of immunoglobulin A (IgA) which is a crucial component of the immune system. In the same process, FOS can increase bacteroides to stimulate IgA production [146,147]. Regarding immunomodulation, prebiotics also improve resistance to allergy and prevent respiratory disease. Relating to the metabolomic effects, prebiotics prevent osteoporosis, diabetes, hypertense, cancer and obesity. Also, prebiotics show therapeutic effects preventing diarrhea, urogenital infections, skin problems and stimulating the synthesis of vitamins [148]. The role of prebiotics in health benefits also includes effects on mental health and bone [144]. Besides these benefits, prebiotics add technological and sensory properties to foods such as texture, mouthfeel, and sugar replacement [149].

Considering dietary prebiotics, they must be resistant to gastric acidity, enzymatic hydrolysis, and gastrointestinal absorption, be metabolized by intestinal microbiota and selectively stimulate the growth and/or activity of gut bacteria associated with health and wellbeing [150]. The most studied and used dietary prebiotics are carbohydrate-based. Benefits of these prebiotics include dietary fiber enrichment, fat, and sugar reduction, extended products shelf-life and enhancement of product quality. Inulin is one of the most recognized and used carbohydrate-based prebiotic. It is commercially processed from Jerusalem artichoke and chicory roots [151]. Inulin has many beneficial effects such as antioxidant and anticancer properties, reduction diarrhea and inflammation period as well as the risk for intestinal bowel problem [152]. Prebiotic carbohydrates are also able to simulate the bifidogenic effect of human milk oligosaccharides, so they can be used to fortify infant milk formulas [149]. In low-fat and low-calorie meat products such as fermented sausages, part of fat is replaced by carbohydrate prebiotics gels. The result is a product with better properties than the conventional ones because the reduced amount of fat, the improvement in physicochemical, and sensorial characteristics, as well as the oxidative stability during storage, reducing the notes of hexanal [153]. However, there is evidence that other substances such as certain proteins and peptides, unsaturated fatty acids, flavonoids, micronutrients, and macronutrients also fit prebiotic definition [154].

The adequate number of prebiotics required to exert the health effect may vary according to the product and the microbial ecosystem of the host. Nevertheless, most of the prebiotics requires a dose above 3 g per day [143,155]. Processing conditions such as high temperatures, low pH values and those that favor Maillard reactions as well as other food components can reduce prebiotics functionalities. As a solution, diverse studies suggest the incorporation of prebiotic ingredients to active edible films and coatings [156]. In this context, agricultural by-products are an important source of prebiotics.

Gómez-García et al. 2021 [157] characterized different fractions of melon peels: solid, rich in carbohydrates; liquid, with a high percentage of ashes and the highest antioxidant activity; and pellet, with the greatest amount of proteins. These three fractions were also rich in nutrients and bioactive substances being suitable for their use in the development of functional foods. In a similar study using pomegranate peel flour, co-product from the juice extraction, the polyphenolic profile was determined revealing high amounts of punicalagin and ellagic acid, important antibacterial agents in the field of food and nutrients [158]. Another interesting source of antioxidant compounds are the by-products generated from olive grove farming, such as olive tree pruning and olive mill leaves. Besides showing a high total phenolic and flavonoid content with notable antioxidant capacity against different radicals, these extracts exhibited antimicrobial capacity corroborating their potential to be used in food and pharmaceutical industries [159].

In the case of pineapple stems and peels, they are interesting by-products to be applied in the food and nutraceutical industries. This biomass presents a moderate content in phenolic compounds such as chlorogenic, caffeic and ferulic acids but a high content in insoluble dietary fiber, mainly cellulose and hemicellulose [160]. The bioavailability, bioaccessibility, and prebiotic potential of pineapple stems and peels by-products have been further studied by Campos *et al.* (2020). As a result, these by-products promoted the growth of beneficial bacterial strains through a synergetic interaction of dietary fiber and polyphenols [161]. Besides that, Xu *et al.* (2019) [162] proved that the incorporation of xylo-oligosaccharides into CNF-reinforced hemicellulose/chitosan-based films improved the mechanical and barrier properties of the film. These xylo-oligosaccharides were released in the intestine and used by the gut microbiota. In other studies, probiotic bacteria growth and activity were stimulated incorporating fructooligosaccharides in starch-based films. It is a cost-effective alternative in food industry due to the cost of the matrix is low and the FOS can be obtained with high productivity [163].

Prebiotics can be added to coatings and films directly or combined with probiotics, which create a symbiotic effect. The choice of the best method depends on the equipment available, costs, efficiency, and application [156]. Direct casting method is the most used technique on a laboratory scale for films production with prebiotics. In this case, prebiotics are incorporated after obtaining a homogeneous and viscous solution and the conditions depend on the type of matrix and the prebiotic added. In any case, solubilization condition, solution mass, plates material, solvent evaporation and thickness must be considered. Regarding to edible coatings incorporated with prebiotics, they can be produced by spread coating, spray coating and dip coating. The last one is the most suitable due to it is simple and requires low time and low-cost equipment. Moreover, these films show greater thickness on fruit and vegetables surfaces and greater resistance to cold injury, color changes and deterioration [164]. Finally, thermoplastic extrusion technology to produce films with prebiotics has technological advantages such as the increase in the content of soluble dietary fiber. It is important to consider the appropriate parameters of extrusion to ensure the stability of prebiotics due to during thermal processing prebiotics can be degraded or hydrolyzed. As an alternative, other methods such as thermoplastic injection and thermo-pressing are used to develop materials incorporated with prebiotics [156].

5. Composites

5.1 Thermoplastics

In today's society there is great dependence on plastic materials, which can be found in all areas of everyday life. Sectors such as agriculture, packaging for the food industry, health, construction, etc. depend to a greater or lesser extent on these materials. This dependence causes an environmental problem, among other reasons because of their low or non-existent biodegradability. Adding natural fillers to thermoplastic compounds can increase their biodegradability, thus having a positive impact on the environment. In addition, if

New Materials for a Circular Economy   Materials Research Forum LLC
Materials Research Foundations 149 (2023) 393-445   https://doi.org/10.21741/9781644902639-11

these bio-based materials come from waste, an economic improvement of the process can be generated.

There are many works in the literature related to the formulation of thermoplastic materials with the addition of bio-based materials coming from agri-food residues, which, depending on the amount, can result in bioplastics. Sarki *et al.* (2011) introduced coconut shells in an epoxy polymeric matrix [165]. In this way it was possible to increase the tensile strength of the polymer as the amount of coconut shell introduced increased. The composite was manufactured by stirring the liquid epoxy polymer, the hardener, and the coconut shells with a particle size of less than 100 μm. These types of materials can be used in the construction of green buildings. Other materials that have been used as fillers in polymer matrices are cork dust and rice husk ashes in proportions of 1, 2 and 5% wt. Although with both materials the flexural strength and toughness were reduced compared to the polymeric matrix used, which in this case was polyester, an increase in impact energy absorption is achieved when the percentage of both fillers is 2.5%, reaching 30% absorption in the case of rice husk ashes [166].

Among the variety of existing thermoplastic polymers, those that are biodegradable are of special interest, although the modification of these materials does not always seek to increase their biodegradability as much as to improve their other characteristics. This is the example of poly(butylene adipate-co-terephthalate) (PBAT), a synthetic polymer with properties comparable to polyethylene, although with higher cost and shorter useful life. The development of a composite with this lignin-reinforced polymer matrix required the addition of zinc stearate and epoxy soybean oil as interfacial modifiers to build dynamic bonds between lignin and PBAT. This material achieved maximum tensile strength of 36.7 MPa and elongation at break of 725.3% at 5% wt. lignin content. When the lignin content was increased to 10% wt. the biocomposite with interfacial modifiers maintained the tensile strength of 35.4 MPa and the elongation at break of 627.8%, being 82% and 31% higher than that of the direct composite without interfacial modifiers, respectively [167].

Another biodegradable polymer widely used as a printing filament for 3D printers is polylactic acid (PLA). The addition of CNF to this polymer improved mechanical properties [168]. These CNF can be obtained by mechanical and/or chemical pretreatment by oxidation using a TEMPO catalyst from cellulose pulps from different raw materials. This composite is prepared by thermal extrusion at CNF concentrations of 1.3 and 5% wt. Characterization of the materials obtained revealed an increase in tensile modulus and strength from 2.9 - to 3.6 GPa and from 58 MPa to 71 MPa for the composite with 5% CNF [169].

PP is a thermoplastic widely used in the manufacture of garden furniture, so it needs be resistant to adverse climatic condition. The integration of peanut shell powder in this PP-matrix can modify its characteristics, improving its resistance to environmental exposure. The PP/peanut shell composite is prepared by melt blending and compression molding with a loading of 0 to 40% wt. These composites were irradiated using a 2.0 MeV electron beam accelerator at a fixed dose of 20 kGy. The properties of irradiated and non-irradiated

New Materials for a Circular Economy   Materials Research Forum LLC
Materials Research Foundations 149 (2023) 393-445   https://doi.org/10.21741/9781644902639-11

composites after natural weathering exposure were compared and characterized by tensile properties, scanning electron microscopy (SEM), carbonyl indices (CI), differential scanning calorimetry (DSC) and weight loss analysis. The results showed that the irradiated composites were more resistant to degradation by natural weathering than the non-irradiated ones. [170].

Due to the wide variety of uses of high-density polyethylene (HDPE) in construction materials, it is interesting to study the mechanical properties obtained by reinforcing it with rice husks, as it is possible to valorize a waste product of one of the most harvested crops in the world, of which up to 4.8% of the total production is wasted [171]. To obtain the integration of this residue into the polymeric matrix, a twin-screw extruder is used, introducing the mixture with different concentrations of rice husk (0, 10, 20, 20, 30 and 40% wt.). Polyethylene with maleic acid is added as a coupling agent. In this way, the tensile and flexural properties are improved by increasing the percentage of residue added at the cost of a decrease in impact strength [172].

On the other hand, the large production of olive oil in countries such as Italy, Greece and Spain makes necessary to valorize/reuse a large amount of waste produced by its processing. An interesting use for the leftover olive pits is to use them in powder form as reinforcement in a thermoplastic such as PLA. Tests have been carried out on the tensile and flexural mechanical properties of the composite obtained, which has been manufactured in proportions of 5, 10 and 15 % at two different particle sizes (30, 60 and 170 mm), integrating it into the matrix by milling. In general, an increase in the tensile modulus and a decrease in the flexural strength is observed [173].

Another material that could have many applications is a thermoplastic with flame-retardant properties. A composite has been designed from high-density aliphatic polyethylene by introducing jute fibers, corn silk and bagasse as reinforcement, ammonium polyphosphate to retard combustion and ferric stearate to activate biodegradability. Samples are prepared in a mixer at 180°C. After characterization, jute composites show the highest mechanical properties of 11 MPa ultimate strength and 8 GPa modulus of elasticity, while corn silk composites have better V-2 performance according to UL94 flame retardancy test. Water absorption of 15% fiber content is reduced to 1% [174].

5.2   Foams

One of the most desirable composite materials today because of their insulating properties and lightweight are foam-type polymers, specifically polyurethanes (PU). PU are very versatile polymers in terms of applications, as they can be used in the foam industry, coatings, elastomers, adhesives and even biomedicine[175–179]. More than 60% of the PU market is represented by foams, which are widely used as bedding, packaging, cushioning, automotive parts, and insulation materials [180]. However, the PU industry is highly dependent on fossil resources as the PU synthesis mostly uses petroleum-derived materials such as polyols and isocyanates. Therefore, many efforts are focused on researching new materials to produce PUs derived from renewable sources.

New Materials for a Circular Economy   Materials Research Forum LLC
Materials Research Foundations 149 (2023) 393-445   https://doi.org/10.21741/9781644902639-11

Reaction 1

HO-R'-OH + OCN-R-NCO ⟶ $\left[\begin{array}{c} \text{-C-N-R-N-C-O-R'-O-} \\ \overset{||}{\text{O}}\ \overset{|}{\text{H}}\quad \overset{|}{\text{H}}\ \overset{||}{\text{O}} \end{array}\right]_n$

Reaction 2

2 H_2O + OCN-R-NCO ⟶ H_2N-R-NH_2 + 2 CO_2

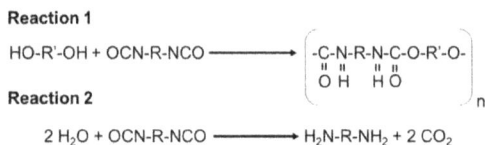

Figure 3. PU foams formation reactions. Taken with permission from [181].

Strategies such as oxypropylation or acid liquefaction of agri-food residues have been used to obtain polyols with high yields and desirable properties [182]. In this sense, when polyols are used for the formulation of foams, the two most important characteristics to consider are the hydroxyl number (I_{OH}) and the viscosity. This is because they directly affect the reaction rate in the foaming and cell formation process of the foams [181]. Two reactions occur during the formation of the foams, (i) the foaming process that competes with the use of isocyanate during (ii) the formation of the cells (*Figure 3*). If the foaming process is faster, foams with higher densities and smaller cells (rigid foams) will be obtained, while if the cell formation is faster, lighter foams with irregular and open cells (flexible foams) will be obtained. The polyols used to produce rigid foams have an I_{OH} between 200 and 800 with viscosities below 300 Pa·s. Lower I_{OH} and viscosities result in flexible foams. In general, the properties of PU foams (PUF) can be controlled during their synthesis depending on the structure and amount of monomers, catalysts, surfactants and blowing agents. Filler, flame retardants, pigments and colorants can also be added [175]. In this context, agri-food residues have been used to produce bio-based polyols subsequently used in the preparation of PUF providing desirable characteristics in a wide variety of applications. *Table 6* lists the agri-food residues used in the last years to obtain polyols, as well as their characteristics.

As can be seen, there is a wide range of possibilities when using agri-food residues to obtain polyols. However, the most used technique for this biomass is liquefaction assisted by acid catalysts such as sulfuric acid, which results in higher yields than other techniques such as base catalyzed-liquefaction or oxypropylation [190]. The liquefaction process involves the thermal deconstruction of biomass in the liquid stage at high temperatures (150-250 °C) in the presence of solvents, the most used being polyhydric alcohols such as polyethylene glycol (PEG) or glycerol. During this process, the biomass is degraded and decomposed into smaller molecules by solvolytic reactions. The polyols obtained are the result of a mixture of different compounds rich in hydroxyl groups that can be used directly to produce PU products. The liquefaction technique has numerous advantages such as obtaining high quality products without the need for pre-fractionation of the biomass [191,192].

Table 6. Agri-food residues to produce liquefied bio-based polyols.

Agri-food residue	Yield (%)	I_{OH} [mg KOH·g]	Viscosity [Pa·s]	Reference
Date seeds	84	586	0.45	[183]
Rapeseed cake	96	395	0.92	
Olive stone	92	496	0.49	
Corncob	91	504	0.55	
Apple pomace	97	428	0.72	
Empty fruit bunch fiber	99	241	113	[184]
Wheat straw	97	604	0.6	[185]
Bay tree pruning	78	224	6.17	[181]
Orange peel	-	470	6.5	[186]
	85	-	-	[187]
Potato peel	>80	293	-	[188]
Olive stone	40	-	6.2	[189]
Olive bagasse	22		4.7	
Grape seeds	24		5.7	
Rice husk	10		4.5	

Agri-food residues-based polyols have been employed to produce PUF. When PUF are produced by blending petroleum-based and liquefied biomass-based polyols, higher thermal stabilities and compressive strengths are obtained than solely petroleum-based PUF. If PUF are synthesized only with bio-based polyols, unique properties are obtained depending on the characteristics of the polyol used. As example, the polyol resulting from the liquefaction of empty fruit bunches (EFB) has been used for the synthesis of rigid PUF with high content of closed cells conferring a high compressive modulus and strength (433.5 and 41.0 MPa, respectively). Moreover, these EFB-PUF present higher amount of residue after thermal decomposition than petroleum-based PUF indicating that they present phenolic complex polymers with higher thermal stability [184]. In a similar study, the polyol resulting from the liquefaction of bay tree pruning waste (BTPW) was used to formulate PUF together with castor oil, achieving an optimal substitution of 50%, studying the effect of the presence or absence of the blowing agent (water) and the nature of the isocyanate used (Methylene diphenyl diisocyanate, MDI, or Toluene-2,4-diisocyanate, TDI). The use of TDI without blowing agent resulted in PUF with high densities (141-150 km/m^3) compared to TDI-PUF with water (41-57 kg/m^3). Therefore, the compressive strength of the latter was considerably higher (120 kPa). The use of MDI resulted in even denser foams with higher compressive strength (250 kPa). In all these foams, a well-defined closed cell morphology was found, indicating that they could be excellent candidates for insulating applications with high thermal resistance due to the use of BTPW-polyol [181]. Similar foams were prepared using wheat straw polyol. These PUF showed excellent biodegradability at 30 and 60 days (5-8% and 9-10%, respectively) in soil media tests, indicating that they could be suitable for agricultural purposes in plant nurseries [185]. The most recent studies in this field have employed the liquefaction product of orange peel for the formation of lightweight PUF with good compressive strength (90 kPa). In addition, the final properties of PUF can be tailored by the choice of additives used in their formation [187]. During the preparation of PUF, the addition of compounds such as dimethyl methyl phosphonate results in materials with flame retardant properties, especially useful for building insulation [186]. Recently, agri-food residues have not only been used for the synthesis of polyols and PUF but have also started to be used as fillers to reinforce the mechanical performance of PUF. Thus, the effect of walnut shells and hazelnut shells as fillers in both rigid and flexible PUF has been tested. In both, such fillers increase the mechanical strength and thermal stability of the materials obtained [193,194].

In summary, agri-food residues represent a promising pathway for use in the preparation of foam-type polymers, either as polyols or as fillers. The wide range of residues already tested for these purposes opens the door to further expanding the range of residues to be used to help achieve a circular bioeconomy.

5.3   Fiberboards

Fiberboards are usually made of wood or lignocellulosic resources and synthetic binder, which are joined together by heat and pressure [195]. There are two different kinds of fiberboards depending on their density. Those fiberboards between 400-900 kg/m^3 in density are known as Medium-Density Fiberboards (MDF), and High-Density Fiberboards (HDF) for densities of more than 900 kg/m^3 [196]. Two processes are possible to manufacture fiberboards which are called wet and dry processes. For wet processing, cellulosic fibers are treated with water to disperse fibers, high temperature to press, and not binders are used to make fiberboards. In a way to avoid binders, self-bonding is used as pretreatment of the lignocellulosic material. However, this technique has disadvantages because of the low-adhesion fiberboards obtained. Besides self-bonding and steam explosion, there are also other pretreatment techniques to improve adhesion and fiber bonding for lignocellulosic fiber from agriculture residues such as, electromagnetic radiation, steam explosion, chemical pre-treatment with acids or alkaline pretreatments, white-rot fungi pre-treatment, thermos-pressing, hot pressing, and enzymatic pre-treatment with lacasses, xylases, and endonucleases [197]. However, in the dry process firstly cellulosic fibers are dried and adhesives are added to be pressed at a determined curing temperature [198]. In this process, the role of adhesives is important to bond fibers, distribute loads of the board, and increase the strength and modulus of the final fiberboard [199]. The most economical and used binders for fiberboards are urea-formaldehyde and phenol-formaldehyde but they have an impact on human health and the environment [200,201]. Long-term exposure is associated with cancer and asthma and, they can be toxic. If people are exposed to large amounts of formaldehyde, it can cause eye, nose, throat, and skin irritation [202]. Besides, they are non-biodegradable and non-recyclable. They are a problem for users, workers, and the environment. Its toxicity is induced by a hydrolysis which is developed during production and because of the reversibility of amino-methylene bonds and instability of the methylene ether bridges during the service time [203,204]. Lignin isolated from lignocellulosic materials can be used as a natural binder in the fiberboard manufacturing. Domínguez-Robles *et al.* isolated alkaline lignin from wheat straw as binder to produce HDF. In addition, the cellulosic fraction of the wheat straw was also used as fiber source for fiberboard. The obtained fiberboards with lignin at 2.5% have a high value of flexural strength. When the amount of lignin is 25%, they have the highest value of flexural modulus. Lignin improves the adhesion of the fibers. Besides, fiberboards with 15% of lignin as composite have a density from 1013 to 1161 kg/m^3. So, lignin can be used as an alternative to binders, and it acquire HDF which could be used for furniture [205]. Mancera *et al.* used commercial lignin from softwood as adhesive for fiberboards from *Vitis vinifera* residues. The cellulosic fiber was obtained from steam explosion. The lignin addition improves the internal bond between fibers, and it is attributed to a high condensation between lignin and fibers. The obtained fiberboards by this pretreatment are HDF and have good mechanical and water resistance properties met standard specifications. Kraft lignin at 5% wt. is enough to get double modulus rupture (MOR) and modulus of elasticity (MOE) in contrast with binderless fiberboards [206].

On the other hand, the use of binders is not a unique problem for the fiberboard industry. The demand for fiber is increasing each year which is contributing to deforestation [207]. Because of this, it is a necessity to find other resources to obtain fiberboards. Agriculture residues are new alternatives that avoid burning those residues and produce greenhouse gases emitted. They are easy to renew and recycle and they are accessible and biodegradable. Furthermore, they would add value to the agricultural and industrial sectors improving the competitiveness of both industries. In the last years, different kinds of agricultural residues are used to manufacture fiberboards such as grapevine branches, coconut husk, banana bunch, wheat and rapeseed straw, bagasse, almond residues, hemp fibers, etc. [208,209].

Giant reed, which is invasive species in many areas, is processed by steam explosion. This pre-treatment consisted of high pressure and then, decompression treatment which disjoins lignocellulosic components without degrading hemicellulose and lignin. This effect promotes bonding between the lignocellulosic components during the fiberboard production and showed optimal final properties. The optimal pre-temperature was 200°C to obtain comparable properties with commercial fiberboards and to achieve HDF [203]. Wheat straw fiberboards with no binders was produced by enzymatic refining (β-1,4-endonuclease). The obtained fibers were used to get HDF, presenting higher mechanical properties and an excellent bonding strength in comparison to commercial fiberboards. The results showed that enzymatic and mechanical treatments can decrease energy used in mechanical refining and obtain fiberboards which are able to use in several applications like panel housing [205]. In another study, wheat straw fiberboards were produced by two independent treatments: i) enzymatic biobeating, and ii) lignocellulose nanofiber addition.

Both treatments showed an increase in the mechanical and physical properties due to better strength and internal bonds of the fibers [210]. Sunflower plants were used to develop thermal insulation fiberboards for construction by thermo-pressing with a heated hydraulic press. A mixture of all lignocellulosic fibers and protein was obtained acting as binder and filler. These fiberboards showed varied density from 500 to 800 kg/m^3 depending on molding temperature and time and pressure [211].

6. Clinical applications

The development of materials for clinical applications is one of the fields of greatest interest in research due to the great importance it has in the healthcare field. Within this field of application, the materials developed must maintain mechanical performance as well as low or even null toxicity, thus serving as drug support or tissue development promoter. As the industrial applications mentioned so far, natural polymers find a place in the pharmaceutical/medical industry. Most of the studies reported in the literature describe the use of polymers such as bacterial cellulose or chitosan due to their high biocompatibility and antimicrobial properties, respectively, which result in favorable mechanisms to control drug release profiles and wound healing processes [212,213]. However, the increase in the generation of agricultural and agro-industrial waste together with the twin goals of reducing food waste and the global carbon footprint, has led researchers to focus their efforts on finding applications of this biomass in the pharmaceutical and biomedical fields.

In this regard, Cui *et al.* used cellulose extracted from durian rind for the manufacture of organohydrogels applicable as antimicrobial wound dressings in medical supplies. The durian is a highly nutritious and health-promoting fruit, but less than half of it is edible, the remaining parts being treated as food waste. The organohydrogels developed, which also contained natural yeast phenolics, exhibited no cytotoxicity when applied as wound dressing on pig skin as a proof of concept [214]. Also considering the cellulosic fraction, cellulose purified from the food by-product okara (soybean residue) has been used to produce biodegradable and non-cytotoxic hydrogels on NIH3T3 cells. These hydrogels demonstrated efficacy as biocompatible and portable deformation sensors, whose further modification could be applied in tissue engineering or biomimetic smart soft materials [215]. As before, the use of cellulose derivatives in this field has also reported promising results. Brown *et al.* prepared nanocrystalline cellulose from wheat straw for the manufacture of fibrin nanocomposites applicable in vascular grafts. These nanocomposites can adapt to the various mechanical properties of native blood vessels [216]. Other cellulose derivatives such as carboxymethyl cellulose (CMC) have been successfully applied as pharmaceutical excipients. CMC prepared from corn waste-holocellulose a single step with a high degree of substitution has been used as pharmaceutical excipient with results comparable to commercial products due to good flow properties [217].

Another relevant cellulose application within the pharmaceutical industry is the design of biocomposite materials for controlled drug delivery. Specifically, cellulose nanoparticles

New Materials for a Circular Economy   Materials Research Forum LLC
Materials Research Foundations 149 (2023) 393-445   https://doi.org/10.21741/9781644902639-11

possess desirable properties for application as drug carrier. Features such as high specific surface area, negative surface charge, their demonstrated decrease in the rate of clearance from the bloodstream and their hydrophilic nature that prevents opsonin protein absorption, allow the binding of large numbers of drugs with optimal dose control. However, the studies carried out in this field so far have used commercial cellulose. More research is still needed on the compatibility of the cellulosic fraction with the human body to scale up this application to the use of agricultural and agri-food residues [218].

Another lignocellulosic fraction that has found a place in the clinical setting is hemicelluloses. Much research has been conducted over the years in the use of hemicellulose-based wound dressing [219,220]. However, interest in this biopolymer is gaining attention as drug-delivery system in recent decades. In this regard, psyllium husk arabinoxylans have been used for the manufacture of films loaded with gentamicin, a model drug. These antibacterial films showed similar results to those of a standard gentamicin solution, being cytocompatible and promising for use as wound dressing [221,222].

Finally, the use of plant biomass-derived compounds in the clinical setting is of interest not only because of their polymeric fraction but also because of the antimicrobial compounds they contain. Several bioactive compounds derived from food crops such as coffee, soybean or pomegranate have been used as antibiotic adjuvants, such as gentamicin, novobiocin or chloramphenicol [223–225]. However, this application still needs to be scaled up to the use of lignocellulosic residues of these crops.

7. Energy

As mentioned during this chapter, lignocellulosic biomass can be used for energy generation through direct combustion or as source for biofuel production, such bioethanol. Its direct combustion converts chemically stored energy into heat and power (*Figure 4*). This conversion process releases CO_2 into the atmosphere but due to the relatively short cycle this process can be considered CO_2 neutral. The heating value of the fuel, the corrosiveness, of the flue gas, the stickiness of the ash, etc. depends on the chemical composition of the feedstock (nutrients, trace elements as heavy metals, structure of the hydrocarbon bonds) and at the same time depends on agricultural species, part of the plant, seasonal and regional variance. The most suitable fuel from a thermochemical perspective comes from seasoned woods, while quick-turnaround energy crops or straw are typically less desirable. The various facets of combustion and systems have been the subject of extensive research since the beginning of the Industrial Revolution, as well as generic texts [226].

A more recently researched concept is "lignocellulosic ethanol" or "cellulosic ethanol". Cellulosic ethanol is defined as the ethanol produced from lignocellulosic material rather than from the plant's seeds or fruit. This kind of fuel uses a high abundant and low-cost raw material such as corn stover, rice and wheat straw, sugarcane bagasse, etc. [227]. Cellulosic bioethanol production as other fuels that derive from lignocellulose, does not imply competition with the food sector and constitutes an alternative use of waste [228,229]. By now, there exists several operational cellulosic ethanol plants in Europe operating at different levels (pilot, demo, and commercial) (*Table 7*).

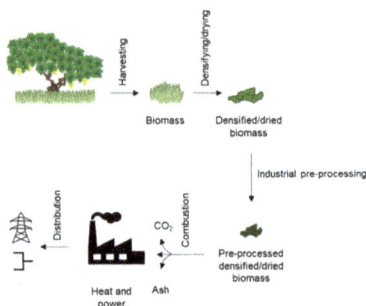

Figure 4. Production of heat and power from solid lignocellulosic biomass.

Table 7. Operational cellulosic ethanol plants in Europe, adapted from [230].

Company	Country	Operation level	Feedstock	Output [t/y]
Aalborg University	Denmark	Pilot	Wheat straw	11
TNO	Netherlands	Pilot	Wheat straw, grass, corn stover, bagasse, wood chips	100
Weyland AS	Norway	Pilot	Various feedstock, mostly spruce and pine	158
SEKAB/EPAB	Sweden	Pilot	Wood chips and agricultural wastes	160
Abengoa	Spain	Demo	Straw and municipal residues	400
ECN	Netherlands	Pilot	Clean wood and demolition wood	346
PROCETHOL 2G	France	Pilot	Woody and agricultural by-products, residues, energy corps	2,700
Inbicon	Denmark	Demo	Wheat straw	4,300
BioAgra	Poland	Demo	Wheat straw and corn stover	50,000
Beta Renewables	Italy	Commercial	Wheat straw	60,000

New Materials for a Circular Economy   Materials Research Forum LLC
Materials Research Foundations 149 (2023) 393-445   https://doi.org/10.21741/9781644902639-11

The complex lignocellulose structure makes it resistant to biological and chemical degradation. Therefore, the production of cellulosic ethanol consists of several steps, mainly (*Fig.5*): 1) pretreatment of the lignocellulosic material, 2) saccharification or hydrolysis, and 3) alcoholic fermentation.

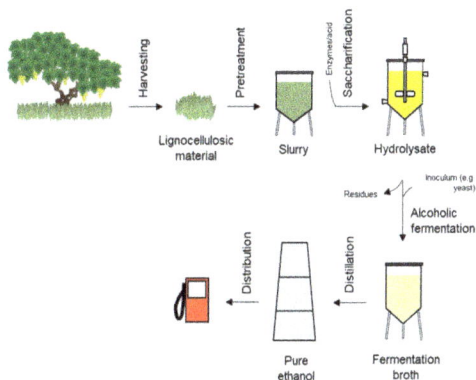

Figure 5. Production of cellulosic bioethanol.

Pretreatment

This is the treatment for disrupting the close inter-component association between main constituents of the plant cell wall. There exist five types of pretreatments: chemical, physical thermal, biological, and combined. One of the most promising pretreatment methods for industrial implementation is the chemical acid-based chemical pretreatment [231]. Mineral acids are typically used, but additional options include organic acids and sulfur dioxide. One of the main inconvenient for most of the pretreatments is the production of inhibitory by-products for downstream biochemical processes (hydrolysis and fermentation) [232]. To minimize the inhibitors effects, some strategies have been attempted: feedstock selection and engineering, detoxification/conditioning, bioabatemen, culturing schemes, selection of microorganism, evolutionary engineering, and genetic/metabolic engineering[231].

Saccharification or hydrolysis

Essential step to convert macromolecules of carbohydrates (celluloses and hemicelluloses) into fermentable sugars (hexoses and pentoses) for ethanol production. This step could be acid or enzymatic in which intervene cellulases and hemicellulases. There are three types of cellulases: endoglucanase, exoglucanase, and β-glucosidase. The first is a glucanase that bonds at random, internal positions in the cellulose chain while exoglucanase progressively cleaves cellobiose from chain ends. β-glucosidases are gluocosidase that converts

New Materials for a Circular Economy   Materials Research Forum LLC
Materials Research Foundations 149 (2023) 393-445   https://doi.org/10.21741/9781644902639-11

cellobiose and small cello-oligosaccharides to glucose [233,234]. The recent discovery of lytic polysaccharide monooxygenases (LPMOs) has revolutionized the perspective on how cellulose is degraded. These enzymes are copper-dependent lytic polysaccharide monooxygenases with a strong preference for cleaving every second glycosidic bond, with low endoglucanase activity but promotes the hydrolytic breakdown of crystalline cellulose [235].

Alcoholic fermentation

Step that converts the hexoses and pentoses in the hydrolysate to ethanol. Fermentation can be carried out by yeasts, bacteria, and filamentous fungus. Among the yeasts employed are *Saccharomyces cerevisiae*, *Candida shehatae*, *Pachysolen tannophilus*, and *Pichia stipitis* [230]. *S. cerevisiae* is the most widely studied yeast, most employed yeast that tolerates a wide range of pH thus making the process less susceptible to infection, able to ferment various lignocellulosic hydrolysates with high ethanol yield, and good tolerance to inhibitors and osmotic pressure [230,236]. *C. shehatae*, *P. tannophilus* and *P. stipitis* are facultative anaerobic yeasts that contrary to *S. cerevisiae*, are able to ferment xylose [237,238]. *Clostridium thermocellum* and *Zymomonas mobilis* are bacteria used for cellulosic ethanol production [230]. *C. thermocellum* is an anaerobic thermophilic bacterium that contributes to either fermentation and hydrolysis (produces cellulases and hemicellulases) while *Z. mobilis* has low biomass yield and a higher ethanol productivity than *S. cerevisiae* [230,236,237].

Conclusions and future aspect

Biorefinery from agro-industrial residues is a response to two of the main objectives of sustainable development. On the one hand, to valorize the enormous amount of lignocellulosic waste generated by the agricultural and agri-food industry, from the cultivation of the biomass to its completion in the food supply chain. On the other hand, to exploit the renewable sources of energy and materials that the application of biorefinery implies, thus breaking with the dependence on fossil resources and their derivatives.

From this point of view, the application of this biorefinery has reached various industrial sectors, which are described in this chapter. These industries range from the most traditional and widely studied, such as papermaking or fine chemicals production, to the most innovative industries which, seeking to make the circular economy and sustainable development part of their sector, find in lignocellulosic biomass an excellent candidate. These latter industries include the food industry (packaging, bioactive compounds, and functional foods), composite materials (thermoplastics, foams, fiberboards), clinical applications, and energy production.

The multidisciplinary approach of lignocellulosic residues biorefinery is evident from the diversity of industries in which it is used- There are still barriers to overcome to achieve a complete transition from out petroleum-based production systems to new bio-based systems. These are due to the inherent challenges to lignocellulosic biomass, such as seasonal biomass availability, specific biomass processing or pretreatment strategies.

These are some of the essential milestones to achieve full biorefinery capacity and are expected to be of relevance in the coming years.

References

[1]   Food and Agriculture Organization of the United Nations (FAO), FAOSTAT, Crops and Livestock Products. (2022).

[2]   E. Espinosa, R.I. Arrebola, I. Bascón-Villegas, M. Sánchez-Gutiérrez, J. Domínguez-Robles, A. Rodríguez, Industrial application of orange tree nanocellulose as papermaking reinforcement agent, Cellulose. 27 (2020) 10781–10797. https://doi.org/10.1007/s10570-020-03353-w

[3]   M. Kissinger, J. Fix, W.E. Rees, Wood and non-wood pulp production: Comparative ecological footprinting on the Canadian prairies, Ecological Economics. 62 (2007) 552–558. https://doi.org/10.1016/j.ecolecon.2006.07.019

[4]   P. Rousu, P. Rousu, J. Anttila, Sustainable pulp production from agricultural waste, Resour Conserv Recycl. 35 (2002) 85–103. https://doi.org/10.1016/S0921-3449(01)00124-0

[5]   E.S. Abd El-Sayed, M. El-Sakhawy, M.A.-M. El-Sakhawy, Non-wood fibers as raw material for pulp and paper industry, Nord Pulp Paper Res J. 35 (2020) 215–230. https://doi.org/10.1515/npprj-2019-0064

[6]   J. Garcia, Pastas celulósicas de materias primas alternativas a las convencionales, in: Caracterización Morfológica de Las Materias Primas, Editorial Gráficas Sol, S.A., 2005: pp. 71–95

[7]   Z. Jahan, M.B.K. Niazi, M.B. Hägg, Ø.W. Gregersen, Decoupling the effect of membrane thickness and CNC concentration in PVA based nanocomposite membranes for CO2/CH4 separation, Sep Purif Technol. 204 (2018) 220–225. https://doi.org/10.1016/J.SEPPUR.2018.04.076

[8]   F. Vargas, Z. González, R. Sánchez, L. Jiménez, A. Rodríguez, Cellulosic pulps of cereal straws as raw material for the manufacture of ecological packaging, Bioresources. 7 (2012) 4161–4170

[9]   L. Jiménez, A. Rodríguez, A. Pérez, A. Moral, L. Serrano, Alternative raw materials and pulping process using clean technologies, Ind Crops Prod. 28 (2008) 11–16. https://doi.org/10.1016/J.INDCROP.2007.12.005

[10]   S. de Lopez, M. Tissot, M. Delmas, Integrated cereal straw valorization by an alkaline pre-extraction of hemicellulose prior to soda-anthraquinone pulping. Case study of barley straw, Biomass Bioenergy. 10 (1996) 201–211. https://doi.org/10.1016/0961-9534(95)00031-3

[11]   F.S. Chakar, A.J. Ragauskas, Review of current and future softwood kraft lignin process chemistry, Ind Crops Prod. 20 (2004) 131–141. https://doi.org/10.1016/J.INDCROP.2004.04.016

[12]   Y. Uraki, Y. Sano, Polyhydric Alcohol Pulping at Atmospheric Pressure: An Effective Method for Organosolv Pulping of Softwoods, Holzforschung. 53 (1999) 411–415. https://doi.org/10.1515/HF.1999.068

[13]   I. Volf, V.I. Popa, 4 - Integrated Processing of Biomass Resources for Fine Chemical Obtaining: Polyphenols, in: V. Popa, I. Volf (Eds.), Biomass as Renewable Raw Material to Obtain Bioproducts of High-Tech Value, Elsevier, 2018: pp. 113–160. https://doi.org/https://doi.org/10.1016/B978-0-444-63774-1.00004-1

[14]   R.A. Sheldon, Green and sustainable manufacture of chemicals from biomass: state of the art, Green Chemistry. 16 (2014) 950–963. https://doi.org/10.1039/C3GC41935E

[15]   H.W. Ryu, D.H. Kim, J. Jae, S.S. Lam, E.D. Park, Y.-K. Park, Recent advances in catalytic co-pyrolysis of biomass and plastic waste for the production of petroleum-like hydrocarbons, Bioresour Technol. 310 (2020) 123473. https://doi.org/https://doi.org/10.1016/j.biortech.2020.123473

[16]   D. Gupta, R. Kumar, K.K. Pant, Hydrotalcite supported bimetallic (Ni-Cu) catalyst: A smart choice for one-pot conversion of biomass-derived platform chemicals to hydrogenated biofuels, Fuel. 277 (2020) 118111. https://doi.org/https://doi.org/10.1016/j.fuel.2020.118111

[17]   P. Ghorbannezhad, M.D. Firouzabadi, A. Ghasemian, P.J. de Wild, H.J. Heeres, Sugarcane bagasse ex-situ catalytic fast pyrolysis for the production of Benzene, Toluene and Xylenes (BTX), J Anal Appl Pyrolysis. 131 (2018) 1–8. https://doi.org/https://doi.org/10.1016/j.jaap.2018.02.019

[18]   A. Alcazar-Ruiz, L. Sanchez-Silva, F. Dorado, Enhancement of BTX production via catalytic fast pyrolysis of almond shell, olive pomace with polyvinyl chloride mixtures, Process Safety and Environmental Protection. 163 (2022) 218–226. https://doi.org/https://doi.org/10.1016/j.psep.2022.05.029

[19]   J. Damay, I.-Z. Boboescu, J.-B. Beigbeder, X. Duret, S. Beauchemin, O. Lalonde, J.-M. Lavoie, Single-stage extraction of whole sorghum extractives and hemicelluloses followed by their conversion to ethanol, Ind Crops Prod. 137 (2019) 636–645. https://doi.org/https://doi.org/10.1016/j.indcrop.2019.05.028

[20]   M. Hijosa-Valsero, A.I. Paniagua-García, R. Díez-Antolínez, Biobutanol production from apple pomace: the importance of pretreatment methods on the fermentability of lignocellulosic agro-food wastes, Appl Microbiol Biotechnol. 101 (2017) 8041–8052. https://doi.org/10.1007/s00253-017-8522-z

[21]   S. Li, W. Deng, Y. Li, Q. Zhang, Y. Wang, Catalytic conversion of cellulose-based biomass and glycerol to lactic acid, Journal of Energy Chemistry. 32 (2019) 138–151. https://doi.org/https://doi.org/10.1016/j.jechem.2018.07.012

[22]   M. Verma, P. Mandyal, D. Singh, N. Gupta, Recent Developments in Heterogeneous Catalytic Routes for the Sustainable Production of Succinic Acid from

Biomass Resources, ChemSusChem. 13 (2020) 4026–4034.
https://doi.org/https://doi.org/10.1002/cssc.202000690

[23]   L.G. Covinich, N.M. Clauser, F.E. Felissia, M.E. Vallejos, M.C. Area, The
       challenge of converting biomass polysaccharides into levulinic acid through
       heterogeneous catalytic processes, Biofuels, Bioproducts and Biorefining. 14 (2020)
       417–445. https://doi.org/https://doi.org/10.1002/bbb.2062

[24]   Y. Su, M. Lu, R. Su, W. Zhou, X. Xu, Q. Li, A 3D MIL-101@rGO composite as
       catalyst for efficient conversion of straw cellulose into valuable organic acid, Chinese
       Chemical Letters. 33 (2022) 2573–2578.
       https://doi.org/https://doi.org/10.1016/j.cclet.2021.08.078

[25]   J.P.A. Silva, J.S.M. Nogueira, C.L. de Aquino Santos, L.M. Carneiro, 9 - 5-
       Hydroxymethylfurfural as a chemical platform for a lignocellulosic biomass
       biorefinery, in: A.K. Chandel, F. Segato (Eds.), Production of Top 12 Biochemicals
       Selected by USDOE from Renewable Resources, Elsevier, 2022: pp. 269–315.
       https://doi.org/https://doi.org/10.1016/B978-0-12-823531-7.00004-4

[26]   F.D. Pileidis, M.-M. Titirici, Levulinic Acid Biorefineries: New Challenges for
       Efficient Utilization of Biomass, ChemSusChem. 9 (2016) 562–582.
       https://doi.org/https://doi.org/10.1002/cssc.201501405

[27]   L. Zhang, G. Xi, J. Zhang, H. Yu, X. Wang, Efficient catalytic system for the
       direct transformation of lignocellulosic biomass to furfural and 5-
       hydroxymethylfurfural, Bioresour Technol. 224 (2017) 656–661.
       https://doi.org/https://doi.org/10.1016/j.biortech.2016.11.097

[28]   A. Kumar, A.S. Chauhan, Shaifali, P. Das, Lignocellulosic biomass and
       carbohydrates as feed-stock for scalable production of 5-hydroxymethylfurfural,
       Cellulose. 28 (2021) 3967–3980. https://doi.org/10.1007/s10570-021-03764-3

[29]   C.B.T.L. Lee, T.Y. Wu, A review on solvent systems for furfural production from
       lignocellulosic biomass, Renewable and Sustainable Energy Reviews. 137 (2021)
       110172. https://doi.org/https://doi.org/10.1016/j.rser.2020.110172

[30]   S. Dutta, S. De, B. Saha, Advances in biomass transformation to 5-
       hydroxymethylfurfural and mechanistic aspects, Biomass Bioenergy. 55 (2013) 355–
       369. https://doi.org/https://doi.org/10.1016/j.biombioe.2013.02.008

[31]   D.A. Iryani, S. Kumagai, M. Nonaka, K. Sasaki, T. Hirajima, Production of 5-
       hydroxymethyl Furfural from Sugarcane Bagasse under Hot Compressed Water,
       Procedia Earth and Planetary Science. 6 (2013) 441–447.
       https://doi.org/https://doi.org/10.1016/j.proeps.2013.01.058

[32]   S. Dutta, S. De, Md.I. Alam, M.M. Abu-Omar, B. Saha, Direct conversion of
       cellulose and lignocellulosic biomass into chemicals and biofuel with metal chloride
       catalysts, J Catal. 288 (2012) 8–15.
       https://doi.org/https://doi.org/10.1016/j.jcat.2011.12.017

[33]   Y. Shao, D.C.W. Tsang, D. Shen, Y. Zhou, Z. Jin, D. Zhou, W. Lu, Y. Long, Acidic seawater improved 5-hydroxymethylfurfural yield from sugarcane bagasse under microwave hydrothermal liquefaction, Environ Res. 184 (2020) 109340. https://doi.org/https://doi.org/10.1016/j.envres.2020.109340

[34]   M. Li, H. Jiang, L. Zhang, X. Yu, H. Liu, A.E.A. Yagoub, C. Zhou, Synthesis of 5-HMF from an ultrasound-ionic liquid pretreated sugarcane bagasse by using a microwave-solid acid/ionic liquid system, Ind Crops Prod. 149 (2020) 112361. https://doi.org/https://doi.org/10.1016/j.indcrop.2020.112361

[35]   Q. Ji, C.P. Tan, A.E.A. Yagoub, L. Chen, D. Yan, C. Zhou, Effects of Acidic Deep Eutectic Solvent Pretreatment on Sugarcane Bagasse for Efficient 5-Hydroxymethylfurfural Production, Energy Technology. 9 (2021) 2100396. https://doi.org/https://doi.org/10.1002/ente.202100396

[36]   Y. Yang, C. Hu, M.M. Abu-Omar, Conversion of carbohydrates and lignocellulosic biomass into 5-hydroxymethylfurfural using AlCl3·6H2O catalyst in a biphasic solvent system, Green Chemistry. 14 (2012) 509–513. https://doi.org/10.1039/C1GC15972K

[37]   H. Gómez Bernal, L. Bernazzani, A.M. Raspolli Galletti, Furfural from corn stover hemicelluloses. A mineral acid-free approach, Green Chemistry. 16 (2014) 3734–3740. https://doi.org/10.1039/C4GC00450G

[38]   L. Yan, N. Liu, Y. Wang, H. Machida, X. Qi, Production of 5-hydroxymethylfurfural from corn stalk catalyzed by corn stalk-derived carbonaceous solid acid catalyst, Bioresour Technol. 173 (2014) 462–466. https://doi.org/https://doi.org/10.1016/j.biortech.2014.09.148

[39]   Z. Xu, W. Li, Z. Du, H. Wu, H. Jameel, H. Chang, L. Ma, Conversion of corn stalk into furfural using a novel heterogeneous strong acid catalyst in γ-valerolactone, Bioresour Technol. 198 (2015) 764–771. https://doi.org/https://doi.org/10.1016/j.biortech.2015.09.104

[40]   L. Liu, X. Yang, Q. Hou, S. Zhang, M. Ju, Corn stalk conversion into 5-hydroxymethylfurfural by modified biochar catalysis in a multi-functional solvent, J Clean Prod. 187 (2018) 380–389. https://doi.org/https://doi.org/10.1016/j.jclepro.2018.03.234

[41]   X.-N. Ye, Q. Lu, X. Wang, T.-P. Wang, H.-Q. Guo, M.-S. Cui, C.-Q. Dong, Y.-P. Yang, Fast Pyrolysis of Corn Stalks at Different Growth Stages to Selectively Produce 4-Vinyl Phenol and 5-Hydroxymethyl Furfural, Waste Biomass Valorization. 10 (2019) 3867–3878. https://doi.org/10.1007/s12649-018-0259-0

[42]   B. Yuan, J. Guan, J. Peng, G. Zhu, J. Jiang, Green hydrolysis of corncob cellulose into 5-hydroxymethylfurfural using hydrophobic imidazole ionic liquids with a recyclable, magnetic metalloporphyrin catalyst, Chemical Engineering Journal. 330 (2017) 109–119. https://doi.org/https://doi.org/10.1016/j.cej.2017.07.058

[43]   B. Nis, B. Kaya Ozsel, Efficient direct conversion of lignocellulosic biomass into biobased platform chemicals in ionic liquid-water medium, Renew Energy. 169 (2021) 1051–1057. https://doi.org/https://doi.org/10.1016/j.renene.2021.01.083

[44]   C. Lucas-Torres, A. Lorente, B. Cabañas, A. Moreno, Microwave heating for the catalytic conversion of melon rind waste into biofuel precursors, J Clean Prod. 138 (2016) 59–69. https://doi.org/https://doi.org/10.1016/j.jclepro.2016.03.122

[45]   S. Kang, J. Fu, G. Zhang, From lignocellulosic biomass to levulinic acid: A review on acid-catalyzed hydrolysis, Renewable and Sustainable Energy Reviews. 94 (2018) 340–362. https://doi.org/https://doi.org/10.1016/j.rser.2018.06.016

[46]   Y.W. Tiong, C.L. Yap, S. Gan, W.S.P. Yap, Conversion of Biomass and Its Derivatives to Levulinic Acid and Levulinate Esters via Ionic Liquids, Ind Eng Chem Res. 57 (2018) 4749–4766. https://doi.org/10.1021/acs.iecr.8b00273

[47]   S. Kang, J. Yu, An intensified reaction technology for high levulinic acid concentration from lignocellulosic biomass, Biomass Bioenergy. 95 (2016) 214–220. https://doi.org/https://doi.org/10.1016/j.biombioe.2016.10.009

[48]   A.S. Khan, Z. Man, M.A. Bustam, A. Nasrullah, Z. Ullah, A. Sarwono, F.U. Shah, N. Muhammad, Efficient conversion of lignocellulosic biomass to levulinic acid using acidic ionic liquids, Carbohydr Polym. 181 (2018) 208–214. https://doi.org/https://doi.org/10.1016/j.carbpol.2017.10.064

[49]   D.B. Bevilaqua, M.K.D. Rambo, T.M. Rizzetti, A.L. Cardoso, A.F. Martins, Cleaner production: levulinic acid from rice husks, J Clean Prod. 47 (2013) 96–101. https://doi.org/https://doi.org/10.1016/j.jclepro.2013.01.035

[50]   N. Ya'aini, N.A.S. Amin, M. Asmadi, Optimization of levulinic acid from lignocellulosic biomass using a new hybrid catalyst, Bioresour Technol. 116 (2012) 58–65. https://doi.org/https://doi.org/10.1016/j.biortech.2012.03.097

[51]   D. Licursi, C. Antonetti, S. Fulignati, S. Vitolo, M. Puccini, E. Ribechini, L. Bernazzani, A.M. Raspolli Galletti, In-depth characterization of valuable char obtained from hydrothermal conversion of hazelnut shells to levulinic acid, Bioresour Technol. 244 (2017) 880–888. https://doi.org/https://doi.org/10.1016/j.biortech.2017.08.012

[52]   X. Zheng, Z. Zhi, X. Gu, X. Li, R. Zhang, X. Lu, Kinetic study of levulinic acid production from corn stalk at mild temperature using FeCl3 as catalyst, Fuel. 187 (2017) 261–267. https://doi.org/https://doi.org/10.1016/j.fuel.2016.09.019

[53]   H. Ji, J.Y. Zhu, R. Gleisner, Integrated production of furfural and levulinic acid from corncob in a one-pot batch reaction incorporating distillation using step temperature profiling, RSC Adv. 7 (2017) 46208–46214. https://doi.org/10.1039/C7RA08818C

[54]   H. Liu, Y. Zhang, T. Hou, X. Chen, C. Gao, L. Han, W. Xiao, Mechanical deconstruction of corn stover as an entry process to facilitate the microwave-assisted production of ethyl levulinate, Fuel Processing Technology. 174 (2018) 53–60. https://doi.org/https://doi.org/10.1016/j.fuproc.2018.02.011

New Materials for a Circular Economy Materials Research Forum LLC
Materials Research Foundations 149 (2023) 393-445 https://doi.org/10.21741/9781644902639-11

[55] K. Lappalainen, N. Vogeler, J. Kärkkäinen, Y. Dong, M. Niemelä, A. Rusanen, A.L. Ruotsalainen, P. Wäli, A. Markkola, U. Lassi, Microwave-assisted conversion of novel biomass materials into levulinic acid, Biomass Convers Biorefin. 8 (2018) 965–970. https://doi.org/10.1007/s13399-018-0334-6

[56] C. Chang, L. Deng, G. Xu, Efficient conversion of wheat straw into methyl levulinate catalyzed by cheap metal sulfate in a biorefinery concept, Ind Crops Prod. 117 (2018) 197–204. https://doi.org/https://doi.org/10.1016/j.indcrop.2018.03.009

[57] Y. Zhang, X. Wang, T. Hou, H. Liu, L. Han, W. Xiao, Efficient microwave-assisted production of biofuel ethyl levulinate from corn stover in ethanol medium, Journal of Energy Chemistry. 27 (2018) 890–897. https://doi.org/https://doi.org/10.1016/j.jechem.2017.06.010

[58] X. Li, R. Xu, Q. Liu, M. Liang, J. Yang, S. Lu, G. Li, L. Lu, C. Si, Valorization of corn stover into furfural and levulinic acid over SAPO-18 zeolites: Effect of Brønsted to Lewis acid sites ratios, Ind Crops Prod. 141 (2019) 111759. https://doi.org/https://doi.org/10.1016/j.indcrop.2019.111759

[59] Q. Guan, T. Lei, Z. Wang, H. Xu, L. Lin, G. Chen, X. Li, Z. Li, Preparation of ethyl levulinate from wheat straw catalysed by sulfonate ionic liquid, Ind Crops Prod. 113 (2018) 150–156. https://doi.org/https://doi.org/10.1016/j.indcrop.2018.01.030

[60] X. Chen, Y. Zhang, J. Mei, G. Zhao, Q. Lyu, X. Lyu, H. Lyu, L. Han, W. Xiao, Ball milling for cellulose depolymerization and alcoholysis to produce methyl levulinate at mild temperature, Fuel Processing Technology. 188 (2019) 129–136. https://doi.org/https://doi.org/10.1016/j.fuproc.2019.02.002

[61] A.P. Pratama, D.U.C. Rahayu, Y.K. Krisnandi, Levulinic Acid Production from Delignified Rice Husk Waste over Manganese Catalysts: Heterogeneous Versus Homogeneous, Catalysts. 10 (2020). https://doi.org/10.3390/catal10030327

[62] E. Rincón, A. Zuliani, A. Jiménez-Quero, F. Vilaplana, R. Luque, L. Serrano, A.M. Balu, Combined Extraction/Purification-Catalytic Microwave-Assisted Conversion of Laurus nobilis L. Pruning Waste Polysaccharides into Methyl Levulinate, ACS Sustain Chem Eng. 8 (2020) 11016–11023. https://doi.org/10.1021/acssuschemeng.0c04161

[63] S. Sid, R.S. Mor, A. Kishore, V.S. Sharanagat, Bio-sourced polymers as alternatives to conventional food packaging materials: A review, Trends Food Sci Technol. 115 (2021) 87–104. https://doi.org/10.1016/j.tifs.2021.06.026

[64] A.D. Tripathi, P.K. Mishra, K. Khosarvi-Darani, A. Agarwal, V. Paul, Hydrothermal treatment of lignocellulose waste for the production of polyhydroxyalkanoates copolymer with potential application in food packaging, Trends Food Sci Technol. 123 (2022) 233–250. https://doi.org/10.1016/j.tifs.2022.03.018

[65] E.L. Sánchez-Safont, A. Aldureid, J.M. Lagarón, J. Gámez-Pérez, L. Cabedo, Biocomposites of different lignocellulosic wastes for sustainable food packaging

applications, Compos B Eng. 145 (2018) 215–225.
https://doi.org/10.1016/j.compositesb.2018.03.037

[66] M. Didone, P. Saxena, E. Brilhuis-Meijer, G. Tosello, G. Bissacco, T.C. Mcaloone, D.C.A. Pigosso, T.J. Howard, Moulded Pulp Manufacturing: Overview and Prospects for the Process Technology, Packaging Technology and Science. 30 (2017) 231–249. https://doi.org/10.1002/pts.2289

[67] W.A. Laftah, W.A. Wan Abdul Rahman, Pulping Process and the Potential of Using Non-Wood Pineapple Leaves Fiber for Pulp and Paper Production: A Review, Journal of Natural Fibers. 13 (2016) 85–102.
https://doi.org/10.1080/15440478.2014.984060

[68] Y. Su, B. Yang, J. Liu, B. Sun, C. Cao, X. Zou, R. Lutes, Z. He, Prospects for Replacement of Some Plastics in packaging with lignocellulose materials: a brief review, Bioresources. 13 (2018) 4550–4576

[69] C. Liu, P. Luan, Q. Li, Z. Cheng, X. Sun, D. Cao, H. Zhu, Biodegradable, Hygienic, and Compostable Tableware from Hybrid Sugarcane and Bamboo Fibers as Plastic Alternative, Matter. 3 (2020) 2066–2079.
https://doi.org/10.1016/j.matt.2020.10.004

[70] S.F. Curling, N. Laflin, G.M. Davies, G.A. Ormondroyd, R.M. Elias, Feasibility of using straw in a strong, thin, pulp moulded packaging material, Ind Crops Prod. 97 (2017) 395–400. https://doi.org/10.1016/j.indcrop.2016.12.042

[71] A. Rahman, J. Fehrenbach, C. Ulven, S. Simsek, K. Hossain, Utilization of wheat-bran cellulosic fibers as reinforcement in bio-based polypropylene composite, Ind Crops Prod. 172 (2021) 114028. https://doi.org/10.1016/j.indcrop.2021.114028

[72] D. Kai, M.J. Tan, P.L. Chee, Y.K. Chua, Y.L. Yap, X.J. Loh, Towards lignin-based functional materials in a sustainable world, Green Chemistry. 18 (2016) 1175–1200. https://doi.org/10.1039/c5gc02616d

[73] C. Liu, P. Luan, Q. Li, Z. Cheng, X. Sun, D. Cao, H. Zhu, Biodegradable, Hygienic, and Compostable Tableware from Hybrid Sugarcane and Bamboo Fibers as Plastic Alternative, Matter. 3 (2020) 2066–2079.
https://doi.org/10.1016/j.matt.2020.10.004

[74] C. V. Lang, J. Jung, T. Wang, Y. Zhao, Investigation of mechanisms and approaches for improving hydrophobicity of molded pulp biocomposites produced from apple pomace, Food and Bioproducts Processing. 133 (2022) 1–15.
https://doi.org/10.1016/j.fbp.2022.02.003

[75] P. Rattanawongkun, N. Kerddonfag, N. Tawichai, U. Intatha, N. Soykeabkaew, Improving agricultural waste pulps via self-blending concept with potential use in moulded pulp packaging, J Environ Chem Eng. 8 (2020) 104320.
https://doi.org/10.1016/j.jece.2020.104320

[76]   F. Vargas, Z. González, R. Sánchez, L. Jiménez, A. Rodríguez, CELLULOSIC PULPS OF CEREAL STRAWS AS RAW MATERIAL FOR THE MANUFACTURE OF ECOLOGICAL PACKAGING, Bioresources. 7 (2012) 4161–4170

[77]   J. Wang, L. Li, H. Xu, Y. Zhang, Y. Liu, F. Zhang, G. Shen, L. Yan, W. Wang, H. Tang, H. Qiu, J.D. Gu, W. Wang, Construction of a fungal consortium for effective degradation of rice straw lignin and potential application in bio-pulping, Bioresour Technol. 344 (2022) 126168. https://doi.org/10.1016/j.biortech.2021.126168

[78]   J.A. Andrade Alves, M.D. Lisboa dos Santos, C.C. Morais, J.L. Ramirez Ascheri, R. Signini, D.M. dos Santos, S.M. Cavalcante Bastos, D.P. Ramirez Ascheri, Sorghum straw: Pulping and bleaching process optimization and synthesis of cellulose acetate, Int J Biol Macromol. 135 (2019) 877–886. https://doi.org/10.1016/j.ijbiomac.2019.05.014

[79]   R.G. Candido, A.R. Gonçalves, Synthesis of cellulose acetate and carboxymethylcellulose from sugarcane straw, Carbohydr Polym. 152 (2016) 679–686. https://doi.org/10.1016/j.carbpol.2016.07.071

[80]   A.M. Das, A.A. Ali, M.P. Hazarika, Synthesis and characterization of cellulose acetate from rice husk: Eco-friendly condition, Carbohydr Polym. 112 (2014) 342–349. https://doi.org/10.1016/j.carbpol.2014.06.006

[81]   W. Zhang, Y. Zhang, J. Cao, W. Jiang, Improving the performance of edible food packaging films by using nanocellulose as an additive, Int J Biol Macromol. 166 (2021) 288–296. https://doi.org/10.1016/j.ijbiomac.2020.10.185

[82]   A. Barhoum, P. Samyn, T. Öhlund, A. Dufresne, Review of recent research on flexible multifunctional nanopapers, Nanoscale. 9 (2017) 15181–15205. https://doi.org/10.1039/c7nr04656a

[83]   L. Urbina, A. Eceiza, N. Gabilondo, M.Á. Corcuera, A. Retegi, Valorization of apple waste for active packaging: multicomponent polyhydroxyalkanoate coated nanopapers with improved hydrophobicity and antioxidant capacity, Food Packag Shelf Life. 21 (2019) 100356. https://doi.org/10.1016/j.fpsl.2019.100356

[84]   K. V. Neenu, C.D. Midhun Dominic, P.M.S. Begum, J. Parameswaranpillai, B.P. Kanoth, D.A. David, S.M. Sajadi, P. Dhanyasree, T.G. Ajithkumar, M. Badawi, Effect of oxalic acid and sulphuric acid hydrolysis on the preparation and properties of pineapple pomace derived cellulose nanofibers and nanopapers, Int J Biol Macromol. 209 (2022) 1745–1759. https://doi.org/10.1016/j.ijbiomac.2022.04.138

[85]   J. Rao, H. Gao, Y. Guan, W. qi Li, Q. Liu, Fabrication of hemicelluloses films with enhanced mechanical properties by graphene oxide for humidity sensing, Carbohydr Polym. 208 (2019) 513–520. https://doi.org/10.1016/j.carbpol.2018.12.099

[86]   E. Bahcegul, H.E. Toraman, N. Ozkan, U. Bakir, Evaluation of alkaline pretreatment temperature on a multi-product basis for the co-production of glucose and hemicellulose based films from lignocellulosic biomass, Bioresour Technol. 103 (2012) 440–445. https://doi.org/10.1016/j.biortech.2011.09.138

[87]   A.N. Arzami, T.M. Ho, K.S. Mikkonen, Valorization of cereal by-product hemicelluloses: Fractionation and purity considerations, Food Research International. 151 (2022) 110818. https://doi.org/10.1016/j.foodres.2021.110818

[88]   S. Sabiha-Hanim, A.M. Siti-Norsafurah, Physical properties of hemicellulose films from sugarcane bagasse, Procedia Eng. 42 (2012) 1390–1395. https://doi.org/10.1016/j.proeng.2012.07.532

[89]   K.M. Haafiz, O.F.A. Taiwo, N. Razak, H. Rokiah, H.M. Hazwan, N.F.M. Rawi, H.P.S.A. Khalil, Development of Green MMT-modified hemicelluloses based nanocomposite film with enhanced functional and barrier properties, Bioresources. 14 (2019) 8029–8047. https://doi.org/10.15376/biores.14.4.8029-8047

[90]   S. Rai, P.K. Dutta, G.K. Mehrotra, Natural Antioxidant and Antimicrobial Agents from Agrowastes: An Emergent Need to Food Packaging, Waste Biomass Valorization. 11 (2020) 1905–1916. https://doi.org/10.1007/s12649-018-0498-0

[91]   L. Ma, Y. Zhu, Y. Huang, L. Zhang, Z. Wang, Strong water-resistant, UV-blocking cellulose/glucomannan/lignin composite films inspired by natural LCC bonds, Carbohydr Polym. 281 (2022) 119083. https://doi.org/10.1016/j.carbpol.2021.119083

[92]   L.R. Mugwagwa, A.F.A. Chimphango, Physicochemical properties and potential application of hemicellulose/pectin/nanocellulose biocomposites as active packaging for fatty foods, Food Packag Shelf Life. 31 (2022) 100795. https://doi.org/10.1016/j.fpsl.2021.100795

[93]   M. Ji, J. Li, F. Li, X. Wang, J. Man, J. Li, C. Zhang, S. Peng, A biodegradable chitosan-based composite film reinforced by ramie fibre and lignin for food packaging, Carbohydr Polym. 281 (2022) 119078. https://doi.org/10.1016/j.carbpol.2021.119078

[94]   J. Xu, R. Xia, L. Zheng, T. Yuan, R. Sun, Plasticized hemicelluloses/chitosan-based edible films reinforced by cellulose nanofiber with enhanced mechanical properties, Carbohydr Polym. 224 (2019) 115164. https://doi.org/10.1016/j.carbpol.2019.115164

[95]   D.A. da S. Rios, M.M. Nakamoto, A.R.C. Braga, E.M.C. da Silva, Food coating using vegetable sources: importance and industrial potential, gaps of knowledge, current application, and future trends, Applied Food Research. 2 (2022) 100073. https://doi.org/10.1016/j.afres.2022.100073

[96]   Y. Zhang, J. Bi, S. Wang, Q. Cao, Y. Li, J. Zhou, B.W. Zhu, Functional food packaging for reducing residual liquid food: Thermo-resistant edible super-hydrophobic coating from coffee and beeswax, J Colloid Interface Sci. 533 (2019) 742–749. https://doi.org/10.1016/j.jcis.2018.09.011

[97]   Z. Tian, R. Zhang, Y. Liu, J. Xu, X. Zhu, T. Lei, K. Li, Hemicellulose-based nanocomposites coating delays lignification of green asparagus by introducing AKD as a hydrophobic modifier, Renew Energy. 178 (2021) 1097–1105. https://doi.org/10.1016/j.renene.2021.06.096

New Materials for a Circular Economy   Materials Research Forum LLC
Materials Research Foundations 149 (2023) 393-445   https://doi.org/10.21741/9781644902639-11

[98]   M. Carlosama Adriana, C. Rodríguez Misael, C. Londoño Guillermo, O. Sánchez Fernando, S. Cock Liliana, Optimization of the reproduction of Weissella cibaria in a fermentation substrate formulated with agroindustrial waste, Biotechnology Reports. 32 (2021) e00671. https://doi.org/10.1016/j.btre.2021.e00671

[99]   R.S. Singh, N. Kaur, J.F. Kennedy, Pullulan production from agro-industrial waste and its applications in food industry: A review, Carbohydr Polym. 217 (2019) 46–57. https://doi.org/10.1016/j.carbpol.2019.04.050

[100] J.V. Carpinelli Macedo, F.F. de Barros Ranke, B. Escaramboni, T.S. Campioni, E.G. Fernández Núñez, P. de Oliva Neto, Cost-effective lactic acid production by fermentation of agro-industrial residues, Biocatal Agric Biotechnol. 27 (2020) 101706. https://doi.org/10.1016/j.bcab.2020.101706

[101] W. Juttuporn, P. Thiengkaew, A. Rodklongtan, M. Rodprapakorn, P. Chitprasert, Ultrasound-Assisted Extraction of Antioxidant and Antibacterial Phenolic Compounds from Steam-Exploded Sugarcane Bagasse, Sugar Tech. 20 (2018) 599–608. https://doi.org/10.1007/s12355-017-0582-y

[102] Q. Wang, Y. Cao, L. Zhou, C.-Z. Jiang, Y. Feng, S. Wei, Effects of postharvest curing treatment on flesh colour and phenolic metabolism in fresh-cut potato products, Food Chem. 169 (2015) 246–254. https://doi.org/10.1016/j.foodchem.2014.08.011

[103] H.F. Wang, S.J. Shao, X.R. Xin, M. Wang, Y. Yao, Z.M. Chen, J.L. Wei, Research on Extraction and Antibacterial Activity of Flavonoids in Potato Peel, Zhongbei Daxue Xuebao (Ziran Kexue Ban)/Journal of North University of China (Natural Science Edition). 38 (2017). https://doi.org/10.3969/j.issn.1673-3193.2017.06.018

[104] L. Sanhueza, R. Melo, R. Montero, K. Maisey, L. Mendoza, M. Wilkens, Synergistic interactions between phenolic compounds identified in grape pomace extract with antibiotics of different classes against Staphylococcus aureus and Escherichia coli, PLoS One. 12 (2017) e0172273. https://doi.org/10.1371/journal.pone.0172273

[105] R.N. Pereira, R.M. Rodrigues, Z. Genisheva, H. Oliveira, V. de Freitas, J.A. Teixeira, A.A. Vicente, Effects of ohmic heating on extraction of food-grade phytochemicals from colored potato, LWT - Food Science and Technology. 74 (2016). https://doi.org/10.1016/j.lwt.2016.07.074

[106] S.R. Dewi, L.A. Stevens, A.E. Pearson, R. Ferrari, D.J. Irvine, E.R. Binner, Investigating the role of solvent type and microwave selective heating on the extraction of phenolic compounds from cacao (Theobroma cacao L.) pod husk, Food and Bioproducts Processing. 134 (2022) 210–222. https://doi.org/10.1016/j.fbp.2022.05.011

[107] B.R. Patra, V.B. Borugadda, A.K. Dalai, Microwave-assisted extraction of sea buckthorn pomace and seed extracts as a proactive antioxidant to stabilize edible oils,

Bioresour Technol Rep. 17 (2022) 100970.
https://doi.org/10.1016/j.biteb.2022.100970

[108] P.K. Sadh, P. Chawla, J.S. Duhan, Fermentation approach on phenolic, antioxidants and functional properties of peanut press cake, Food Biosci. 22 (2018) 113–120. https://doi.org/10.1016/j.fbio.2018.01.011

[109] P. Chandra, D.S. Arora, Production of Antioxidant Bioactive Phenolic Compounds by Solid-state Fermentation on Agro-residues Using Various Fungi Isolated from Soil, Asian Journal of Biotechnology. 8 (2016) 8–15. https://doi.org/10.3923/ajbkr.2016.8.15

[110] D. Dursun, A.C. Dalgıç, Optimization of astaxanthin pigment bioprocessing by four different yeast species using wheat wastes, Biocatal Agric Biotechnol. 7 (2016) 1–6. https://doi.org/10.1016/j.bcab.2016.04.006

[111] D.L. Abd Razak, N.Y. Abd Rashid, A. Jamaluddin, S.A. Sharifudin, A. Abd Kahar, K. Long, Cosmeceutical potentials and bioactive compounds of rice bran fermented with single and mix culture of Aspergillus oryzae and Rhizopus oryzae, Journal of the Saudi Society of Agricultural Sciences. 16 (2017) 127–134. https://doi.org/10.1016/j.jssas.2015.04.001

[112] H. Mohd Zaini, J. Roslan, S. Saallah, E. Munsu, N.S. Sulaiman, W. Pindi, Banana peels as a bioactive ingredient and its potential application in the food industry, J Funct Foods. 92 (2022) 105054. https://doi.org/10.1016/j.jff.2022.105054

[113] A.M. Aboul-Enein, Z.A. Salama, A.A. Gaafar, H.F. Aly, F.A. Bou-Elella, H.A. Ahmed, Identification of phenolic compounds from banana peel (Musa paradaisica L.) as antioxidant and antimicrobial agents, J Chem Pharm Res. 8 (2016)

[114] J. Saavedra, A. Córdova, R. Navarro, P. Díaz-Calderón, C. Fuentealba, C. Astudillo-Castro, L. Toledo, J. Enrione, L. Galvez, Industrial avocado waste: Functional compounds preservation by convective drying process, J Food Eng. 198 (2017) 81–90. https://doi.org/10.1016/j.jfoodeng.2016.11.018

[115] J.A. Poveda-Giraldo, C.A. Cardona Alzate, A biorefinery for the valorization of marigold (Calendula officinalis) residues to produce biogas and phenolic compounds, Food and Bioproducts Processing. 125 (2021) 91–104. https://doi.org/10.1016/j.fbp.2020.10.015

[116] E. Ekeledo, S. Latif, A. Abass, J. Müller, Antioxidant potential of extracts from peels and stems of yellow-fleshed and white cassava varieties, Int J Food Sci Technol. 56 (2021) 1333–1342. https://doi.org/10.1111/ijfs.14814

[117] Y.A. Rodríguez-Restrepo, P. Ferreira-Santos, C.E. Orrego, J.A. Teixeira, C.M.R. Rocha, Valorization of rice by-products: Protein-phenolic based fractions with bioactive potential, J Cereal Sci. 95 (2020) 103039. https://doi.org/10.1016/j.jcs.2020.103039

[118] S. Punia, K.S. Sandhu, S. Grasso, S.S. Purewal, M. Kaur, A.K. Siroha, K. Kumar, V. Kumar, M. Kumar, Aspergillus oryzae Fermented Rice Bran: A Byproduct with

Enhanced Bioactive Compounds and Antioxidant Potential, Foods. 10 (2020) 70. https://doi.org/10.3390/foods10010070

[119] R. Zheng, S. Su, H. Zhou, H. Yan, J. Ye, Z. Zhao, L. You, X. Fu, Antioxidant/antihyperglycemic activity of phenolics from sugarcane (Saccharum officinarum L.) bagasse and identification by UHPLC-HR-TOFMS, Ind Crops Prod. 101 (2017) 104–114. https://doi.org/10.1016/j.indcrop.2017.03.012

[120] A. de J. Cenobio-Galindo, G. Díaz-Monroy, G. Medina-Pérez, M.J. Franco-Fernández, F.E. Ludeña-Urquizo, R. Vieyra-Alberto, R.G. Campos-Montiel, Multiple Emulsions with Extracts of Cactus Pear Added in A Yogurt: Antioxidant Activity, In Vitro Simulated Digestion and Shelf Life, Foods. 8 (2019) 429. https://doi.org/10.3390/foods8100429

[121] C. Caleja, L. Barros, A.L. Antonio, M. Carocho, M.B.P.P. Oliveira, I.C.F.R. Ferreira, Fortification of yogurts with different antioxidant preservatives: A comparative study between natural and synthetic additives, Food Chem. 210 (2016) 262–268. https://doi.org/10.1016/j.foodchem.2016.04.114

[122] J. Bonilla, T. Poloni, R. v. Lourenço, P.J.A. Sobral, Antioxidant potential of eugenol and ginger essential oils with gelatin/chitosan films, Food Biosci. 23 (2018) 107–114. https://doi.org/10.1016/j.fbio.2018.03.007

[123] M.A. Giaconia, S. dos P. Ramos, C.F. Pereira, A.C. Lemes, V.V. de Rosso, A.R.C. Braga, Overcoming restrictions of bioactive compounds biological effects in food using nanometer-sized structures, Food Hydrocoll. 107 (2020) 105939. https://doi.org/10.1016/j.foodhyd.2020.105939

[124] L.M. Reyes, M. Landgraf, P.J.A. Sobral, Gelatin-based films activated with red propolis ethanolic extract and essential oils, Food Packag Shelf Life. 27 (2021) 100607. https://doi.org/10.1016/j.fpsl.2020.100607

[125] H.H. Lau, R. Murney, N.L. Yakovlev, M. V. Novoselova, S.H. Lim, N. Roy, H. Singh, G.B. Sukhorukov, B. Haigh, M. V. Kiryukhin, Protein-tannic acid multilayer films: A multifunctional material for microencapsulation of food-derived bioactives, J Colloid Interface Sci. 505 (2017) 332–340. https://doi.org/10.1016/j.jcis.2017.06.001

[126] E.M.C. Alexandre, R.V. Lourenço, A.M.Q.B. Bittante, I.C.F. Moraes, P.J. do A. Sobral, Gelatin-based films reinforced with montmorillonite and activated with nanoemulsion of ginger essential oil for food packaging applications, Food Packag Shelf Life. 10 (2016) 87–96. https://doi.org/10.1016/j.fpsl.2016.10.004

[127] A.C.K. Bierhalz, M.A. da Silva, T.G. Kieckbusch, Fundamentals of two-dimensional films and membranes, in: Biopolymer Membranes and Films, Elsevier, 2020: pp. 35–66. https://doi.org/10.1016/B978-0-12-818134-8.00002-X

[128] H. Zhao, Z. Yang, L. Guo, Nacre-inspired composites with different macroscopic dimensions: strategies for improved mechanical performance and applications, NPG Asia Mater. 10 (2018) 1–22. https://doi.org/10.1038/s41427-018-0009-6

[129] C.P. Boeira, D.C.B. Flores, J. dos S. Alves, M.R. de Moura, P.T.S. Melo, C.M.B. Rolim, D.R. Nogueira-Librelotto, C.S. da Rosa, Effect of corn stigma extract on physical and antioxidant properties of biodegradable and edible gelatin and corn starch films, Int J Biol Macromol. 208 (2022) 698–706. https://doi.org/10.1016/j.ijbiomac.2022.03.164

[130] P. Yilmaz, E. Demirhan, B. Ozbek, Development of Ficus carica Linn leaves extract incorporated chitosan films for active food packaging materials and investigation of their properties, Food Biosci. 46 (2022) 101542. https://doi.org/10.1016/j.fbio.2021.101542

[131] V.D. Medeiros Silva, M.C. Coutinho Macedo, C.G. Rodrigues, A. Neris dos Santos, A.C. de Freitas e Loyola, C.A. Fante, Biodegradable edible films of ripe banana peel and starch enriched with extract of Eriobotrya japonica leaves, Food Biosci. 38 (2020) 100750. https://doi.org/10.1016/j.fbio.2020.100750

[132] N. Alqahtani, T. Alnemr, S. Ali, Development of low-cost biodegradable films from corn starch and date palm pits (Phoenix dactylifera), Food Biosci. 42 (2021) 101199. https://doi.org/10.1016/j.fbio.2021.101199

[133] E. Jamróz, J. Tkaczewska, J. Kopeć, A. Cholewa-Wójcik, Shelf-life extension of salmon using active total biodegradable packaging with tea ground waste and furcellaran-CMC double-layered films, Food Chem. 383 (2022) 132425. https://doi.org/10.1016/j.foodchem.2022.132425

[134] W.G. Sganzerla, G.B. Rosa, A.L.A. Ferreira, C.G. da Rosa, P.C. Beling, L.O. Xavier, C.M. Hansen, J.P. Ferrareze, M.R. Nunes, P.L.M. Barreto, A.P. de Lima Veeck, Bioactive food packaging based on starch, citric pectin and functionalized with Acca sellowiana waste by-product: Characterization and application in the postharvest conservation of apple, Int J Biol Macromol. 147 (2020) 295–303. https://doi.org/10.1016/j.ijbiomac.2020.01.074

[135] P. Thivya, Y.K. Bhosale, S. Anandakumar, V. Hema, V.R. Sinija, Development of active packaging film from sodium alginate/carboxymethyl cellulose containing shallot waste extracts for anti-browning of fresh-cut produce, Int J Biol Macromol. 188 (2021) 790–799. https://doi.org/10.1016/J.IJBIOMAC.2021.08.039

[136] Y. Jiang, H. Yin, X. Zhou, D. Wang, Y. Zhong, Q. Xia, Y. Deng, Y. Zhao, Antimicrobial, antioxidant and physical properties of chitosan film containing Akebia trifoliata (Thunb.) Koidz. peel extract/montmorillonite and its application, Food Chem. 361 (2021) 130111. https://doi.org/10.1016/j.foodchem.2021.130111

[137] D. Wang, Y. Dong, X. Chen, Y. Liu, J. Wang, X. Wang, C. Wang, H. Song, Incorporation of apricot (Prunus armeniaca) kernel essential oil into chitosan films displaying antimicrobial effect against Listeria monocytogenes and improving quality indices of spiced beef, Int J Biol Macromol. 162 (2020) 838–844. https://doi.org/10.1016/j.ijbiomac.2020.06.220

[138] L. and welfare Ministry of Health, Food with health claims, food for special dietary uses and nutrition labeling. , Https://Www.Mhlw.Go.Jp/English/Topics/Foodsafety/Fhc/02.Html . (n.d.)

[139] European Comission, Food Safety. Nutrition Claims, Https://Ec.Europa.Eu/Food/Safety/Labelling-and-Nutrition/Nutrition-and-Health-Claims/Nutrition-Claims_en. (n.d.)

[140] D. Granato, F.J. Barba, D. Bursać Kovačević, J.M. Lorenzo, A.G. Cruz, P. Putnik, Functional Foods: Product Development, Technological Trends, Efficacy Testing, and Safety, Annu Rev Food Sci Technol. 11 (2020) 93–118. https://doi.org/10.1146/annurev-food-032519-051708

[141] D.M. Martirosyan, J. Singh, A new definition of functional food by FFC: what makes a new definition unique?, Functional Foods in Health and Disease. 5 (2015) 209. https://doi.org/10.31989/ffhd.v5i6.183

[142] T.C. Pimentel, B.B. Torres de Assis, C. dos Santos Rocha, V.A. Marcolino, M. Rosset, M. Magnani, Prebiotics in non-dairy products: Technological and physiological functionality, challenges, and perspectives, Food Biosci. 46 (2022) 101585. https://doi.org/10.1016/j.fbio.2022.101585

[143] G.R. Gibson, R. Hutkins, M.E. Sanders, S.L. Prescott, R.A. Reimer, S.J. Salminen, K. Scott, C. Stanton, K.S. Swanson, P.D. Cani, K. Verbeke, G. Reid, Expert consensus document: The International Scientific Association for Probiotics and Prebiotics (ISAPP) consensus statement on the definition and scope of prebiotics, Nat Rev Gastroenterol Hepatol. 14 (2017) 491–502. https://doi.org/10.1038/nrgastro.2017.75

[144] E.M. Dewulf, P.D. Cani, S.P. Claus, S. Fuentes, P.G. Puylaert, A.M. Neyrinck, L.B. Bindels, W.M. de Vos, G.R. Gibson, J.-P. Thissen, N.M. Delzenne, Insight into the prebiotic concept: lessons from an exploratory, double blind intervention study with inulin-type fructans in obese women, Gut. 62 (2013) 1112–1121. https://doi.org/10.1136/gutjnl-2012-303304

[145] D. Vandeputte, G. Falony, S. Vieira-Silva, J. Wang, M. Sailer, S. Theis, K. Verbeke, J. Raes, Prebiotic inulin-type fructans induce specific changes in the human gut microbiota, Gut. 66 (2017) 1968–1974. https://doi.org/10.1136/gutjnl-2016-313271

[146] S. Hachimura, M. Totsuka, A. Hosono, Immunomodulation by food: impact on gut immunity and immune cell function, Biosci Biotechnol Biochem. 82 (2018) 584–599. https://doi.org/10.1080/09168451.2018.1433017

[147] Y. Nakanishi, K. Murashima, H. Ohara, T. Suzuki, H. Hayashi, M. Sakamoto, T. Fukasawa, H. Kubota, A. Hosono, T. Kono, S. Kaminogawa, Y. Benno, Increase in Terminal Restriction Fragments of *Bacteroidetes* -Derived 16S rRNA Genes after Administration of Short-Chain Fructooligosaccharides, Appl Environ Microbiol. 72 (2006) 6271–6276. https://doi.org/10.1128/AEM.00477-06

[148] S. Manzoor, S.M. Wani, S. Ahmad Mir, D. Rizwan, Role of probiotics and prebiotics in mitigation of different diseases, Nutrition. 96 (2022) 111602. https://doi.org/10.1016/j.nut.2022.111602

[149] H.S. Arruda, G.A. Pereira, M.E.F. Almeida, Neri-Numa. Iramaia A., R.A.S. Sancho, G. Molina, G.M. Pastore, Current Knowledge and Future Perspectives of Oligosaccharides Research, in: Atta-ur-Rahman (Ed.), Frontiers in Natural Product Chemistry Volume 3, BENTHAM SCIENCE PUBLISHERS, 2017: pp. 91–175. https://doi.org/10.2174/9781681085340117030005

[150] K.P. Scott, R. Grimaldi, M. Cunningham, S.R. Sarbini, A. Wijeyesekera, M.L.K. Tang, J.C. -Y. Lee, Y.F. Yau, J. Ansell, S. Theis, K. Yang, R. Menon, J. Arfsten, S. Manurung, V. Gourineni, G.R. Gibson, Developments in understanding and applying prebiotics in research and practice—an ISAPP conference paper, J Appl Microbiol. 128 (2020) 934–949. https://doi.org/10.1111/jam.14424

[151] M.C. Rosa, M.R.S. Carmo, C.F. Balthazar, J.T. Guimarães, E.A. Esmerino, M.Q. Freitas, M.C. Silva, T.C. Pimentel, A.G. Cruz, Dairy products with prebiotics: An overview of the health benefits, technological and sensory properties, Int Dairy J. 117 (2021) 105009. https://doi.org/10.1016/j.idairyj.2021.105009

[152] T.J. Ashaolu, Immune boosting functional foods and their mechanisms: A critical evaluation of probiotics and prebiotics, Biomedicine & Pharmacotherapy. 130 (2020) 110625. https://doi.org/10.1016/j.biopha.2020.110625

[153] M. Glisic, M. Baltic, M. Glisic, D. Trbovic, M. Jokanovic, N. Parunovic, M. Dimitrijevic, B. Suvajdzic, M. Boskovic, D. Vasilev, Inulin-based emulsion-filled gel as a fat replacer in prebiotic- and <scp>PUFA</scp> -enriched dry fermented sausages, Int J Food Sci Technol. 54 (2019) 787–797. https://doi.org/10.1111/ijfs.13996

[154] T.J. Ashaolu, B. Saibandith, C.T. Yupanqui, S. Wichienchot, Human colonic microbiota modulation and branched chain fatty acids production affected by soy protein hydrolysate, Int J Food Sci Technol. 54 (2019) 141–148. https://doi.org/10.1111/ijfs.13916

[155] M. Roberfroid, G.R. Gibson, L. Hoyles, A.L. McCartney, R. Rastall, I. Rowland, D. Wolvers, B. Watzl, H. Szajewska, B. Stahl, F. Guarner, F. Respondek, K. Whelan, V. Coxam, M.-J. Davicco, L. Léotoing, Y. Wittrant, N.M. Delzenne, P.D. Cani, A.M. Neyrinck, A. Meheust, Prebiotic effects: metabolic and health benefits, British Journal of Nutrition. 104 (2010) S1–S63. https://doi.org/10.1017/S0007114510003363

[156] A.F.S. Paulo, T.R. Baú, E.I. Ida, M.A. Shirai, Edible coatings and films with incorporation of prebiotics —A review, Food Research International. 148 (2021) 110629. https://doi.org/10.1016/j.foodres.2021.110629

[157] R. Gómez-García, D.A. Campos, A. Oliveira, C.N. Aguilar, A.R. Madureira, M. Pintado, A chemical valorisation of melon peels towards functional food ingredients:

Bioactives profile and antioxidant properties, Food Chem. 335 (2021).
https://doi.org/10.1016/j.foodchem.2020.127579

[158] B. Gullon, M.E. Pintado, J.A. Pérez-Álvarez, M. Viuda-Martos, Assessment of polyphenolic profile and antibacterial activity of pomegranate peel (Punica granatum) flour obtained from co-product of juice extraction, Food Control. 59 (2016) 94–98. https://doi.org/10.1016/j.foodcont.2015.05.025

[159] B. Gullón, P. Gullón, G. Eibes, C. Cara, A. de Torres, J.C. López-Linares, E. Ruiz, E. Castro, Valorisation of olive agro-industrial by-products as a source of bioactive compounds, Science of the Total Environment. 645 (2018) 533–542. https://doi.org/10.1016/j.scitotenv.2018.07.155

[160] D.A. Campos, T.B. Ribeiro, J.A. Teixeira, L. Pastrana, M.M. Pintado, Integral valorization of pineapple (Ananas comosus L.) By-products through a green chemistry approach towards Added Value Ingredients, Foods. 9 (2020). https://doi.org/10.3390/foods9010060

[161] D.A. Campos, E.R. Coscueta, A.A. Vilas-Boas, S. Silva, J.A. Teixeira, L.M. Pastrana, M.M. Pintado, Impact of functional flours from pineapple by-products on human intestinal microbiota, J Funct Foods. 67 (2020). https://doi.org/10.1016/j.jff.2020.103830

[162] J. Xu, R. Xia, T. Yuan, R. Sun, Use of xylooligosaccharides (XOS) in hemicelluloses/chitosan-based films reinforced by cellulose nanofiber: Effect on physicochemical properties, Food Chem. 298 (2019) 125041. https://doi.org/10.1016/j.foodchem.2019.125041

[163] G.T. Bersaneti, S. Garcia, S. Mali, M.A. Pedrine Colabone Celligoi, Evaluation of the prebiotic activities of edible starch films with the addition of nystose from Bacillus subtilis natto, LWT. 116 (2019) 108502. https://doi.org/10.1016/j.lwt.2019.108502

[164] S.R. Kanatt, M.S. Rao, S.P. Chawla, A. Sharma, Effects of chitosan coating on shelf-life of ready-to-cook meat products during chilled storage, LWT - Food Science and Technology. 53 (2013) 321–326. https://doi.org/10.1016/j.lwt.2013.01.019

[165] J. Sarki, S.B. Hassan, V.S. Aigbodion, J.E. Oghenevweta, Potential of using coconut shell particle fillers in eco-composite materials, J Alloys Compd. 509 (2011) 2381–2385. https://doi.org/10.1016/J.JALLCOM.2010.11.025

[166] P.N.B. Reis, J.A.M. Ferreira, P.A.A. Silva, Mechanical behaviour of composites filled by agro-waste materials, Fibers and Polymers. 12 (2011) 240–246. https://doi.org/10.1007/S12221-011-0240-Z

[167] W. Li, J. Huang, W. Liu, X. Qiu, H. Lou, L. Zheng, Lignin modified PBAT composites with enhanced strength based on interfacial dynamic bonds, J Appl Polym Sci. (2022). https://doi.org/10.1002/APP.52476

[168] Z. Yang, X. Li, J. Si, Z. Cui, K. Peng, Morphological, Mechanical and Thermal Properties of Poly(lactic acid) (PLA)/Cellulose Nanofibrils (CNF) Composites

Nanofiber for Tissue Engineering, Journal of Wuhan University of Technology-Mater. Sci. Ed. 2019 34:1. 34 (2019) 207–215. https://doi.org/10.1007/S11595-019-2037-7

[169] M. Jonoobi, J. Harun, A.P. Mathew, K. Oksman, Mechanical properties of cellulose nanofiber (CNF) reinforced polylactic acid (PLA) prepared by twin screw extrusion, Compos Sci Technol. 70 (2010) 1742–1747. https://doi.org/10.1016/J.COMPSCITECH.2010.07.005

[170] N.F. Zaaba, H. Ismail, Comparative study of irradiated and non-irradiated recycled polypropylene/peanut shell powder composites under the effects of natural weathering degradation, Bioresources. 13 (2018) 487–505. https://doi.org/10.15376/BIORES.13.1.487-505

[171] A.P. Gupte, M. Basaglia, S. Casella, L. Favaro, Rice waste streams as a promising source of biofuels: feedstocks, biotechnologies and future perspectives, Renewable and Sustainable Energy Reviews. 167 (2022) 112673. https://doi.org/10.1016/J.RSER.2022.112673

[172] J.Y. Tong, N.R.R. Royan, Y.C. Ng, M.H. Ab Ghani, S. Ahmad, Study of the mechanical and morphology properties of recycled hdpe composite using rice husk filler, Advances in Materials Science and Engineering. 2014 (2014). https://doi.org/10.1155/2014/938961

[173] A.F. Koutsomitopoulou, J.C. Bénézet, A. Bergeret, G.C. Papanicolaou, Preparation and characterization of olive pit powder as a filler to PLA-matrix bio-composites, Powder Technol. 255 (2014) 10–16. https://doi.org/10.1016/J.POWTEC.2013.10.047

[174] R. Elgamsy, A. Allah Abo Elmagd, A. Elrahman Mokhtar, I. Khalid, N. Taha, S. Sadek, M.L. Tawfic, T. Attia, A. Elsabbagh, Developing fire retardant composites of biodegradable polyethylene reinforced with agricultural wastes, Ain Shams Engineering Journal. 13 (2022). https://doi.org/10.1016/J.ASEJ.2022.101768

[175] J. Peyrton, L. Avérous, Structure-properties relationships of cellular materials from biobased polyurethane foams, Materials Science and Engineering: R: Reports. 145 (2021) 100608. https://doi.org/https://doi.org/10.1016/j.mser.2021.100608

[176] A. Noreen, K.M. Zia, M. Zuber, S. Tabasum, M.J. Saif, Recent trends in environmentally friendly water-borne polyurethane coatings: A review, Korean Journal of Chemical Engineering. 33 (2016) 388–400. https://doi.org/10.1007/s11814-015-0241-5

[177] F. Xie, T. Zhang, P. Bryant, V. Kurusingal, J.M. Colwell, B. Laycock, Degradation and stabilization of polyurethane elastomers, Prog Polym Sci. 90 (2019) 211–268. https://doi.org/https://doi.org/10.1016/j.progpolymsci.2018.12.003

[178] Y. Chen, H. Zhang, Z. Zhu, S. Fu, High-value utilization of hydroxymethylated lignin in polyurethane adhesives, Int J Biol Macromol. 152 (2020) 775–785. https://doi.org/https://doi.org/10.1016/j.ijbiomac.2020.02.321

[179] S. Wendels, L. Avérous, Biobased polyurethanes for biomedical applications, Bioact Mater. 6 (2021) 1083–1106. https://doi.org/https://doi.org/10.1016/j.bioactmat.2020.10.002

[180] H. Li, C. (Charles) Xu, Z. Yuan, Q. Wei, Synthesis of bio-based polyurethane foams with liquefied wheat straw: Process optimization, Biomass Bioenergy. 111 (2018) 134–140. https://doi.org/https://doi.org/10.1016/j.biombioe.2018.02.011

[181] E. Rincón, A.M. Balu, R. Luque, L. Serrano, Insulating rigid polyurethane foams from laurel tree pruning based polyol, J Appl Polym Sci. 138 (2021) 49789. https://doi.org/https://doi.org/10.1002/app.49789

[182] R. Briones, L. Serrano, R. ben Younes, I. Mondragon, J. Labidi, Polyol production by chemical modification of date seeds, Ind Crops Prod. 34 (2011) 1035–1040. https://doi.org/https://doi.org/10.1016/j.indcrop.2011.03.012

[183] R. Briones, L. Serrano, J. Labidi, Valorization of some lignocellulosic agro-industrial residues to obtain biopolyols, Journal of Chemical Technology & Biotechnology. 87 (2012) 244–249. https://doi.org/https://doi.org/10.1002/jctb.2706

[184] U.A. Amran, S. Zakaria, C.H. Chia, R. Roslan, S.N.S. Jaafar, K.M. Salleh, Polyols and rigid polyurethane foams derived from liquefied lignocellulosic and cellulosic biomass, Cellulose. 26 (2019) 3231–3246. https://doi.org/10.1007/s10570-019-02271-w

[185] L. Serrano, E. Rincón, A. García, J. Rodríguez, R. Briones, Bio-Degradable Polyurethane Foams Produced by Liquefied Polyol from Wheat Straw Biomass, Polymers (Basel). 12 (2020). https://doi.org/10.3390/polym12112646

[186] C. Zhang, S. Bhoyate, M. Ionescu, P.K. Kahol, R.K. Gupta, Highly flame retardant and bio-based rigid polyurethane foams derived from orange peel oil, Polym Eng Sci. 58 (2018) 2078–2087. https://doi.org/https://doi.org/10.1002/pen.24819

[187] I. Domingos, J. Ferreira, L. Cruz-Lopes, B. Esteves, Polyurethane foams from liquefied orange peel wastes, Food and Bioproducts Processing. 115 (2019) 223–229. https://doi.org/https://doi.org/10.1016/j.fbp.2019.04.002

[188] R. Galhano dos Santos, P. Ventura, J.C. Bordado, M.M. Mateus, Direct and efficient liquefaction of potato peel into bio-oil, Environ Chem Lett. 15 (2017) 453–458. https://doi.org/10.1007/s10311-017-0620-8

[189] J.A.D. Condeço, S. Hariharakrishnan, O.M. Ofili, M.M. Mateus, J.M. Bordado, M.J.N. Correia, Energetic valorisation of agricultural residues by solvent-based liquefaction, Biomass Bioenergy. 147 (2021) 106003. https://doi.org/https://doi.org/10.1016/j.biombioe.2021.106003

[190] H. Li, S. Feng, Z. Yuan, Q. Wei, C.C. Xu, Highly efficient liquefaction of wheat straw for the production of bio-polyols and bio-based polyurethane foams, Ind Crops Prod. 109 (2017) 426–433. https://doi.org/https://doi.org/10.1016/j.indcrop.2017.08.060

[191] S. Hu, Y. Li, Two-step sequential liquefaction of lignocellulosic biomass by crude glycerol for the production of polyols and polyurethane foams, Bioresour Technol. 161 (2014) 410–415. https://doi.org/https://doi.org/10.1016/j.biortech.2014.03.072

[192] J.-Y. Kim, H.W. Lee, S.M. Lee, J. Jae, Y.-K. Park, Overview of the recent advances in lignocellulose liquefaction for producing biofuels, bio-based materials and chemicals, Bioresour Technol. 279 (2019) 373–384. https://doi.org/https://doi.org/10.1016/j.biortech.2019.01.055

[193] S. Członka, A. Strąkowska, A. Kairytė, Effect of walnut shells and silanized walnut shells on the mechanical and thermal properties of rigid polyurethane foams, Polym Test. 87 (2020) 106534. https://doi.org/https://doi.org/10.1016/j.polymertesting.2020.106534

[194] A. Bryśkiewicz, M. Zieleniewska, K. Przyjemska, P. Chojnacki, J. Ryszkowska, Modification of flexible polyurethane foams by the addition of natural origin fillers, Polym Degrad Stab. 132 (2016) 32–40. https://doi.org/https://doi.org/10.1016/j.polymdegradstab.2016.05.002

[195] G. Wuzella, A.R. Mahendran, T. Bätge, S. Jury, A. Kandelbauer, Novel, binder-free fiber reinforced composites based on a renewable resource from the reed-like plant Typha sp., Ind Crops Prod. 33 (2011) 683–689. https://doi.org/10.1016/j.indcrop.2011.01.008

[196] J. Domínguez-Robles, Q. Tarrés, M. Delgado-Aguilar, A. Rodríguez, F.X. Espinach, P. Mutjé, Approaching a new generation of fiberboards taking advantage of self lignin as green adhesive, Int J Biol Macromol. 108 (2018) 927–935. https://doi.org/10.1016/j.ijbiomac.2017.11.005

[197] M. Nasir, D.P. Khali, M. Jawaid, P.M. Tahir, R. Siakeng, M. Asim, T.A. Khan, Recent development in binderless fiber-board fabrication from agricultural residues: A review, Constr Build Mater. 211 (2019) 502–516. https://doi.org/10.1016/j.conbuildmat.2019.03.279

[198] N.M. Stark, Z. Cai, C. Carll, Chapter 11 - Wood-Based Composite Materials Panel Products , Glued-Laminated Timber , Structural Materials, Wood Handbook - Wood as an Engineering Material. (2010) 1–28

[199] C.R. Frihart, C.G. Hunt, Adhesives with Wood Materials- Bond Formation and Performance, Wood Handbook: Wood as an Engineering Material. (2010) 10.1-10.24

[200] D. Theng, G. Arbat, M. Delgado-Aguilar, F. Vilaseca, B. Ngo, P. Mutjé, All-lignocellulosic fiberboard from corn biomass and cellulose nanofibers, Ind Crops Prod. 76 (2015) 166–173. https://doi.org/10.1016/j.indcrop.2015.06.046

[201] A.M. El-Kassas, A.H.I. Mourad, Novel fibers preparation technique for manufacturing of rice straw based fiberboards and their characterization, Mater Des. 50 (2013) 757–765. https://doi.org/10.1016/j.matdes.2013.03.057

[202] G. McGwin, J. Lienert, J.I. Kennedy, Formaldehyde exposure and asthma in children: A systematic review, Environ Health Perspect. 118 (2010) 313–317. https://doi.org/10.1289/ehp.0901143

[203] F. Vitrone, D. Ramos, V. Vitagliano, F. Ferrando, J. Salvadó, All-lignocellulosic fiberboards from giant reed (Arundo donax L.): Effect of steam explosion pre-treatment on physical and mechanical properties, Constr Build Mater. 319 (2022). https://doi.org/10.1016/j.conbuildmat.2021.126064

[204] J. Domínguez-Robles, Q. Tarrés, M. Alcalà, N.E. El Mansouri, A. Rodríguez, P. Mutjé, M. Delgado-Aguilar, Development of high-performance binderless fiberboards from wheat straw residue, Constr Build Mater. 232 (2020). https://doi.org/10.1016/j.conbuildmat.2019.117247

[205] J. Domínguez-Robles, Q. Tarrés, M. Alcalà, N.E. el Mansouri, A. Rodríguez, P. Mutjé, M. Delgado-Aguilar, Development of high-performance binderless fiberboards from wheat straw residue, Constr Build Mater. 232 (2020) 117247. https://doi.org/10.1016/J.CONBUILDMAT.2019.117247

[206] C. Mancera, N.E. El Mansouri, M.A. Pelach, F. Francesc, J. Salvadó, Feasibility of incorporating treated lignins in fiberboards made from agricultural waste, Waste Management. 32 (2012) 1962–1967. https://doi.org/10.1016/j.wasman.2012.05.019

[207] D. de Castro Sales, A.E. Cabral, M.S. Medeiros, Development of fiberboard panels manufactured from reclaimed cement bags, Journal of Building Engineering. 34 (2021). https://doi.org/10.1016/j.jobe.2020.101525

[208] E. Espinosa, Q. Tarrés, D. Theng, M. Delgado-Aguilar, A. Rodríguez, P. Mutjé, Effect of enzymatic treatment (endo-glucanases) of fiber and mechanical lignocellulose nanofibers addition on physical and mechanical properties of binderless high-density fiberboards made from wheat straw, Journal of Building Engineering. 44 (2021). https://doi.org/10.1016/j.jobe.2021.103392

[209] P. Hýsková, Š. Hýsek, O. Schönfelder, P. Šedivka, M. Lexa, V. Jarský, Utilization of agricultural rests: Straw-based composite panels made from enzymatic modified wheat and rapeseed straw, Ind Crops Prod. 144 (2020). https://doi.org/10.1016/j.indcrop.2019.112067

[210] E. Espinosa, Q. Tarrés, D. Theng, M. Delgado-Aguilar, A. Rodríguez, P. Mutjé, Effect of enzymatic treatment (endo-glucanases) of fiber and mechanical lignocellulose nanofibers addition on physical and mechanical properties of binderless high-density fiberboards made from wheat straw, Journal of Building Engineering. 44 (2021). https://doi.org/10.1016/j.jobe.2021.103392

[211] P. Evon, V. Vandenbossche, P.Y. Pontalier, L. Rigal, New thermal insulation fiberboards from cake generated during biorefinery of sunflower whole plant in a twin-screw extruder, Ind Crops Prod. 52 (2014) 354–362. https://doi.org/10.1016/j.indcrop.2013.10.049

[212] J.A. Okolie, S. Nanda, A.K. Dalai, J.A. Kozinski, Chemistry and Specialty Industrial Applications of Lignocellulosic Biomass, Waste Biomass Valorization. 12 (2021) 2145–2169. https://doi.org/10.1007/s12649-020-01123-0

[213] Matica, Aachmann, Tøndervik, Sletta, Ostafe, Chitosan as a Wound Dressing Starting Material: Antimicrobial Properties and Mode of Action, Int J Mol Sci. 20 (2019) 5889. https://doi.org/10.3390/ijms20235889

[214] X. Cui, J. Lee, K.R. Ng, W.N. Chen, Food Waste Durian Rind-Derived Cellulose Organohydrogels: Toward Anti-Freezing and Antimicrobial Wound Dressing, ACS Sustain Chem Eng. 9 (2021) 1304–1312. https://doi.org/10.1021/acssuschemeng.0c07705

[215] X. Cui, J.J.L. Lee, W.N. Chen, Eco-friendly and biodegradable cellulose hydrogels produced from low cost okara: towards non-toxic flexible electronics, Sci Rep. 9 (2019) 18166. https://doi.org/10.1038/s41598-019-54638-5

[216] E.E. Brown, D. Hu, N. Abu Lail, X. Zhang, Potential of Nanocrystalline Cellulose–Fibrin Nanocomposites for Artificial Vascular Graft Applications, Biomacromolecules. 14 (2013) 1063–1071. https://doi.org/10.1021/bm3019467

[217] Md.S. Rahman, Md.I. H. Mondal, Mst.S. Yeasmin, M.A. Sayeed, M.A. Hossain, M.B. Ahmed, Conversion of Lignocellulosic Corn Agro-Waste into Cellulose Derivative and Its Potential Application as Pharmaceutical Excipient, Processes. 8 (2020) 711. https://doi.org/10.3390/pr8060711

[218] A. Karimian, H. Parsian, M. Majidinia, M. Rahimi, S.M. Mir, H. Samadi Kafil, V. Shafiei-Irannejad, M. Kheyrollah, H. Ostadi, B. Yousefi, Nanocrystalline cellulose: Preparation, physicochemical properties, and applications in drug delivery systems, Int J Biol Macromol. 133 (2019) 850–859. https://doi.org/10.1016/j.ijbiomac.2019.04.117

[219] D. Melandri, A. de Angelis, R. Orioli, G. Ponzielli, P. Lualdi, N. Giarratana, V. Reiner, Use of a new hemicellulose dressing (Veloderm®) for the treatment of split-thickness skin graft donor sites: A within-patient controlled study, Burns. 32 (2006) 964–972. https://doi.org/10.1016/J.BURNS.2006.03.013

[220] L.M. Ferreira, L. Blanes, A. Gragnani, D.F. Veiga, F.P. Veiga, G.B. Nery, G.H.H.R. Rocha, H.C. Gomes, M.G. Rocha, R. Okamoto, Hemicellulose dressing versus rayon dressing in the re-epithelialization of split-thickness skin graft donor sites: a multicenter study, J Tissue Viability. 18 (2009) 88–94. https://doi.org/10.1016/J.JTV.2009.06.001

[221] N. Ahmad, D. Tayyeb, I. Ali, N. K. Alruwaili, W. Ahmad, A. ur Rehman, A.H. Khan, M.S. Iqbal, Development and Characterization of Hemicellulose-Based Films for Antibacterial Wound-Dressing Application, Polymers (Basel). 12 (2020) 548. https://doi.org/10.3390/polym12030548

[222] N. Ahmad, M.M. Ahmad, N.K. Alruwaili, Z.A. Alrowaili, F.A. Alomar, S. Akhtar, O.A. Alsaidan, N.A. Alhakamy, A. Zafar, M. Elmowafy, M.H. Elkomy, Antibiotic-Loaded Psyllium Husk Hemicellulose and Gelatin-Based Polymeric Films for Wound

Dressing Application, Pharmaceutics. 13 (2021) 236.
https://doi.org/10.3390/pharmaceutics13020236

[223] M. Taniguchi, R. Aida, K. Saito, A. Ochiai, S. Takesono, E. Saitoh, T. Tanaka, Identification and characterization of multifunctional cationic peptides from traditional Japanese fermented soybean Natto extracts, J Biosci Bioeng. 127 (2019) 472–478. https://doi.org/10.1016/J.JBIOSC.2018.09.016

[224] B.S.F. Bazzaz, B. Khameneh, M.R.Z. Ostad, H. Hosseinzadeh, In vitro evaluation of antibacterial activity of verbascoside, lemon verbena extract and caffeine in combination with gentamicin against drug-resistant Staphylococcus aureus and Escherichia coli clinical isolates, Avicenna J Phytomed. 8 (2018) 246–253

[225] P.N. Phatthalung, S. Chusri, S.P. Voravuthikunchai, Thai ethnomedicinal plants as resistant modifying agents for combating Acinetobacter baumannii infections, BMC Complement Altern Med. 12 (2012) 56. https://doi.org/10.1186/1472-6882-12-56

[226] M. Mandø, Direct combustion of biomass, in: Biomass Combustion Science, Technology and Engineering, Elsevier, 2013: pp. 61–83. https://doi.org/10.1533/9780857097439.2.61

[227] G. de Bhowmick, A.K. Sarmah, R. Sen, Lignocellulosic biorefinery as a model for sustainable development of biofuels and value added products, Bioresour Technol. 247 (2018) 1144–1154. https://doi.org/10.1016/j.biortech.2017.09.163

[228] A. Tesfaw, F. Assefa, Current Trends in Bioethanol Production by *Saccharomyces cerevisiae* : Substrate, Inhibitor Reduction, Growth Variables, Coculture, and Immobilization, Int Sch Res Notices. 2014 (2014) 1–11. https://doi.org/10.1155/2014/532852

[229] J.K. Saini, R. Saini, L. Tewari, Lignocellulosic agriculture wastes as biomass feedstocks for second-generation bioethanol production: concepts and recent developments, 3 Biotech. 5 (2015) 337–353. https://doi.org/10.1007/s13205-014-0246-5

[230] P. Karagoz, R.M. Bill, M. Ozkan, Lignocellulosic ethanol production: Evaluation of new approaches, cell immobilization and reactor configurations, Renew Energy. 143 (2019) 741–752. https://doi.org/10.1016/j.renene.2019.05.045

[231] L.J. Jönsson, C. Martín, Pretreatment of lignocellulose: Formation of inhibitory by-products and strategies for minimizing their effects, Bioresour Technol. 199 (2016) 103–112. https://doi.org/10.1016/j.biortech.2015.10.009

[232] L.N. Jayakody, J. Ferdouse, N. Hayashi, H. Kitagaki, Identification and detoxification of glycolaldehyde, an unattended bioethanol fermentation inhibitor, Crit Rev Biotechnol. 37 (2017) 177–189. https://doi.org/10.3109/07388551.2015.1128877

[233] S.T. Merino, J. Cherry, Progress and Challenges in Enzyme Development for Biomass Utilization, in: 2007: pp. 95–120. https://doi.org/10.1007/10_2007_066

[234] G. Müller, A. Várnai, K.S. Johansen, V.G.H. Eijsink, S.J. Horn, Harnessing the potential of LPMO-containing cellulase cocktails poses new demands on processing

conditions, Biotechnol Biofuels. 8 (2015) 187. https://doi.org/10.1186/s13068-015-0376-y

[235] T. v. Vuong, B. Liu, M. Sandgren, E.R. Master, Microplate-Based Detection of Lytic Polysaccharide Monooxygenase Activity by Fluorescence-Labeling of Insoluble Oxidized Products, Biomacromolecules. 18 (2017) 610–616. https://doi.org/10.1021/acs.biomac.6b01790

[236] H. Chacón-Navarrete, C. Martín, J. Moreno-García, Yeast immobilization systems for second-generation ethanol production: actual trends and future perspectives, Biofuels, Bioproducts and Biorefining. 15 (2021) 1549–1565. https://doi.org/10.1002/bbb.2250

[237] H. Zabed, J.N. Sahu, A.N. Boyce, G. Faruq, Fuel ethanol production from lignocellulosic biomass: An overview on feedstocks and technological approaches, Renewable and Sustainable Energy Reviews. 66 (2016) 751–774. https://doi.org/10.1016/j.rser.2016.08.038

[238] P. Yuvadetkun, A. Reungsang, M. Boonmee, Comparison between free cells and immobilized cells of Candida shehatae in ethanol production from rice straw hydrolysate using repeated batch cultivation, Renew Energy. 115 (2018) 634–640. https://doi.org/10.1016/j.renene.2017.08.033

New Materials for a Circular Economy
Materials Research Foundations 149 (2023) 446-483

Materials Research Forum LLC
https://doi.org/10.21741/9781644902639-12

Chapter 12

Polymeric Electrolytes in Fuel Cells: A Sustainable Approach

N. Ureña[1]*, M.T. Pérez-Prior[1]

[1]Department of Materials Science and Engineering and Chemical Engineering, Universidad Carlos III de Madrid, 28911 Leganes, Spain

murena@ing.uc3m.es

Abstract

Concerns are increasing over the thoughtless uses of fossil fuels in transport and electricity generation due to their harmful health and environmental impacts. As an alternative, in recent years, renewable energy has received an unprecedented worldwide interest. One of the most important points is the search for sustainable electrochemical energy generation and storage devices. The environmental impact of the use of such devices, in sectors such as transport, is indisputable. The use of vehicles that contain fuel cells, compared to those that use fossil fuels, drastically reduces the CO_2 emissions into the atmosphere. In this context, fuel cells are electrochemical devices with a low carbon footprint. However, fuel cell components are based on scarce and expensive materials. Concretely, in the hydrogen fuel cells, two of the main components, platinum as the electrode and Nafion® as the electrolyte, respectively, are high-priced and difficult to obtain. This limits the industry implementation of fuel cell technology. In this chapter, conventional component recovery methods are summarized showing their limitations. Furthermore, promising approaches for component recovery are discussed. These approaches pave the way to new, and more eco-friendly ways for a circular economy.

Keywords

Renewable Energy, Polymer Electrolyte, Hydrogen, Fuel Cell, Circular Economy

Contents

1.   Introduction

The greenhouse effect has received more attention in recent years due to the huge amount of gasses, such as CO_2, that are emitted into the atmosphere. In fact, the amount of CO_2 emissions in the United States only decreased by 8% between 1990 and 2020 (Fig. 1).

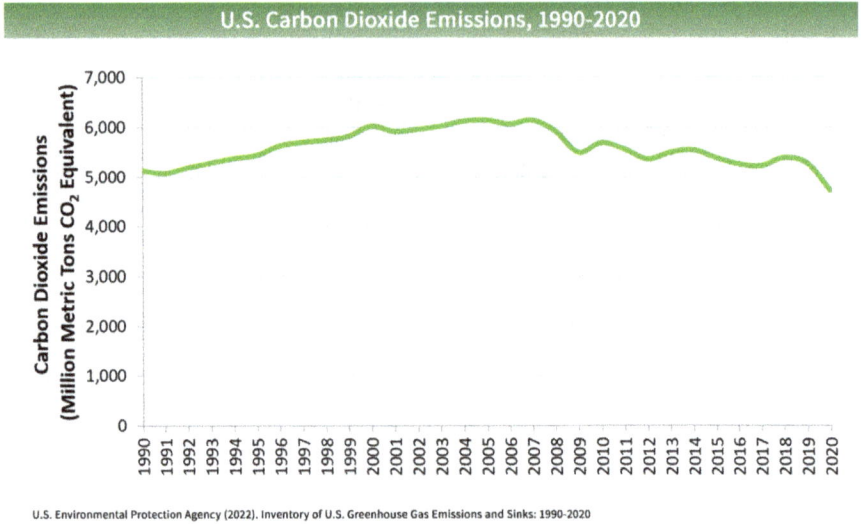

U.S. Environmental Protection Agency (2022). Inventory of U.S. Greenhouse Gas Emissions and Sinks: 1990-2020

Figure 1. U.S Carbon dioxide emissions estimate [1].

Human activities notably increase CO_2 emissions, and it will continue to increase and further warm the climate unless substantial efforts are undertaken to reduce them worldwide. Significant opportunities to mitigate anthropogenic emissions of these gasses exist, although some will be easier to exploit than others [2]. Therefore, to reduce the generation of this type of greenhouse gas (GHG), it is necessary to efficiently store electricity from renewable energies that contain little or no carbon. In this context, the design of mobile electricity generation devices to reduce urban pollution is of great interest. These devices include electrochemical energy storage equipment such as batteries, superconductors, and power generation devices such as fuel cells (FCs), which are currently the best solution for mitigating urban pollution and reducing CO_2 emissions. From this side, the European Union (EU) seeks to fully engage citizens in this unprecedented challenge of reducing GHGs with different proposals. One of the most significant proposals is the use of vehicles that do not emit polluting gasses into the atmosphere, such as those that use FCs. With these initiatives, a decrease in CO_2 emissions of more than 30% is expected by 2030 [3]. Furthermore, FCs have been proposed as a power source for several applications such as transportation, stationary power, or portable devices [4,5]. In this way, in addition to reducing emissions of pollutants into the atmosphere, the use of fossil fuels would be reduced [6].

In addition to the sustainability problem, the unavailability of fossil fuels makes it essential to reduce their use. These resources are found only in certain localized areas of the world. Areas not always accessible, with unstable political situations that watch over their economic interests. This situation directly influences the geopolitical relations between exporting and importing countries. The high dependence on exporting countries results in an unprecedented framework of energy insecurity. Recently, this scenario proposes, especially in the most developed countries, to substantially reinforce the system of renewable energy networks and create a system based on a circular economy (CE).

The current model is a linear economy (LC) created to mass produce and sell products (flat line, the so-called open cycle) (Fig. 2). The main issue is that we can not afford, in the long run, to continue with this production system based on irrational usage, and the abuse of resources and energy. Conversely, CE is far more sustainable because it is an economy with the main purpose of capitalizing on the materials used while reducing or even eliminating waste as much as possible. In this way, the product of today is considered to be the resource of tomorrow's product [7]. When products are reused, repaired, or recycled, employment is generated, and when waste from one process is used as an input into others, efficiency and productivity gains are achieved (Porter Hypothesis) [8].

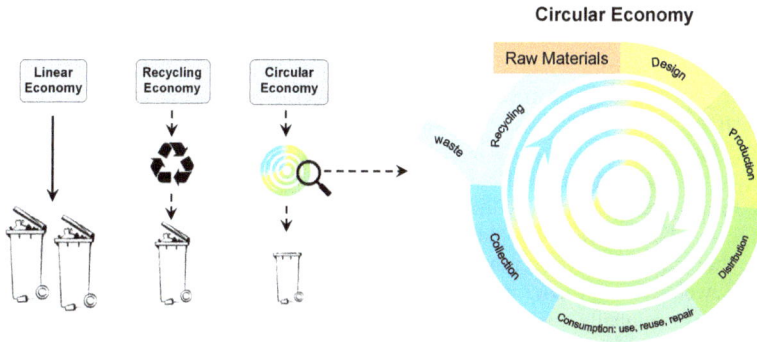

Figure 2. Linear economy (LC) versus circular economy (CE).

Today's society is based on a LC model where mobility and consumption prevail producing enormous amounts of CO_2 that end up in the atmosphere causing global warming of the earth. It is urgent to move to a CE model and harness and store energy from renewable sources, in order to achieve the sustainable development goals set by the EU. In this sense, much more rigorous requirements are expected, than those we have seen so far, in terms of fuel economy and mandatory fuel efficiency standards. Through this shift toward a CE, by 2030, as estimated by Ellen MacArthur Foundation, the European Union would be able to decrease their net spending on resources by 600 billion euros, to increase annual resource productivity by 3%, and thus to generate a total of 1.8 trillion euros of annual net benefit

[7]. In spite of these very encouraging economic advantages, up till this point, it is estimated that only 6% of all materials used in the worldwide economy make up part of a closed loop recycling model. The EU economy is estimated to have twice the degree of circularity compared to the global average. An estimated 13% of materials processed are recycled, which can still be considered very low [9].

In relation to the sustainability and availability of fossil fuel, much attention has been devoted to devices for electrochemical energy storage and power generation. However, in general, there is no unique energy technology that can be generalized to govern the global energy market. Therefore, a combination of energy producing and storing technologies exists, based on the availability of resources in different countries, and on the different policies of these countries concerning importing energy-producing and storing resources. As for the power-generation systems, fuel cells are gaining special interest. One of the main reasons is because, in terms of sustainability and availability, they use hydrogen as fuel, which is "clean" and is the most abundant element on the planet. It makes up 75% of matter by mass and 90% of matter by number of atoms. Due to the fast-pace development of hydrogen technology on one hand, and a continuously growing energy demand on another, many countries have been lately focusing on defining a domestic hydrogen roadmap. Moreover, since hydrogen and fuel cells are able to support this continuous global growth and demand, many countries include the development of hydrogen as part of their national strategy and push the promotion of the fuel cell industry. A good example here is Japan, where the government has considered hydrogen energy as an essential part of their national energy strategy. This includes the development of the industrialization and marketing of hydrogen technology, leading to a high production capacity of fuel cell vehicles (estimated 10k units of Toyota Mirai), household fuel cell cogeneration systems (estimated 30k sets of Ene-farm systems), and more than 100 hydrogen refueling stations [10-12]. In the end, it is expected that fuel cells will play an important role in the energy transition. This would involve the generation of green hydrogen with water electrolysis by means of electricity from renewable energy sources. Despite being a device close to a high degree of implementation globally, there are many aspects which need further efforts. In the context of sustainability and availability, the Fig. 3 shows the global supply and the risk of raw materials necessary for fuel cells. The raw materials are located in specific zones [13]. Therefore, and in order to reduce the environmental impact during production and transportation, recycling and limiting consumption is necessary.

In this chapter, a fuel cell technology evaluation is presented. The description of the principal components, their main applications, availability and sustainability are detailed. The current state of their conventional recovery methods for a circular economy is described. Additionally, alternative methods for component recovery are discussed, highlighting the main advantages and limitations from an environmental point of view for their implementation in the industry. The bases for the assessment of the environmental impact for the fuel cell are discussed.

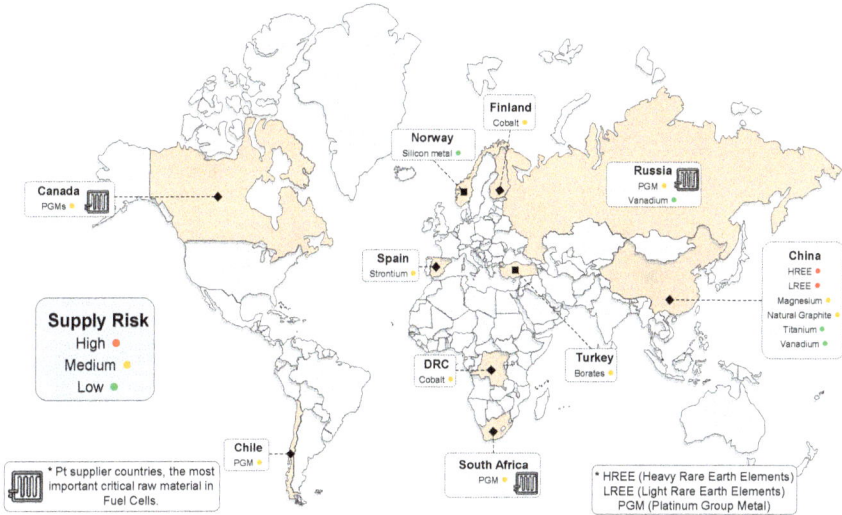

Figure 3. Global supply and risk of raw materials for fuel cells.

2.   Fuel cells – A general overview

A fuel cell is a device that converts chemical energy into electric energy, heat, and water by means of an electrochemical oxidation-reduction (redox) reaction. During the redox reaction, a fuel is oxidized and a comburent is reduced. A fuel cell is composed of an anode or negative electrode, a cathode or positive electrode, and the electrolyte. The basic electrochemical unit being referred to is the "monocell" (Scheme 1). In a monocell, the two electrodes are separated by the electrolyte and the system is connected to an external circuit. While the term "fuel cell" is often used, the basic electrochemical unit being referred to is the "monocell". A fuel cell consists of one or more of these monocells electrically connected.

The electrodes are fed by a gas or a liquid flow such as hydrogen and oxygen. Oxidation reaction of the fuel (hydrogen) takes place at the anode. While at the cathode, the reduction reaction of the comburent (oxygen, which comes, in most cases, from the air) takes place. Both the anode and the cathode have a complex structure. However, it is not the focus of this chapter (information has been published about this subject; see references listed from [12,14-27]. In general terms, it is composed of carbon, Teflon®, and as the catalyst, platinum (see Section 4).

The electrolyte is an ionic conductor which provides the medium for transfer of charge, as ions, inside the system between the electrodes. However, it does not allow electron conduction to pass.

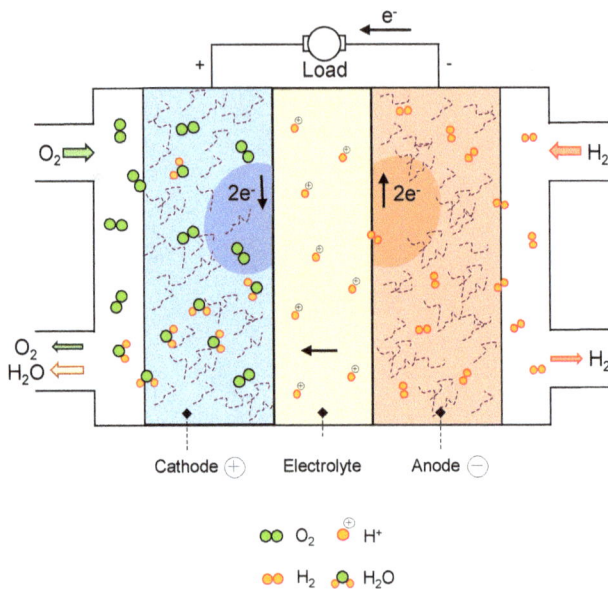

Scheme 1. Polymeric membrane fuel cell (modified from [6]).

Thus, H_2 is oxidized in the presence of the catalyst to give rise to protons with the release of electrons. The electrons flow through an external circuit to the cathode creating a current that can be harnessed before it reaches the cathode. The generated protons pass through the electrolyte until they reach the cathode where, when combined with O_2, water is formed. Both the oxidation and reduction half-reactions and the overall reaction are shown below.

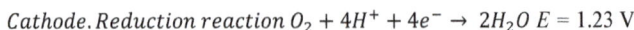

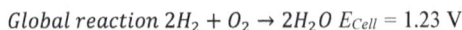

$Cathode. Reduction\ reaction\ O_2 + 4H^+ + 4e^- \rightarrow 2H_2O\ E = 1.23\ V$

$Anode. Oxidation\ reaction\ H_2 \rightarrow 2H^+ + 2e^-\ E = 0.00\ V$

$Global\ reaction\ 2H_2 + O_2 \rightarrow 2H_2O\ E_{Cell} = 1.23\ V$

New Materials for a Circular Economy
Materials Research Foundations 149 (2023) 446-483

Materials Research Forum LLC
https://doi.org/10.21741/9781644902639-12

There are several types of fuel cells. The classification is made based on the type of electrolyte used in each of them. Thus, polymeric membrane fuel cells (PEM) have a solid polymeric electrolyte, direct methanol fuel cells (DMFC) have polymeric membranes and use methanol which is directly oxidized to carbon dioxide, alkaline fuel cells (AFC) contain a basic solution as electrolyte, phosphoric acid fuel cells (PAFC) are consisted of an acid solution, molten carbonate fuel cells (MCFC) contain a salt of molten carbonate, and solid oxide fuel cells (SOFC) have a ceramic electrolyte. They also differ in the type of mobile ion that is transferred through the electrolyte and in the operating temperature. From a general point of view, all of them use hydrogen as fuel, except DMFC which use methanol.

This chapter is focused on the proton exchange membrane fuel cells (PEMFCs). The main advantages of these fuel cells compared to the others can be summarized as follows: (i) their operating temperature is relatively low, they operate below 100 °C, (ii) they have a compact structure which allows the generation of more energy per volume unit, and (iii) they reach a high efficiency (60%). Another advantage to be highlighted is the use of a solid electrolyte versus a liquid electrolyte which reduces corrosion problems. It achieves greater durability, and a safer sealing of the device, reducing manufacturing costs. The main disadvantage is associated with the heat that the device dissipates, which makes it necessary to include cooling in the system [28].

These devices can be used for portable devices, as a stationary power source (3-7 kW), and in the construction sector (250 kW) [29]. Moreover, they are used as a power source in electric vehicles, which is one of the most important applications [30]. The Fig. 4 shows the trend in the fuel types of new cars for the 2017-2022 period in the EU. Although those cars that run on petrol and diesel have been at the forefront of the market since 2017, important changes can be seen in just three years. In particular, sales of diesel-powered cars have dropped by almost half (from 44% to 28%). This fact has given way to other alternative cars that have increased their presence significantly such as electrically-chargeable vehicles (5% for battery electric or BEV, and 5% for plug-in hybrids vehicles or PHEV), or hybrids (hybrid electric vehicles, HEV), which account for 12%. And this trend is expected to continue in the next few years.

Even though the market for BEV is lately undergoing more increase, the limitation of lithium-ion batteries in terms of energy density restricts the range of BEVs. This creates a limitation in the maximum weight of batteries carried by a vehicle and the space within the vehicles to store the batteries. On the other hand, FCEV (fuel cell electric vehicles, included as part of "alternative fuels" in Fig. 4) offer consumers a wider range, due to its capacity to deliver high energy density.

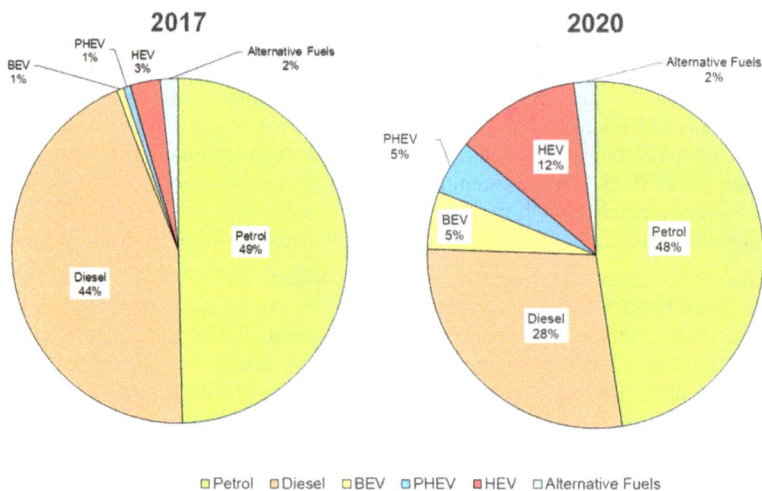

Figure 4. Fuel types of new passenger cars in the EU.

For this reason, developing fuel cells for transportation vehicles has attracted automakers over the last 30 years. Companies such as Volkswagen, Honda, Ford, Hyundai, and Toyota have significantly invested in order to advance the fuel cell technology for FCEVs. In the case of passenger cars, the outcome of these investments has yielded a relatively disappointing reality. In 2021 only two automakers, Toyota and Hyundai, have FCEV on the market, with records of fewer than 10,000 cars sold during 2020.

On the contrary, BEVs that have also started developing during the same time as FCEVs, have a record of more than 3,000,000 vehicles sold in 2020. This pushes fuel cells from the primary market of passenger cars, yet remains present as a primary axis of sustainable energy development for major European countries. The European hydrogen energy strategy, for the transport sector, consists of its implementation in urban buses, taxis, and sections of the railway system that cannot be electrified.

3. Proton exchange membrane fuel cells: Components and materials

The monocell consists of an assembly of bipolar plates, a separator (normally made of Teflon® or even silicone), electrodes, and a solid electrolyte (polymeric membrane). Usually, monocells are connected in an appropriate series arrangement (stack) to provide the required operating voltage and current levels [28,30]. In this way, the stacks are a widely used practical power source. The stack is a set of, what is commonly known as, membrane electrode assemblies (MEA) and bipolar plates. The plates are electrically

New Materials for a Circular Economy   Materials Research Forum LLC
Materials Research Foundations 149 (2023) 446-483   https://doi.org/10.21741/9781644902639-12

connected in series through one or more MEAs. They are the current collectors, and those through which the reactants involved in the reactions are supplied to the electrodes. In terms of energy, a monocell generates a potential difference of 0.7 V [28]. Therefore, to generate an appreciable potential difference, this arrangement is necessary. The power achieved in a fuel cell stack generally ranges between 50 and 250 kW [31]. Regarding power density, it is possible to reach 350 mW cm^{-2} with this type of device [32].

In the next section, the main recycling methods for a circular economy of the principal components of the fuel cell is described.

4.   Electrodes

Generally, electrodes are composed of two layers namely, the gas diffusion layer (GDL) made of carbon black and Teflon®, and the catalyst layer (CL) with platinum, carbon, and Teflon® [12,14]. In turn, GDL can be divided into two layers, (i) a macroporous layer (MPL) of carbon powder and hydrophobic agent (usually Teflon®) and (ii) a microporous layer composed of carbon paper or carbon cloth (Fig. 5). Although some researchers consider that an electrode is composed of three layers and different structures have been discussed (see references [14-24] for further details). The GDL provides channels for gas and electrons, discharges water products, connects bipolar plates with the CL and acts as a support for the latter [25]. It is made up of carbon, alcohol, water and polytetrafluoroethylene (PTFE), or in some cases another hydrophobic substance. The use of PTFE helps promote the transportation of water and gas in flooding conditions [26,27]. The catalyst layer on the other hand is where reactions of electrochemical nature take place which in turn converts hydrogen and oxygen into water and electricity. The thickness of the CL usually ranges between 5 μm to 100 μm with a porosity varying between 40% to 70%. The catalyst particle size varies between 1 nm to 10 nm with and must be well dispersed in the CL [33]. It is worth mentioning that a nafion solution can also be present, acting as a proton conductor between the catalyst layer and the electrolyte membrane.

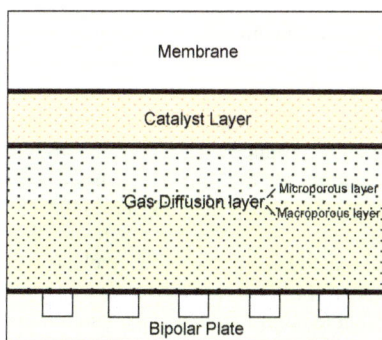

Figure 5. Electrode's structure of PEMFC (modified from [26]).

The most commonly used catalysts include carbon supported platinum and platinum/ruthenium for the best catalyst dispersion and utilization for PEM fuel cells. Platinum and ruthenium are part of the platinum group metal (PGM), rare precious metals which are being used in a diverse range of industrial applications. Another less common catalyst are non-platinum based where organometals and non-noble metals are considered [34]. However, Pt remains the most effective out of the other catalysts, for its use in PEMFC, and therefore is the most used and studied in the field.

From the several components of the electrodes, this section focuses on the catalyst, since Pt is one of the raw materials registered by the European Commission (see below).

Platinum rarely exists as a native metal but commonly forms a wide variety of alloys with other metals, notably with iron, and less commonly with tin, copper, lead, mercury, and silver. In other platinum-group minerals the PGM are bonded to sulfur, arsenic, antimony, tellurium, bismuth, and selenium.

The platinum-group elements (PGE), also more commonly referred to as platinum-group metals (PGM), refer to six chemical elements which are ruthenium (Ru), rhodium (Rh), palladium (Pd), osmium (Os), iridium (Ir), and platinum (Pt). Interestingly, minerals that have an essential part made of one or more PGMs are referred to as platinum-group minerals, and also share the same abbreviation PGM. However, in this chapter PGM will be used only to refer to platinum-group metals.

Modified-platinum based and non-platinum based catalysts, even though in theory, are considered to be highly promising substitutes for platinum, at present date they are still far from being considered in manufacturing and industrial applications. Several candidate materials have been tested for their applicability in fuel cells and their capability to render good work efficiency, however nothing promising has come through as an eventual substitute for Pt. Sludge biochar-based catalysts (SBCs) are an interesting example to mention where they have been used as an electrode in microbial fuel cells (MFCs) [35]. Finally, it is worthy to mention that some of these alternative catalysts have recovery processes similar to Pt that will be mentioned in the coming sections.

4.1 Applications

In recent decades, the rate of use of PGM has increased significantly, with more than 80% of their total global cumulative production. The main reasons for this change are the increment of the global population, and the spread of prosperity across the world. The PGM are rare precious metals, although unlike gold, they are used in a diverse range of industrial applications. For instance, new technologies, such as those needed for modern communication and computing, and those that produce clean energy, require considerable quantities of numerous metals [36]. Generally, they are known for their uses as catalysts, one of the most important applications. Specifically, platinum is predominantly used in autocatalysts in diesel vehicles to transform pollutants such as CO_2, NO_x, hydrocarbon (HC), in less harmful products (about 35% of CO, 30% of HC, and 25% of NO_x produced into the atmosphere is from the transportation sector) [37]. Being the main contribution to

Pt demand in Europe where around 28% of the passenger cars had a diesel engine in 2020 (Fig. 4). Other sectors such as jewelry, glass, biomedical, and chemical are also part of the network of applications of this metal [38-41]. Eventually, the use of Pt in electrolyzers and FCs for stationary and transportation applications grows, and that demand is expected to increase significantly [42]. Nowadays, the FC stacks of a vehicle contain between 25 and 35 g of platinum (0.25-0.35 g kW^{-1}) [43]. And only in the range of 2-3% of the total PGM uses is set aside for these applications. One of the assumptions concerning the grams of Pt per KW in a vehicle estimates that now till 2030, the quantity of Pt will decrease until 0.07 g. This will depend on global interests because nowadays, the current generation of FCs in vehicles are more concerned about durability, and the stack durability is highly correlated with Pt use. Nonetheless, these technologies are becoming crucial from the environmental point of view, in particular in the automotive industry. Therefore, on one hand if the hydrogen infrastructure starts to grow to some extent, consequently, the percentage of the total PGM uses would increase significantly. On the other hand, however, the growth of the BEV market will eventually decrease the demand for platinum, rhodium, palladium, whose recycling is estimated to be between 50% to 60% [3].

4.2 Abundance and pollution data for its extraction

The PGM elements are one of the least abundant elements. Within PGMs, Pt and Pd are the (relatively) most common whereas Rh is considered to be the rarest [44]. Moreover, global PGMs resources are small compared with more common metals such as copper or iron [41]. In particular, the main natural deposits of platinum are found in Russia, Canada, and South Africa. 90% of all known Pt in the world can be found in the Bushveld Complex in South Africa, and its production accounts to more than 75% of the worldwide platinum production.

The mining of Pt is very costly and has a very high impact on the environment because of the low concentration of Pt per cubic meter of land [45-49]. In order to produce 1 kg of platinum, it is estimated that 18,860 MJ to 254,860 MJ is required. Moreover, for the extraction of a ton of platinum an estimated 100,000 m^3 to 1,200,000 m^3 of water is required. On the other hand, recycled Pt only requires 1400MJ to 3400 MJ and 3000 m^3 to 6000 m^3 per kg [3]. Lastly, the sulfide minerals found in platinum ores can lead to extremely hazardous fumes, along with tons of CO_2 emitted, in the process of smelting and refining [47].

The positive point is that consumption is also small. Hence, according to data reported by USGS, the known Pt resources are enough to supply the demand for at least the next 200 years [50]. From this point of view it is clear that there is no imminent risk of PGM supply. However, from the geopolitical and the market perspectives previously discussed (Section 1 and 4.1, respectively), in the short run supply risk situation may occur [45,51-53]. With this scenario where PGMs are i) important for high technology products and emerging innovations, ii) difficult to replace in specific devices, iii) relevant for fighting climate change and for improving the environment, EU commission denoted PGMs as Critical Raw Materials (CRM) [54]. This means that, according to the EU commission, PGMs exceed

New Materials for a Circular Economy   Materials Research Forum LLC
Materials Research Foundations 149 (2023) 446-483   https://doi.org/10.21741/9781644902639-12

the threshold of supply risk and economic value. Therefore, in 2020, the EU approved new targets for recycling rates for its industrial use allowing a circular economy [3].

4.3   Recovery technologies of Pt from spent catalysts

In general, the metals retain their value quite well even after recycling. Moreover, Pt can be recycled many times with no change in their functions. The advantages about recycling metals, and specifically Pt, are numerous. Recycling helps reduce negative environmental impact. The energy required to recycle metals is much less than the energy required to mine new metals. Moreover, by recycling, the strain on limited and scarce natural resources can be reduced. Furthermore, recycling directly leads to the decrease of waste. Finally, recycling is a plus from an economic point of view because it reduces the costs which in turn increases revenue. For these reasons, Pt recycling and recovery is important in order to meet the worldwide demand of this limited metal, via a sustainable approach, and an economically viable manner.

4.3.1 Conventional technologies

The obtainment of metals varies depending on the homogeneity, type of material acting as support, percentage of metal employed, and the loading percentage of PGM and other compounds. There are two main conventional approaches to obtain recycled Pt as well as other recycled PGMs. They are pyrometallurgical and hydrometallurgical recycling technologies. Actually, they are the most used techniques since they profit from the established facilities of big companies which use platinum, other PGMs as catalysts, or other manufactured materials [55,56].

The pyrometallurgical recycling process is a thermal treatment, normally in a furnace, of the used Pt mixed with other materials. The volatile compounds such as support material, plastics, oxides, and others will be eliminated by means of this technique and the Pt metallic phase will be enriched. The hydrometallurgical recycling process consists of leaching metals with the presence of highly oxidizing species. By combining both techniques, a higher percentage of recovery is obtained. However, high hazardous liquid and gaseous wastes are generated, especially with pyrometallurgical recycling processes. Then, the separation or precipitation step is required in order to retrieve Pt from the resulting mix.

4.3.1.1 Pyrometallurgical recovery process

This method permits, through thermal treatment, to obtain the metals in a higher percentage from PGM sources at end of life. It is one of the main processes to recover Pt from spent catalysts [57,58]. Three main sub-categories are found; i) chlorination, ii) smelting collection, and iii) sintering processes [59,60]. In contrast to other methods such as hydrometallurgy, pyrometallurgical recovery process does not require a specific pre-treatment of the Pt source, which means, it is a very efficient process [59]. On the contrary, this technique is highly energy demanding due to the temperature of operation above 1000 °C. Moreover, it goes against all the efforts that are being made from the environmental point of view due to the high amount of harmful liquids and gasses this process emits. Furthermore, an additional process of treating the wastes is required [61,62].

In this category, chlorination or carbochlorination allows the extraction of Pt and other metal species in their respective chloride salts from spent catalysts at high temperatures. The secondary step is the separation i) depending on the nature of the metal chlorides, ii) through several cycles of wash, and iii) by the presence of an activated carbon layer [63]. Through chlorination, highly pure metals are obtained, and progress has been made in terms of efficiency and environmental impact [64,65]. However, disadvantages like the corrosion of the furnace and its components, the presence of hazardous gases such as Cl_2 which causes health and environmental risks, and the energetic cost makes this process uninteresting from an industrial perspective [66].

Another method to obtain Pt is by smelting. Here, platinum waste is mixed with a collector, flux, and reducing agents and then smelted at very high temperatures. Usually this is done in electric, inductive, or plasma ovens with temperatures going up to 1000 °C. It is worthy to mention that in this method pretreatment is necessary. Therefore, as part of this process, and in order to obtain Pt in higher percentage, taking apart the components, burning of non-metallic substances, calcination, and reduction with other metals can take place. In presence of supported materials, such as carbon, the melting process by using specific fluxes are required. After that, the resulting slag should present low viscosity. The separation of Pt enriched metal phase followed by the purification of Pt containing alloy are carried out [67]. One of the oldest method for recovering PGMs from spent catalysts is the metal lead collection method [68]. Copper collection process [69] or iron collection process [70] are other alternative methods used for spent catalysts. With the latter, the recovery rates for Pt reported are up to 98% in the literature. Copper collection method is relatively low in cost, very efficient, and has a low smelting temperature which makes it interesting from an industrial perspective [59,66].

The sintering process uses plasma by reducing the Pt compounds. In particular, a thermal plasma process was developed in order to recover spent alumina-supported platinum catalysts (Pt/Al_2O_3) by platinum oxide (PtO_2) [71]. This remarkable process has been extensively implemented in the petrochemical industry. In addition, Bronshtein et al. demonstrated that Pt recovery from spent VCCs can be very efficient when using a sintering process with chloride salts [72]. It is important to remark that non-desirable species are employed or generated during the process.

Briefly, close to 100% of recovery Pt is obtained by means of pyrometallurgy. One of the main reasons why it is still the more common industrial process to recover Pt. Consequently, their scalability secures lower costs. However, the elevated energy consumption and the large amount of hazardous wastes (such as SO_2, CO_2), and the new environmental regulations force us to look for more eco-friendly alternatives [57].

New Materials for a Circular Economy   Materials Research Forum LLC
Materials Research Foundations 149 (2023) 446-483   https://doi.org/10.21741/9781644902639-12

4.3.1.2 Hydrometallurgical recovery process

The present process uses acid or bases solution with an oxidant to leach the metal followed by separation or concentration. Finally, the recovery is obtained in a metal or salt form. Compared to pyrometallurgy, softer work conditions in terms of temperature are required. Therefore, lower emissions and harmful wastes are emitted to the environment. On the contrary, depending on the nature of the metal to recover, a chemical or physical pretreatment is usually carried out to i) reduce the amount of Pt oxides produced during their previous life cycle, ii) limit the presence of organic compounds, or iii) reduce the particle size of the pristine sample [73,74]. The literature shows the importance of an oxidizing agent, such as chlorine, hydrogen peroxide, and bromine, in allowing Pt to be dissolved by dropping the potential energy barrier enough to perform leaching [75-80]. The treatment of catalysts with aqua regia or other acidic oxidants, although can lead to the emission of hazardous gases, creates a metal containing solution that will require additional treatment to separate the Pt [81]. Jafarifar *et al.* studied two innovative techniques (reflux and irradiation) for recycling catalysts from the petrochemical industry point of view with Pt recovery rates ranging between 96% up to 98% [82]. Others investigated ways to limit the environmental impact due to aqua regia use by proposing other leaching agents in the presence of Cl^- as a Pt complexing agent. When using H_2O_2 combined with HCl, a recory rate of 95% has been reported [83]. Moreover, a study demonstrated that by using ozone with concentrated chloride can yield a Pt recovery rate of up to 90% [84]. Recently, a solvomettalurgic process using inorganic lixiviants reported Pt recovery rates around 80% without the need to use aggressive elements or the generation of harmful wastes [85]. Finally, the use of microwave radiation by Suoranta *et al.* [86] in order to assist leaching, made it faster, and therefore less expensive than conventional leaching all while reporting a Pt recovery rate of 90%.

4.3.2 Alternative technologies

In recent years, the development of alternative methods from pyrometallurgical and hydrometallurgical recovery processes which are within the framework of the latest government regulations are receiving more attention. The new technologies can allow recycling not only Pt and the family of PGMs, but also other significant materials such as support materials or reactor membranes.

4.3.2.1 Selective electrochemical dissolution

Platinum is the most widely used metal as an electrocatalyst in fuel cell devices and hydrolyzers [87]. Recently, greater attention is being given to developing, at lab-scale, Pt electrochemical dissolution as an alternative method to recover the metal [88,89]. For instance, Sharma *et al.* recovered Pt from spent low temperature PEMFC electrodes through dissolution in HCl with an oxidizer and the recycling of platinum through redeposition of the dissolved Pt on carbon. The recycled Pt/C showed electrochemical performance comparable to that of the commercial 20 wt % Pt/C. Pavlišič and coworkers studied the effect of voltage scan rate on Pt dissolution from benchmark Pt/C fuel cell

New Materials for a Circular Economy   Materials Research Forum LLC
Materials Research Foundations 149 (2023) 446-483   https://doi.org/10.21741/9781644902639-12

electrocatalysts. A variety of electrochemical treatments were carried out in order to better understand this methodology.

Electrochemical dissolution methodology recovers Pt catalyst highly pure at optimal conditions (pH, temperature, and voltage). Therefore, it can be proposed as an alternative eco-friendlier and cheaper process. However, considerable effort is required in optimizing the methodology to achieve large scale reproducibility.

4.3.2.2 Bioleaching

In general, bioleaching is a process described as *the dissolution of metals from their mineral sources by certain naturally occurring microorganisms* [90,91]. In this biometallurgical process the microorganisms are usually bacteria. Two main types of bacteria are used, mesophilic and heterotrophic bacteria. The former, which are autotrophic and chemolithotrophic, are able to leach metal ions when iron or sulfur is present. The latter are able to leach precious metal ions due to the presence of cyanide produced by the oxidative decarboxylation of some substrates [92]. For instance, chromobacterium violaceum has been employed to recover rare metals [93]. Bacteria involved in bioleaching demand a favorable environment in order to grow and have good bioleaching abilities. For example, pH plays an important role because it influences the growth of the bacteria population and also impacts the rate of leaching [94].

It is worth to note that bioleaching is still considered to be rather understudied. There is still a lot of room for research necessary to better understand the role of all the parameters involved and necessary for its optimization in order for this phenomena to achieve as promising results as better understood methods and processes such as hydrometallurgy and pyrometallurgy.

4.3.3 Others

In recent years, research has shed light on several techniques that can be considered as more environmentally friendly methods of recovering platinum [95,96]. In this context, a good example is the use of magnetic separation in order to concentrate platinum after which leaching with Ni electro-less or $FeCl_2$ vapor deposition is used [97,98]. Platinum was alloyed by forming magnetic compounds isolated from the remaining catalyst. By doing so, the sample quantity that is required to be further treated decreases, along with the consumption of energy and reagents to do so. In the case of pyrometallurgy, Martinez *et al.* put forward a different chlorination method for VCCs [99]. Molten salts were used as a solvent in order to have adequate rates of recovery that were recorded to reach up to 50%. Here the temperature is clearly kept lower than conventional techniques, however, and in spite of lower consumption of energy, the problem encountered is in the form of chlorine derivatives which are considered to be environmentally hazardous. Ding *et al.* also used a method with low temperatures compared to the conventional 2000 °C range. In their work, Iron-Platinum alloys were formed in an efficient iron collector with temperatures ranging between 1300 °C and 1400 °C by adjusting the composition of the slag and thus acquiring a platinum recovery rate of 99.25%. Even lower temperatures (950 °C) were reported by

Moracali that avoided the use of a plasma furnace. Here the use of a flux made of B_2O_3, Na_2O, and FeS_2 achieved a platinum recovery rate of 99% [100]. It is interesting to note that even though some reactants used by Moracali have a very high cost, this technique demonstrates a real environmental advantage because of its low energy consumption and the use of less hazardous species. Sasaki *et al.*, by using zinc as a pretreatment, alloyed platinum in VCCs. [101,102]. Since the platinum-zinc alloys are very dissolvable in aqua regia, this enhances the rate of recovery (almost 100%), decreases the overall process time, and minimizes the amount of oxidant required. Sasaki *et al.* also showed that aqua regia, when diluted, works as efficiently as undiluted when considering the extraction of platinum in a sample treated by zinc. In the field of metallurgy, in order to perform controlled and efficient heating, typically microwaves are used [82]. Some researchers however have used microwaves in order to pretreat samples before the extraction of platinum. In both a traditional oven and a microwave, Suoranata *et al.* have performed leaching with aqua regia and HCl using a range of temperatures between 90 °C and 210 °C [86]. The authors report that the microwave oven, compared to a traditional oven, promises a more energy efficient process. When the former is used, both aqua regia and HCl are able to leach, reaching a recovery rate reported to be up to 91%, in a relatively shorter time. According to the authors, this is due to a mechanism of local superheating. Platinum and other PGM particles could be selectively heated by microwave ovens. Spooren *et al.* also worked with microwave ovens in order to heat the VCC in the presence of several oxidants ($NaClO_3$, $NaHSO_4\cdot H_2O$, and $KHSO_4$) to change the Pt into Pt^{2+} and Pt^{4+} [103]. Then the authors performed a microwave acidic leaching method with 1M HCl which yielded a platinum recovery rate of 85% in only 30 minutes. By having such short process time, their method is considered to be very efficient from an energy consumption point of view. Microwave-based leaching has also been investigated for industry purposes by companies like VITO and other stakeholders of the PLATIRUS project that are ready to start testing in a prototype phase [104]. Finally, in an attempt to use a leaching process which is more environmentally friendly, authors such as Chen *et al.*, Trinh *et al.*, and Nogueira *et al.* have considered using agents that produce less hazardous emissions such as $CuCl_2$ and $NaClO_3$ by considering roasting and oxidation cycles as pretreatments [105-107]. Moreover, Trinh *et al.* recycled all the components of a VCC by roasting-assisted leaching.

5. Conventional electrolyte: Nafion®

Within the PEM-type fuel cells, the most widely used are the fuel cells that contain proton exchange membranes. PEMFCs use a solid polymer membrane (commonly with a thickness range between 127 to 254 μm) which is permeable to protons but it does not conduct electrons [30]. One of the most common and commercially available PEM materials used as an electrolyte is Nafion® from Dupont (perfluorosulfonic acid (PFSA) ionomer) whose chemical structure is shown in Fig. 6 [108,109].

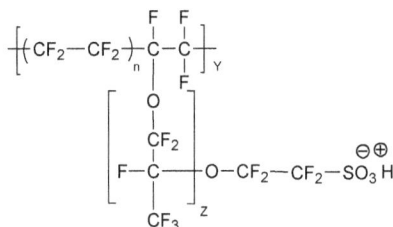

Figure 6. Chemical structure of Nafion®.

Nafion® is made up of a main hydrophobic polymer chain based on polytetrafluoroethylene (PTFE), which provides good mechanical stability, and side chains attached by ester bonds with hydrophilic sulfonic groups responsible for ionic conductivity. Due to the super acid groups, commercial Nafion® membranes have high conductivity values and high stability against oxidation, which allows achieving high performance and durability in the fuel cell [110,111]. The main disadvantages of this material are, on the one hand, its high cost, and on the other hand, the decrease in ionic conductivity that it presents at high temperatures due to hydration problems.

Besides the use of Nafion® in electrochemical devices such as FCs [112], these membranes are also involved in chloro-alkali electrolysis [113], and photoelectrochemical systems [114]. In 2021, the expansion of PEMFCs mainly in the transportation and power supplies markets is growing at a steady rate and the global Nafion® market is anticipated to rise at a considerable rate between 2022 and 2030 [115]. Consequently, it will be necessary to implement protocols in the industry for the reprocessing of these materials (due to their end-of-life, failed membranes).

Nowadays, there are three procedures to process polymer wastes, i) burial in landfills, ii) incineration, and iii) mechanical recycling. Taking into account that burial presents an extreme issue related to the timeline degradation of polymer wastes and a null recovery percentage, and incineration is associated with a high amount of emission of harmful gases into the environment and a low energy recovery efficiency, both techniques seem to be out of date with respect to the new global initiatives. Instead mechanical recycling is considered as a transitory solution. Mechanical recycling of polymer based products can be summarized by a process of sorting, washing, drying, and melting in order to obtain new polymer based materials [116,117].

In the following section, alternative Nafion® recovery methods used so far are summarized. However, it is an area that needs more research to improve recovery percentage, cost reduction, etc., which would make the methodologies practical on an industrial scale. This also indirectly impairs the employment of material recovery techniques of the rest of the device. For instance, pyrometallurgical recovery processing of Pt (methodology described in Section 4.3.1.1) used as electrocatalyst in FC devices is not viable at large scale due to the presence of fluorine atoms in the MEA. In fact, Koehler *et al.* proposed a pyrometallurgical recovery process in which the MEAs were treated with a medium present in the supercritical state to obtain the fluorine containing constituents from precious metals. However, this process employed a high energy consumption and hazardous chemicals [118]. Specifically, there are fluoropolymers in the membrane, the treated GDL using Teflon® (tetrafluoroethylene, PTFE) [119], and the catalyst layer's binding agent. This process, however, is practical because the generated ash can be evaluated and assayed and then commercially exchanged. However, the incineration of this kind of polymers allows the formation of toxic compounds such as hydrogen fluoride (HF). In the same way, hydrometallurgically has been carried out for MEA recycling. However, the presence of big volumes of strong acids and oxidants used, such as HCl, H_2SO_4, HNO_3, and H_2O_2, can react vigorously with the fluorinated compounds generating harmful gases. Devaluing these methods due to its harmful effect on the environment [120-122].

5.1   Recovering Nafion® Membranes

An alternative to conventional processes for treating polymer waste could be the recycling and regeneration of membranes. It can strengthen both the economic system in the industry and the environmental state [123,124]. In the methodology, the end of life polymeric electrolyte is separated from the rest of the MEA components. The membrane is then dissolved in an appropriate solvent. Once the resulting ionomer is obtained, it can be recast into a film [123]. Generally, this polymer is not soluble in alcohol and water at room temperature. However, it is possible to dissolve by increasing the temperature at high pressure in a mixture of ethanol and water [123,125-132]. During the last years, different Nafion® recycling methods have been thoroughly studied by different researchers to create cleaner and environmentally friendly processes. For instance, Xu *et al.* [122] used concentrated sulfuric acid solution to treat catalyst-coated membranes. As products they obtained Nafion® solution and catalyst. Afterwards, the remaining product was reused in fuel cells. However, they reported the formation of harmful SO_2 gas obtained during the process. Shore *et al.* proposed the immersion of the MEA in alcohol and water solution in order to weaken the adhesive force between membranes and electrodes [133]. They investigated the influence of using alkyl alcohol solutions at different concentrations. Xu *et. al* used isopropyl alcohol and water solution to separate the membrane from the rest of the components of the MEA at boiling temperature. After, the membrane was redissolved in dimethyl sulfoxide (DMSO) and cast. The drawback observed was the deterioration of the mechanical properties post-recast [123]. Other authors tried to dissolve MEAs in alcohol/water solvent in autoclave, followed by the separation of the Pt particles from the Nafion® ionomer solution by filtration in different steps. This methodology turned out to be complex and high-consumption energy [134]. Similarly, Moghaddam *et al.* applied three methods to dissolve the membranes, and then the resulting solutions were recast and characterized. They showed the recast Nafion membranes had appropriate water uptake, proton conductivity, and chemical and mechanical stability compared with the commercial Nafion® membrane [135]. Oki *et al.* developed a recovery process to obtain membranes from several MEAs used in mobile equipment simultaneously. The separation of the

New Materials for a Circular Economy   Materials Research Forum LLC
Materials Research Foundations 149 (2023) 446-483   https://doi.org/10.21741/9781644902639-12

electrolyte from the electrode was carried out by using milder organic solvents and minimizing the deformation of the membranes. Nevertheless, and for process validation purposes, the recovery membrane should be further characterized [136]. Silva *et al.* treated perfluorosulfonate ionomer dispersions in three organic solvents (ethylene glycol, dimethyl sulfoxide (DMSO), and dimethylformamide (DMF)) to prepare solution-cast membranes. The dispersions were obtained by dissolution of Nafion® 112 membranes in a reaction using water/ethanol solution. The results showed lower chemical stability for all cast membranes compared with commercial ones. Moreover, only membranes cast from DMF-based dispersions gave conductivity performance comparable to those of Nafion® 112 and 115 [130]. In 2000, Laporta *et al.* obtained Nafion® water dispersion from several procedures with Nafion® 117. The morphology of the resulting dispersions was compared with Nafion® commercial dispersion. The prepared Nafion® dispersions solution were casted obtaining membranes with a range thickness from 5 to 20 µm. The results exhibited that the thin membranes could absorb more water than the commercial ones. Consequently, they also showed equal and even higher proton conductivity [126]. Valente *et al.* studied strengths, weaknesses, opportunities, and threats (SWOT analysis) with the current methods of recycling and alcohol dissolution. As a conclusion, the authors claim to have a process with low environmental impact, low cost of operation, high rate of recovery, and the possibility to recycle more than one material. Nevertheless, additional work is necessary in order to have a high level of maturity of the technology. This will depend on technical aspects of the fuel cell on one hand, and on regulatory affairs and the level of product deployment on the other [137].

5.2   Alternative electrolytes

To overcome the limitations offered by Nafion®, different polymeric materials have been developed. From a sustainability point of view, one of the principal aspects to keep in mind for the choice of alternative membranes, is to find polymers with absence of fluorine atoms in the structure, since this would avoid the generation of toxic gases. Another important alternative is the utilization of biodegradable polymers, and the use of engineering plastics which can be repurposed into new value-added materials. Nonetheless, these research fields remain relatively immature. Nowadays, durability and ionic conductivity are the main focuses for researchers and also for the industry [138]. The most common alternative polymeric matrices are based on polybenzimidazole (PBI) [139-141], polystyrene (PS) [142-144], polyether ether ketones (PEEK) [145-147] or polysulfone (PSU) [148]. These membranes have a dual role of being a proton conduction medium and a separator of the gases. The degradation of PEMs has a significant impact on the durability of PEMFCs. The failure of the PEMs can lead to a high hydrogen and oxygen crossover which directly impacts the performance of the FC [149]. For these reasons, the durability of the membranes and the high costs are one of the main barriers that limit the commercialization of PEMFCs [150,151]. As a consequence, huge amounts of polymer wastes have been generated during the last decades [152,153].

For this reason, in general terms, synthetic polymers designed for faster degradability and recyclability are becoming more and more interesting [117]. They are used in a large number of applications [154,155] and their production has reportedly grown from 15 million tons in the 1960's to 311 million tons in 2014. This number is expected to triple by 2050 [156]. Consequently, the implementation of these materials in our daily life would enormously contribute to the development of the global circular economy.

In particular, one interesting example is the one of Jones *et al.* They described a repurposing method of polycarbonate (PC) to obtain polysulfone (PSU) with an average molecular weight of 11.000 g mol^{-1}, whose characteristic could be ideal to be reused for FC applications, although it needs further characterization. This practical method is based on the decomposition of PC under alkaline conditions generating bisphenolate and CO_2. With a carbonate salt, the resulting bisphenolate can take part in the polycondensation reaction to obtain the PSU without extra steps of isolation and purification [157].

6.   Other components of a fuel cell

Up till this point, we have mainly focused on recovering the main components of the MEA. However, it is worth having an overview on the possibility of recovery, reuse, or recyclability of the remaining components of a fuel cell. These components include but are not limited to the mechanical components, joints, plastic components, electronics, hoses, compression plates, bipolar plates, current collectors, and hydrogen tanks.

Components such as joints (usually made of silicon or rubber) and plastic are not reusable because of the wearing they undergo, however they are easily recycled. Compressor plates (carbon fiber or steel), bipolar plates (metallic or graphite), and current collectors (copper plate) rarely undergo wearing and therefore can be reused in new fuel cells. In a worst case scenario, these materials are perfectly recyclable. The recyclability of the electronics used within fuel cell systems is similar to typical electronics.

Hydrogen tanks are however a little bit more particular because they are considered to be more complicated to recycle. This is mainly because of the complex composition of the inner cylinder that has to withstand high safety standards. Here composite materials are favorably used, where materials such as carbon fiber, optic fiber, and epoxy resin can be found. Due to the complexity of recycling here, the strategy of hydrogen tank manufacturing aims to provide a very long end-of-life for these products. Typically the lifetime of hydrogen tanks can range between 20-30 years. In some cases, manufacturers produce hydrogen tanks with non-limited life that can easily be used for up to 60 years. In this case a regular inspection (every 5 years) could be required in accordance with safety standards.

New Materials for a Circular Economy   Materials Research Forum LLC
Materials Research Foundations 149 (2023) 446-483   https://doi.org/10.21741/9781644902639-12

7.   Life cycle assessment (LCA) of fuel cell technology: A powerful tool to evaluate environmental impacts of fuel cell components

Life Cycle Assessment (LCA) of electrochemical devices such as FCs and batteries has emerged currently as a powerful tool to evaluate the environmental impacts associated with these technologies [158,159], and specifically, with regard to the main components of these devices. Focusing on polymer electrolyte membrane fuel cells, some works published in recent years reveal the environmental impact associated with the materials used to develop the FC components [13,160,161].

Mori *et al.* [13] performed a quite complete and clear compilation of the relevant data on materials that are the most commonly used in FC and hydrogen technologies. Table 1 shows the materials used to prepare the key components of a PEMFC. The vast majority of the materials used in PEMFCs have been classified as non-hazardous.

Table 1. Materials used in the fabrication of the main components of a PEMFC (Extracted from [13]).

Component	Material
Catalyst layer	Platinum and Pt-alloys
	Carbon black (catalyst support)
	PTFE-Teflon (hydrophobic agent)
	Perfluorosulphonic acid (PFSA-Nafion)
Electrolyte	Sulfonated polyether ether ketone (s-PEEK)
	Polybenzimidazole (PBI) doped with H_3PO_4 (HT-PEM)
GDL	Carbon fibres like Polyacrylonitrile Fibres (PAN)
	Metallic mesh (Steel)
Interconnect	Graphite
	Stainless steel
Sealant	Thermoplastic (PTFE)

The environmental impact indicators to produce 1 g of the material used in the development of a fuel cell vary as a function of the material [13]. Thus, the environmental impact of platinum is several orders of magnitude higher than other materials. In addition, polymer electrolytes such as Nafion® have a relevant impact on the environment. As an example, it has been estimated a value for the Global Warming Potential of Pt and Nafion® of 3.30×10^1 and 8.31×10^{-1} kg CO_2 eq (100 years) while for graphite is 2.26×10^{-5} kg CO_2 eq.

In the same line, the life cycle analysis of a PEMFC for electric vehicles carried out by Usai and coworkers [162] highlights the impact that some of the main components of this

New Materials for a Circular Economy   Materials Research Forum LLC
Materials Research Foundations 149 (2023) 446-483   https://doi.org/10.21741/9781644902639-12

system have on the environment and the importance of taking measures to reduce it. These authors have applied a cradle-to-gate life cycle assessment for a PEMFC which implies the assessment of life cycle from the resource to the factory gate. The FC stack, the FC auxiliaries and the storage tanks are the main components of the electrochemical device that are considered to perform the LCA. The electrical energy generated from the redox reaction depends on the FC stack which in turn is made up of several sub-components such as membrane, bipolar plates, catalyst ink, or current collectors among others. Fuel cell auxiliaries like air, water, heat, or fuel management, and electronics have the function of maintaining adequate working conditions. And finally, the storage tanks store gaseous hydrogen that is supplied to the stack. These tanks contain carbon fiber.

Each of these sub-components exerts an effect on the environment which is analyzed. Thus, Table 2 summarizes the contribution (in percentage) of the main FC components on environmental categories obtained by Usai *et al.* [162].

*Table 2. Contribution of **fuel cell components** (approximate percentage) on environmental categories. (Extracted from [162]).*

Fuel Cell Component	Impact Category[a]							
	GWP	FDP	FETP	HTP	MEP	MDP	PMFP	TAP
	[kg CO$_2$ eq]	[kg oil eq]	[kg 1,4-DB eq]	[kg 1,4-DB eq]	[kg N eq]	[kg Fe eq]	[kg PM$_{10}$ eq]	[kg SO$_2$ eq]
Catalyst	20-25	30-35	35-40	55-60	30-35	70-75	60-65	70-75
H$_2$ Tanks	40-45	60-70	80-90	12-14	50-55	2-3	12-15	89-91
Membrane	2-3	0	0	0	0	0	0	0
Auxiliaries	15-20	10-15	50-60	25-30	9-12	25-27	15-17	4-6
Bipolar Plates	15-20	10-15	1-3	3-4	4-6	1-2	7-9	3-4
Other stacks	4-5	5-7	3-5	3-4	3-4	1-2	1-2	1-2

[a]GWP: Global Warming Potential; FDP: Fossil Depletion Potential; FETP: Freshwater Ecotoxicity Potential; HTP: Human Toxicity Potential; MEP: Marine Eutrophication Potential; MDP: Metal Depletion Potential; PMFP: Particulate Matter Formation Potential; TAP; Terrestrial Acidification Potential.

The Global warming potential (denoted as GWP) of the FC components described above varies from one to another. Usai *et al.* observed that the contribution of both catalyst and tanks is around 53% of the GWP. The preparation of catalysts contributes a percentage of 24% of the GWP impacts (approximately 1.1 ton CO$_2$ eq). The mining activities as well as the preparation of the powder used as a catalyst cause most of the environmental impacts. A percentage of around 17% of the GWP impacts are associated with the fuel cell auxiliaries. Regarding the polymer electrolyte, Nafion® has a contribution to the GWP lower than 2%. The production of the catalyst, the auxiliaries, and the tanks have high impact on other categories studied observing a similar tendency as GWP. However, the membrane used as a solid electrolyte as well as the other components of the stack exhibit lower environmental impact.

So, in view of these results Usai *et al.* proposed some strategies to reduce the environmental impacts of FC components such as the reduction of the platinum loading used to prepare the electrodes, and the use of gold-coated steel bipolar plates instead of carbon-coated titanium plates. These changes can contribute to reducing the footprint of this technology.

8.   Fuel cells and international agreements of sustainability

In the last few years, numerous efforts have been made to limit human environmental impacts, one of which is based on the involvement of fuel cells. In fact, with the Kyoto Protocol, which is an international agreement, the reduction of the emissions of six gases of greenhouse effect: CO_2, methane gas (CH_4), nitrous oxide (N_2O), and three fluorinated greenhouse gases (hydrofluorocarbons (HFCs), perfluorocarbons (PFCs), sulfur hexafluoride (SF_6)) was targeted from 2008 to 2012. In this agreement, the reduction of the emission of this kind of gases is fixed to a minimum of 5%. A second Kyoto commitment was later established aiming to reduce, by 2020, the emission of GHGs to at least 18% lower than the levels recorded in 1990 [163]. After that, the Paris agreement took place in 2015, and 195 countries signed an international accord concerning climate change. The countries agreed to limit the worldwide temperature increase to maximum of 1.5 °C above pre-industrial level, to offer international help to developing countries in order to better adapt to new weather changes, to reduce as much as possible the effect of global warming, to reduce the emissions in 2030 by at least 40%, and to invest until 2020 around 100 billion U.S. dollars for this cause [164]. In addition, the European Union has recently launched a roadmap for global climate action to attain 80% reduction in GHG emissions by 2050 with respect to 1990 according to the goals set in the Paris Agreement [165].

9.   Current status of circular economy for fuel cells and future prospects

In the energy sector, and in the context of green technology and sustainability, fuel cells are without a doubt part of the future. Throughout the years, huge investments have been made from both the private and public sector in order to further understand and improve fuel cell systems. Moreover, in recent years, the focus has not only been centered on improving the efficiency of these systems, but has also been focused on the environmental impact of all the system's components and their respective carbon footprint. Indeed, in order to stay coherent with the principle of producing clean energy, one must also question the entire chain of product life cycle, in terms of sustainability and carbon footprint. In order to stay aligned with this coherence, the circular economy must also be taken into consideration as an ideal model for the commercialization and industrialization of fuel cells in all sectors, in particular energy and transportation.

Even though fuel cells are not quite there in the race to transition to a circular economy, today's research and studies are pushing it to this direction. Piece by piece, study by study, there is hope that one day in the future the technology maturity we are looking for will take shape. In this ideal future, one can imagine a purely circular economy for fuel cells with two important pillars. A primary pillar, which is more of technical nature, where the

New Materials for a Circular Economy   Materials Research Forum LLC
Materials Research Foundations 149 (2023) 446-483   https://doi.org/10.21741/9781644902639-12

industrialization of recycling can be found. In this pillar, the dismantling of fuel cell stacks, the implementation and validation of recycling technologies, and the reassembly of quality controlled recycled product stands. The secondary pillar is more of a management nature, with a focus on business cases that consider cost/benefit analysis based on circular economy. Moreover, strategic assessments and product life cycle analysis are necessary foundations to be taken into consideration. Finally, standardization, ecolabelling, ecodesiging, and working without regulatory frames are all key elements to complete the picture of a coherent, sustainable, and green energy source: Fuel Cells.

References

[1] Information on https://www.epa.gov

[2] S.A. Montzka, E.J. Dlugokencky, J.H. Butler, Non-CO2 greenhouse gases and climate change, Nature 476 (2011) 43-50. https://doi.org/10.1038/nature10322

[3] Information on https://ec.europa.eu/info/index_en

[4] M.Z. Jacobson, W.G. Colella, D.M. Golden, Cleaning the air and improving health with hydrogen fuel cell vehicles, Science 308 (2005) 1901-1905. https://doi.org/10.1126/science.1109157

[5] M.A. Hickner, H.Ghassemi, Y.S. Kim, B.R. Einsla, J.E. McGrath, Alternative polymer systems for proton exchange membranes (PEMs), Chem. Rev. 104 (2004) 4587-4612. https://doi.org/10.1021/cr020711a

[6] L. Carrete, K.A. Friedrich, U. Stimming, Fuel cells-Fundamentals and applications, Fuel Cells 1 (2001) 5-39. https://doi.org/10.1002/1615-6854(200105)1:1<5::AID-FUCE5>3.0.CO;2-G

[7] Ellen Macarthur Foundation, Towards the Circular Economy. Economic and Business Rationale for an Accelerated Transition, first ed., Cower, UK, 2010, pp. 21-34.

[8] M. Wagner, The porter hypothesis revisited: a literature review of theoretical models and empirical tests, Centre for Sustainability Management, Univesitat Luneburg, Lüneburg, Germany, 2003.

[9] W. Haas, F. Krausmann, D. Wiedenhofer, M. Heinz, How circular is the global economy?: an assessment of material flows, waste production, and recycling in the European union and the world in 2005, J. Ind. Ecol. 19 (2015) 765-777. https://doi.org/10.1111/jiec.12244

[10] H.S. Oh, J.G. Oh, S. Haam, K. Arunabha, B. Roh, I. Hwang, H. Kim, On-line mass spectrometry study of carbon corrosion in polymer electrolyte membrane fuel cells, Electrochem. Commun. 10 (2008) 1048-1051. https://doi.org/10.1016/j.elecom.2008.05.006

[11] Y. Wang, D.F. Ruiz Diaz, K.S. Chen, Z. Wang, X.C. Adroher, Materials, technological status, and fundamentals of PEM fuel cells-A review, Mater. Today 32 (2020) 178-203. https://doi.org/10.1016/j.mattod.2019.06.005

[12] L. Fan, Z. Tu, S.H. Chan, Recent development of hydrogen and fuel cell technologies: A review, Energy Rep. 7 (2021) 8421-8446. https://doi.org/10.1016/j.egyr.2021.08.003

[13] M. Mori, R. Stropnik, M. Sekavčnik, A. Lotrič, Criticality and life-cycle assessment of materials used in fuel-cell and hydrogen technologies, Sustainability 13 (2021) 3565. https://doi.org/10.3390/su13063565

[14] S. Park, J.-W. Lee, B.N. Popov, A review of gas diffusion layer in PEM fuel cells: materials and designs, Int. J. Hydrog. Energy 37 (2012) 5850-5865. https://doi.org/10.1016/j.ijhydene.2011.12.148

[15] S. Park, B.N. Popov, Effect of a GDL based on carbon paper or carbon cloth on PEM fuel cell performance, Fuel 90 (2011) 436-440. https://doi.org/10.1016/j.fuel.2010.09.003

[16] M. Han, S. Chan, S.P. Jiang, Development of carbon-filled gas diffusion layer for polymer electrolyte fuel cells, J. Power Sources 159 (2006) 1005-1014. https://doi.org/10.1016/j.jpowsour.2005.12.003

[17] A. Arvay, E. Yli-Rantala, C.-H. Liu, X.-H. Peng, P. Koski, L. Cindrella, P. Kauranen, P.M. Wilde, A. M. Kannan, Characterization techniques for gas diffusion layers for proton exchange membrane fuel cells-a review, J. Power Sources 213 (2012) 317-337. https://doi.org/10.1016/j.jpowsour.2012.04.026

[18] M. Han, J. Xu, S. Chan, S.P. Jiang, Characterization of gas diffusion layers for PEMFC, Electrochim. Acta 53 (2008) 5361-5367. https://doi.org/10.1016/j.electacta.2008.02.057

[19] T. Kitahara, T. Konomi, H. Nakajima, Microporous layer coated gas diffusion layers for enhanced performance of polymer electrolyte fuel cells, J. Power Sources 195 (2010) 2202-2211. https://doi.org/10.1016/j.jpowsour.2009.10.089

[20] R. Anderson, M. Blanco, X. Bi, D.P. Wilkinson, Anode water removal and cathode gas diffusion layer flooding in a proton exchange membrane fuel cell, Int. J. Hydrog. Energy 37 (2012) 16093-16103. https://doi.org/10.1016/j.ijhydene.2012.08.013

[21] E.-D. Wang, P.-F. Shi, C.-Y. Du, Treatment and characterization of gas diffusion layers by sucrose carbonization for PEMFC applications, Electrochem. Commun. 10 (2008) 555-558. https://doi.org/10.1016/j.elecom.2008.01.031

[22] C.-H. Liu, T.-H. Ko, J.-W. Shen, S.-I. Chang, S.-I. Chang, Y.-K. Liao, Effect of hydrophobic gas diffusion layers on the performance of the polymer exchange membrane fuel cell, J. Power Sources 191 (2009) 489-494. https://doi.org/10.1016/j.jpowsour.2009.02.017

[23] N. Zamel, X. Li, Effective transport properties for polymer electrolyte membrane fuel cells-with a focus on the gas diffusion layer, Progr. Energy Combust. Sci. 39 (2013) 111-146. https://doi.org/10.1016/j.pecs.2012.07.002

[24] A. Tamayol, M. Bahrami, Water permeation through gas diffusion layers of proton exchange membrane fuel cells, J. Power Sources 196 (2011) 6356-6361. https://doi.org/10.1016/j.jpowsour.2011.02.069

[25] J.F. Lin, J. Wertz, R. Ahmad, M. Thommes, A.M. Kannan, Effect of carbon paper substrate of the gas diffusion layer on the performance of proton exchange membrane fuel cell, Electrochim. Acta 55 (2010) 2746-2751. https://doi.org/10.1016/j.electacta.2009.12.056

[26] S.E. Iyuke, A.B. Mohamad, A.A.H. Kadhum, W.R.W. Daud, C. Rachid, Improved membrane and electrode assemblies for proton exchange membrane fuel cells, J. Power Sources 114 (2003) 195-202. https://doi.org/10.1016/S0378-7753(03)00016-8

[27] S. Litster, G. McLean, PEM fuel cell electrodes, J. Power Sources 130 (2004) 61-76. https://doi.org/10.1016/j.jpowsour.2003.12.055

[28] B.C.H. Steele, Material science and engineering: the enabling technology for the commercialisation of fuel cell systems, J. Mater. Sci. 36 (2001) 1053-1068. https://doi.org/10.1023/A:1004853019349

[29] A.J. Leo, H. Ghezel-Ayagh, R. Sanderson, Ultra high efficiency hybrid direct fuel cell/turbine power plant. ASME. Turbo expo: power for land, sea, and air, 2000, Volume 2: Coal, biomass and alternative fuels; combustion and fuels; oil and gas applications. https://doi.org/10.1115/2000-GT-0552

[30] A. Kraytsberg, Y. Ein-Eli, Review of advanced materials for proton exchange membrane fuel cells, Energy fuels 28 (2014) 7303-7330. https://doi.org/10.1021/ef501977k

[31] P.R. Behera, R. Dash, S.M. Ali, K.K. Mohapatra, A review on fuel cell and its applications, Int. J. Res. Eng. Technol. 3 (2014) 562-565. https://doi.org/10.15623/ijret.2014.0303105

[32] D.J. Kim, M.J. Jo, S.Y. Nam, A review of polymer-nanocomposite electrolyte membranes for fuel cell application, J. Ind. Eng. Chem. 21 (2015) 36-52. https://doi.org/10.1016/j.jiec.2014.04.030

[33] D. Van Dao, G. Adilbish, I.-H. Lee, Y.-T. Yu, Enhanced electrocatalytic property of Pt/C electrode with double catalyst layers for PEMFC, Int. J. Hydrog. Energy 44 (2019) 24580-24590. https://doi.org/10.1016/j.ijhydene.2019.07.156

[34] S. Zhang, X.Z. Yuan, J.N.C. Hin, H. Wang, K.A. Friedrich, M. Schulze, A review of platinum-based catalyst layer degradation in proton exchange membrane fuel cells, J. Power Sources 194 (2009) 588-600. https://doi.org/10.1016/j.jpowsour.2009.06.073

[35] H.-J. Huang, T. Yang, F.-Y. Lai, G.-Q. Wu, Co-pyrolysis of sewage sludge and sawdust/rice straw for the production of biochar, J. Anal. Appl. Pyrolysis 125 (2017) 61-68. https://doi.org/10.1016/j.jaap.2017.04.018

[36] G. Gunn, Critical metals handbook, in G. Gunn (Ed.), John Wiley and Sons. Hoboken, NJ, USA, 2014, pp. 284. https://doi.org/10.1002/9781118755341.ch12

[37] S. Dey, N.S. Mehta, Automobile pollution control using catalysis, Resour. Environ. Sustain. 2 (2020) 100006. https://doi.org/10.1016/j.resenv.2020.100006

[38] T. Biggs, S.S. Taylor, E. Lingen, The hardening of platinum alloys for potential jewellery application, Platinum Metals Rev. 49 (2005) 2-15. https://doi.org/10.1595/147106705X24409

[39] C. Courdec, Platinum group metals in glass making, Platinum Metals Rev. 54 (2010) 186-191. https://doi.org/10.1595/147106710X514012

[40] A. Cowley, B. Woodward, A healthy future: platinum in medical applications, Platinum Metals Rev. 55 (2011) 98-107. https://doi.org/10.1595/147106711X566816

[41] A.E. Hughes, N. Haque, S.A. Northey, S. Giddey, Platinum group metals: a review of resources, production and usage with a focus on catalysts, Resources 10 (2021) 93. https://doi.org/10.3390/resources10090093

[42] J. Hou, M. Yang, C. Ke, G. Wei, C. Priest, Z. Qiao, G. Wu, J. Zhang, Platinum-group-metal catalysts for proton exchange membrane fuel cells: From catalyst design to electrode structure optimization. Energy Chem. 2 (2020) 100023. https://doi.org/10.1016/j.enchem.2019.100023

[43] D. Banham, J. Zou, S. Mukerjee, Z. Liu, D. Yang, Y. Zhang, Y. Peng, A. Dong, Ultralow platinum loading proton exchange membrane fuel cells: Performance losses and solutions, J. Power Sources 490 (2021) 229515. https://doi.org/10.1016/j.jpowsour.2021.229515

[44] H.E. Suess, Abundances of the elements, Rev. Mod. Phys. 28 (1956) 53-73. https://doi.org/10.1103/RevModPhys.28.53

[45] L. Grandell, A. Lehtilä, M. Kivinen, T. Koljonen, S. Kihlman, L.S. Lauri, Role of critical metals in the future markets of clean energy technologies, Renew. Energy 95 (2016) 53-62. https://doi.org/10.1016/j.renene.2016.03.102

[46] A.N. Løvik, C. Hagelüken, P. Wäger, Improving supply security of critical metals: current developments and research in the EU, Sustain. Mater. Technol. 15 (2018) 9-18. https://doi.org/10.1016/j.susmat.2018.01.003

[47] J. Burlakovs, Z. Vincevica-Gaile, M. Krievans, Y. Jani, M. Horttanainen, K.M. Pehme, E. Dace, R.H. Setyobudi, J. Pilecka, G. Denafas, I. Grinfelde, A. Bhatnagar, V. Rud, V. Rudovica, R.L. Mersky, O. Anne, M. Kriipsalu, R. Ozola-Davidane, T. Tamm, M. Klavins, Platinum group elements in geosphere and anthroposphere:

Interplay among the global reserves, urban ores, markets and circular economy, Minerals 10 (2020) 558. https://doi.org/10.3390/min10060558

[48] P. Sahu, M.S. Jena, N.R. Mandre, R. Venugopal, Platinum group elements mineralogy, beneficiation, and extraction practices-an overview, Miner. Process. Extr. Metall. Rev. 42 (2020) 1-14. https://doi.org/10.1080/08827508.2020.1795848

[49] B. J. Glaister, G.M. Mudd, The environmental costs of platinum-PGM mining and sustainability: Is the glass half-full or half-empty?, Miner. Eng. 23 (2010) 438-450. https://doi.org/10.1016/j.mineng.2009.12.007

[50] Information on https://www.usgs.gov

[51] A.L. Gulley, N.T. Nassar, S. Xun, China, the United States, and competition for resources that enable emerging technologies, Proc. Natl. Acad. Sci. USA 115 (2018) 4111-4115. https://doi.org/10.1073/pnas.1717152115

[52] N.T. Nassar, J. Brainard, A. Gulley, R. Manley, G. Matos, G. Lederer, L.R. Bird, D. Pineault, E. Alonso, J. Gambogi, S.M. Fortier, Evaluating the mineral commodity supply risk of the U.S. manufacturing sector, Sci. Adv. 6 (2020) 8647. https://doi.org/10.1126/sciadv.aay8647

[53] Y. Yuan, M. Yellishetty, G.M. Mudd, M.A. Muñoz, S.A. Northey, T.T. Werner, Toward dynamic evaluations of materials criticality: A systems framework applied to platinum, Resour. Conserv. Recycl. 152 (2020) 104532. https://doi.org/10.1016/j.resconrec.2019.104532

[54] Information on https://eur-lex.europa.eu

[55] F.K. Crundwell, M.S. Moats, V. Ramachandran , T.G. Robinson, W.G. Davenport, Extractive Metallurgy of Nickel, Cobalt and Platinum Group Metals, first ed., Elsevier, Oxford, UK, 2011, pp. 610. https://doi.org/10.1016/B978-0-08-096809-4.10038-3

[56] N. Ritschel, J. Taylor, T. England, B. Peters, F. Stoffner, C. Röhlich, S. Voss, H. Winkler, Heraeus Deutschland GmbH and Co, Process for the Production of a PGM-Enriched Alloy, U.S. Patent 10,202,669 (2019).

[57] M.K. Jha, J.C. Lee, M.S. Kim, J. Jeong, B.S. Kim, V. Kumar, Hydrometallurgical recovery/recycling of platinum by the leaching of spent catalysts: A review, Hydrometallurgy 133 (2013) 23-32. https://doi.org/10.1016/j.hydromet.2012.11.012

[58] H. Dong, J. Zhao, J. Chen, Y. Wu, B. Li, Recovery of platinum group metals from spent catalysts: A review, Int. J. Miner. Process 145 (2015) 108-113. https://doi.org/10.1016/j.minpro.2015.06.009

[59] S.K. Padamata, A.S. Yasinskiy, P.V. Polyakov, E.A. Pavlov, D.Y. Varyukhin, Recovery of noble metals from spent catalysts: A review, Metall. Mater. Trans. B 51 (2020) 2413-2435. https://doi.org/10.1007/s11663-020-01913-w

[60] Z. Peng, Z. Li, X. Lin, H. Tang, L. Ye, Y. Ma, M. Rao, Y. Zhang, G. Li, T. Jiang, Pyrometallurgical recovery of platinum group metals from spent catalysts, JOM 69 (2017) 1553-1562. https://doi.org/10.1007/s11837-017-2450-3

[61] T. Havlik, D. Orac, M. Petranikova, A. Miskufova, F. Kukurugya, Z. Takacova, Leaching of copper and tin from used printed circuit boards after thermal treatment, J. Hazard. Mater. 183 (2010) 866-873. https://doi.org/10.1016/j.jhazmat.2010.07.107

[62] M.W. Ojeda, E. Perino, M.del C. Ruiz, Gold Extraction by chlorination using a pyrometallurgical process, Miner. Eng. 22 (2009) 409-411. https://doi.org/10.1016/j.mineng.2008.09.002

[63] R.J. Allen, P.C. Foller, J. Giallombardo, Two step method for recovery of dispersed noble metals, EU Patent 0492691A1 (1992).

[64] C. Horike, K. Morita, T.H. Okabe, Effective dissolution of platinum by using chloride salts in recovery process. Metall. Mater. Trans. B Process, Metall. Mater. Process. Sci. 43 (2012) 1300-1307. https://doi.org/10.1007/s11663-012-9746-z

[65] C.H. Kim, S.I. Woo, S.H. Jeon, Recovery of platinum-group metals from recycled automotive catalytic converters by carbochlorination, Ind. Eng. Chem. Res. 39 (2000) 1185-1192. https://doi.org/10.1021/ie9905355

[66] K. Staszak, Chemical and petrochemical industry, Phys. Sci. Rev. 3 (2018) 1-28. https://doi.org/10.1016/j.revip.2017.11.001

[67] Y. Kayanuma, T.H. Okabe, M. Maeda, Metal vapor treatment for enhancing the dissolution of platinum group metals from automotive catalyst scrap, Metall. Mater. Trans. B 35 (2004) 817-824. https://doi.org/10.1007/s11663-004-0075-8

[68] M. Benson, C.R. Bennett, J.E. Harry, M.K. Patel, M. Cross, The recovery mechanism of platinum group metals from catalytic converters in spent automotive exhaust systems, Resour. Conserv. Recycl. 31 (2000) 1-7. https://doi.org/10.1016/S0921-3449(00)00062-8

[69] G. Kolliopoulos, E. Balomenos, I. Giannopoulou, I. Yakoumis, D. Panias, Behavior of platinum group during their pyrometallurgical recovery from spent automotive catalysts, O. A. Lib. J. 1 (2014) 1-9. https://doi.org/10.4236/oalib.1100736

[70] X. He, Y. Li, X. Wu, Y. Zhoo, H. Weng, W. Liu, Study on the process of enrichment platinum group metals by plasma melting technology, Precious Met. 37 (2016) 1-5.

[71] K.C. Chiang, K.L. Chen, C.Y. Chen, J.J. Huang, Y.H. Shen, M.Y. Yeh, F.F. Wong, Recovery of spent alumina-supported platinum catalyst and reduction of platinum oxide via plasma sintering technique, J. Taiwan Inst. Chem. Eng. 42 (2011) 158-165. https://doi.org/10.1016/j.jtice.2010.05.003

[72] I. Bronshtein, Y. Feldman, S. Shilstein, E. Wachtel, I. Lubomirsky, V. Kaplan, Efficient chloride salt extraction of platinum group metals from spent catalysts, J. Sustain. Metall. 4 (2018) 103-114. https://doi.org/10.1007/s40831-017-0155-z

[73] D.J. De Aberasturi, R. Pinedo, I.R. De Larramendi, J.I.R. De Larramendi, T. Rojo, Recovery by hydrometallurgical extraction of the platinum-group metals from car catalytic converters, Miner. Eng. 24 (2011) 505-513. https://doi.org/10.1016/j.mineng.2010.12.009

[74] A. Fornalczyk, M. Saternus, Removal of platinum group metals from the used autocatalytic converter, Metalurgija 48 (2009) 133-136.

[75] T.N. Angelidis, Development of a laboratory scale hydrometallurgical procedure for the recovery of Pt and Rh from spent automotive catalysts, Top. Catal. 16 (2001) 419-423. https://doi.org/10.1023/A:1016641906103

[76] K. Han, X. Meng, Recovery of platinum group metals and rhenium from materials using halogen reagents, U.S. Patent 5,542,957A (1995).

[77] J.S. Yoo, Metal recovery and rejuvenation of metal-loaded spent catalysts, Catal. Today 44 (1998) 27-46. https://doi.org/10.1016/S0920-5861(98)00171-0

[78] S. Harjanto, Y. Cao, A. Shibayama, I. Naitoh, T. Nanami, K. Kasahara, Y. Okumura, K. Liu, T. Fujita, Leaching of Pt, Pd and Rh from automotive catalyst residue in various chloride based solutions, Mater. Trans. 47 (2006) 129-135. https://doi.org/10.2320/matertrans.47.129

[79] R. Panda, M.K. Jha, D.D. Pathak, Commercial processes for the extraction of platinum group metals (PGMs), in H. Kim et. al (Eds.), Rare metal technology. The Minerals, Metals & Materials Series. Springer, Cham, Switzerland, 2018, pp. 119-130. https://doi.org/10.1007/978-3-319-72350-1_11

[80] C. Saguru, S. Ndlovu, D. Moropeng, A review of recent studies into hydrometallurgical methods for recovering PGMs from used catalytic converters, Hydrometallurgy 182 (2018) 44-56. https://doi.org/10.1016/j.hydromet.2018.10.012

[81] M. Massucci, S.L. Clegg, P. Brimblecombe, Equilibrium partial pressures, thermodynamic properties of aqueous and solid phases, and Cl2 production from aqueous HCl and HNO3 and their mixtures, J. Phys. Chem. A 103 (1999) 4209-4226. https://doi.org/10.1021/jp9847179

[82] D. Jafarifar, M.R. Daryanavard, S. Sheibani, Ultra fast microwave-assisted leaching for recovery of platinum from spent catalyst, Hydrometallurgy 78 (2005) 166-171. https://doi.org/10.1016/j.hydromet.2005.02.006

[83] E. Kizilaslan, S. Aktaş, M.K. Şeşen, Towards environmentally safe recovery of platinum from scrap automotive catalytic converters, Turkish J. Eng. Environ. Sci. 33 (2009) 83-90.

[84] R. Torres, G.T. Lapidus, Platinum, palladium and gold leaching from magnetite ore, with concentrated chloride solutions and ozone, Hydrometallurgy 166 (2016) 185-194. https://doi.org/10.1016/j.hydromet.2016.06.009

[85] V.T. Nguyen, S. Riaño, E. Aktan, C. Deferm, J. Fransaer, K. Binnemans, Solvometallurgical recovery of platinum group metals from spent automotive catalysts, ACS Sustain. Chem. Eng. 9 (2021) 337-350. https://doi.org/10.1021/acssuschemeng.0c07355

[86] T. Suoranta, O. Zugazua, M. Niemelä, P. Perämäki, Recovery of palladium, platinum, rhodium and ruthenium from catalyst materials using microwave-assisted leaching and cloud point extraction, Hydrometallurgy 154 (2015) 56-62. https://doi.org/10.1016/j.hydromet.2015.03.014

[87] S. Sui, X. Wang, X. Zhou, Y. Su, S. Riffat, C.-J. Liu, A comprehensive review of Pt electrocatalysts for the oxygen reduction reaction: Nanostructure, activity, mechanism and carbon support in PEM fuel cells, J. Mater. Chem. A 5 (2017) 1808-1825. https://doi.org/10.1039/C6TA08580F

[88] R. Sharma, S. Gyergyek, P.B. Lund, S.M. Andersen, Recovery of Pt and Ru from spent low-temperature polymer electrolyte membrane fuel cell electrodes and recycling of Pt by direct redeposition of the dissolved precursor on carbon, ACS Appl. Energy Mater. 4 (2021) 6842-6852. https://doi.org/10.1021/acsaem.1c00964

[89] A. Pavlišič, P. Jovanovič, V.S. Šelih, M. Šala, N. Hodnik, M. Gaberšček, Platinum dissolution and redeposition from Pt/C fuel cell electrocatalyst at potential cycling, J. Electrochem. Soc. 165 (2018) F3161-F3165. https://doi.org/10.1149/2.0191806jes

[90] C.L. Brierley, N.W. Le Roux, Bacterial leaching, CRC Crit. Rev. Microbiol. 6 (1978) 273-284. https://doi.org/10.3109/10408417809090623

[91] D.G. Lundgren, E.E. Malouf, Microbial extraction and concentration of metals, ADV. Biotechnol. Process. 1 (1983) 223-249.

[92] H. Brandl, S. Lehmann, M.A. Faramarzi, D. Martinelli, Biomobilization of silver, gold, and platinum from solid waste materials by HCN-forming microorganisms, Hydrometallurgy 94 (2008) 14-17. https://doi.org/10.1016/j.hydromet.2008.05.016

[93] V.A. Pham, Y.P. Ting, Gold Bioleaching of electronic waste by cyanogenic bacteria and its enhancement with bio-oxidation, Adv. Mater. Res. 71-73 (2009) 661-664. https://doi.org/10.4028/www.scientific.net/AMR.71-73.661

[94] T.D. Chi, J.C. Lee, B.D. Pandey, K. Yoo, J. Jeong, Bioleaching of gold and copper from waste mobile phone pcbs by using a cyanogenic bacterium, Proc. Miner. Eng. 24 (2011) 1219-1222. https://doi.org/10.1016/j.mineng.2011.05.009

[95] H.B. Trinh, J.-C. Lee, Y.-J. Suh, J. Lee, A review on the recycling processes of spent auto-catalysts: Towards the development of sustainable metallurgy, Waste Manag. 114 (2020) 148-165. https://doi.org/10.1016/j.wasman.2020.06.030

[96] R. Granados-Fernández, M.A. Montiel, S. Díaz-Abad, M.A. Rodrigo, J. Lobato, Platinum recovery techniques for a circular economy, Catalysts 11 (2021) 937. https://doi.org/10.3390/catal11080937

[97] Y.-K. Taninouchi, T.H. Okabe, Recovery of platinum group metals from spent catalysts using iron chloride vapor treatment, Metall. Mater. Trans. B 49 (2018) 1781-1793. https://doi.org/10.1007/s11663-018-1269-9

[98] Y.-K. Taninouchi, T. Watanabe, T.H. Okabe, Recovery of platinum group metals from spent catalysts using electroless nickel plating and magnetic separation, Proc. Mater. Trans. 58 (2017) 410-419. https://doi.org/10.2320/matertrans.M-M2017801

[99] A.M. Martinez, K.S. Osen, A. Støre, Recovery of platinum group metals from secondary sources by selective chlorination from molten salt media, Proc. Miner. Met. Mater. Ser. (2020) 221-233. https://doi.org/10.1007/978-3-030-36758-9_21

[100] M.H. Morcali, A new approach to recover platinum-group metals from spent catalytic converters via iron Matte, Resour. Conserv. Recycl. 159 (2020) 104891. https://doi.org/10.1016/j.resconrec.2020.104891

[101] H. Sasaki, M. Maeda, Enhanced dissolution of Pt from Pt-Zn intermetallic compounds and underpotential dissolution from Zn-rich alloys, J. Phys. Chem. C 117 (2013) 18457-18463. https://doi.org/10.1021/jp405184e

[102] H. Sasaki, M. Maeda, Zn-Vapor Pretreatment for acid leaching of platinum group metals from automotive catalytic converters, Hydrometallurgy 147-148 (2014) 59-67. https://doi.org/10.1016/j.hydromet.2014.04.019

[103] J. Spooren, T. Abo Atia, Combined microwave assisted roasting and leaching to recover platinum group metals from spent automotive catalysts, Miner. Eng. 146 (2020) 106153. https://doi.org/10.1016/j.mineng.2019.106153

[104] G. Nicol, E. Goosey, D.S. Yildiz, E. Loving, V.T. Nguyen, S. Riaño, I. Yakoumis, A.M. Martinez, A. Siriwardana, A. Unzurrunzaga, J. Spooren, T. Abo Atia, B. Michielsen, X. Dominguez-Benetton, O. Lanaridi, Platinum group metals recovery using secondary raw materials (PLATIRUS): Project overview with a focus on processing spent autocatalyst, Johns. Matthey Technol. Rev. 65 (2021) 127-147. https://doi.org/10.1595/205651321x16057842276133

[105] S. Chen, S. Shen, Y. Cheng, H. Wang, B. Lv, F. Wang, Effect of O2, H2 and CO pretreatments on leaching Rh from spent auto-catalysts with acidic sodium chlorate solution, Hydrometallurgy 144-145 (2014) 69-76. https://doi.org/10.1016/j.hydromet.2014.01.018

[106] H.B. Trinh, J.-C. Lee, R.R. Srivastava, S. Kim, Total recycling of all the components from spent auto-catalyst by NaOH roasting-assisted hydrometallurgical route, J. Hazard. Mater. 379 (2019) 120772. https://doi.org/10.1016/j.jhazmat.2019.120772

[107] C.A. Nogueira, A.P. Paiva, M.C. Costa, A.M. Rosa da Costa, Leaching efficiency and kinetics of the recovery of palladium and rhodium from a spent auto-catalyst in HCl/CuCl2 media, Environ. Technol. 41 (2020) 2293-2304. https://doi.org/10.1080/09593330.2018.1563635

[108] D.J. Connolly, W.F. Gresham, Fluorocarbon vinyl ether polymers, U.S. Patent 3,282,875 (1966).

[109] S.J. Peighambardoust, S. Rowshanzamir, M. Amjadi, Review of the proton exchange membranes for fuel cell applications, Int. J. Hydrog. Energy 35 (2010) 9349-9384. https://doi.org/10.1016/j.ijhydene.2010.05.017

[110] S. Motupally, A.J. Becker, J.W. Beidner, Diffusion of water in Nafion-115 membranes, J. Electrochem. Soc. 147 (2000) 3171-3177. https://doi.org/10.1149/1.1393879

[111] P. Choi, N.H. Jalani, R. Datta, Thermodynamics and proton transport in Nafion II. Proton diffusion mechanisms and conductivity, J. Electrochem. Soc. 152 (2005) 123-130. https://doi.org/10.1149/1.1859814

[112] M. Vinothkannan, A.R. Kim, D.J. Yoo, Potential carbon nanomaterials as additives for state-of-the-art Nafion electrolyte in proton-exchange membrane fuel cells: a concise review, RSC Adv. 11 (2021) 18351-18370. https://doi.org/10.1039/D1RA00685A

[113] F. Mohammadi, A. Rabiee, Solution casting, characterization, and performance evaluation of perfluorosulfonic sodium type membranes for chlor-alkali application, J. Appl. Polym. Sci. 120 (2011) 3469-3476. https://doi.org/10.1002/app.33526

[114] X.L. Zheng, J.P. Song, T. Ling, Z.P. Hu, P.F. Yin, K. Davey, X.W. Du, S.Z. Qiao, Strongly coupled Nafion molecules and ordered porous CdS networks for enhanced visible-light photoelectrochemical hydrogen evolution, Adv. Mater. 28 (2016) 4935-4942. https://doi.org/10.1002/adma.201600437

[115] Information on https://www.factmr.com

[116] Information on http:// www.plasticsrecyclers.eu

[117] M. Hong, E.Y.X. Chen, Chemically recyclable polymers: a circular economy approach to sustainability, Green Chem. 19 (2017) 3692-3706. https://doi.org/10.1039/C7GC01496A

[118] J. Koehler, R. Zuber, B. Matthias, V. Baenisch, M. Lopez, Process for recycling fuel cell components containing precious metals, WO Patent 024507A1 (2006).

[119] a) W. Grot, (DuPont) U.S. Patent 3,969,285 (1976)

b) W. Grot, (DuPont) U.S. Patent 4,026,783 (1977)

c) W. Grot, (DuPont) U.S. Patent 4,030,988 (1977).

[120] R. Wittstock, A. Pehlken, M. Wark, Challenges in automotive fuel cells recycling, Recycling 1 (2016) 343-364. https://doi.org/10.3390/recycling1030343

[121] L. Duclos, L. Svecova, V. Laforest, G. Mandil, P.X. Thivel, Process development and optimization for platinum recovery from PEM fuel cell catalyst, Hydrometallurgy 160 (2016) 79-89. https://doi.org/10.1016/j.hydromet.2015.12.013

[122] F. Xu, S. Mu, M. Pan, Recycling of membrane electrode assembly of PEMFC by acid processing, Int. J. Hydrog. Energy 35 (2010) 2976-2979. https://doi.org/10.1016/j.ijhydene.2009.05.087

[123] H. Xu, X. Wang, Z. Shao, I. Hsing, Recycling and regeneration of used perfluorosulfonic membranes for polymer electrolyte fuel cells, J. Appl. Electrochem. 32 (2002) 1337-1340. https://doi.org/10.1023/A:1022636000888

[124] C. Handley, N. Brandon, R. Van Der Vorst, Impact of the European Union vehicle waste directive on end-of-life options for polymer electrolyte fuel cells, J. Power Sources 106 (2002) 344-352. https://doi.org/10.1016/S0378-7753(01)01019-9

[125] P.A. Cirkel, T. Okada, S. Kinugasa, Equilibrium aggregation in perfluorinated ionomer solutions, Macromolecules 32 (1999) 531-533. https://doi.org/10.1021/ma981421s

[126] M. Laporta, M. Pegoraro, L. Zanderighi, Recast Nafion-117 thin film from water solution, Macromol. Mater. Eng. 282 (2000) 22-29. https://doi.org/10.1002/1439-2054(20001001)282:1<22::AID-MAME22>3.0.CO;2-#

[127] R.B. Moore, C.R. Martin, Procedure for preparing solution-cast perfluorosulfonate ionomer films and membranes, Anal. Chem. 58 (1986) 2569-2570. https://doi.org/10.1021/ac00125a046

[128] F. Mura, R. Silva, A. Pozio, Study on the conductivity of recast Nafion/montmorillonite and Nafion/TiO2 composite membranes, Electrochim. Acta 52 (2007) 5824-5828. https://doi.org/10.1016/j.electacta.2007.02.081

[129] A. Pozio, A. Cemmi, F. Mura, A. Masci, R. F. Silva, Study on the durability of recast Nafion/montmorillonite composite membranes in low humidification conditions, Int. J. Electrochem. Sci. 2011 (2010) 1-5. https://doi.org/10.4061/2011/252031

[130] R.F. Silva, M. De Francesco, A. Pozio, Solution-cast Nafion ionomer membranes: preparation and characterization, Electrochim. Acta 49 (2004) 3211-3219. https://doi.org/10.1016/j.electacta.2004.02.035

[131] R.F. Silva, S. Passerini, A. Pozio, Solution-cast Nafion/montmorillonite composite membrane with low methanol permeability, Electrochim. Acta 50 (2005) 2639-2645. https://doi.org/10.1016/j.electacta.2004.11.008

[132] A. Tsatsas, W. Risen, Studies on solution cast perfluorocarbonsulfonic acid ionomers, J. Polym. Sci., Part B: Polym. Phys. 31 (1993) 1223-1227. https://doi.org/10.1002/polb.1993.090310917

[133] a) L. Shore, B.R. Arthur, H.S. Shulman, M.L. Fall, Process for recycling components of a PEM fuel cell membrane electrode assembly, U.S. Patent WO2006/115684 (2006)

b) L. Shore, Process for recycling components of a PEM fuel cell membrane electrode assembly, U.S. Patent 8124261B2 (2012).

[134] S. Grot, W. Grot, Recycling of used perfluorosulfonic acid membranes, U.S. Patent 7,255,798 (2007).

[135] J.A. Moghaddam, M.J. Parnian, S. Rowshanzamir, Preparation, characterization, and electrochemical properties investigation of recycled proton exchange membrane for fuel cell applications, Energy 161 (2018) 699-709. https://doi.org/10.1016/j.energy.2018.07.123

[136] T. Oki, T. Katsumata, K. Hashimoto, M. Kobayashi, Recovery of platinum catalyst and polymer electrolyte from used small fuel cells by particle separation technology, Mater. Trans. 50 (2009) 1864-1870. https://doi.org/10.2320/matertrans.M-M2009812

[137] A. Valente, D. Iribarren, J. Dufour, End of life of fuel cells and hydrogen products: from technologies to strategies, Int. J. Hydrog. Energy 44 (2019) 20965-20977. https://doi.org/10.1016/j.ijhydene.2019.01.110

[138] P. Pei, H. Chen, Main factors affecting the lifetime of Proton Exchange Membrane fuel cells in vehicle applications: a review, Appl. Energy 125 (2014) 60-75. https://doi.org/10.1016/j.apenergy.2014.03.048

[139] X. Glipa, B. Bonnet, B. Mula, D. Jones, J. Rozière, Investigation of the conduction properties of phosphoric and sulfuric acid doped polybenzimidazole, J. Mater. Chem. 9 (1999) 3045-3049. https://doi.org/10.1039/a906060j

[140] M. Kawahara, J. Morita, M. Rikukawa, K. Sanui, N. Ogata, Synthesis and proton conductivity of thermally stable polymer electrolyte: poly(benzimidazole) complexes with strong acid molecules, Electrochim. Acta 45 (2000) 1395-1398. https://doi.org/10.1016/S0013-4686(99)00349-7

[141] P. Staiti, M. Minutoli, Influence of composition and acid treatment on proton conduction of composite polybenzimidazole membranes, J. Power Sources 94 (2001) 9-13. https://doi.org/10.1016/S0378-7753(00)00597-8

[142] J. Yu, B. Yi, D. Xing, F. Liu, Z. Shao, Y. Fu, H. Zhang, Degradation mechanism of polystyrene sulfonic acid membrane and application of its composite membranes in fuel cells, Phys. Chem. Chem. Phys. 5 (2003) 611-615. https://doi.org/10.1039/b209020a

[143] Y. A. Elabd, E. Napadensky, C. W. Walker, K. I. Winey, Transport properties of sulfonated poly(styrene-b-isobutylene-b-styrene) triblock copolymers at high ion-exchange capacities, Macromolecules 39 (2006) 399-407. https://doi.org/10.1021/ma051958n

[144] F. Göktepe, A. Bozkurt, Ş. T. Günday, Synthesis and proton conductivity of poly(styrene sulfonic acid)/heterocycle-based membranes, Polym. Int. 57 (2008) 133-138. https://doi.org/10.1002/pi.2335

[145] S. Vetter, B. Ruffmann, I. Buder, S. P. Nunes, Proton conductive membranes of sulfonated poly(ether ketone ketone), J. Membr. Sci. 260 (2005) 181-186. https://doi.org/10.1016/j.memsci.2005.02.036

[146] T. Maharana, A.K. Sutar, N. Nath, A. Routaray, Y.S. Negi, B. Mohanty, Polyetheretherketone (PPEK) membrane for fuel cell applications, in: A. Tiwari, S. Valyukh (Eds.), Adv. Energy Mater.Wiley, New Jersey, 2014, pp. 433-464. https://doi.org/10.1002/9781118904923.ch11

[147] P. Xing, G.P. Robertson, M.D. Guiver, S.D. Mikhailenko, K. Wang, S. Kaliaguine, Synthesis and characterization of sulfonated poly(ether ether ketone) for proton exchange membranes, J. Membr. Sci. 229 (2004) 95-106. https://doi.org/10.1016/j.memsci.2003.09.019

[148] H.L. Wu, C.C.M. Ma, F.Y. Liu, C.Y. Chen, S.J. Lee, C.L. Chiang, Preparation and characterization of poly(ether sulfone)/sulfonated poly(ether ether ketone) blend membranes, Eur. Polym. J. 42 (2006) 1688-1695. https://doi.org/10.1016/j.eurpolymj.2006.01.018

[149] Z. Wang, H. Tang, H. Zhang, M. Lei, R. Chen, P. Xiao, M. Pan, Synthesis of Nafion/ CeO2 hybrid for chemically durable proton exchange membrane of fuel cell, J. Membr. Sci. 421 (2012) 201-210. https://doi.org/10.1016/j.memsci.2012.07.014

[150] S. Xiao, H. Zhang, C. Bi, Y. Zhang, Y. Zhang, H. Dai, Z. Mai, X. Li, Degradation location study of proton exchange membrane at open circuit operation, J. Power Sources 195 (2010) 5305-5311. https://doi.org/10.1016/j.jpowsour.2010.03.010

[151] L. Zhang, S.R. Chae, Z. Hendren, J.S. Park, M.R. Wiesner, Recent advances in proton exchange membranes for fuel cell applications, Chem. Eng. J. 204 (2012) 87-97. https://doi.org/10.1016/j.cej.2012.07.103

[152] J.R. Jambeck, R. Geyer, C. Wilcox, T.R. Siegler, M. Perryman, A. Andrady, R. Narayan, K.L. Law, Marine pollution. Plastic waste inputs from land into the ocean, Science 347 (2015) 768-771. https://doi.org/10.1126/science.1260352

[153] C.J. Moore, Synthetic polymers in the marine environment: a rapidly increasing, long-term threat, Environ. Res. 180 (2008) 131-139. https://doi.org/10.1016/j.envres.2008.07.025

[154] R.A. Gross, B. Kalra, Biodegradable polymers for the environment, Science 297 (2002) 803-807. https://doi.org/10.1126/science.297.5582.803

[155] B. Rieger, A. Künkel, G.W. Coates, R. Reichardt, E. Dinjus, T.A. Zevaco, Synthetic biodegradable polymers, Springer, Berlin, Heidelberg, 2012, pp. 49. https://doi.org/10.1007/978-3-642-27154-0

[156] Information on https://ellenmacarthurfoundation.org/

[157] G.O. Jones, A. Yuen, R.J. Wojtecki, J.L. Hedrick, J.M. García, Computational and experimental investigations of one-step conversion of poly(carbonate)s into value-

added poly(aryl ether sulfone)s, Proc. Natl. Acad. Sci. U. S. A., 113 (2016) 7722-7726. https://doi.org/10.1073/pnas.1600924113

[158] M.S. Koroma, D. Costa, M. Philippot, G. Cardellini, M. S. Hosen, T. Coosemans, M. Messagie, Life cycle assessment of battery electric vehicles: Implications of future electricity mix and different battery end-of-life management, Sci. Total Environ. 831 (2022) 154859. https://doi.org/10.1016/j.scitotenv.2022.154859

[159] F. Arshad, J. Lin, N. Manurkar, E. Fan, A. Ahmad, M-u-N. Tariq, F. Wu, R. Chen, L. Li, Life cycle assessment of Lithium-ion Batteries: A critical review, Resour. Conserv. Recycl. 180 (2022) 106164. https://doi.org/10.1016/j.resconrec.2022.106164

[160] S. Evangelisti, C. Tagliaferri, J.L.B. Dan, P. Lettieri, Life cycle assessment of a polymer electrolyte membrane fuel cell system for passenger vehicles, J. Clean. Prod. 142 (2017) 4339-4355. https://doi.org/10.1016/j.jclepro.2016.11.159

[161] R. Stropnik, A. Lotrič, A.B. Montenegro, M. Sekavčnik, M. Mori, Critical materials in PEMFC systems and a LCA analysis for the potential reduction of environmental impacts with EoL strategies, Energy Sci. Eng. 7 (2019) 2519-2539. https://doi.org/10.1002/ese3.441

[162] L. Usai, C.R. Hung, F. Vásquez, M. Windsheimer, O.S. Burheim, A.H. Strømman, Life cycle assessment of fuel cell systems for light duty vehicles, current state-of-the-art and future impacts, J. Clean. Prod. 280 (2021) 125086. https://doi.org/10.1016/j.jclepro.2020.125086

[163] S.G. Poulopoulos, Atmospheric Environment, in: S.G. Poulopoulos, V.J. Inglezakis (Eds.), Environment and Development, Elsevier, 2016, pp. 45-136. https://doi.org/10.1016/B978-0-444-62733-9.00002-2

[164] United Nations. Adoption of the Paris Agreement; United Nations Framework Convention on Climate Change: Paris, France, 2015.

[165] European Commission. The European Green Deal; European Commission: Brussels, Belgium, 2019; Volume 53.

Keyword Index

About the Editors

Dr. Alberto García-Peñas has developed an intense research activity in different research centers and universities, such as Institute of Materials Science of Madrid (Madrid, Spain), Institute of Polymer Science and Technology (Madrid, Spain), University of Lisbon (Lisbon, Portugal), University of Karlsruhe (Karlsruhe, Germany), Shenzhen University (Shenzhen, China), the Fudan University (Shanghai, China), or University Carlos III of Madrid (Leganés, Spain). Recently, he carried out a short stay at the University of Cape Town in South Africa (ERASMUS + program).

His profile has been enriched by different research lines, and through the numerous research groups where he has worked which are totally independent from each other, as can be deduced from his publications and projects. This fact has contributed to a multidisciplinary training that has motivated a high scientific production as can be observed through his 87 works, including 53 JCR publications, 1 book, 5 book chapters, 5 popular articles, 6 scientific technical reports, and 17 scientific contributions in various areas such as thermo-responsive materials, polyolefins, smart polymers, and composites. His articles have been published mostly in reputed journals such as Chemosphere, Chemical Engineering Journal, Nano Research, Carbohydrate Polymers, etc... (Q1 = 75%), and a small proportion in the second quartile (Q2 = 25%), where 46% of papers have five or fewer authors. Furthermore, he is the first or last author of a large number of articles (50%), and the corresponding author (25%) of some of these publications.

He is very active in searching funding, research grants and cooperation networks. He has participated in a European project (COST Action), a Latin American project (CYTED), and he was the leader (Principal Investigator, PI) in two Asian projects (People's Government of Guangdong Province, 50620171375 and Postdoctoral Science Foundation, 2018M633119). Nowadays, he has started in January 2022 to manage an interdisciplinary project as coordinator and PI that deals with chemical recycling of polyolefins (CICLAPNER-CM-UC3M), granted by the Government of the Community of Madrid.

Dr. García-Peñas has received 10 awards and distinctions, among which stands out the world "Borealis Student Innovation Award", the recognition of the Spanish National Research Council for his doctoral thesis or the awards from the specialized groups of the Real Sociedad Española de Química and the Real Sociedad Española de Física to his work in the field of correlation between microstructural features and final properties of polymeric materials. In December 2018, he was invited to the Orthopedic National Conference (China) to present his work in the field of materials with antitumor activity, where he received the recognition from the Xi'an Hong Hui Hospital. He obtained 10 scholarships for different scientific events and participated in 21 specialized courses. He was delegate of the China-Spain Researchers Network, as a reward for organizing the First Meeting of Ibero-American Researchers in southern China, and the Night of European Researchers in Shenzhen (EURAXESS), co-organized by the European Commission and the Embassy of Spain in Beijing. In addition, he co-founded the first contact office for Spanish-speaking researchers in China. He is contributing to open access journals as part of the Topical Advisory Panel of Polymers (Q1), and Crystals (Q2).

He is Coordinator of the master's degree in Circular Engineering, Secretary for Academic Affairs of "Álvaro Alonso Barba" Institute of Chemistry and Materials Technology, Founder of the Cirmat Symposium, Computer Council Member, and Laboratory Safety Officer at University Carlos III of Madrid. Furthermore, he is currently supervisor of a doctoral thesis, three master's degree projects and a degree project at the Carlos III University of Madrid. In addition, Dr. García-Peñas is mentor of the Euraxess Mentoring Programme.

Dr. Gaurav Sharma

International Research Centre of Nanotechnology for Himalayan Sustainability (IRCNHS), Shoolini University, Solan, 173229, Himachal Pradesh, India

Email: Gaurav.541@shooliniuniversity.com

Dr. Gaurav Sharma research activity started in 2009 at Shoolini University (India) as a master of philosophy student, and then, he continued his research work as PhD student with the preparation and characterization of diverse multifunctional nanomaterials, and their composites, specially focused on potential applications in environmental remediation (as photocatalysts and adsorbents). For four years he worked as assistant professor in the School of chemistry at Shoolini University (India), where he carried out diverse research lines, interrelated to each other based on synthesis and characterization of nanocomposites, hydrogels, bi and trimetallic nanoparticles, ion exchangers, adsorbents and photocatalysts etc. Moreover, he performed and taught different courses as nanochemistry, polymer chemistry, spectroscopy and natural products, among others. On the other hand, he supervised 3 PhD, 5 Master of Philosophy, and more than 25 Master and Bachelors students. He established collaborative research with various professors in countries as Finland, Saudi Arabia, China, Spain and South Africa. In this context, he was invited as visiting research professor from University of KwaZuklu-Natal (South Africa) in 2017 and 2019. In 2017, he joined as postdoctoral fellow at college of materials science and engineering, Shenzhen University. He got project from china postdoctoral science foundation in 2018. The outcome of his research work was depicted in more than 150 publications, in various journals such as Renewable and Sustainable

Energy Reviews, Chemical Engineering Journal, Journal of Cleaner Production, Carbohydrate Polymers, ACS Applied Materials and Interfaces, Journal of Hazardous Materials, Applied Catalysis B, and International Journal of Biological Macromolecules etc, 9 book chapters and 7 edited books. He is also serving as Director, International Research Centre of Nanotechnology for Himalayan Sustainability (IRCNHS), Shoolini University, India. He is a Highly Cited Researcher -2020, 2021 Crossfield (Web of Science); and also Ranked among the World top 2% Scientists (Current year 2019, 2020, &2021 category) as per Stanford.

His h-index is 66, citations: more than 11000 (web of science); Google Scholar: h-index is 68, citations: more than 11000. He is Associate Editor of the International Journal of Environmental Science and Technology (Springer). Editorial Board member of Current Organic Chemistry, Current analytical Chemistry, Materials-MDPI Innovations in Corrosion and Materials Science, Journal of Nanostructure in Chemistry, Nanotechnology for Environmental Engineering(Springer), Letters in Applied NanoBioScience etc, and Academic Editor of Journal of Nanomaterials, Advances in Polymer Technology.